Sunspots: Theory and Observations

NATO ASI Series

Advanced Science Institutes Series

A Series presenting the results of activities sponsored by the NATO Science Committee, which aims at the dissemination of advanced scientific and technological knowledge, with a view to strengthening links between scientific communities.

The Series is published by an international board of publishers in conjunction with the NATO Scientific Affairs Division

A Life Sciences **B Physics**	Plenum Publishing Corporation London and New York
C Mathematical and Physical Sciences **D Behavioural and Social Sciences** **E Applied Sciences**	Kluwer Academic Publishers Dordrecht, Boston and London
F Computer and Systems Sciences **G Ecological Sciences** **H Cell Biology** **I Global Environmental Change**	Springer-Verlag Berlin, Heidelberg, New York, London, Paris and Tokyo

NATO-PCO-DATA BASE

The electronic index to the NATO ASI Series provides full bibliographical references (with keywords and/or abstracts) to more than 30000 contributions from international scientists published in all sections of the NATO ASI Series.
Access to the NATO-PCO-DATA BASE is possible in two ways:

– via online FILE 128 (NATO-PCO-DATA BASE) hosted by ESRIN, Via Galileo Galilei, I-00044 Frascati, Italy.

– via CD-ROM "NATO-PCO-DATA BASE" with user-friendly retrieval software in English, French and German (© WTV GmbH and DATAWARE Technologies Inc. 1989).

The CD-ROM can be ordered through any member of the Board of Publishers or through NATO-PCO, Overijse, Belgium.

Series C: Mathematical and Physical Sciences - Vol. 375

Sunspots: Theory and Observations

edited by

John H. Thomas

Department of Mechanical Engineering,
Department of Physics and Astronomy
and C.E.K. Mees Observatory,
University of Rochester,
Rochester, NY, U.S.A.

and

Nigel O. Weiss

Department of Applied Mathematics and Theoretical Physics,
University of Cambridge,
Cambridge, U.K.

Kluwer Academic Publishers

Dordrecht / Boston / London

Published in cooperation with NATO Scientific Affairs Division

Proceedings of the NATO Advanced Research Workshop on
The Theory of Sunspots
Cambridge, U.K.
September 22–27, 1991

Library of Congress Cataloging-in-Publication Data

NATO Advanced Research Workshop on the Theory of Sunspots (1991 : Cambridge, England)
Sunspots : theory and observations / edited by John H. Thomas and Nigel O. Weiss.
p. cm. -- (NATO ASI series. Series C, Mathematical and physical sciences ; vol. 375)
"Proceedings of the NATO Advanced Research Workshop on the Theory of Sunspots, Cambridge, England, 22-27 September 1991."
"Published in cooperation with NATO Scientific Affairs Division."
Includes bibliographical references and index.
ISBN 0-7923-1852-8 (alk. paper)
1. Sunspots--Congresses. I. Thomas, John H., 1941- . II. Weiss, N. O. (Nigel Oscar) III. North Atlantic Treaty Organization. Scientific Affairs Division. IV. Title. V. Series: NATO ASI series. Series C, Mathematical and physical sciences ; no. 375.
QB525.N38 1991
523.7'4--dc20 92-20041

ISBN 0-7923-1852-8

Published by Kluwer Academic Publishers,
P.O. Box 17, 3300 AA Dordrecht, The Netherlands.

Kluwer Academic Publishers incorporates the publishing programmes of
D. Reidel, Martinus Nijhoff, Dr W. Junk and MTP Press.

Sold and distributed in the U.S.A. and Canada
by Kluwer Academic Publishers,
101 Philip Drive, Norwell, MA 02061, U.S.A.

In all other countries, sold and distributed
by Kluwer Academic Publishers Group,
P.O. Box 322, 3300 AH Dordrecht, The Netherlands.

Printed on acid-free paper

Printed in the Netherlands

TABLE OF CONTENTS

PREFACE

This volume contains the invited papers presented at the NATO Advanced Research Workshop on the Theory of Sunspots, held in Cambridge, England, 22-27 September 1991. The idea of holding this Workshop first arose during the Solar Optical Telescope workshop on Theoretical Problems in High-Resolution Solar Physics in Munich in 1985. At that meeting, separate discussion groups were formed to consider specific topics in solar physics. The discussion group on sunspots recommended that there be a meeting devoted to theoretical problems associated with sunspots, the motivation being the consensus that theory seemed to lag behind the observational evidence in our quest for a satisfactory understanding of the physics of sunspots. This recommendation was warmly received and the two of us were designated to organize the Workshop.

Although the Workshop eventually took place later than originally envisioned, the delay turned out to be fortunate and the timing of the Workshop was ideal for a number of reasons. There have been remarkable improvements in high-resolution observations of sunspots in the past few years, and many important new observational results were presented for the first time at this Workshop (by groups working at the Lockheed Palo Alto Research Laboratories, the Swedish and German telescopes in the Canary Islands, and the U.S. National Solar Observatory). Vector magnetographs and Stokes polarimetry have at last given us reliable measurements of the vector magnetic fields in sunspots. Recent discoveries, such as the absorption of solar p-modes by sunspots, pose new and important theoretical problems. Impressive advances in numerical simulations of processes such as compressible magnetoconvection allow us to construct more realistic theoretical models of a sunspot. Also, coincidentally, this Workshop took place exactly ten years after the last meeting devoted exclusively to sunspots, the workshop on the Physics of Sunspots held at Sacramento Peak Observatory in 1981 (which was itself preceded by a meeting held in Florence in 1964).

The response to invitations to the Workshop was very positive. The vast majority of theoreticians working on the physics of sunspots attended the Workshop, along with many of the best observers. In all, there were 56 participants.

The program for the Workshop was planned carefully in consultation with the Scientific Organizing Committee. After the introductory talks, the topics of specific interest divided naturally into four scientific sessions, each under the control of a session chairman. The titles of these four sessions and their respective chairmen were: Overall Structure of Sunspots (D. J. Galloway); Fine Structure of Sunspots (H. U. Schmidt); Waves and Oscillations in Sunspots (D. O. Gough); and The Relation of Sunspots to the Global Solar Magnetic Field (R. Rosner). This volume contains sections corresponding to these four session topics, followed by E. N. Parker's stimulating summary. In addition, we as editors have written an extensive introductory chapter to provide an overall perspective of the subject and make this book more accessible to the nonspecialist.

The expressed purposes of the Workshop were to establish what is known observationally and theoretically about sunspots, to formulate the key theoretical problems, and to identify the most promising approaches to their solution. While we generally maintained this emphasis on theory throughout the Workshop, the many important new observational results presented at the Workshop in themselves constitute a major contribution. To reflect this, we have decided to use the title Sunspots: Theory and Observations for these published proceedings.

We are grateful to the members of the Scientific and Local Organizing Committees (listed below) and the session chairmen for their important contributions to the success of the Workshop. We thank the Department of Applied Mathematics and Theoretical Physics, University of Cambridge, for providing the meeting room and other facilities at no cost and Keith Moffatt for welcoming participants to the Department. We also thank Clare College, Cambridge, where all of the participants were comfortably housed and well fed. We are grateful to the Royal Astronomical Society for a grant from the Victor Nadarov Fund which enabled us to support some participants from non-NATO countries. We thank Karen Stringer and John Turner for their help with administrative matters. We are especially grateful to Louise Hindle for her secretarial assistance before and during the workshop and in the preparation of this book.

Finally, on behalf of all the participants, we wish to express our gratitude to the NATO Scientific Affairs Division for their generous financial support of this workshop.

JHT and NOW

Cambridge, January 1992

Scientific Organizing Committee

P. Maltby	(Norway)	University of Oslo
F. Moreno-Insertis	(Spain)	Instituto de Astrofisica de Canarias
R. Muller	(France)	Observatoire du Pic du Midi
E. N. Parker	(USA)	University of Chicago
H. U. Schmidt	(Germany)	Max-Planck-Institut für Astrophysik
J. H. Thomas (Director)	(USA)	University of Rochester
N. O. Weiss (Director)	(UK)	University of Cambridge
C. Zwaan	(Netherlands)	Rijksuniversiteit te Utrecht

Local Organizing Committee

D. P. Brownjohn
D. J. Galloway
D. W. Hughes
P. C. Matthews
M. R. E. Proctor
N. O. Weiss (Chairman)

LIST OF PARTICIPANTS

Aballe Villero, M.A.	(Spain)	Instituto de Astrofisica de Canarias, Tenerife
Bogdan, T.	(USA)	High Altitude Observatory, Boulder
Brownjohn, D.P.	(UK)	University of Cambridge
Bumba, V.	(Czechoslov.)	Ondrejov Observatory
Bünte, M.	(Switzerland)	ETH, Zurich
Byrne, P.B.	(UK)	Armagh Observatory
Chitre, S.M.	(India)	Tata Institute, Bombay
Choudhuri, A. Rai	(India)	Indian Institute of Science, Bangalore
Cram, L.E.	(Australia)	University of Sydney
Degenhardt, D.	(Germany)	University of Göttingen
Dialetis, D.	(Greece)	National Observatory, Athens
Evans, D.J.	(UK)	Queen Mary & Westfield College, London
Fox, P.A.	(USA)	Yale University
Frank, Z.A.	(USA)	Lockheed Palo Alto Research Laboratories
Galloway, D.J.	(Australia)	University of Sydney
Golub, L.	(USA)	Center for Astrophysics, Cambridge, MA
Gough, D.O.	(UK)	University of Cambridge
Hasan, S.S.	(USA)	Center for Astrophysics, Cambridge, MA
Hughes, D.W.	(UK)	University of Leeds
Isaak, G.R.	(UK)	University of Birmingham
Jahn, K.	(Poland)	Warsaw University Observatory
Lawrence, J.K.	(USA)	California State University, Northridge
Lites, B.W.	(USA)	High Altitude Observatory, Boulder
Livingston, W.C.	(USA)	National Solar Observatory, Tucson
Low, B.C.	(USA)	High Altitude Observatory, Boulder
Maltby, P.	(Norway)	University of Oslo
Matthews, P.C.	(UK)	University of Cambridge
Montesinos, B.	(UK)	University of Oxford
Moreno-Insertis, F.	(Spain)	Instituto de Astrofisica de Canarias, Tenerife
Muller, R.	(France)	Observatoire du Pic du Midi
Parker, E.N.	(USA)	University of Chicago
Pizzo, V.J.	(USA)	High Altitude Observatory, Boulder
Proctor, M.R.E.	(UK)	University of Cambridge
Roberts, B.	(UK)	University of St Andrews
Rosenthal, C.S.	(USA)	University of Colorado
Rosner, R.	(USA)	University of Chicago
Ryutova, M.P.	(USSR)	Institute of Nuclear Physics, Novosibirsk
Scharmer, G.B.	(Sweden)	Swedish Solar Observatory, La Palma
Schatten, K.	(USA)	NASA Goddard SFC, Greenbelt
Schmidt, H.U.	(Germany)	Max-Planck-Institut für Astrophysik, Munich

Schmidt, W.	(Germany)	Kiepenheuer Institut für Sonnenphysik
Shine, R.A.	(USA)	Lockheed Palo Alto Research Laboratories
Similon, P.L.	(USA)	Yale University
Skumanich, A.	(USA)	High Altitude Observatory, Boulder
Solanki, S.K.	(Switzerland)	ETH, Zurich
Spruit, H.C.	(Germany)	Max-Planck-Institut für Astrophysik, Munich
Staude, J.	(Germany)	Zentralinstitut für Astrophysik, Potsdam
Strous, L.H.	(Netherlands)	Rijksuniversiteit te Utrecht
Theobald, M.L.	(USA)	Yale University
Thomas, J.H.	(USA)	University of Rochester
Title, A.M.	(USA)	Lockheed Palo Alto Research Laboratories
Watson, P.G.	(UK)	University of Cambridge
Weiss, N.O.	(UK)	University of Cambridge
Wilson, P.R.	(Australia)	University of Sydney
Zhugzhda, Yu.D.	(USSR)	IZMIRAN, Troitsk
Zwaan, C.	(Netherlands)	Rijksuniversiteit te Utrecht

I. INTRODUCTION

"Our knowledge of the sun's action is but fragmentary, and the publication of speculations on the nature of his spots would be a very precarious adventure."
(Carrington 1858)

Before the discovery of magnetic fields in sunspots any such speculations would indeed have been premature. Subsequent advances, both observational and theoretical, have enormously increased our knowledge: within the last 10 years our understanding of sunspots has been transformed, by a combination of high-resolution observations, detailed modelling and numerical computations. We therefore decided to make this volume as close to a monograph as possible (allowing, of course, for differences in opinion), rather than a mere record of proceedings. We have included invited reviews only and all of them have been refereed. Many new results were presented at the meeting in contributed papers and we have tried to mention them wherever possible. This book does not supersede the classic monograph by Bray and Loughhead (1964) but does attempt to bring it up to date. Readers are also referred to the proceedings of a workshop at Sunspot, New Mexico, 10 years ago, edited by Cram and Thomas (1981). Suggestions for further reading are given overleaf.

The layout of this book follows the structure of the meeting. We have, however, replaced our opening remarks with a chapter that aims to provide a comprehensive review of our current understanding of the theory of sunspots; this chapter is meant to serve as an introduction and guide to those that follow, but we hope that it can also stand alone as an up-to-date introduction to the subject for the student or nonspecialist. It is followed by 16 authoritative reviews which cover major developments in the field. They are grouped into four sections, each preceded by introductory remarks from the editors. The concluding chapter summarizes the state of the subject, emphasizing the many challenges that still remain.

Taken together, these chapters provide a book that can stand on its own. We hope that it will be read by students embarking on research: they will find that the study of sunspots is making rapid progress and that there are many exciting problems to be solved.

References

Bray, R.J., and Loughhead, R.E. 1964. *Sunspots* (London: Chapman and Hall).

Carrington, R.C. 1858. On the distribution of the solar spots in latitude since the beginning of the year 1854. *Mon. Not. Roy. Astron. Soc.* **19**, 1–3.

Cram, L.E., and Thomas, J.H. (eds) 1981. *The Physics of Sunspots* (Sunspot, NM: Sacramento Peak Observatory).

Further reading

Anyone interested in learning about sunspots in a broader context should start with the excellent account in

Noyes, R.W. 1982. *The Sun, Our Star* (Cambridge, MA: Harvard University Press),

which is aimed at a general reader. More details are provided in the following recent books on solar physics:

Durrant, C.J. 1988. *The Atmosphere of the Sun* (Bristol: Adam Hilger);

Foukal, P. 1990. *Solar Astrophysics* (New York: Wiley);

Stix, M. 1989. *The Sun: An Introduction* (Berlin: Springer-Verlag).

The standard reference on hydromagnetic theory relevant to the Sun is

Priest, E.R. 1982. *Solar Magnetohydrodynamics* (Dordrecht: Reidel).

There are also a number of useful review articles on particular topics that are referred to in the following chapters.

THE THEORY OF SUNSPOTS

JOHN H. THOMAS
Department of Mechanical Engineering, Department of Physics and Astronomy, and C.E.K. Mees Observatory
University of Rochester
Rochester, New York 14627
USA

and

NIGEL O. WEISS
Department of Applied Mathematics and Theoretical Physics
University of Cambridge
Silver Street
Cambridge CB3 9EW
UK

ABSTRACT. This review covers the present state of our theoretical understanding of the physics of sunspots, along with the principal observational results that need to be explained. The topics covered range from the detailed structure of an individual sunspot to the broad connection between sunspots and the global solar magnetic field and the solar cycle. Our aim is to give a critical discussion of the theoretical ideas and models without presenting mathematical details. After outlining the historical development of the basic concepts associated with the magnetohydrodynamic theory of sunspots, we discuss recent treatments of their properties and structure, placing special emphasis on developments that have occurred within the last ten years. There have been remarkable improvements in the theoretical modelling of sunspots, led by new ideas and by more elaborate and realistic numerical simulations. At the same time, new observations have raised new theoretical questions or caused old ones to be reconsidered. In particular, measurements of oscillations in and around sunspots have opened up the new field of sunspot seismology, while recent high-resolution observations have forced us to rethink the structure of a sunspot penumbra.

1. Introduction

Although sunspots have been observed through telescopes since the time of Galileo, and their basic magnetic character has been known since shortly after the beginning of this century, we still do not have an adequate theoretical understanding of their basic structure and evolution. Sunspots pose a major challenge not only to solar physicists but also to plasma physicists and astrophysicists in general, for they provide the best testbed for cosmical magnetohydrodynamics. Moreover, it is only by studying the physics of these spots that we shall be able to describe the prominent starspots that appear on other much more active stars.

Individual sunspots often have a fairly regular appearance, making the theoretician's circular spot a reasonable idealization of the photospheric observations. So the problems

J. H. Thomas and N. O. Weiss (eds.), Sunspots: Theory and Observations, 3–59.

of explaining the existence of an umbra and penumbra, and predicting the subphotospheric structure of a spot, seem ideally suited for theoretical study. It is sobering that, after all these years, they have not yet been solved. If we cannot explain such simple phenomena, how can we hope to model accretion discs or active galactic nuclei? It seems, however, that unresolved objects can be relatively easy to explain: as detailed observations accrue they may actually make such problems harder. Our task would be much simpler if we only had to deal with an axisymmetric structure, like that of an ideal pore – but real sunspots are complicated by the presence of a filamentary penumbra.

The study of sunspots has entered an exciting phase, for it is only in the last few years that high-resolution observations have become sufficiently detailed to allow us to formulate the key theoretical problems precisely and to make a start in answering them. In solar physics, theory is usually led by observations. In the words of Pascal, written only a few years after Galileo's death,

"L'imagination ... se lassera plutôt de concevoir que la nature de fournir"

(The imagination tires sooner of conceiving than Nature does of providing—*Pensées*, XV, No.199) or, as Eugene Parker puts it, "Nature is cleverer than we are". Of the many problems that we shall discuss, some, like the cooling of sunspots or the origin of the Evershed outflow, are relatively old. Others, like the varying inclination of the penumbral magnetic field or the absorption of p-modes by a sunspot, are very new.

Theoretical progress has depended mainly on an improved understanding of the underlying physics and its relation to the observations (see e.g. Priest 1982, Stix 1989), coupled with detailed modelling of umbral and penumbral atmospheres. Magnetoconvective processes play a vital part in energy transport and they are essentially nonlinear. To understand them we rely, on the one hand, on recent advances in nonlinear dynamics and, on the other, on large-scale numerical experiments and simulations.

Meanwhile it has become possible to observe structures on much finer scales: whereas Beckers and Schröter (1969) had a spatial resolution of $0.6''$ (400 km) in the continuum, white-light observations at the Pic du Midi, La Palma, Tenerife, Sacramento Peak or Big Bear can now achieve a resolution of $0.3''$, while spectroscopic measurements resolve magnetic fields and velocities on a scale of $0.5''$. Sophisticated techniques have been developed, notably at the Lockheed Palo Alto Research Laboratory, for collecting and processing the data digitally; these techniques are particularly valuable for studying time-dependent behaviour. In the future we may expect improvements from adaptive optics and, eventually, as a result of observations made from space.

Solar research has also been stimulated by observations of magnetic activity in other late-type stars. Photometric variations are interpreted as due to starspots that may occupy up to 60% of the surface area of an active star, as described by Byrne (1992) in this volume. This solar-stellar connection lends increased importance to modelling spots on the sun, which may be smaller but can easily be resolved.

We begin this paper with a brief historical review. Then, in Section 3, we discuss the overall structure of a sunspot, and models of the magnetic field underneath its surface. Fine structure of the umbra and penumbra is described in Section 4, and related to underlying convective processes. In Section 5 we consider umbral oscillations and penumbral waves, together with developments in the new field of sunspot seismology. Section 6 is devoted to the global issue of how sunspot magnetic fields are related to the solar dynamo and the final section looks towards developments that may be expected in the future. Our aim is to provide a general theoretical overview, without repeating details that are presented elsewhere in this volume.

2. Historical Perspective

The first telescopic observations, by Fabricius, Galileo and Scheiner, were made in 1611. Both Galileo and Scheiner recognized the distinction between the umbra and penumbra, which appears clearly in the engraving by Hevelius (1647) reproduced in Figure 1*a*. Aside from the discovery of the Wilson depression in 1769 there was little further progress until the first large telescopes were developed by William Herschel (1801), who described penumbral filaments ("Shallows have no Corrugations, but are tufted", in his idiosyncratic nomenclature). By the mid-nineteenth century, Secchi (1870) was able to discern this filamentary structure and to reproduce it in drawings like that shown in Figure 1*b*.

(*a*)

(*b*)

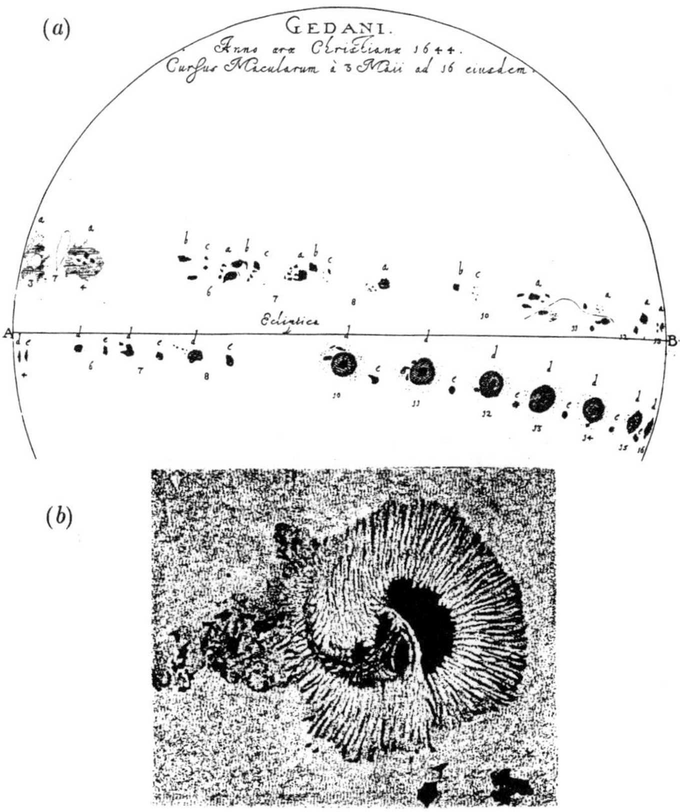

Figure 1. Early sunspot observations. (*a*) Engraving by Hevelius (1647), showing spots observed on successive days, as they rotated with the Sun. The umbra and penumbra are clearly distinguished. (*b*) Drawing by Secchi (1870), showing the filamentary structure of the penumbra, a light bridge, pores and the surrounding granulation.

Galileo had established that sunspots rotate with the surface of the Sun. In the eighteenth century both Wilson and William Herschel described spots as dark holes in the bright exterior of the Sun ("Places where the luminous Clouds of the Sun are Removed"). John Herschel later suggested that they were sites of cyclonic motion and this notion of tornadoes (associated with upward or downward motion) dominated nineteenth-century discussions. The meteorological analogy was developed by Russell (1921) and Rosseland (1926), and eventually led to Bjerknes's (1926) model of the solar cycle: he envisaged azimuthal vortex tubes, transported by a meridional circulation within the convective zone, which popped through the photosphere to form sunspot pairs.

The modern age of solar physics began shortly after 1900 with rapid advances in atomic physics and the application of spectroscopy to the understanding of the solar atmosphere. In 1908 Hale discovered that sunspots contain strong magnetic fields by measuring the Zeeman splitting of magnetically sensitive absorption lines (Hale 1908b). This was the first discovery of an extraterrestrial magnetic field and as such ushered in a whole new branch of astrophysics. In the late nineteenth century there had been a good deal of speculation (mostly based on arguments that later proved incorrect) that perhaps all large, rotating bodies are magnetic. In 1891 Schuster, in an address to the British Royal Institution, remarked that the form of the solar corona suggests that the Sun has a magnetic field. Zeeman's discovery of the influence of a magnetic field on spectral lines opened the way for testing this idea. Hale undertook to look for magnetic fields in sunspots because he was convinced that sunspots were closely analogous to terrestrial tornadoes and that electrons moving in the sunspot's vortex would induce a magnetic field (see Hale 1908a).

Hale's discovery of the magnetic field in a sunspot was followed quickly by the discovery of a persistent radial outflow in the penumbra by Evershed in 1909. Using the spectrograph at Kodaikanal Observatory, Evershed found radial velocities, away from the centre of the sunspot, in weak photospheric spectral lines and radial inflow in stronger chromospheric lines, such as the dark central core of the Ca II H and K lines. Subsequent measurements of the Evershed flow in over 500 different spectral lines by St. John (1913) at Mt. Wilson showed a systematic vertical shear in the Evershed velocity. Kinman (1952) later showed that the photospheric Evershed velocity increases outward in the the penumbra.

The discovery of strong magnetic fields in sunspots led Larmor (1919) to propose an axisymmetric model in which the field was locally generated from a small seed field by induction, owing to meridional flow of highly conducting gas across the field lines. Cowling (1934) showed that such a local dynamo was impossible, and suggested that sunspots appeared as a result of a pre-existing tube of magnetic flux being carried upwards and breaking through the surface. These ideas were developed during the 1930s in the course of a remarkable correspondence between Cowling and Biermann (Cowling 1985). They agreed that there had to be a balance between magnetic pressure inside the spot and gas pressure outside. Then Biermann suggested that the coolness of sunspots could be explained in terms of restriction of convection by a magnetic field; in his reply, Cowling began by being sceptical but then estimated the relative values of the magnetic and Reynolds stresses and finally agreed that Biermann was correct. Their correspondence was interrupted by the war but in 1941 Biermann published his conjecture that convection was inhibited by the strong magnetic fields in sunspots. Biermann's suggestion, considering the great influence it has had on our theoretical understanding of sunspots, was remarkably brief and is perhaps worth quoting (our translation from the German):

... The second point concerns the cooling mechanism of sunspots. It is clear that the mechanism proposed by Russell (adiabatic cooling of elements ascending through a stably stratified region) no longer functions if the spots owe their origin to a convection zone that extends to just below the photosphere. Now, however, because of the altered conditions there necessarily exists a new cooling mechanism of the following sort. According to the results of theory, an extended convection zone is only to be expected if convection is responsible for the energy transport in the convection zone right up to its outer boundary. Wherever there is a strong magnetic field, the field tries to suppress the convection through braking of the eddy motions. Whether this succeeds depends, according to an investigation by Cowling (1933), on the ratio of the magnetic energy density to the energy density of the turbulent motions. The former amounts to $10^{5.6} \cdot (H/3200)^2$ erg cm^{-3}, the latter to $3/2\ \rho v^2 = 10^{4.2} \cdot (\rho/10^{-6.5})$. Up to densities ρ of the order of 10^{-5}, therefore, the convective energy transport beneath a sunspot must be suppressed and the heat transport accordingly reduced to the fraction that would be carried by radiation alone. The energy flow that cannot reach the surface in the sunspot must accordingly appear in the surroundings of the spot, where the faculae and chromospheric eruptions also arise preferentially. This interpretation considers the magnetic field as the primary cause of sunspots. A comparison of magnetic with thermal energy density makes this idea appear entirely plausible. So does a survey of the present state of the theory of solar magnetism (Cowling 1934), which, however, will not be carried out here.*

Biermann's remarks might well have been forgotten if Cowling (1953, 1957) had not drawn attention to them and elaborated on them in subsequent reviews. Later discussions of sunspot cooling by Hoyle (1949) and Cowling (1953) pointed out the possibility that the magnetic field may only reduce, rather than fully suppress, convective heat transport

* Der zweite Punkt betrifft den Abkühlungsmechanismus der Flecken. Es ist klar, daß der von Russell vorgeschlagene Mechanismus (adiabatische Abkühlung durch ein stabil geschichtetes Gebiet hindurch aufsteigender Elemente) nicht mehr funktioniert, wenn die Flecken ihre Entstehung einer Konvektionszone verdanken, die bis dicht unter die Photosphäre reicht. Nun ergibt sich aber zwangsläufig aus der geänderten Sachlage ein neuer Abkühlungsmechanismus der folgenden Art. Nach den Ergebnissen der Theorie ist eine ausgedehnte Konvektionszone nur dann zu erwarten, wenn der Energietransport in der Konvektionszone bis an ihre äußere Grenze durch die Konvektion besorgt wird. Wenn sich nun irgendwo ein starkes Magnetfeld befindet, so sucht dieses durch Wirbelstrombremsung an eben dieser Stelle die Konvektion zu unterdrücken. Ob ihm dieses gelingt, hängt nach einer Untersuchung von Cowling (1933) ab von dem Verhältnis der magnetischen Energiedichte zu derjenigen der turbulenten Bewegung. Die erste beträgt $10^{5.6} \cdot (H/3200)^2$erg/cm^3, die letztere $3/2\ \rho v^2 = 10^{4.2} \cdot (\rho/10^{-6.5})$. Bis zu Dichten ρ der Ordnung 10^{-5} hinauf muß daher unterhalb der Flecken der konvektive Energietransport unterdrückt sein und der Strahlungsstrom demgemäß reduziert sein auf den Bruchteil, der von der Strahlung allein transportiert wird. Der Energiestrom, der in den Flecken nicht zur Oberfläche gelangen kann, muß demgemäß in der Umgebung der Flecken auftreten. Hier entstehen ja auch bevorzugt die Fackeln und chromosphärischen Eruptionen. Diese Deutung sieht das Magnetfeld als das Primäre im Sonnenfleck an. Ein Vergleich der magnetischen mit der thermischen Energiedichte läßt jedoch diese Vorstellung als durchaus plausibel erscheinen. Das gleiche ergibt ein Überblick über den gegenwärtigen Stand der Theorie des Sonnenmagnetismus (Cowling 1934), wie indessen an dieser Stelle nicht ausgeführt werden soll.

within a sunspot. Hoyle argued that the magnetic field will suppress motions across but not along the field, so that heat will be convected preferentially along field lines in a strong magnetic field. Thus the fanning out with height of the sunspot magnetic field near the solar surface will cool the sunspot, simply by spreading the heat flux over a greater area.

Although some important results concerning the behaviour of magnetic fields in conducting fluids had been obtained before 1940, the new science of magnetohydrodynamics really began in earnest in the 1940s following two important theoretical discoveries by Alfvén (1942). The first, now generally known as Alfvén's theorem (although Alfvén did not provide a formal proof), was the fact that in a highly conducting fluid the magnetic lines of force must follow the motion of the fluid, i.e. are "frozen in". The second discovery was the existence of the transverse waves which have come to be known as Alfvén waves and are one of the most immediate consequences of Alfvén's theorem. Alfvén (1943) also recognized the magnetohydrostatic balance in a sunspot and went on to formulate a theory of sunspots based on waves travelling along field lines of the Sun's overall magnetic field; this theory has since proved incorrect.

Cowling (1946) calculated the electrical conductivity of plasma near the photosphere and estimated that the Ohmic decay time for a sunspot would be around a century. Since spots appear within a few days and have lifetimes ranging from days to months it became clear that they must be formed from pre-existing flux. Cowling envisaged toroidal flux tubes running as girdles around the Sun (like a magnetic analogue of Bjerknes's vortex tubes) and suggested that loops were carried upwards by convection to emerge as bipolar sunspot pairs.

The magnetic buoyancy of an isolated magnetic flux tube, another fundamental idea in solar magnetohydrodynamics, was first described independently by Parker (1955a) and by Jensen (1955). Parker demonstrated that an isolated magnetic flux tube in a stratified gas under gravity must be buoyant, provided that its temperature is not too different from that of its surroundings. Jensen's paper, which has often been overlooked, puts forth clearly the basic ideas of the inherent buoyancy of an isolated magnetic flux tube, the limiting maximum field strength (for a fully evacuated flux tube), and the buoyant rise of a toroidal flux tube. Parker also showed how the buoyancy of toroidal flux tubes in the Sun could explain the general features of the appearance of sunspots at the solar surface, such as their occurrence in bipolar pairs and their nearly east-west orientation; in addition, he related buoyant flux tubes to a model of the solar dynamo, which explained both the migration in latitude of the sunspot zones and the polarity reversal of the magnetic cycle (Parker 1955b). These theoretical ideas were absorbed into Babcock's (1961) phenomenological model of the solar cycle.

The study of magnetoconvection was initiated by Walén (1949), who compared the buoyancy force on a fluid element rising through a superadiabatically stratified layer with the curvature force exerted by the distorted field lines (Cowling 1953; Weiss 1991). Cowling (1957) developed this physical argument into a linear stability analysis valid in the absence of any dissipation. This provided a quantitative criterion that could be applied to sunspots. However, the problem is complicated by double-diffusive effects, which had meanwhile been investigated by Thompson (1951) and Chandrasekhar (1952): they showed that instability would set in as overstable oscillations, rather than as monotonically growing modes, if the ratio of the magnetic to the thermal diffusivity is sufficiently small. This is the situation at or below the photosphere, where radiative transport of heat is more effective than the molecular resistivity; so sunspot umbrae are expected to be unstable to oscillatory modes, which are relatively ineffective at transporting heat. The results of these linear investigations are contained in Chandrasekhar's (1961) treatise – in which sunspots are never mentioned.

Meanwhile, observers were struggling to overcome the distortion caused by atmospheric interference. The Stratoscope balloon flights of 1959 marked the beginning of a new era. They yielded images of sunspots with unprecedented resolution (Danielson 1961a), limited only by the aperture of the telescope. Figure 2 shows a detail from a Stratoscope picture of penumbral filaments (Danielson 1966), which were clearly resolved for the first time, so that their widths and lifetimes could be measured. Moreover, this paper was followed immediately by Danielson's (1961b) theoretical study of convection in the inclined magnetic field of the penumbra. These results, along with many others, are covered in the classic monograph by Bray and Loughhead (1964), which remains an essential reference for the study of sunspots.

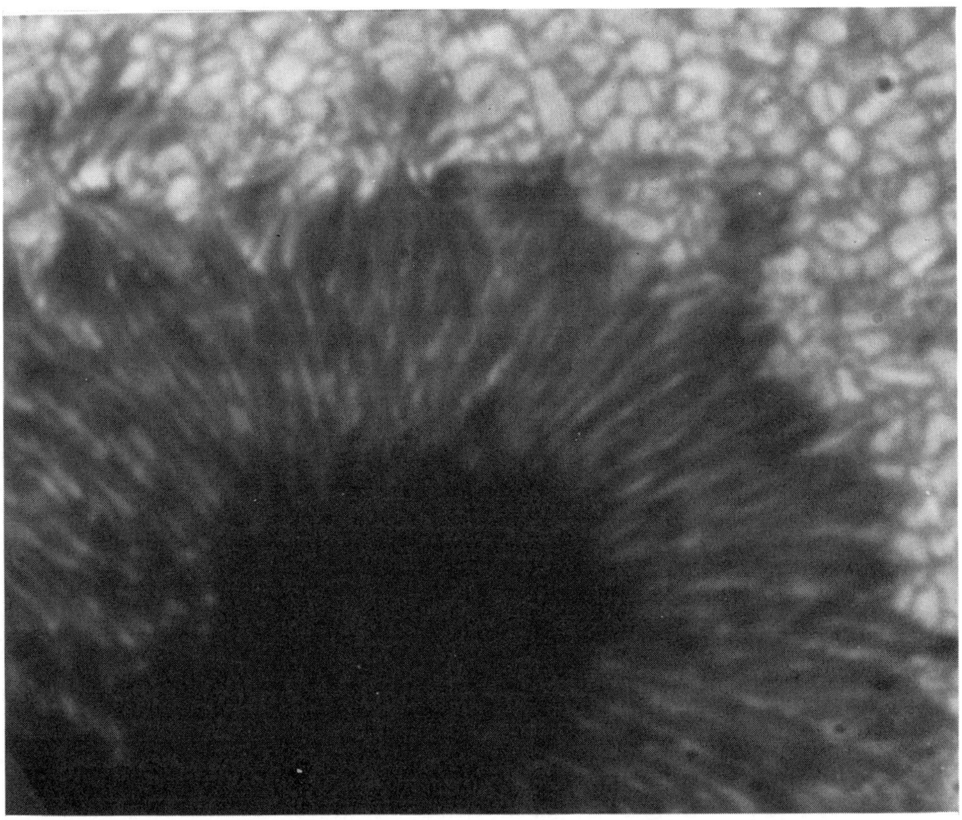

Figure 2. Stratoscope image of a sunspot, showing granulation and fine-scale penumbral structure. (Courtesy of the late R. E. Danielson.)

3. Overall Structure of Sunspots

During the past three decades observers have produced white-light images of ever-increasing clarity, from space as well as from the ground. The striking image in Figure 3 was obtained from the ground at the Pic du Midi Observatory by Muller (1973a). It shows the distinction between the umbra, which is more or less uniformly dark, and the penumbra, with its irregular filamentary striations. The umbra only emits energy at about 20% of the rate at which it is radiated from the normal photosphere, while the average penumbral intensity ratio is 75%, varying from 60% to 95% between dark and bright filaments (Muller 1973b). The upper part of the spot in Figure 3 is fairly regular but the lower part shows an extended penumbra surrounded by a string of pores, presumably associated with a complicated field structure.

Sunspots are observed with radii ranging from 1800 km to 25000 km and even, in one case, up to 65000 km. Fortunately, all well-developed symmetrical sunspots (i.e. roughly circular spots with a radius $R \gtrsim 10000$ km) have a similar structure: the umbral radius is approximately $0.4R$, while the azimuthally averaged intensity and magnetic field are functions of the normalized radius (s/R) only, where s is the radial distance from the central axis. (Of course, the radius of the flux tube beneath a sunspot decreases with depth and its vertical structure depends on the stratification of the convection zone as well as on the surface radius.) Thus we are able to discuss the overall structure of a typical sunspot. In this section we consider thermal properties first, before turning to the magnetic field.

3.1. THERMAL PROPERTIES

Measurements of the umbral intensity of sunspots at various wavelengths go back over 150 years, but it has only been in the past 25 years that improved instruments and accurate methods of correcting for scattered light have produced truly reliable results (Zwaan 1968). Measurements of the centre-to-limb intensity variation at a fixed wavelength and of the intensity variation with wavelength at a fixed sunspot position both give important information about the vertical temperature stratification in the umbra. These measurements have made it possible to construct semi-empirical models of the umbral atmosphere with nearly the same level of sophistication as in models of the quiet solar atmosphere. The first such model was the Sunspot sunspot model (Avrett 1981), which was later modified to give an improved representation of the deepest layers based on infrared measurements (Maltby *et al.* 1986). These and other semi-empirical sunspot models are discussed by Maltby (1992) in his review in this volume. Such models are an important input to theoretical work on sunspot structure and on waves and oscillations.

The most surprising result to have emerged from recent observations of umbral intensities is the discovery by Albregtsen and Maltby (1978, 1981) that the umbral intensity of sunspots varies significantly over the solar cycle. At a wavelength of 1.67 μm the ratio of umbral intensity to normal photospheric intensity increases nearly linearly from about 0.5 at solar minimum to about 0.6 at solar maximum (see Fig. 2 in Maltby's review). This discovery must have important implications for the relationship between sunspots and the solar dynamo. It may be that the umbral brightness depends upon the age of the subphotospheric flux tube that forms the sunspot (Schüssler 1980), or the depth at which the flux tube first formed (Yoshimura 1983).

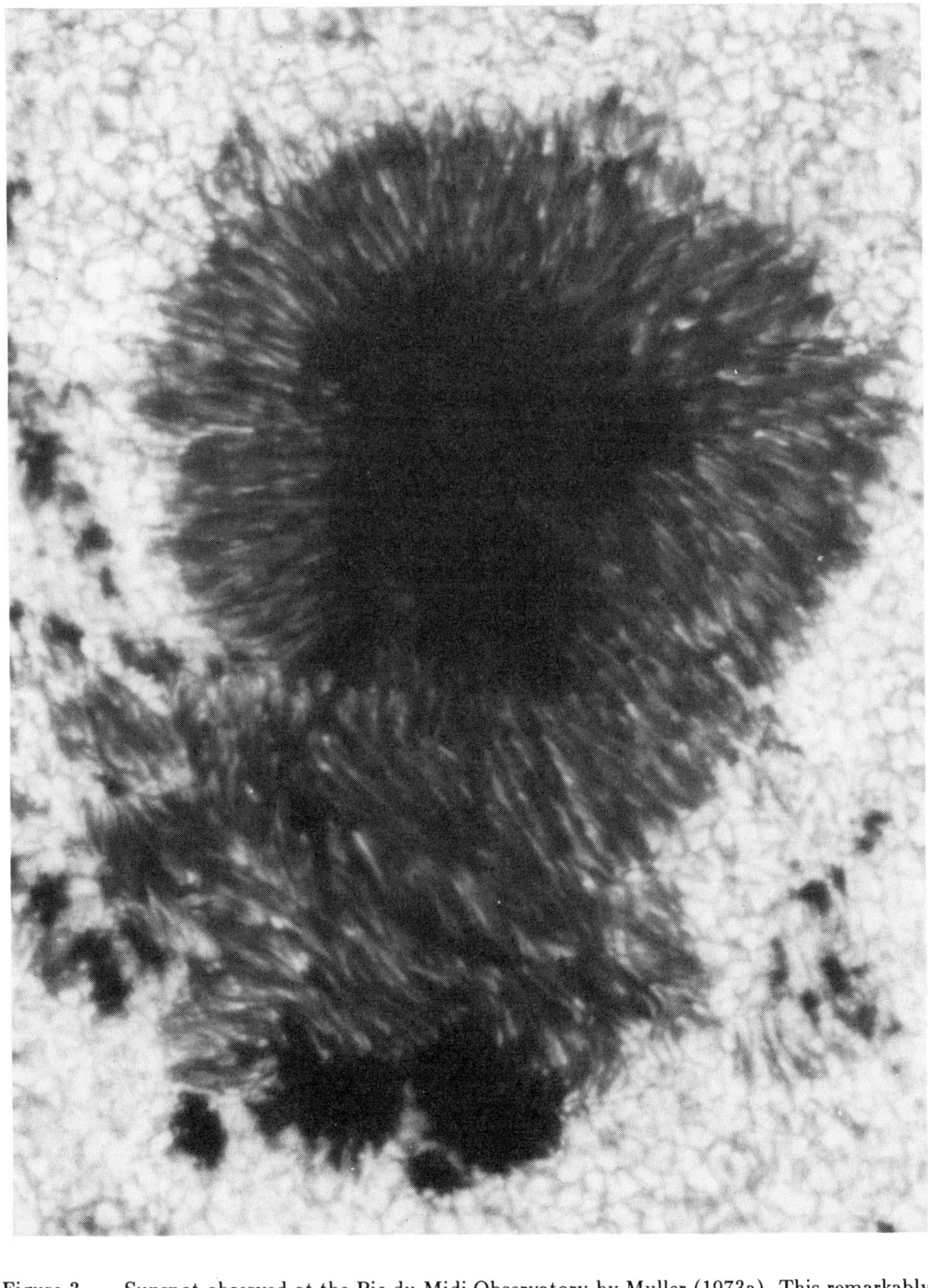

Figure 3. Sunspot observed at the Pic du Midi Observatory by Muller (1973a). This remarkably clear image shows penumbral filaments with elongated bright grains.

Biermann's (1941) suggestion, that sunspots are relatively cool owing to the inhibition of convection by their magnetic field, is now generally considered to be basically correct, although it is realized that significant convective transport of energy must occur in order to maintain the brightness of the umbra. In calculating the thermal structure of a sunspot, the partial suppression of convection can easily be represented in a parametrized form by reducing the value of the mixing length used for a standard model of the convection zone (Chitre 1963; Deinzer 1965; Chitre and Shaviv 1967; Yun 1970; Jahn 1989).

It is surprising that after Biermann's comments over 30 years elapsed before anyone got around to examining theoretically the fate of the heat flux that is diverted by a sunspot as a consequence of the inhibition mechanism. Parker (1974) reopened the subject by arguing, on the basis of a simple steady-state thermal model of a shallow sunspot, that the diverted heat flux would show up in the form of an intense bright ring immediately around the sunspot, which is not observed. He proposed instead that the spot must be cooled by a flux of Alfvén waves (or other MHD waves; see Roberts 1976), an idea that had been explored in several earlier theoretical papers (Danielson 1965; Zwaan 1965; Musman 1967; Savage 1969). This proposal was severely criticized by Cowling (1976) on both hydromagnetic and thermodynamic grounds. Eschrich and Krause (1977) then showed that the bright ring problem can be resolved in a steady-state thermal model that assumes anisotropic heat transport in the sunspot. Spruit (1977) demonstrated, also by means of a steady-state thermal model, that lateral convective heat transport in the convection zone is likely to be so efficient that the diverted heat flux is spread over a very large region of the sun and no detectable bright ring is formed. Subsequent calculations by Clark (1979) further elucidated the fate of the diverted heat flux and lent yet more support to inhibition of convection as a viable cooling mechanism. In the meantime, the observations of Beckers and Schneeberger (1977) showed that the upward flux of wave energy was insufficient to cool a sunspot, and Thomas (1978) confirmed that significant cooling by upward-propagating Alfvén waves would not be expected on theoretical grounds because the waves are reflected

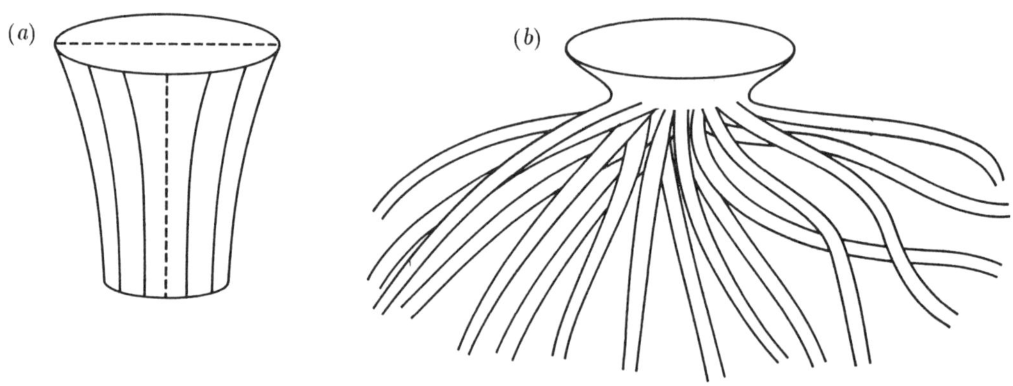

Figure 4. Extreme models of a sunspot. (*a*) Monolithic column and (*b*) jellyfish. Neither model is compatible with observations.

strongly downward in the umbral atmosphere. Later calculations (Scheuer and Thomas 1981) showed that downward-propagating compressible MHD waves suffer fairly strong upward reflection in a sunspot, posing further problems for the wave-cooling mechanism.

Although the various steady-state thermal models mentioned above were important in demonstrating that inhibition of convection is a plausible mechanism for sunspot cooling, in retrospect we now realize that these models are somewhat irrelevant because of the inherent transient nature of the process. The lifetime of a sunspot (a few months at most) is very much shorter than the time scale for thermal readjustments in the convection zone, so a steady state is never reached. This was first pointed out by Spruit (1982a,b) and by Foukal, Fowler, and Livshits (1983). Their results show that the energy flux blocked by the sunspot, instead of reappearing immediately at the surface outside the sunspot, is stored in the convection zone over a time of about 10^5 years, the time scale for thermal adjustment of the convection zone. Thus, the 'missing flux' of a sunspot is truly missing, in the sense that it does not reappear at the surface on any time scale of observational interest. The influence of the various relevant time scales on the fate of the heat flux blocked by a sunspot is beautifully explained in the chapter by Spruit (1992) in this volume.

3.2. MAGNETIC STRUCTURE

Owing to the Wilson depression the photospheric magnetic field is measured at different geometrical levels in the atmosphere. It is generally agreed that the azimuthally averaged field strength in a large spot has a maximum value approaching 3000 G at the centre of the umbra and that it decreases monotonically outwards. Estimates of the field at the outer edge of the penumbra range from 1300–1500 G (Beckers and Schröter 1969; Beckers 1981), sufficient to balance the external gas pressure, down to 700 G (Lites and Skumanich 1990) or even less, as described by Skumanich (1992) in this volume. At the spot boundary the photospheric field drops to zero over a distance that is too small to be resolved. The field inclination is more controversial (Beckers 1981; Skumanich 1992). Early measurements were consistent with a mean inclination to the vertical that increased linearly with radius from zero at the centre to 90° at the edge of the penumbra (Hale and Nicholson 1938; Beckers and Schröter 1969). However, Wittmann (1974) concluded that the field was much more nearly horizontal over most of the penumbra. More recent measurements (Kawakami 1983; Adam 1990; Lites and Skumanich 1990) all agree that the mean inclination at the edge of the spot is only 70°. Superimposed on this are the fluctuations described by Title *et al.* (1992) in this volume.

The subphotospheric structure of the magnetic field cannot be determined directly from observations. Cowling (1946) originally considered two possibilities: a monolithic flux tube extending deep into the convective zone or a local dipolar structure maintained by currents near the surface. Of these, he clearly preferred the monolithic structure sketched in Figure 4*a* (Cowling 1957). Parker (1975), on the other hand, favoured a configuration like that in Figure 4*b*, where the constricted field splits into many independent flux tubes immediately below the surface, so that the structure resembles a magnetic jellyfish. We now know that both these pictures are too naive. Radiative transport alone is insufficient to supply the energy emitted from the umbra (Schlüter and Temesváry 1958) and convection must take over just below the visible surface. To be effective, convection must distort the field, so the monolithic plug of Figure 4*a* should be replaced by a coherent but inhomogeneous structure, like that sketched in Figure 5*a*. There are several arguments against the constricted tube of Figure 4*b*. First, the surface field would not fall off monotonically with increasing radius; secondly, there is no means of stabilizing it against flute instabilities at the waist (Parker 1975; Meyer, Schmidt and Weiss 1977); furthermore, such a configuration would not be able to sustain coherent oscillations, such as the umbral oscillations that have been observed (see the reviews by Lites (1992) and Chitre (1992) in this volume); in addition, the systematic properties of sunspots (rotation rates, polarities and inclination of spot groups) become difficult to explain. In its modern form, the jellyfish has developed into a tight bundle of isolated flux tubes, separated by field-free plasma, as shown in Figure 5*b* (Obridko 1974; Parker 1979a,b,c; Spruit 1981a,c; Zwaan 1981). Each of the configurations in Figure 5 has its advocates and each is compatible with the available observations (cf. Moreno-Insertis 1987). To decide between them would require a detailed, quantitative theory of nonlinear magnetoconvection, which cannot yet be provided.

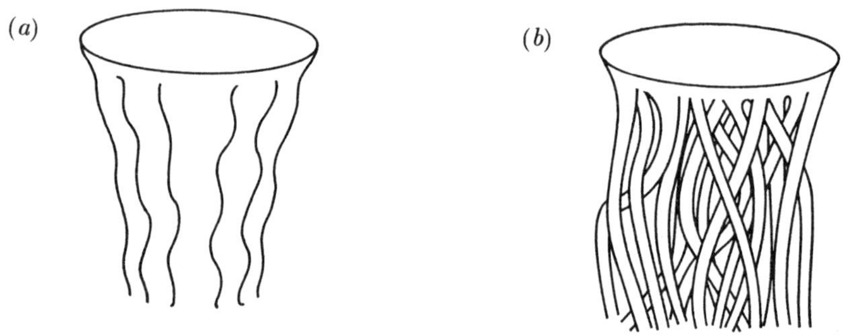

Figure 5. Sunspot models that are compatible with observations. (*a*) Coherent flux tube and (*b*) tight cluster.

The lifetime of a sunspot is much longer than the dynamical timescale (the Alfvén transit time is only an hour) so a spot is manifestly in a state of stable magnetohydrostatic equilibrium. The development of increasingly sophisticated equilibrium models is described in Jahn's (1992) chapter in this volume. In an idealized sunspot the field is axisymmetric and purely meridional. If the spot is assumed to be horizontally stratified it follows that the field within it must be force-free, and therefore current-free. Hence we consider an isolated flux tube with a potential field $\mathbf{B}$, referred to cylindrical polar co-ordinates (s, ϕ, z), that is maintained by azimuthal surface currents at its boundary. At this boundary – the magnetopause – the magnetic field drops discontinuously to zero but the normal stress must be continuous, so that $B^2 = 8\pi\Delta P$, where $\Delta P(z)$ is the difference between the external and internal gas pressures. In the thin flux tube approximation the radial variation of B_z is ignored; other illustrative models can be obtained by taking potential fields derived from well-known solutions of Laplace's equation (e.g. dipole or Bessel function potentials) and matching ΔP and its derivative at the surface (Simon and Weiss 1970; Spruit 1976; Simon, Weiss and Nye 1983). To construct a proper solution it is necessary to choose the radius $R(z)$ of the flux tube so that the normal stress is continuous at the magnetopause. Wegmann (1981) developed an elegant method for solving this free boundary problem, with appropriate boundary conditions at the top and bottom of the flux tube, and this method was implemented by Schmidt and Wegmann (1983; see Schmidt 1991). This approach yields valid models for a pore but it fails to describe penumbral structure in a sunspot. The photosphere in the umbra is depressed by 500–700 km below its level outside the spot and the pressure difference ΔP at the level of the Wilson depression corresponds to a field strength of about 5000 G, far greater than the measured value at the centre of a spot. So it is necessary to allow electric currents within the flux tube, with an associated radial gradient in gas pressure that is balanced by curvature of the magnetic field lines. Indeed, such currents had already appeared in the spot models of Schlüter and Temesváry (1958), Chitre (1963), Deinzer (1965), Yun (1970) and Pizzo (1986).

As we have already mentioned, a flux tube may be liable to interchange (or fluting) instabilities if the magnetopause is concave towards the field-free plasma (Parker 1975). Magnetic buoyancy provides a stabilizing mechanism: the fall in gas pressure across the magnetopause produces a drop in density, so that the configuration tends to be stable if the magnetized plasma floats above the field-free region. (This is the converse of Rayleigh-Taylor instability.) Whether a flux tube is stable or not depends on the competition between these two effects. For the case of a potential field immersed in an adiabatically stratified layer the energy principle of Bernstein *et al.* (1958) can be used to show that the magnetopause is stable if $d|B_s|/dz > 0$ on the boundary $s = R$ (Meyer, Schmidt and Weiss 1977). The flux tube flares outwards, owing to the decrease in gas pressure with height z, and the stability condition requires that the increase in the inclination, χ, to the vertical should be more important than the reduction in field strength. An alternative expression of the stability criterion, which explicitly reveals the roles of curvature and inclination, is that the ratio of the pressure scale-height to the radius of curvature of the magnetopause should be less than $\frac{1}{2}\sin\chi$. This local condition is satisfied at depths less than 2000 km for simple sunspot models, provided the magnetic flux is sufficiently large.

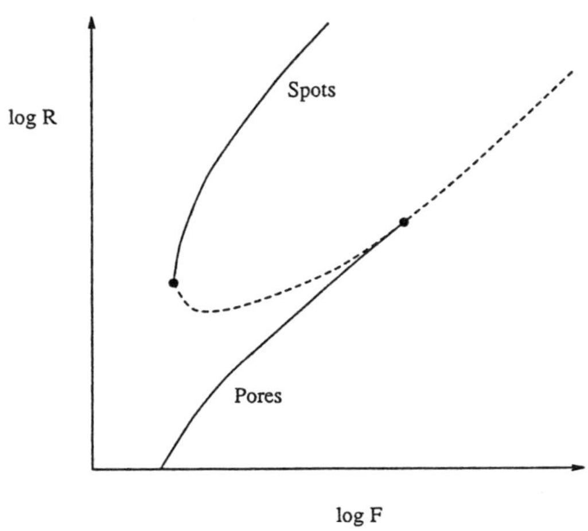

Figure 6. Pore-spot bifurcation diagram in $F - R$ plane. Solid and broken lines indicate stable and unstable solutions, respectively.

In these models the magnetopause becomes increasingly inclined as the magnetic flux F within the tube is increased. In principle, it should be possible to construct an infinite sequence of models, depending on F as a parameter, that are in magnetohydrostatic and thermal equilibrium. These would correspond to pores rather than to sunspots. The observations tell us that, for radii greater than about 3500 km, corresponding to fluxes of 4×10^{20} mx, these models are unstable to convective modes that lead to the formation of a sunspot with a thick penumbra (Bray and Loughhead 1964; Simon and Weiss 1970). Since the smallest sunspots have radii of about 1800 km, corresponding to fluxes of 10^{20} mx, the bifurcation must be subcritical. The associated bifurcation structure is sketched in Figure 6, which shows a branch of unstable spot configurations that acquires stability in a saddle-node bifurcation at the turning point, so allowing the observed hysteresis. To compute such a sequence of models would, however, require a detailed theory of penumbral convection.

4. Fine Structure of Sunspots

4.1. MAGNETOCONVECTION IN THE UMBRA

Semi-empirical umbral models, based on observations, show that radiation is responsible for energy transport at the visible surface of the umbra, where the continuum optical depth $\tau_{0.5} \approx 1$, but that the fraction of the energy that is carried by radiation drops rapidly for $\tau_{0.5} \gtrsim 3$ (Maltby 1992). Theoretical models of sunspots confirm that convective transport takes over within a few scale-heights below the photosphere. At that level there must therefore be horizontal variations in temperature. The only direct evidence of convection comes from umbral dots, whose interpretation still remains unclear. Their properties are discussed by Muller (1992) elsewhere in this volume. Although isolated "central" umbral dots (Grossman-Doerth, Schmidt and Schröter 1986), like those visible in Figure 7, seem to be present in all large sunspots, Maltby (1992) finds that spots with umbral radii greater than 3000 km have dark patches in their cores within which no variation of intensity can be detected. If so, the horizontal temperature gradients needed to drive convection must be smoothed out owing to the efficiency of radiative transfer in the photosphere.

Theoretical studies have established that the imposition of a strong vertical magnetic field impedes convection in an electrically conducting fluid. It follows intuitively that sunspots should achieve an equilibrium in which they are cooler than the normal photosphere; this can be verified by constructing model umbral atmospheres, whose properties depend also on the variation of opacity with density and temperature. Attempts to understand convective processes in sunspots have inspired detailed studies of magnetoconvection, which has become a significant subject in its own right, as can be seen from the recent review by Weiss (1991) and the chapter by Proctor (1992) in this volume.

Early investigations were restricted to the Boussinesq approximation. The linear theory developed by Chandrasekhar (1961) was extended by Danielson (1961b) and Weiss (1964a); Musman (1967) and Savage (1969) later explored the effects of allowing penetration into a stably stratified region (see earlier reviews by Danielson 1965, 1966 and Weiss 1969, 1977). The key feature of this theory is the tendency to favour oscillatory convection if ζ, the ratio of the magnetic to the thermal diffusivity, is less than unity. In a sunspot umbra, $\zeta \approx 10^{-3}$

at the photosphere, where radiative diffusion is more effective than molecular resistivity; hence we would expect to find oscillatory convection, in narrow elongated cells with motion that is predominantly vertical (Cowling 1976). This provides an obvious explanation for umbral dots (Danielson 1964; Knobloch and Weiss 1984). However, the opacity increases rapidly with depth, owing to ionization of H and He, leading to a sharp reduction in the radiative diffusivity, and so there is a range of depths between 2000 and 20000 km where $\zeta > 1$ (Meyer *et al.* 1974).

Nonlinear Boussinesq magnetoconvection has been extensively investigated (Galloway and Moore 1979; Weiss 1981a,b; Proctor and Weiss 1982). For $\zeta < 1$ and a sufficiently strong field there is a branch of periodic solutions, corresponding to thermally destabilized hydromagnetic oscillations. As the superadiabatic gradient is increased, nonlinear oscillations eventually give way to steady overturning convection. Magnetic flux is then confined to stagnant sheets or tubes, so that the field is segregated from the motion. For $\zeta \geq 1$, however, this segregation is no longer found.

Compressible convection is more complicated. The modes that are destabilized depend on the plasma beta, $\beta = 8\pi P/B^2$. In the Boussinesq limit, where $\beta \gg 1$, the overstable oscillations correspond to trapped slow magnetoacoustic waves. When $\beta \ll 1$, the slow magnetoacoustic mode is again destabilized but now it corresponds to sound waves travelling along the field lines (Syrovatsky and Zhugzhda 1967; Zhugzhda 1970). There is a complicated transition for $\beta \approx 1$ (Cowling 1976; Obridko 1979), where the fast and slow modes are linked by the stratification and even a subadiabatically stratified layer may become unstable (Antia and Chitre 1979; Cattaneo 1984). Once again, overstable oscillations only appear through double-diffusive effects, which introduce phase differences between the thermal and magnetic perturbations. In the absence of any diffusion there is only a direct instability, leading to exponentially growing modes (Moreno-Insertis and Spruit 1989).

Recent work has emphasized the nonlinear compressible regime. Most published results have been for idealized two-dimensional model calculations. Steady solutions were computed by Hurlburt and Toomre (1988) and numerical experiments by Hurlburt *et al.* (1989) have revealed a wealth of time-dependent behaviour, involving standing waves, travelling waves, modulated waves and mixed-mode oscillatory solutions. The chapter by Proctor (1992) in this volume summarizes these results. Work on three-dimensional magnetoconvection is now in progress; indeed, Nordlund and Stein (1989) have already reported some preliminary results obtained in an ambitious attempt to simulate three-dimensional convection in a sunspot umbra.

Any attempt to explain the pattern of convection in a sunspot requires a bold extrapolation from the available numerical results. It is clear that magnetic inhibition of convection is most effective within 1500 km of the photosphere, where the field is dynamically strong, the stratification is highly superadiabatic and $\zeta \ll 1$. It is here that the magnetic throttle, which limits energy transport to the umbra, is located. The temperature difference, ΔT, between the sunspot umbra and the mean undisturbed atmosphere at the level of the Wilson depression is about 9000 K but ΔT decreases rapidly with depth, z, below the normal photosphere. The estimated value of ΔT falls to 500 K at $z = 2000$ km and then to 25 K at $z = 6000$ km (Meyer *et al.* 1974): a spot would be a shallow phenomenon if it were defined by its thermal properties alone.

Following Meyer *et al.* (1974), we can distinguish schematically between three regions. In the topmost layer (500 km $\lesssim z \lesssim$ 2000 km), where $\zeta < 1$, convective transport is inefficient; oscillations are favoured locally and energy is transported vertically rather than across the field. Then comes a region (2000 km $\lesssim z \lesssim$ 10000 km) where $\zeta > 1$. These circumstances favour steady overturning convection, which is likely to become time-dependent in a highly nonlinear regime. Numerical experiments, both for stratified layers and in the Boussinesq approximation, show no tendency for the formation of isolated flux tubes, with the field segregated from the convective motion, when $\zeta > 1$ (Knobloch and Weiss 1984; Weiss *et al.* 1990). In this region there may be significant lateral transport, bringing energy in from the ambient plasma. However, the sunspot models constructed by Jahn and Schmidt (1992; Jahn 1992) suggest that the energy supplied to the umbra may be channelled along field lines, as originally suggested by Hoyle (1949). Moreover, the systematic variation in umbral temperature with the solar cycle suggests that the inner part of the flux tube must be thermally insulated from its surroundings. Finally, there is a deep layer, with $z \gtrsim 10000$ km, where ζ gradually increases. External convection in this region may provide the collar that is needed to suppress the residual, weakly unstable interchanges.

It is not obvious how overturning convection in the layer with $\zeta > 1$ is coupled to oscillatory convection immediately below the photosphere (Obridko 1974; Knobloch and Weiss 1984). Weiss *et al.* (1990) approached this problem by investigating two-dimensional nonlinear magnetoconvection in a stratified layer across which the density increased by a factor of 11, with a corresponding increase in the value of ζ, such that $\zeta < 1$ in the upper half and $\zeta > 1$ in the lower half of the region. Although convection sets in as a direct mode, leading to a branch of steady solutions, subsequent bifurcations lead to spatially modulated periodic solutions (see Fig. 8 of Proctor 1992). The sense of motion does not vary at the base of the layer but the flow reverses at the top as alternate plumes force their way through to the surface. Moreover, the preferred horizontal scale is controlled by the upper part of the region, where the magnetic field is dynamically most important. These model calculations suggest that umbral convection should have a small horizontal scale, with localized hot plumes, associated with weakened magnetic fields, appearing as spasmodic bursts. The numerical results are consistent with the appearance of central umbral dots, though local variations of intensity are apparently too slight to be detected in the umbral cores.

Episodic behaviour in the region immediately below the visible surface has to be superimposed on a more ordered pattern of convection at greater depths. Is convection there sufficiently vigorous to produce an intermittent magnetic structure, with many isolated flux tubes within the spot? Idealized numerical experiments at Rayleigh numbers, Ra, that are only moderately supercritical ($Ra \leq 10^5$, where convection sets in at $Ra \approx 8 \times 10^3$) show no sign of such a segregation if $\zeta > 1$ throughout the layer. In a spot, convection is much more vigorous and will certainly become time-dependent, with substantial variations in field strength. However, it does not follow that fluid elements are perpetually field-free or else attached to strong fields. Instead, they will experience irregular variations in field strength, and it may be possible to estimate the filling factor for strong fields at any instant. In one limit, convection can be modelled by assuming that there are isolated flux tubes separated by field-free plasma and that convective transport is confined to the field-free regions, while the flux tubes are heated by radiative transport only (Parker 1979a,c; Choudhuri 1986). Choudhuri (1992) and Parker (1992), in their chapters in this volume, claim that this extreme model is exact: they argue that oscillatory convection in the field-free plasma penetrates to just below the photosphere, giving rise to umbral dots, while the field below is permanently confined to a tight cluster of individual flux tubes. As yet, however, there is no nonlinear calculation that supports their picture.

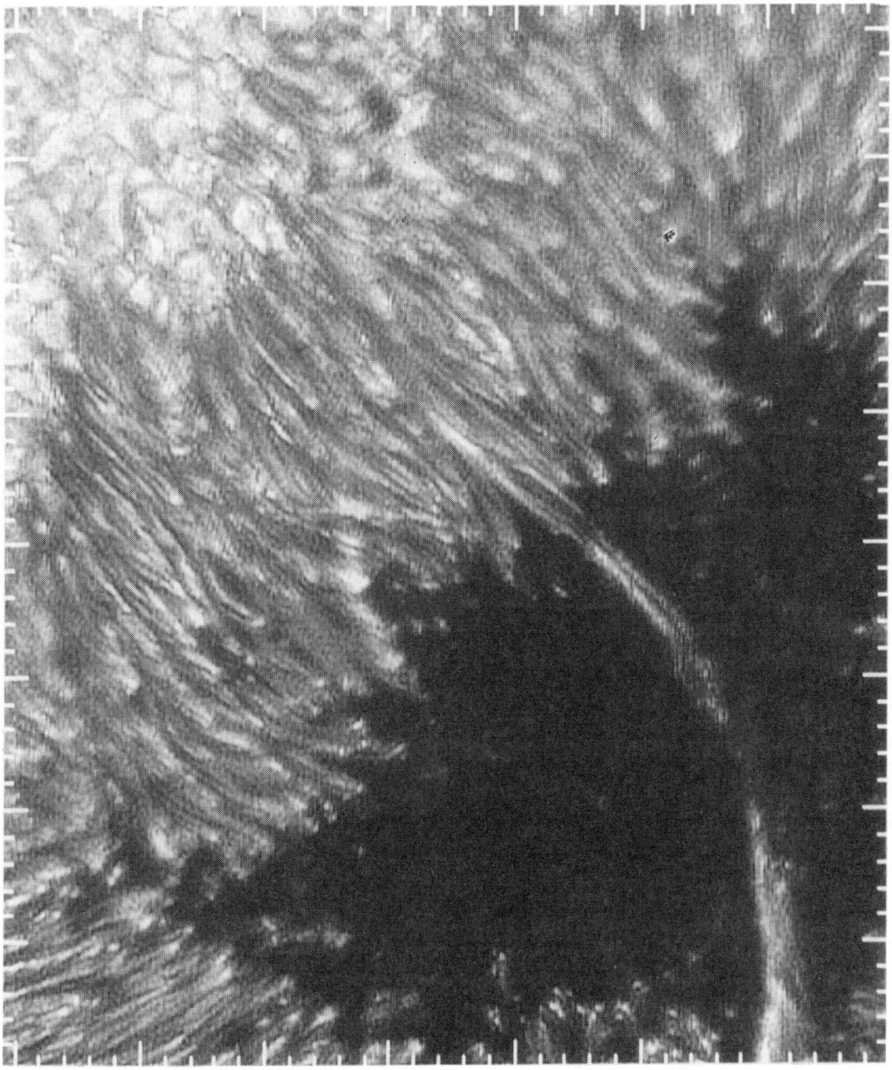

Figure 7. High-resolution image of part of the umbra and penumbra of a sunspot, taken during exceptional seeing with the Swedish Solar Telescope on La Palma and processed digitally at the Lockheed Palo Alto Research Laboratories. The tick marks are at 1 $''$ intervals and structures as small as 0.22 $''$ can be detected. (Courtesy of Solar and Optical Physics Department of LPARL.)

4.2. PENUMBRAL FINE STRUCTURE

Our understanding of penumbral structure has developed as observations have been obtained with ever-increasing resolution. The radial striations recorded by Secchi in 1870 (see Figure 1*b*) are revealed as alternating bright and dark filaments in the Stratoscope image reproduced in Figure 2. Danielson (1961a) found that the bright filaments have lengths of about 5000 km, widths of about 300 km and lifetimes of around 45 min. The exceptionally clear white-light picture from the Pic du Midi in Figure 3 shows yet more detail. As Muller (1992) explains in his chapter in this volume, the bright filaments are composed of elongated grains which migrate towards the umbra (Muller 1973a, 1976). Figure 7 displays the fine structure with even greater clarity: this remarkable picture was obtained using the Swedish Solar Telescope on La Palma. Digital processing techniques, originally developed at the Lockheed Palo Alto Research Laboratories in connection with the Spacelab-2 mission in 1985, make it possible to correct both for the telescope function and for atmospheric distortion. The image in Figure 7 has been processed so as to emphasize fine structure both in the penumbra and in the umbra, where a light bridge and several isolated umbral dots are visible.

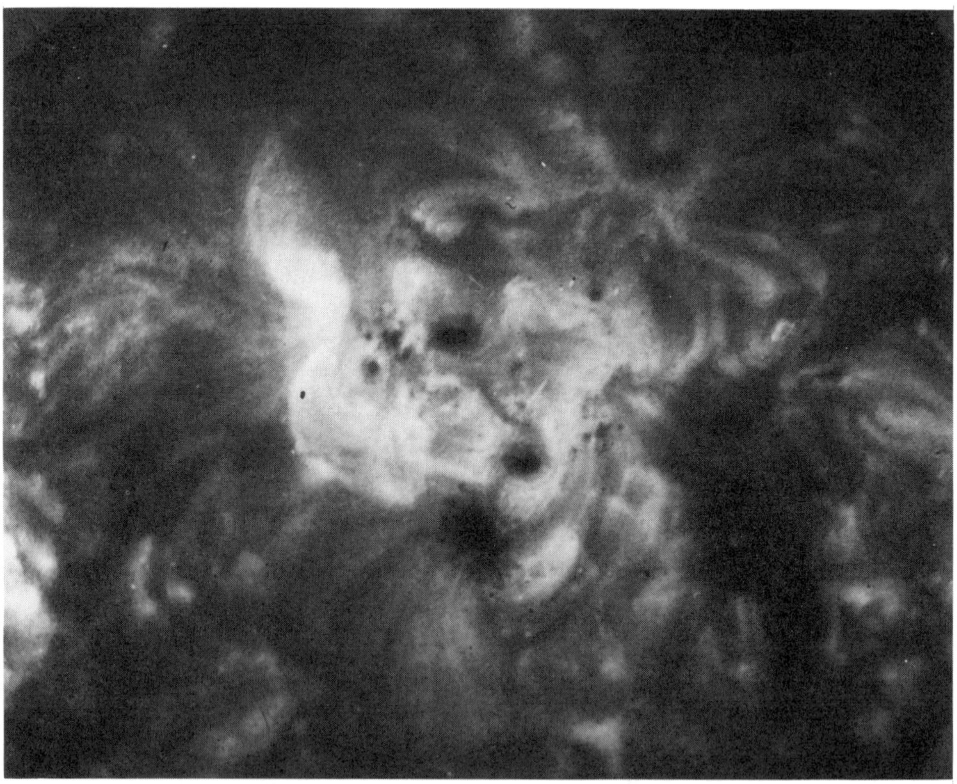

Figure 8. NIXT Combined X-ray and white-light image, showing magnetic flux tubes emerging from the penumbra into the corona and extending for more than 10^5 km. (Courtesy of IBM Research and SAO.)

At the outer edge of the penumbra the dark filaments appear to lie above the granulation (Moore 1981), though they are probably depressed below the normal photosphere (Schmidt, Spruit and Weiss 1986). The filamentary structure that is so prominent in Hα images is quite different: the so-called superpenumbra lies higher in the atmosphere and extends to distances of twice the spot radius or more (Bray and Loughhead 1974; Stix 1989). Filamentary features in the corona extend even further. The photograph in Figure 8, obtained from a rocket with the Normal Incidence X-Ray Telescope, shows coronal structures, visible in soft X-ray emission, superposed exactly on a white-light photospheric image. Several sunspots can be seen. The X-ray filaments that emerge from the inner penumbra of the large spot follow field lines that curve round over distances of 3×10^5 km before returning to the surface of the Sun (Sams, Golub and Weiss 1992).

What is the geometry of the magnetic field associated with penumbral filaments? This has been a controversial issue for over 20 years. Observations by Beckers and Schröter (1969; Beckers 1981) already suggested that the field in bright filaments is significantly inclined to that in dark filaments, which is almost horizontal and parallel to the Evershed flow. This structure has at last been confirmed by the high-resolution observations that are described in the chapter contributed by Title *et al.* (1992) to this volume. It is now clear that the magnetic field varies rapidly in inclination – but not in magnitude – in the azimuthal direction. At the outer edge of the penumbra the field in dark filaments is almost horizontal, while that in bright filaments is elevated by about 35° above the horizontal, so that the mean field is inclined at 15–20° (Title *et al.* 1991, 1992). The connection between this structure and the Evershed flow is discussed in Section 4.3 below. Near the inner boundary of the penumbra the mean field is inclined at about 45° with fluctuations of ±15–20° between bright and dark filaments. Thus the magnetic field has an interlocking comb-like structure, as illustrated in Figure 9. Furthermore, the X-ray observations in Figure 8 show that a significant fraction of the magnetic flux emerging from bright filaments in the inner penumbra is connected to footpoints at great distances from the spot.

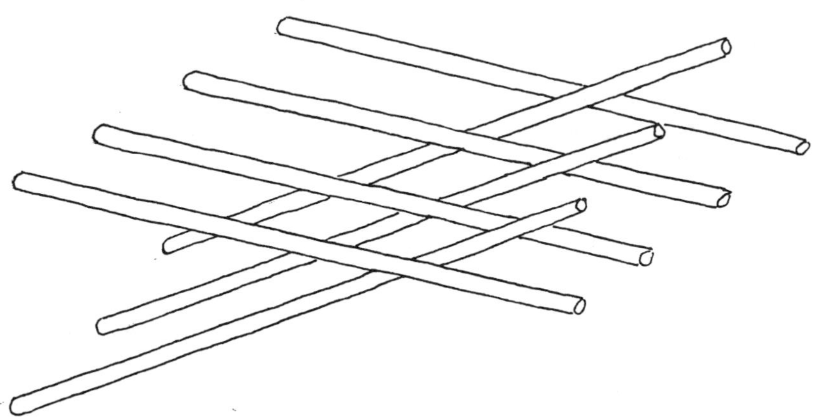

Figure 9. Sketch showing the interlocking comb-like structure of the magnetic field in bright and dark penumbral filaments.

It is important to gain a proper theoretical understanding of the filamentary penumbra, not only for its own sake but also as the key to explaining the overall equilibrium of a sunspot. Although it is obvious in general terms that radial filaments must be caused by the strong inclined magnetic field, which resists lateral distortion, the intricate penumbral structure is exceptionally difficult to model. In a shallow layer with an inclined magnetic field convection first sets in as rolls with their axes parallel to the horizontal component of the field (Chandrasekhar 1961). Danielson (1961b) and Pikel'ner (1961) therefore suggested that penumbral filaments correspond to convection rolls aligned with a mean field that is almost horizontal. Danielson assumed that these rolls have a depth of 300 km, comparable with their width, and applied linear Boussinesq theory to show that steady convection could occur. This picture implies that the penumbra is extremely shallow – indeed, the bright filaments might even be produced by penetration from beneath the magnetopause. Shallow penumbral models have tempted other theoreticians too (Schmidt, Spruit and Weiss 1986) but such models are demonstrably inconsistent (Schmidt 1987, 1991; Weiss 1990). The field at the penumbral boundary has a strength of about $\frac{1}{3}B_0$, where B_0 is the field at the centre of the umbra. If $\mathbf{B}$ is horizontal at the spot boundary ($s = R$) then we can suppose that the radial field is constant over a hemisphere of radius R, so that the total flux $F \approx \frac{2}{3}\pi R^2 B_0$. Now the umbra has a radius $r_u \approx 0.4R$ and therefore only contributes a flux $F_u < 0.16\pi R^2 B_0$. Hence more than three-quarters of the flux must emerge through the penumbra, which is impossible if the field there is horizontal. In fact, we now know that the mean field at the outer edge of the penumbra is inclined at 20° to the horizontal, so the total flux $F \approx 0.44\pi R^2 B_0$ and is made up of roughly equal contributions emerging through the umbra ($r < r_u$), the inner penumbra ($r_u < r < 0.7R$) and the outer penumbra ($0.7R < r < R$). This is only possible if the magnetopause is inclined and it follows that the penumbra must therefore be deep. More specifically, we expect the magnetopause to be at a depth of order 3000 km at the middle of the penumbra.

There is some evidence to suggest that bright filaments are moving upwards relative to dark filaments but the observational results are inconclusive (Beckers 1981; Muller 1992). On the theoretical side, there have been attempts to extend Danielson's model (Galloway 1975; Busse 1987) but no one has succeeded in describing a convective interaction that produces a two-component field. Moreover, linear and weakly nonlinear studies of compressible convection in an inclined magnetic field suggest that waves travelling horizontally along the field are the favoured mode (Matthews *et al.* 1992) and these nonlinear travelling waves have been demonstrated in numerical experiments (Hurlburt *et al.* 1992).

Heat transport in the penumbra is more likely to be dominated by the inclined magnetopause. Magnetic buoyancy, which stabilizes this boundary, also limits the extent to which the magnetopause is penetrated by convection from below. A simple estimate, obtained by balancing the kinetic energy of a convecting plume against the potential energy gained (Schmidt, Spruit and Weiss 1986) suggests that corrugations of the magnetopause may have an amplitude of around 300 km. These distortions will trigger motions within the penumbra. Spruit (1981b, 1987) proposed that there is an exchange between flux tubes, or sheets, lying approximately parallel to the magnetopause and moving in the perpendicular direction (see Fig. 2 of Jahn 1992). In this model of penumbral convection there is a continuous interchange between hot rising and cold sinking flux sheets (Schmidt 1991; Jahn 1992). An individual flux tube gains thermal energy in the vicinity of the magnetopause and moves upwards and inwards, so that its intersection with the photosphere migrates radially inwards. This provides an explanation for the proper motion of bright grains (Muller 1973a,

1976). After radiating its energy the flux tube cools and becomes more nearly horizontal at the photosphere (cf. Wentzel 1992); then it drifts back towards the magnetopause. Superimposed on this motion is the Evershed flow, which opposes the inflow in bright filaments and augments the outward proper motion of dark filaments (Spruit 1981b). This general picture of the convective process in the outer penumbra is certainly consistent with the available observations. Unfortunately, it is not easy to devise a well-posed model problem that would allow a quantitative treatment. The geometry is awkward and the boundary conditions are uncertain. Several effects are simultaneously involved and it is hard to isolate them individually; on the other hand, a full three-dimensional simulation would be extremely complicated and difficult to interpret. For the moment, then, this process has to be represented by parametrized models.

Jahn (1989) has constructed axisymmetric sunspot models with a thick penumbra. In these models, which are described elsewhere in this volume (Jahn 1992), there is a continuous distribution of azimuthal volume currents in the penumbra, in addition to the surface current at the magnetopause. The magnetopause itself is a free surface, adjusted to obtain a model that is in magnetohydrostatic equilibrium, with a surface field that matches observations. The resulting configuration is illustrated in Figure 1*b* of Jahn (1992). As expected, the penumbra tapers gradually downwards but it extends to a depth of order 10^4 km. More recently, Jahn and Schmidt (1992) have constructed a tripartite model; instead of following the peripatetic motions of flux tubes in the penumbra they introduce an inner current sheet (the 'peripatopause') that separates the umbra from the penumbra, which is current-free. The umbra, the penumbra and the external plasma are each horizontally stratified and in thermal as well as magnetohydrostatic equilibrium. In these models it turns out that most of the energy flux through the penumbra has to be drawn through the magnetopause from the external plasma, while the umbra is heated directly from below (Jahn 1992).

Behaviour near the umbra-penumbra boundary raises several complicated issues. The bright grains move inwards, with apparent velocities around 0.5 km s^{-1} (Muller 1973a, 1992), although no corresponding Doppler shift has been detected (Zirin and Wang 1991). Either the proper motion is a travelling wave or, more likely, the inward motion is at least partially cancelled by the horizontal component of the Evershed flow along the inclined field lines (Spruit 1981b; Schmidt, Spruit and Weiss 1986). These bright features extend into the dark umbra, as can be seen in Figures 2, 3 and 7. The Stratoscope photographs showed changes over an interval of 10 min (Danielson 1961a) and Scharmer (1991) has produced a dramatic video sequence showing bright filamentary features, like those in Figure 7, which move inward across the umbra-penumbra boundary and then break up into several smaller grains, which resemble umbral dots. It seems likely that this is the origin of "peripheral" umbral dots (Grossman-Doerth, Schmidt and Schröter 1986). Indeed, Livingston (1991) has argued that *all* umbral dots lie on filaments connected to the penumbra. These features should probably be regarded as penumbral intrusions in the umbra, presumably associated with magnetic fields whose inclination to the vertical is greater than that around them. Conversely, the X-ray images in Figure 8 suggest that there are umbral inclusions in the penumbra; these inclusions have fields that are more nearly vertical and they do not take part in the convective interchanges (Sams, Golub and Weiss 1992).

Our present understanding of energy transport in a sunspot is summarized in Figure 10. Near the axis, energy is carried vertically to the umbra by convective processes that are magnetically dominated and become increasingly inefficient near the surface. This energy is either transported from great depths or drawn from the thermal energy of the flux tube at depths of order 5×10^4 km (such that the relevant Kelvin-Helmholtz time is much longer than 11 years – cf. Spruit 1992). The systematic variation of umbral intensity during the solar cycle makes it unlikely that energy is brought in laterally to supply the umbral core (though it is conceivable that there are systematic variations in the magnetic throttle). The penumbra is supplied predominantly from outside: although the penumbral intensity also varies with the solar cycle, the relative amplitude of the variation is much less than for the umbra (Albregtsen and Maltby 1978), suggesting that only 10% of the energy is actually carried vertically up the flux tube, in agreement with the models of Jahn and Schmidt (1992). The rest relies on convective interchanges that bring individual flux tubes into contact with the magnetopause. We may speculate that those filaments that extend into the umbra have been in contact with the external plasma and that this is still true of the isolated flux tubes that give rise to central umbral dots. It is even possible that peripheral umbral dots are associated with field-free plasma that crosses the magnetopause, though it remains unlikely that any such plasma can penetrate to the umbral core.

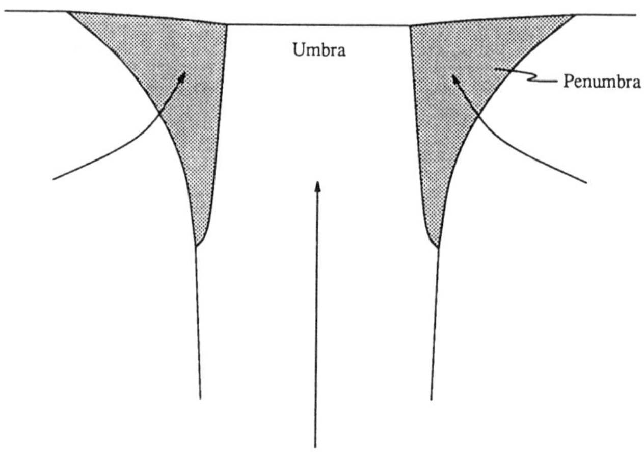

Figure 10. Sketch illustrating energy transport in the umbra and penumbra.

4.3. THE EVERSHED FLOW

Perhaps the most readily apparent dynamical phenomenon in a sunspot is the horizontal Evershed outflow in the penumbral photosphere, observed as a Doppler shift and asymmetry of photospheric spectral lines in sunspots away from disc centre, as described by Muller (1992) in this volume. The asymmetry in the line profile is caused by the vertical gradient in the Evershed flow speed and by differences in flow speed between the bright and dark filaments. Low-frequency wave motions can also contribute to the asymmetry; indeed, Maltby and Eriksen (1967) once suggested that the Evershed effect might be due entirely to wave motions (see also Lamb 1975), but this seems to have been disproved by observations at higher spatial resolution. The Evershed flow is a common feature of all fully developed sunspots. It is clearly an important characteristic of sunspot structure and it demands a satisfactory theoretical explanation, but this has proved difficult. The Evershed flow and the filamentary structure of the penumbra are intimately related and must be considered together in a complete theoretical model.

The speed of the overall photospheric Evershed flow is typically 1–2 km s^{-1} when measured at a low spatial resolution of a few arcseconds. Measurements at somewhat higher spatial resolution (one arcsecond) begin to show horizontal inhomogeneities in the flow, with speeds reaching as high as 6 km s^{-1}. The average flow speed decreases with height (i.e., with increasing line strength) and reverses above a certain height to form the reversed Evershed inflow in the chromosphere (St. John 1913; Kinman 1952). At a fixed height the flow speed increases outward across the penumbra, reaching a maximum at or near the outer penumbral boundary. The radial position of the maximum speed varies systematically with the height of formation of the spectral line observed, moving outward for lines formed higher up (Ichimoto 1987). The Evershed effect disappears rather abruptly at the outer penumbral boundary (see, for example, Wiehr *et al.* 1986), not because the flow stops abruptly (which would have observable dynamical consequences), but because the gas flowing along field lines slightly inclined to the horizontal becomes suddenly transparent to the observed radiation (Makita 1963; Zwaan 1968; Thomas 1981).

The earliest observations with sub-arcsecond resolution showed that the photospheric Evershed flow is largely concentrated into the dark penumbral filaments (Beckers 1968, 1969) where the magnetic field is more nearly horizontal (Beckers and Schröter 1969). Some observers have disagreed with this result (see, for example, Wiehr and Stellmacher 1989; Lites, Scharmer and Skumanich 1990), finding no particular spatial relation between the Evershed flow and continuum intensity variations. However, the recent high-resolution observations reported by Title *et al.* (1992) in this volume confirm that the Evershed flow is structured on the same spatial scale as the penumbral filaments, with the flow occurring primarily in the dark penumbral filaments where the field is more nearly horizontal (but not necessarily stronger) than average. Shine *et al.* (1990) find that the flow occurs in intermittent clouds which recur at intervals of about 10 min. These clouds expand as they move outwards, becoming wider than the filaments.

The reversed, inward Evershed flow in the chromospheric superpenumbra, observed in Hα, is similarly concentrated into the dark fibrils (Beckers 1964; Haugen 1967; Maltby 1975). The average flow speed in the superpenumbra above the outer boundary of the photospheric penumbra is about 7 km s^{-1}, but the flow speed in individual filaments can be 20 km s^{-1} or more (Haugen 1969; see also Alissandrakis *et al.* 1988). The overall geometry and strength of the chromospheric Evershed flow is consistent with the siphon flow model of Meyer and Schmidt (1968), which is discussed below. In some filaments the flow is observed to slow abruptly at a location of enhanced line emission (Maltby 1975), strongly suggestive of the standing shock wave in a supersonic siphon flow.

Early theoretical interpretations of the Evershed flow focused on convectively driven motions. Sweet (1955) considered the photospheric Evershed flow to be part of a large-scale convective motion driven by the excess temperature below the sunspot due to the buildup of the blocked solar heat flux. In this model, the Evershed flow is partly responsible for carrying the blocked heat flux and thus avoiding a bright ring around the spot. Our current understanding of the fate of the blocked heat flux – see the article by Spruit (1992) in this volume – and the fine structure of the Evershed flow rules out Sweet's model. Other convective mechanisms for the Evershed flow have been proposed (e.g., Wilson 1968; Chitre 1968). Danielson (1961b) proposed that the penumbral filaments themselves are a convective phenomenon; on the basis of linear stability theory for the onset of convection in a magnetic field, he argued that the bright and dark penumbral filaments are produced by long convective rolls aligned with a nearly horizontal mean magnetic field in the penumbra. Danielson's linear theory does not imply a radial outflow, but Galloway (1975) showed how the convective rolls could produce such a flow as a nonlinear effect. The convective rolls concentrate the magnetic flux into the dark (downwelling) filaments and then an outward radial motion is driven in the dark filaments by the Lorentz force due to the distorted magnetic field lines. On the other hand, Busse (1987) ascribed the Evershed effect to Reynolds stresses produced by three-dimensional convection in an inclined magnetic field, as demonstrated by a highly idealized, weakly nonlinear model. Although a convective mechanism for the photospheric Evershed flow cannot be ruled out, no one has yet produced a convective model that produces light and dark filaments inclined at 40 degrees to each other and drives a flow along the dark filaments. (On the other hand, no siphon-flow model has yet included the effects of the convective interchange motions that are clearly taking place in the penumbra.)

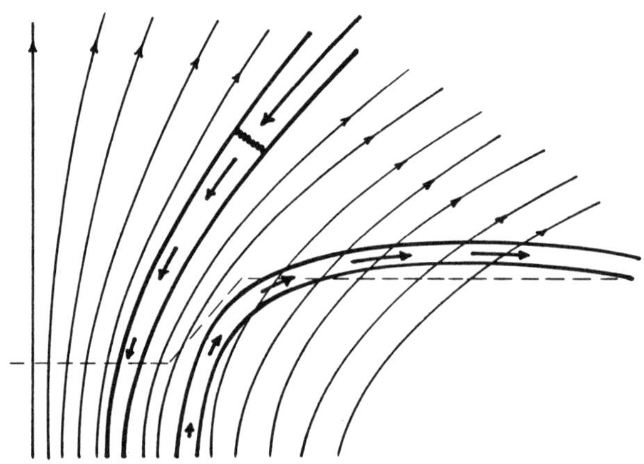

Figure 11. Sketch of the siphon-flow model of the Evershed flow. The photospheric Evershed flow consists of outward siphon flows in nearly horizontal penumbral flux tubes. The reversed, chromospheric Evershed flow is due to inward siphon flows along flux tubes emanating from the umbra. These chromospheric flows may be supersonic, in which case they will decelerate abruptly to subsonic speed at a standing shock wave (as shown).

Meyer and Schmidt (1968) proposed that the photospheric and the reversed chromospheric Evershed flow are both due to siphon flows along arched magnetic flux tubes, driven by pressure differences between the footpoints of the flux tubes. Their original paper on siphon flows is remarkably concise and rich in new ideas. Consider an arched magnetic flux tube with two footpoints at the same geometric height. For higher arched flux tubes, with one footpoint in the umbra and the other footpoint somewhere outside the sunspot in the quiet photosphere, the gas pressure will be lower at the umbral footpoint (where the magnetic pressure is higher), so the flow along the arch will be directed inward toward the umbra. This flow corresponds to the reversed Evershed flow observed at chromospheric heights. For lower arched flux tubes originating in the penumbra, where the magnetic field strength is of the order of 1000 G, and connecting to another footpoint with a higher magnetic field strength (in the umbra of another sunspot, for example), the flow will be outward away from the umbra, as is the case in the observed photospheric Evershed flow. This configuration is sketched in Figure 11.

Meyer and Schmidt's siphon-flow model was proposed before the discovery, in the early 1970s, that nearly all of the magnetic flux outside sunspots is concentrated into small magnetic elements with field strengths of 1200 to 1500 G (see section 6.2). That discovery nicely clarified Meyer and Schmidt's explanation of the normal and reversed Evershed flow, as pointed out by Spruit (1981b). If we imagine for simplicity that all the flux tubes emanating from a sunspot have as their opposite footpoint a magnetic element of field strength 1500 G, then the flow will be outward for low-lying tubes with footpoints in the penumbra (where the field strength is less than 1500 G) and inward for higher-reaching tubes with footpoints in the umbra (where the field strength is greater than 1500 G).

Following Meyer and Schmidt's paper, there were several theoretical studies of siphon flows in coronal loops (see the review by Priest 1981). All of these studies, including that of Meyer and Schmidt, were based on the assumption that the shape of the magnetic flux tube is rigid and is unaffected by the siphon flow. This assumption is justified in the chromosphere and corona, where the plasma beta is small, but not in the photosphere, where the plasma beta of the flux tube is of order unity. Thomas (1984a, 1988) investigated the effects of relaxing this assumption and allowing the cross-sectional area of the flux tube to change in response to variations in the internal gas pressure induced by the siphon flow. In this case the critical speed becomes the tube speed $c_t = [c^2a^2/(c^2+a^2)]^{\frac{1}{2}}$, where c is the sound speed and a is the Alfvén speed, instead of the sound speed c as in the case of an effectively rigid flux tube in the limit of low plasma beta. (The tube speed is the speed of propagation of a longitudinal, 'sausage' wave along a thin, isolated magnetic flux tube (Defouw 1976).)

Siphon flows in arched, isolated magnetic flux tubes, corresponding to conditions in the solar photosphere and subphotosphere, have been studied in several recent papers (Thomas 1988; Montesinos and Thomas 1989, 1992; Thomas and Montesinos 1990, 1991; Degenhardt 1989, 1991). These calculations produce flow speeds in good agreement with those measured for the photospheric Evershed flow, but the horizontal extent of the arched flux tubes is too short to correspond to a penumbral filament. (This is a consequence of having to balance the magnetic buoyancy with magnetic curvature forces.) However, Thomas and Montesinos (1992) have recently shown that when there is an ambient magnetic field in the surrounding atmosphere, weaker by a few hundred gauss than the field strength of the flux tube, suitably long arches containing siphon flows of the correct strength do exist. The addition of the ambient magnetic field to the model is suggested by the recent evidence that the penumbra is a much deeper structure than hitherto imagined (see the discussion in section 4.2). The nearly horizontal flux tubes carrying the photospheric Evershed flow do not lie above field-free gas, but instead are embedded within an inclined magnetic field coming up through the penumbral photosphere (as shown in Figure 11). The new picture of a deep penumbra may solve another puzzle that was associated with the shallow penumbra: why a magnetic flux tube overlying nearly normal photosphere should appear dark (see Cram, Nye, and Thomas 1981). The inclined magnetic field outside the horizontal filaments reduces the heat transport to the dark filaments and increases the density (and hence the opacity) within the dark filaments, compared to the case of a shallow penumbra overlying field-free gas.

Overall, then, it seems that the siphon-flow model is in reasonable agreement with the principal observed features of the photospheric and chromospheric Evershed flow, although further refinement of the theoretical models and more detailed comparisons with observations are needed to confirm this.

5. Waves and Oscillations in Sunspots

The various types of oscillations in sunspots are interesting phenomena in their own right but are especially important because of the clues they offer concerning sunspot structure. Research on sunspot oscillations began in 1969 with the discovery of the periodic umbral flashes in the Ca II H and K lines by Beckers and Tallant (1969). Then, in the single year 1972, three new kinds of wave motion in sunspots were reported: the 3-minute umbral velocity and intensity oscillations in the umbral photosphere and chromosphere by Giovanelli (1972), Bhatnagar and Tanaka (1972), and Beckers and Schultz (1972); 5-minute oscillations in the umbral photosphere by Bhatnagar, Livingston and Harvey (1972); and running penumbral waves in Hα by Zirin and Stein (1972) and Giovanelli (1972). Theoretical modelling of sunspot oscillations began with the work of Nye and Thomas (1974, 1976) on running penumbral waves and of Uchida and Sakurai (1975) on 3-minute umbral oscillations. In 1982, Thomas, Cram and Nye (1982) pointed out that observations of the interaction of a sunspot with solar p-mode oscillations could be used as a probe of the structure of the sunspot beneath the visible surface of the Sun. Interest in the possibilities of this concept of 'sunspot seismology' greatly increased in 1987 with the discovery by Braun, Duvall and Labonte (1987, 1988) that sunspots absorb a significant fraction of the energy of incident solar p-modes.

The subject of sunspot oscillations is covered thoroughly in this volume. Lites (1992) presents a comprehensive review of the state of our observational knowledge of sunspot oscillations and the implications for theoretical models. Roberts (1992) gives an overview of the theory of magnetohydrodynamic waves in structured magnetic fields, which underlies all of the theoretical modelling of sunspot oscillations. Chitre (1992) discusses specific theoretical models of umbral oscillations and penumbral waves and the possible mechanisms for exciting them. Bogdan (1992) gives a thorough account of the results of the new and active subject of sunspot seismology. Other useful, older reviews of sunspot oscillations include those by Moore (1981), Thomas (1981), Moore and Rabin (1985), and Thomas (1985).

5.1. 3-MINUTE UMBRAL OSCILLATIONS

There are apparently two different types of oscillations in sunspot umbrae, with periods around 5 minutes and 3 minutes respectively. The 5-minute oscillations may be interpreted as the passive response of the sunspot to forcing by the p-mode oscillations in the surrounding convection zone (Thomas 1981), whereas the 3-minute oscillations represent a resonant oscillation of the sunspot itself. There are two resonant cavities for magneto-atmospheric waves in the stratified umbral atmosphere: a cavity for resonant fast modes in the low photosphere and subphotosphere (Uchida and Sakurai 1975; Antia and Chitre 1979; Scheuer and Thomas 1981; Thomas and Scheuer 1982; Cally 1983; Abdelatif 1990; Hasan 1991), and a cavity for resonant slow modes in the umbral chromosphere (Zhugzhda, Locans and Staude 1983; Zhugzhda, Staude and Locans 1984; Gurman and Leibacher 1984). (The terms 'fast' and 'slow' have only local significance in a stratified magnetic atmosphere, as exemplified here by the fact that an oscillation with a 3-minute period is a fast mode in the low photosphere and a slow mode in the chromosphere.) The early work focused on one or the other of these resonant cavities as competing explanations for the observed oscillations, but in more recent discussions a unified model has emerged in which the contributions of both cavities and their coupling are recognized (Thomas 1984b, 1985; Zhugzhda, Locans and Staude 1987).

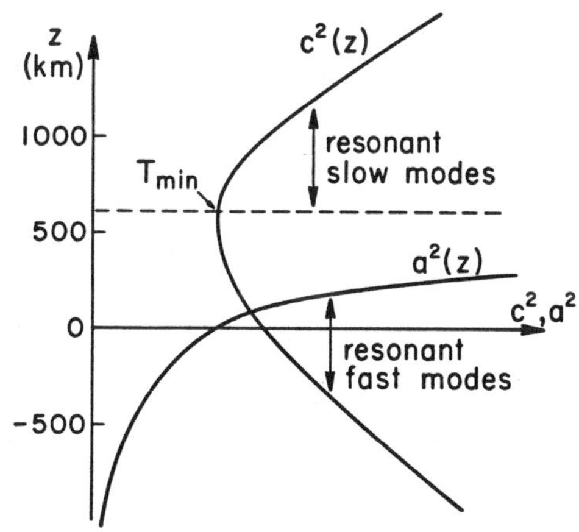

Figure 12. Schematic plot of the variation of the square of the sound speed, c^2, and the square of the Alfvén speed, a^2, with height z in a sunspot umbra, illustrating the coupled resonant cavities for 3-minute umbral oscillations. (From Thomas 1985.)

The nature of these two resonant cavities is illustrated schematically in Figure 12. The propagation of magneto-atmospheric waves is largely determined by the distribution of the square of the adiabatic sound speed, $c^2(z)$, and the square of the Alfvén speed, $a^2(z)$, with height z in the atmosphere (see Thomas 1983). In the umbral atmosphere, $c^2(z)$ roughly follows the temperature distribution $T(z)$ with its minimum in the upper photosphere, while $a^2(z) = B^2/4\pi\rho$ increases nearly exponentially with height because of the nearly exponential decrease of the density ρ and the relatively slow decrease of magnetic field strength B. The lower cavity for 3-minute fast modes is formed by the increase in sound speed with depth in the subphotosphere and the increase in Alfvén speed with height in the photosphere. In the chromosphere, where $a^2 >> c^2$, a slow mode is nearly a pure acoustic mode propagating along the vertical magnetic field lines; the chromospheric cavity is formed by the temperature minimum below (giving a cutoff frequency for propagation) and the increase in sound speed with height above. Oscillations excited in the lower cavity can tunnel through the layer between the two cavities in the upper photosphere where the wave is evanescent; in this way, the two cavities are coupled.

The 3-minute umbral oscillations are excited by some mechanism operating below the visible surface of the umbra that excites oscillations in the subphotospheric cavity; energy is then transmitted to the chromospheric cavity through tunnelling. Although the amplitude of the 3-min oscillations is much smaller in the photosphere than in the chromosphere, the energy density is much greater in the photosphere because of the much greater mass density there (Lites and Thomas 1985; Abdelatif, Lites and Thomas 1986), and it is clear that the photospheric oscillations drive the chromospheric oscillations and not the reverse. But the precise mechanism for the excitation of the photospheric resonance has not been definitely established. One school of thought is that the 3-minute oscillations are excited by overstable, oscillatory convection just below the umbral photosphere (Moore 1973; Mullan and Yun 1973; Knobloch and Weiss 1984); perhaps a better way to view this is that the oscillations are self-excited (overstable) in the superadiabatic layers just below the umbral surface (Antia and Chitre 1979). Further theoretical support for the self-excitation mechanism is provided by Chitre (1992) in his chapter in this volume. Another possibility is that the oscillations are excited by the broad-band, high-frequency tail of the spectrum of the solar p-mode oscillations in the surrounding convection zone (Moore and Rabin 1985). Although there is some observational evidence against this (e.g. Lites and Thomas 1985), Lites (1992) concludes that this possibility cannot be ruled out on observational grounds, especially in view of new evidence concerning the concentration of high-frequency acoustic emission (just above the minimum cutoff frequency of 5.5 mHz) from the convection zone around active regions (Brown *et al.* 1992).

As our theoretical understanding of 3-minute umbral oscillations improves, it may become possible to probe the structure of the umbral atmosphere using simultaneous measurements of umbral oscillations in a number of spectral lines formed at different heights. Several such observations already exist (Giovanelli, Harvey and Livingston 1978; von Uexküll, Kneer and Mattig 1983; Thomas, Cram and Nye 1984; Lites and Thomas 1985; Lites 1986; Thomas *et al.* 1987) and show that the oscillations are sensitive to atmospheric structure (see also the chapter by Lites (1992) in this volume). However, the umbral oscillations clearly exhibit nonlinear behaviour in the chromosphere, so present theoretical models based on linear wave theory are inadequate for probing the structure of the umbral chromosphere. (This contrasts with subsurface sunspot seismology discussed next, for which linear theory may well be adequate for most purposes.)

5.2. 5-MINUTE UMBRAL OSCILLATIONS AND SUNSPOT SEISMOLOGY

A sunspot floats in the solar convection zone and is buffetted and squeezed by the resonant *p*-mode waves that reside there. The *p*-modes have most of their energy at periods around 5 minutes; thus, it is not surprising that we should observe 5-minute oscillations inside a sunspot too (Thomas 1981). The detailed interaction between the sunspot and the *p*-modes in the surrounding convection zone must depend on the configuration of the spot's subsurface magnetic field and the temperature and density within it. This suggests that observations of 5-minute oscillations in and around sunspots may reveal something about the subsurface structure of the spot. Furthermore, since different individual *p*-modes have their energy concentrated at different depths, by decomposing this interaction into individual modes we can in principle detect variations in sunspot structure with depth. These possibilities of 'sunspot seismology' were first pointed out by Thomas, Cram and

Nye (1982) on the basis of their measurements of the temporal power spectrum of umbral oscillations and a very simple model for the interaction.

Observations of 5-minute umbral oscillations are complicated by the strong possibility of contamination of the signal by stray light from the surrounding photosphere. Methods of getting around this problem have involved the use of molecular lines or low-excitation lines formed only in the cool umbra or the use of the Stokes V profile of a magnetically sensitive line for measuring the velocity. As a result of careful measurements by a number of observers, summarized by Lites (1992) in this volume, we now have a consistent picture of these oscillations. There is a broad peak of oscillatory power in the 5-minute band (2.5–4.5 mHz) with a typical rms amplitude of 75 m s^{-1}, which is a factor of two or three lower than the rms amplitude in this frequency band outside sunspots. This suppression of oscillation amplitude within the umbra and its dependence on frequency and wavelength offer further useful information for sunspot seismology. Abdelatif, Lites and Thomas (1986) compared the spatial properties of 5-minute oscillations in the umbra and the surrounding photosphere and showed that not only is the oscillatory power reduced in the umbra, but it is also shifted to smaller wavenumbers (longer wavelengths). This effect was interpreted theoretically by Abdelatif and Thomas (1987) as being due to the fact that an acoustic wave outside the sunspot is transformed into a fast magnetoacoustic wave upon penetrating the umbra. The greater propagation speed of the fast magnetoacoustic wave (for a fixed frequency) gives a longer wavelength in the umbra. Moreover, the umbra acts as a selective filter in admitting some acoustic p-modes more readily than others, because of the enhanced (resonant) transmission of waves for which an integral number of wavelengths fit across the umbral diameter.

A very significant advance is sunspot seismology came with the observations of Braun, Duvall and LaBonte (1987, 1988) showing that a sunspot absorbs an appreciable fraction of the energy of incident p-modes. They decomposed the p-modes in a circular annulus surrounding a sunspot into cylindrical waves (of moderate azimuthal order, $-5 \leq m \leq 5$) propagating radially inward and outward and found a deficit in the energy of outward-propagating waves. They defined the deficit in terms of an absorption coefficient, which they found to increase with increasing horizontal wavenumber and to reach values as high as 0.5 for wavenumbers greater than about 7 Mm^{-1}. The work of Braun, Duvall and LaBonte also introduced a new approach to sunspot seismology, namely, to view the problem as a classical scattering problem and look only at the ingoing and outgoing wave fields in the surroundings and not at oscillations in the sunspot itself. This approach had been suggested independently in a theoretical paper by Bogdan and Zweibel (1987).

The discovery of the absorption of p-modes by sunspots has been followed by a flurry of efforts by theoreticians to explain this effect using idealized models. Hollweg (1988) investigated the possibility of resonance absorption of the incoming waves in the surface transition layer (or magnetopause) of the sunspot magnetic flux tube (see also Sakurai, Goossens and Hollweg 1991). Lou (1990) considered the possible enhancement of resonance absorption by viscous effects, and Chitre and Davila (1991) showed that resonance absorption is enhanced when there is an azimuthal twist in the sunspot flux tube. Goossens and Poedts (1992) considered resistive effects and found them to be similar to viscous effects; they show that the efficiency of resonance absorption is independent of the actual dissipation mechanism. Rosenthal (1991, 1992) considered resonance absorption in more general equilibrium configurations, both with and without twist, including viscous dissipation. In a different approach, Bogdan (1989) considered the possible conversion of the incoming acoustic waves into magnetohydrodynamic wave modes of the sunspot flux tube, or its constituent fibrils in a cluster model. This approach was extended by Bogdan and Knölker (1991) to include the effects of radiative transfer and the coupling to radiation-diffusion modes. Another attractive possibility is that the *p*-modes are converted into slow modes that propagate downward along the sunspot magnetic flux tube and carry energy away from the surface regions of the sunspot (Spruit 1991; Spruit and Bogdan 1992). All of these proposed mechanisms (and more) are discussed in considerable detail in the chapter by Bogdan (1992) in this volume. At this point, it is difficult to assess them and to choose a favourite. The identification of the actual mechanism for *p*-mode absorption in sunspots will have to await more realistic theoretical models and further observations with higher spatial resolution.

5.3. PENUMBRAL WAVES

The running penumbral waves discovered by Zirin and Stein (1972) and Giovanelli (1972) are observed as dark wavefronts in Hα in the form of concentric arcs that depart from the umbra-penumbra boundary, propagate radially outward across the penumbra at speeds of 10 to 20 km s^{-1}, and repeat with periods in the range 200–300 s. The wavefront arcs often extend azimuthally most of the way around the penumbra (i.e. are almost complete circles). The velocity amplitude of the waves is about 1 km s^{-1} in the chromosphere and about 200 m s^{-1} at the temperature minimum. Penumbral waves have also been detected in the low photosphere (Musman, Nye and Thomas 1976) with even smaller amplitude (100 m s^{-1}) but these waves were more intermittent and had higher propagation speeds (by a factor of two) than the chromospheric penumbral waves, so the connection between the two is not clear. A detailed summary of the observations of running penumbral waves is given by Lites (1992) in this volume.

There have been basically two kinds of theoretical models for penumbral waves. In one model the wave mode is a fast magneto-atmospheric wave that is vertically trapped in the photosphere by the increase in sound speed down into the convection zone and the increase in Alfvén speed up into the chromosphere (Nye and Thomas 1974, 1976; Antia, Chitre and Gokhale 1978; Cally and Adam 1983). In the other model the wave is an interfacial wave trapped at the magnetopause, i.e. the interface between the field-free convection zone and the overlying penumbral magnetic field (Small and Roberts 1984; Roberts 1992). These models are actually related; whether a wave of the right period exists as a body wave trapped in a shallow cavity or as an interfacial wave trapped on the magnetopause depends on the numerical values of the parameters one assumes in the equilibrium configuration. In both of these models the wave energy is largely confined to the trapping region and the wave is vertically evanescent in the chromosphere (where it is observed in Hα); however, the wave amplitude is actually much larger in the chromosphere than in the photosphere because of the greatly reduced mass density in the chromosphere.

Both of these models are based on the concept of a shallow penumbra, that is, a nearly horizontal penumbral magnetic field overlying a field-free convection zone and penetrating only a short distance below the visible surface. In the light of all the recent evidence for a deep penumbra with a component of magnetic field inclined as much as 40 degrees to the horizontal (see section 4.2), these theoretical models of penumbral waves need to be reconsidered. The picture of a shallow, nearly horizontal magnetopause in the penumbra is no longer valid. However, it may be that the penumbral waves are carried only by the dark, nearly horizontal penumbral filaments; this is suggested by the observation that the wavefronts appear more and more ragged as they propagate outward across the penumbra. But other models based on the newly emerged picture of a deep penumbra with oblique light and dark filaments should be considered. One possibility is that the penumbral waves are a manifestation of the travelling-wave form of magnetoconvection that is seen to occur in numerical simulations of magnetoconvection in an inclined overall magnetic field (see the review by Proctor (1992) in this volume). This possibility has the added advantage that it would account naturally for the excitation of the penumbral waves, through convective overstability. A simple magnetoconvective theory of penumbral waves was discussed by Galloway (1978), on the basis of standing waves that appeared in his numerical simulations of Boussinesq magnetoconvection in a vertical cylinder with a vertical applied magnetic field. More realistic models of this sort would be well worth considering.

6. Sunspots and Active Regions

So far we have only discussed the physics of an isolated spot but, however prominent sunspots may be, they are not the only products of solar magnetic activity. We have to relate them to other manifestations of this activity, to the flux tubes that erupt through the photosphere to form active regions and to the global magnetic field within the Sun. This section is concerned with behaviour at and just below the photosphere, which leads naturally to consideration of the large-scale field in Section 7. A typical active region starts with the emergence of magnetic flux, which rapidly organizes itself into pores and then into spots. Although most spots are short-lived, some survive for weeks or months in magnetohydrostatic equilibrium as they gradually decay. Our survey of the evolution of sunspots, in the first part of this section, is relatively brief, since Zwaan's (1992) chapter in this book provides a comprehensive review of the relevant observations and their significance. Outside pores and sunspots, magnetic flux is confined to isolated flux tubes with intense magnetic fields. These flux tubes are small compared with the scale of the granulation, whereas the horizontal scale of convection in the umbra of a spot is much less than the umbral radius. Theoretical studies of the formation and structure of thin flux tubes raise many interesting issues, which are discussed in the second part of this section.

6.1. THE EVOLUTION OF SUNSPOTS

The systematic properties of the magnetic fields in active regions – their scale, orientation and polarity – argue that they are produced by a coherent bundle of toroidal flux that rises from deep in the convective zone and breaks through the photosphere. When it first emerges, this bundle is shredded, giving rise to small flux concentrations, which amalgamate to form pores at junctions in the network. The pores stream outwards towards the east-west boundaries of the region, where spots are likely to form; indeed, a growing spot may form a penumbra on its limbward side while it is still acquiring more flux at its inner edge. These processes, first demonstrated by Vrabec (1974) in a famous movie, have been confirmed by high-resolution observations at La Palma.

To explain these results we suppose that a large ∩-shaped flux tube rises upwards through the convection zone and that the horizontal portion becomes frayed and shredded on the journey (see the chapter by Moreno-Insertis 1992). Thus flux appears in a fragmented form.

Once the whole flux tube has emerged, the fragments can coalesce to form sunspots, which are held together by magnetic buoyancy, like a bunch of tethered balloons. This changing pattern has been compared to a rising tree (Vrabec 1974) but the analogy is misleading, for matter flows downwards along the field and does not rise above the photosphere.

Small flux concentrations form pores, or isolated umbrae, with almost uniform fields of around 1900 G (Brants and Zwaan 1982). If the radius of a pore exceeds 2000 km it generally develops a penumbra, on a rapid dynamical timescale of about 2 hr. However, large pores, with radii of up to 3500 km, are occasionally found, while the smallest sunspots have radii of only 1750 km, with rudimentary penumbrae (Bray and Loughhead 1964). As the magnetic flux in a pore increases, its boundary becomes increasingly inclined (Simon and Weiss 1970). Eventually the configuration must become unstable to modes which develop into a filamentary penumbra. As indicated in Section 3.2, the bifurcation is apparently subcritical (cf. Figure 6). This transition deserves more detailed theoretical investigation.

Where pre-existing pores (or smaller spots) amalgamate to form a sunspot, seams may persist, as in a ball of putty. These inhomogeneities preserve a memory of the spot's formation and contribute to its identity: in H. U. Schmidt's words, each sunspot has a face that can be recognized. Moreover, the seams can develop into light bridges, which look like intrusions of external plasma from the penumbra, or even from the field-free exterior, into the dark umbra. These lines of weakness facilitate the eventual breakup of a spot (Garcia de la Rosa 1987). However, they do not seem to be an essential feature of its structure.

A large isolated spot may recur for a few solar rotations. Such a spot must be in magnetohydrostatic and thermal equilibrium, as discussed in Sections 3 and 4. Although the ratio β of the external gas pressure to the magnetic pressure is of order unity at the surface, it has to be far greater at the base of the convection zone. The estimated value of the field strength there is around 10^5 G (see Section 7.2 below), giving $\beta \approx 2 \times 10^5$. Now the value of β remains constant for a thin vertical flux tube in thermal equilibrium with its surroundings (Parker 1979a). Hence the field strength decreases upwards as the square root of the external pressure until the flux tube becomes significantly cooler than the ambient plasma. We know, however, that the field cannot drop below 3000 G, so we can estimate the depth of this transition by noting that a field of 3000 G with $\beta = 2 \times 10^5$ yields a pressure of 6×10^{10} dyne cm^{-2}, corresponding to a depth of 20000 km. Above this level, the superadiabatic gradient required to supply energy to the umbra is enhanced, and the spot becomes significantly cooler than its surroundings.

The pattern of photospheric convection is abnormal in active regions. The diameters of mesogranules and supergranules are reduced owing to the presence of magnetic fields (cf. Stein, Brandenburg and Nordlund 1992) but pores still tend to form at junctions in the network, which correspond to regions of converging flow. Flux tubes that are sufficiently small are apparently swept aside by the convective motion but sunspots react back on the supergranular flow around them and control it. Large, isolated spots are surrounded by annular moat cells and even small, newly-formed spots may be partially enclosed. The moat cell seems to be a large and long-lived supergranule, stabilized by the sunspot that occupies its core and with a diameter that may be twice that of the spot (Sheeley 1972; Harvey and Harvey 1973; Vrabec 1974; Zwaan 1992). The surface flow is radially outward and we may conjecture that the hollow rising plume feeds energy into the penumbra. At greater depths the flow may converge on the flux tube and so provide it with a stabilizing collar (Meyer *et al.* 1974).

The moat is swept clean of magnetic fields, apart from isolated magnetic features (of both polarities) that move outwards from the spot. These moving magnetic features, with a typical flux of 10^{19} mx, appear to be responsible for a net loss of flux that leads to the gradual decay of a sunspot. Now the area of a long-lived spot decays at an approximately constant rate, and it follows that the rate of decay of the magnetic flux is likewise constant (Cowling 1946; Bray and Loughhead 1964; Gokhale and Zwaan 1972). At first sight it might appear that this slow decay occurs as small flux tubes are torn away from the periphery of the spot. Such a process would, however, produce a decay rate proportional to the spot radius, which contradicts the observations. There have been two attempts to resolve this difficulty. Gokhale and Zwaan (1972) considered a simple model of a sunspot with a uniform umbral field surrounded by a vertical current sheet, with a thickness proportional to the spot radius. They found that a combination of diffusion and a radial inflow could maintain a self-similar structure as the spot decayed. (They assumed a laminar diffusivity and required a current sheet that was implausibly thin; adopting a turbulent diffusivity would allow a thick penumbra.) Meyer *et al.* (1974) assumed that flux loss was controlled by turbulent diffusion, acting throughout the spot at a depth where the magnetopause was vertical. In their model solution the field profile $B(s)$ was Gaussian and there was a critical field strength B_c such that flux was carried away wherever $B < B_c$: this led to a constant decay rate but, although the field profile remained self-similar, the central field strength decreased with time. Given our present understanding of penumbral structure (as described in Section 4.2), it now seems likely that the dominant diffusive process is produced by convective exchanges in a deep penumbra, which occupies most of the spot area. Further progress clearly requires a two-dimensional formulation of the problem.

6.2. INTENSE FLUX TUBES OUTSIDE SUNSPOTS

Improved high-resolution observations over the past twenty-five years have revealed that nearly all of the magnetic flux threading the photosphere outside sunspots and pores is in the form of localized, intense magnetic elements with field strengths in the range 1000-1500 G and diameters less than 200 km at the solar surface. These intense flux elements, which are closely related to sunspots, have been the subject of several recent review articles (Spruit and Roberts 1983; Solanki 1987, 1990; Stenflo 1989; Thomas 1990; Schüssler 1990, 1991).

The standard theoretical model of one of these intense flux elements is a thin, isolated magnetic flux tube emerging from the solar surface. In equilibrium the flux tube will be in lateral pressure balance with the surrounding field-free gas, so that $p + B^2/8\pi = p_e$, where p is the internal gas pressure, p_e is the external gas pressure, and $B^2/8\pi$ is the magnetic pressure for a field strength B. For a fixed external pressure, the maximum field strength is that of a totally evacuated flux tube ($p = 0$), for which $B = (8\pi p_e)^{1/2} \equiv B_p$. At the solar surface (at optical depth $\tau_{0.5} = 1.0$) this maximum value B_p is about 1700 G. Since the observed field strengths approach this limiting value, the flux tubes must in fact be substantially evacuated. (Observed field strengths greater than 1700 G have occasionally been reported, but they are not necessarily inconsistent with the model if the gas within the flux tube is more transparent than the surrounding atmosphere so that one is seeing to a greater depth in the flux tube. The value of B_p increases to 2000 G at a depth of only some 60 km below the solar surface.)

In the convection zone just below the solar surface the combined effects of advection and diffusion of magnetic lines of force lead to the expulsion of magnetic flux from the interior of convective cells or eddies and the concentration of the magnetic field into sheets or tubes at the boundaries of the cells or eddies (Parker 1963; Weiss 1966). This basic process of flux expulsion has been studied extensively, both for a Boussinesq fluid (see Parker 1979a, chapter 10, and the review by Proctor and Weiss 1982) and for fully compressible convection (see Nordlund 1983, Hughes and Proctor 1988, and Proctor 1992). The field strength in isolated sheets or tubes is limited by diffusion (in the kinematic regime) or dynamically, in the case of vigorous convection (Galloway and Weiss 1981). Simple physical arguments suggest that the process of flux expulsion and concentration by convection is capable of creating magnetic flux tubes with a field strength up to the 'equipartition' value $B_e = (4\pi\rho v^2)^{1/2}$, given by a balance between the magnetic pressure $B^2/8\pi$ and the dynamical pressure $\rho v^2/2$. In Boussinesq magnetoconvection the peak value is increased by a factor proportional to the square root of the ratio of the viscous to the magnetic diffusivity (Galloway, Proctor and Weiss 1978). Compressible magnetoconvection allows a much wider range of possibilities, as a result of buoyancy braking and evacuation of the flux tube (Hurlburt and Toomre 1988).

At the solar surface the equipartition field strength B_e corresponding to observed granular motions is roughly 200 G (for $\rho = 3 \times 10^{-7}$ gm cm^{-3} and $v = 1$ km s^{-1}); at a depth of 1000 km (with $\rho = 3 \times 10^{-6}$ gm cm^{-3}) B_e rises to 600 G. So expulsion and concentration of flux alone are not sufficient to create flux tubes with the observed magnetic field strength. Considerable effort has therefore been devoted to identifying some other mechanism that will partially evacuate the flux tube.

One such mechanism, first proposed by Parker (1978) and Zwaan (1978), is the convective collapse of the flux tube where it passes through the superadiabatic temperature gradient in the external atmosphere just below the solar surface. This mechanism has been studied extensively. The early work consisted primarily of linear stability analyses for adiabatic perturbations within the thin flux tube approximation (Webb and Roberts 1978; Spruit and Zweibel 1979; Unno and Ando 1979). The basic mechanism of the instability that leads to the convective collapse is illustrated in Figure 13. The tube is assumed to be initially in thermal equilibrium with the surrounding convection zone, in which the mean temperature profile is superadiabatic just below the solar surface (solid curve). The temperature profile inside the tube is initially the same as the external temperature profile. However, the temperature of a vertically displaced parcel of fluid within the tube will follow an adiabat (dashed line) because of the inhibiting effect of the magnetic field on small-scale convective heat transport within the tube. For a downward displacement, the parcel will be cooler than its surroundings and, provided the magnetic pressure is not too high, lateral pressure balance implies that the parcel will also be more dense than its surroundings. Thus, the displacement will be accelerated and the tube will collapse and concentrate. If the initial magnetic field strength is high enough, however, the lateral pressure balance will be maintained by an increase in magnetic pressure and the gas pressure and density of the parcel may be less than in the surroundings, providing a restoring buoyancy force. This argument suggests that there is a critical initial magnetic field strength above which the flux tube is stable and below which it is unstable to collapse. This is indeed confirmed by linear stability analysis, which also gives a critical surface field strength of around 1350 G (Spruit and Zweibel 1979), in good agreement with observations.

Spruit (1979) argued that the final state of the flux tube following the instability is a stable, collapsed equilibrium state with higher magnetic field strength. This collapsed state is overstable if radiative heat exchange and viscosity are included. The nonlinear development of the convective collapse has been studied numerically with models of varying complexity by Venkatakrishnan (1983, 1985), Hasan (1984, 1985), and Massaglia, Bodo, and Rossi (1989). The overstable oscillations about the collapsed state have periods greater than 1000 s. However, other processes, such as the birth and decay of convective granules, have shorter time scales and may be the limiting factor in determining the lifetime of the collapsed flux tube.

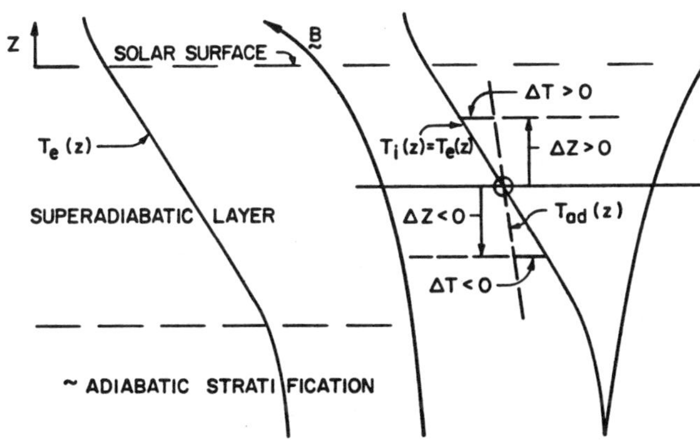

Figure 13. Schematic illustration of the basic mechanism of convective collapse of a magnetic flux tube in thermal equilibrium in the upper solar convection zone.

It should be mentioned that the various theoretical analyses of the convective collapse mechanism mentioned above, both for the linear instability and for its nonlinear development, are quite sensitive to the upper and lower boundary conditions assumed. A critical discussion of the influence of various boundary conditions has been given by Schüssler (1990).

On the Sun the processes of flux expulsion and convective collapse are occurring simultaneously and continually. In fact, they are different aspects of the same phenomenon (Hughes and Proctor 1988). Numerical experiments on steady two-dimensional magneto-convection in a fully compressible layer (Hurlburt and Toomre 1988) reveal an asymmetry between rising and falling plumes, owing to the combined effects of stratification and buoyancy braking. Strong fields, with strengths exceeding B_e and approaching B_p, are formed in the narrow, rapidly sinking plumes, which are partially evacuated. The structure of isolated flux tubes embedded in a turbulent convecting layer (represented using the mixing-length formalism) has been investigated in considerable detail by Schüssler (1990, 1991) and his colleagues (Deinzer *et al.* 1984; Knölker, Schüssler and Weisshaar 1988; Knölker and Schüssler 1988; Grossmann-Doerth *et al.* 1989).

All this is illustrated in Nordlund's (1983, 1986) three-dimensional numerical simulation of granular convection, in the anelastic approximation but including radiative transfer and a magnetic field. The magnetic flux is concentrated into the cool intergranular lanes containing downdrafts. This concentrated magnetic field inhibits heat transport into the flux tube, and the tube cools and collapses even further because of radiative losses at the solar surface. (The final state is affected by strong numerical diffusion.) The magnetic flux is rearranged as the pattern of granular convection changes, and the lifetime of an individual flux tube is comparable to the lifetime of an individual granule. Work on fully compressible three-dimensional magnetoconvection is now in progress (Nordlund and Stein 1989; Stein, Brandenburg and Nordlund 1992; see Proctor 1992).

Another mechanism that can contribute to the formation of intense flux elements is a siphon flow along an arched flux tube (Thomas 1988; Montesinos and Thomas 1989; Thomas and Montesinos 1990). Thomas and Montesinos (1991) pointed out that the observational signature of this mechanism would be a pair of magnetic elements of opposite polarity, separated by a distance of from one to several arcseconds, with an upflow in one element (the upstream footpoint of the arch) and a downflow and somewhat greater magnetic field strength in the other element (the downstream footpoint). The magnetic field will be slightly inclined to the vertical in each element of the pair, in directions consistent with their being connected by an arched flux tube. Recent observations by Rüedi, Solanki and Rabin (1992) have tentatively confirmed the existence of this mechanism in flux tubes arching across a magnetic neutral line on the Sun.

The ubiquitous nature of intense magnetic flux tubes on the Sun leads naturally to the question of the possible wave modes of these flux tubes and their relation to the long-standing problem of the heating of the solar chromosphere and corona. This is an extensive subject: we shall not attempt to cover it here but instead shall only refer to the review articles by Spruit (1981d), Spruit and Roberts (1983), Thomas (1985), Roberts (1986), Ryutova (1990) and the chapter by Roberts (1992) in this volume.

7. Sunspots and the Solar Dynamo

It was by studying sunspots that the principal properties of the solar cycle – the 11-year periodicity of solar activity, the equatorward drift of the activity zones and the alternating magnetic polarities of sunspot pairs in the two hemispheres – were originally established. This section is concerned with the relationship between sunspots and the global magnetic field. We begin with the origins of the solar cycle and review the current state of dynamo theory. Next, we consider the formation of flux tubes in a magnetic layer at the interface between the radiative and convective zones, and discuss their subsequent rise through the convection zone to emerge in active regions at the solar surface.

7.1. DYNAMO THEORY: WHERE WE STAND

The toroidal magnetic field that emerges in active regions is antisymmetric about the equator and reverses its direction at sunspot minimum. Any theory therefore has to explain the origin of the 22-year magnetic cycle, which actually seems to be aperiodic and is modulated by grand minima with a characteristic timescale of 200 yr. Only two types of mechanism have been put forward: oscillators and dynamos.

In an oscillator, an underlying poloidal field is wound up by differential rotation to give a toroidal field which grows until it is strong enough to reverse the motion. Thus kinetic energy is transformed into magnetic energy and back again on an Alfvénic timescale, without significant dissipation. The mechanism has the advantage of simplicity, but no-one has yet explained how such oscillations can be maintained. Moreover, the only variations of angular velocity observed at the surface have a period of 11 yr, rather than 22 yr, and the polar fields reverse near sunspot maximum. For these reasons, it is generally accepted that a dynamo is required (Cowling 1981; Hughes 1992; Rosner and Weiss 1992). However, some measurements of the rotation rate deep in the solar interior have been interpreted by helioseismologists as indicating a cyclic variation in angular velocity (Goode and Dziembowski 1991) and this has led to a revival of the oscillator theory (Goode, Gough and McIntyre 1992).

Although Larmor (1919) had suggested that large-scale fields in the Earth and Sun, as well as individual sunspots, are maintained by self-exciting dynamo action, Cowling's (1934) theorem raised doubts as to whether such a mechanism is viable. These doubts were dispelled in the 1950s and since then dynamo theory has developed into a flourishing activity (Moffatt 1978; Parker 1979a; Krause and Rädler 1980; Zel'dovich, Ruzmaikin and Sokoloff 1983). Aspects of dynamo theory relevant to the origin of sunspots are discussed in the chapter by Hughes (1992) and have frequently been reviewed (e.g. Weiss 1989; Belvedere 1990; Hoyng 1990; Brandenburg and Tuominen 1991; Stix 1991; Rosner and Weiss 1992), so we need only summarize them here.

In a self-exciting stellar dynamo the magnetic field develops as an instability, generated by the combination of turbulent convection and differential rotation. So the first stage is to study linear (kinematic) models, in order to identify velocity fields that can act as dynamos. Next come nonlinear dynamos, where the Lorentz force alters the motion so as to limit growth of the magnetic field. Gilman (1983) and Glatzmaier (1985) have constructed fully self-consistent nonlinear dynamo models in spherical geometry. The rotating fluid is confined within a spherical annulus and convection is driven by heating from below. The magnetic fields that are generated reproduce certain key features of the solar cycle. There are also discrepancies, notably in the direction in which the activity zones migrate, but this is not surprising, given the simplifications that were necessary in order to make the computations possible. What matters is that these calculations demonstrate that the dynamo process actually works.

Much of the work in this subject has been concerned with mean-field dynamo theory, which describes the evolution of fields obtained by averaging over small-scale turbulent structures. The ideas originally put forward by Parker (1955b) have since been extended and elaborated (Moffatt 1978; Parker 1979a; Krause and Rädler 1980). Although the mean-field dynamo equations can be rigorously justified in certain circumstances, they can only be regarded as a physically plausible parametrized model of the solar dynamo, which can yield valuable insights into properties of the solar cycle (cf. Stix 1989) and its relationship to the behaviour of more active stars. In particular, mean-field dynamo models provide a means of exploring the range of qualitative behaviour (including aperiodic modulation) that might be found in the nonlinear regime.

Current research is focused on two aspects of dynamo theory. The first is concerned with fast dynamos (Zel'dovich, Ruzmaikin and Sokoloff 1983). The issue here is how to construct a kinematic dynamo in which the growth rate of the field tends to some positive limit as the magnetic diffusivity tends to zero (and the magnetic Reynolds number becomes infinite). This problem raises issues that have fascinated mathematicians (Childress 1992). It is also important for astrophysical applications, since the magnetic Reynolds number is very large in a star like the Sun and the existence of fast dynamos is implicitly assumed in mean-field dynamo theory. Fortunately, a recent calculation by Galloway and Proctor (1992) has provided convincing numerical evidence that fast dynamos do exist.

The second range of problems involves detailed studies of turbulent dynamos, relying heavily on numerical experiments. Here it is important to ensure that solutions are not significantly affected by truncation error, and that dynamo action is genuine and not a transient effect (Cattaneo, Hughes and Weiss 1991). Turbulent dynamo action has been convincingly demonstrated by Nordlund *et al.* (1992), who modelled convection in a stratified, rotating layer. The motion is dominated by tornado-like sinking plumes and the magnetic field is highly intermittent, with flux tubes wound around the vortices. Such a fibril dynamo (Schüssler 1983) may be responsible for the disordered fields that emerge as ephemeral active regions and produce X-ray bright points, but cannot so readily explain the systematic features of the solar cycle.

Helioseismological evidence shows that there is a strong radial gradient in the angular velocity immediately below the convection zone (Dziembowski, Goode and Libbrecht 1989; Goode 1991), making that the most likely site for the formation of a strong toroidal field. Other arguments, based on flux expulsion and inhibition of magnetic buoyancy, had already suggested that the dynamo is located in a region of weak convective overshoot at the interface between the radiative and convective zones (Spiegel and Weiss 1980; Golub *et al.* 1981; Van Ballegooijen 1982). These results have led to a new phenomenological model of the solar cycle: in this picture there is a shell dynamo, with the strong toroidal field confined to a thin magnetic layer, which encroaches upon the convection zone at sunspot maximum and retreats at sunspot minimum.

7.2. FORMATION AND EMERGENCE OF MAGNETIC FLUX TUBES

If this picture of a dynamo acting just beneath the base of the convection zone is accepted, the next task is to investigate the formation of isolated flux tubes and to describe their evolution as they rise upwards through the convection zone. These topics are authoritatively reviewed in the two chapters by Hughes (1992) and Moreno-Insertis (1992) in this volume. It is well-known that a stratified layer in magnetohydrostatic equilibrium, with a depth-dependent, horizontal magnetic field, is prone to instabilities driven by magnetic buoyancy (Parker 1979a; Hughes and Proctor 1988; Hughes 1991). A continuously varying field typically becomes unstable if the field-strength decreases sufficiently rapidly upwards (Newcomb 1961; Parker 1966; Gilman 1970; Thomas and Nye 1975; Acheson 1978, 1979; Hughes 1985a,b). More relevant in this context is the instability of a layer in which the magnetic field drops discontinuously to zero (Kruskal and Schwarzschild 1954; Parker 1979a). The most unstable modes have a much longer wavelength along the field than in the transverse direction; so it is easy to imagine that they develop so as to form isolated loops of magnetic flux which can enter the convection zone and rise up to the surface.

Modelling this nonlinear process is less straightforward. Weakly nonlinear behaviour was discussed by Hughes (1987) but so far the only fully nonlinear calculations have been confined to two-dimensional geometries, which give a cross-section of the structure. Numerical experiments by Cattaneo and Hughes (1988) provide a graphic demonstration of the breakup of a finite magnetic layer embedded in a compressible atmosphere. The field, which is purely toroidal, does not, however, evolve to form isolated flux tubes; instead, it gradually spreads throughout the region. Subsequent computations with a sheared magnetic field look more promising, for they show the instability swelling and developing into a mushroom-like structure, which detaches itself and rises upwards (Cattaneo, Chiueh and Hughes 1990a,b). To go further evidently requires a treatment that is fully three-dimensional.

Suppose now that the processes outlined above have injected a flux tube into the convection zone. In the simplest case an axisymmetric tube will float upwards owing to magnetic buoyancy, for an isolated flux tube will be less dense than its surroundings if they are in thermal equilibrium (Parker 1955a, 1979a; Jensen 1955; Weiss 1964b). The restriction to axisymmetry is, of course, unrealistic. In particular, the flux tube, as it rises, will become unstable to non-axisymmetric kink modes (Spruit and Van Ballegooijen 1982; Van Ballegooijen 1983). Then we expect gas to drain downwards from the crests to the valleys (Parker 1966), so encouraging loops to rise until they reach the photosphere. This process was carefully modelled by Moreno-Insertis (1986, 1987; see also Choudhuri 1989), who computed shapes and rise times for tubes with different field strengths and fluxes (see Moreno-Insertis 1992); he found that a loop reaches the surface after one or two months and that the field at the crest falls so that, for instance, an initial field of 10^5 G is reduced to the local equipartition value ($B_e \approx 3000$ G) at a depth of about 15000 km.

This development is complicated by other effects. The rising tube moves through a vigorously convecting layer. Where the magnetic field is strong, it will dominate the pattern of convection in its neighbourhood but near the surface the flux tube will be affected by supergranular convection; the latter interaction has been illustrated in simple models (Schmidt, Simon and Weiss 1985; Moreno-Insertis 1987). The numerical results of Cattaneo, Chiueh and Hughes (1990a) suggest that the rising tube will also be flattened in cross-section. Such a configuration is liable to dynamical instabilities that will lead to fragmentation (Schüssler 1979; Tsinganos 1980) and the fragments will be further distorted by turbulent convection as they approach the surface.

The rise time of a kink-unstable flux tube is comparable with the rotation period of the Sun, so the Coriolis force cannot be neglected. Its consequences were studied by Choudhuri and Gilman (1987; see also Choudhuri 1989; D'Silva and Choudhuri 1991). The most important effect is that flux tubes are displaced parallel to the rotation axis, so that they emerge at higher latitudes and not in the sunspot belts at all. To overcome this, the initial field strength has to be sufficiently large. For this reason, and also to ensure that the field remains greater than B_e until relatively near the surface – where β decreases rapidly, as we saw in Section 6.1 – Moreno-Insertis (1992; see also Petrovay 1991) concludes that the magnetic layer must contain fields of order 10^5 G (an order of magnitude greater than the value of B_e at the base of the convection zone). His arguments are compelling. If we accept them we may still suppose that the field strength in the magnetic layer is typically around 10^4 G and that the layer is about 10^4 km thick (Galloway and Weiss 1981; Parker 1987). Where the field rises locally to about 10^5 G it becomes unstable to buoyancy-driven interchanges, releasing isolated loops of flux which rise upwards to emerge in active regions.

8. The Way Ahead

The study of sunspots has a venerable history. Yet it is only in the last 50 years that we have begun to understand them and much of our knowledge has been acquired within the last two decades. We have described the major theoretical problems raised by observations of sunspots and outlined our current understanding of them. Further detailed discussions of specific issues are provided by experts in the chapters that follow in this book.

The rapid progress that has taken place recently can be ascribed to several causes. Precise measurements have led to the construction of accurate atmospheric models. High-resolution observations have provided images that reveal fine structure with much greater clarity, while digital processing has yielded time sequences that display dynamical behaviour. Most of the new results have come from ground-based observatories but they have been augmented by observations from balloons, from rockets and from space. Technical improvements in spectroscopy, such as the use of tunable filters and the development of Stokes polarimetry, have made it possible to measure velocities and the vector magnetic field with enhanced precision. The detection of p-mode oscillations has opened up the whole new field of sunspot seismology. Finally, there is the solar-stellar connection, including observations of starspots.

Meanwhile, there have been great advances in theory, as we have explained. These advances have depended first on gaining a fuller understanding of the specific physical processes that are involved and, secondly, on improvements in computing power and numerical techniques. There has been enormous progress in the theory of hydromagnetic instabilities, of waves in stratified atmospheres and structured magnetic fields and of magnetoconvection, helped by developments in nonlinear dynamics. Computational models of sunspots in magnetohydrostatic equilibrium have become increasingly elaborate, aspects of magnetoconvection have been clarified by idealized numerical experiments and there have been ambitious simulations of the interaction between convection and magnetic fields in the Sun. From all these investigations a new physical picture of sunspots has emerged. This picture is far from complete but it does provide a firm basis for future developments.

We conclude, therefore, by listing some topics where we expect to see substantial progress in the near future. First, we require accurate measurements of the field profile at the surface, which can be used to construct global models of spots in magnetohydrostatic equilibrium and so to determine their deep structure. Then we can turn to multicomponent models. The key to detailed modelling of the structure of a sunspot is a proper understanding of the penumbra. We need precise treatments, first of the instability responsible for the transition from pores to spots, and then of the nonlinear convective interchanges that lead to the formation of a filamentary penumbra. A full description of the Evershed effect should emerge naturally from such a model. We also need an accurate, quantitative description of nonlinear convection in the umbra that can settle the cluster-monolith controversy. Given an understanding of these magnetoconvective processes, it will be possible both to describe the fine structure of penumbral filaments and to construct global sunspot models that can be compared with observations. Sunspot seismology is still in its infancy but we can expect to gain a satisfactory understanding of the absorption of p-modes in a spot and there is the tantalising possibility of inverting measurements to obtain information about subsurface structure in the flux tube beneath a sunspot. Finally, we may expect a fuller understanding of compressible convection and its interaction with rotation, which will lead to a better description of the solar dynamo. Only then will it be possible to model the generation of isolated flux loops and to follow their consequent emergence to form sunspots. We are confident that these projects will provide enough challenges and excitement for a new generation of students working on the theory of sunspots.

Acknowledgements

We are grateful to the NATO Scientific Affairs Division, whose support for the Advanced Research Workshop on *The Theory of Sunspots* made it possible for us to write this review. We have been helped by many discussions with colleagues at the Workshop and we especially thank F. Moreno-Insertis, H. U. Schmidt and C. Zwaan for suggestions and comments on this chapter. JHT was supported in part by NASA grants NAGW-2123 and NAGW-2444 and NOW held a SERC Senior Fellowship while this work was carried out.

References

Abdelatif, T. E. 1990. Magneto-atmospheric waves. *Solar Phys.* **129**, 201-219.

Abdelatif, T. E., Lites, B. W. and Thomas, J. H. 1986. The interaction of solar *p*-modes with a sunspot. I. Observations. *Astrophys. J.* **311**, 1015-1024.

Abdelatif, T. E. and Thomas, J. H.. 1987. The interaction of solar *p*-modes with a sunspot. II. Simple theoretical models. *Astrophys. J.* **320**, 884-892.

Acheson, D. J. 1978. On the instability of toroidal magnetic fields and differential rotation in stars. *Phil. Trans. Roy. Soc. Lond. A* **289**, 459-500.

Acheson, D. J. 1979. Instability by magnetic buoyancy. *Solar Phys.* **62**, 23-50.

Adam, M.G. 1990. An observational examination of models for sunspot magnetic field configurations. *Solar Phys.* **125**, 37-44.

Albregtsen, F. and Maltby, P. 1978. New light on sunspot darkness and the solar cycle. *Nature* **274**, 41-42.

Albregtsen, F. and Maltby, P. 1981. Solar cycle variation of sunspot intensity. *Solar Phys.* **71**, 269-283.

Alfvén, H. 1942. On the existence of electromagnetic-hydromagnetic waves. *Arkiv. f. Mat. Astron. Fysik* **29B**, No. 2, pp 1-7.

Alfvén, H. 1943. On the effect of a vertical magnetic field in a conducting atmosphere. *Arkiv. f. Mat. Astron. Fysik* **29A**, No. 11, pp. 1-6.

Alissandrakis, C. E., Dialetis, D., Mein, P., Schmieder, B. and Simon, G. 1988. The Evershed flow in the solar photosphere, chromosphere, and chromosphere-corona transition region. *Astron. Astrophys.* **201**, 339-349.

Antia, H. M. and Chitre, S. M. 1979. Waves in the sunspot umbra. *Solar Phys.* **63**, 67-78.

Antia, H. M., Chitre, S. M. and Gokhale, M. H. 1978. Waves in the sunspot penumbra. *Solar Phys.* **60**, 31-46.

Avrett, E. H. 1981. Reference model atmosphere calculation: The Sunspot sunspot model. In *The Physics of Sunspots*, ed. L. E. Cram and J. H. Thomas (Sunspot: Sacramento Peak Observatory), pp. 235-255.

Babcock, H. W. 1961. The topology of the Sun's magnetic field and the 22-year cycle. *Astrophys. J.* **133**, 572-587.

Beckers, J. M. 1964. A study of fine structures in the solar chromosphere. AFCRL Environmental Research Paper No. 49, Sacramento Peak Observatory.

Beckers, J. M. 1968. High-resolution measurements of photosphere and sunspot velocity and magnetic fields using a narrow-band birefringent filter. *Solar Phys.* **3**, 258-268.

Beckers, J. M. 1969. The microstructure of sunspots. In *Plasma Instabilities in Astrophysics*, ed. D. G. Wentzel and D. A. Tidman (New York: Gordon and Breach), pp. 139-152.

Beckers, J.M. 1981. Dynamics of the solar photosphere. In *The Sun as a Star*, ed. S. Jordan (Washington: NASA SP-450), pp. 11-64.

Beckers, J. M. and Schröter, E. H. 1969. The intensity, velocity, and magnetic structure of a sunspot region. IV. Properties of a unipolar sunspot. *Solar Phys.* **10**, 384-403.

Beckers, J. M. and Schneeberger, T. J. 1977. Alfvén waves in the corona above sunspots. *Astrophys. J.* **215**, 356-363.

Beckers, J. M. and R. B. Schultz, R. B. 1972. Oscillatory motions in sunspots. *Solar Phys.* **27**, 61-70.

Beckers, J. M. and Tallant, P. E. 1969. Chromospheric inhomogeneities in sunspot umbrae. *Solar Phys.* **7**, 351-365.

Belvedere, G. 1990. Solar and stellar cycles. In *Inside the Sun* ed. M. Cribier and G. Berthomieu (Dordrecht: Kluwer), pp. 371-386.

Bernstein, I. B., Frieman, E. A., Kruskal, M. D. and Kulsrud, R. M. 1958. An energy principle for hydromagnetic stability problems. *Proc. Roy. Soc. Lond. A* **244**, 17-40.

Bhatnagar, A., Livingston, W. C. and Harvey, J. W. 1972. Observations of sunspot umbral velocity oscillations. *Solar Phys.* **27**, 80-88.

Bhatnagar, A. and K. Tanaka. 1972. Intensity oscillation in Hα fine structure. *Solar Phys.* **24**, 87-97.

Biermann, L. 1941. Der gegenwärtige Stand der Theorie konvektiver Sonnenmodelle. *Vierteljahrsschrift Astron. Ges.* **76**, 194-200.

Bjerknes, V. 1926. Solar hydrodynamics. *Astrophys. J.* **64**, 93-121.

Bogdan, T. J. 1989. On the resonance scattering of sound waves by slender magnetic flux tubes. *Astrophys. J.* **345**, 1042-1049.

Bogdan, T. J. 1992. Sunspot seismology: the interaction of a sunspot with solar p-modes. (This volume)

Bogdan, T. J. and Knölker, M. 1991. Scattering of acoustic waves from a magnetic flux tube embedded in a radiating fluid. *Astrophys. J.* **369**, 219-236.

Bogdan, T. J. and Zweibel, E. G. 1987. Propagation of compressive waves through fibril magnetic fields. *Astrophys. J.* **312**, 444-456.

Brandenburg, A. and Tuominen, J. 1991. The solar dynamo. In *The Sun and Cool Stars: Activity, Magnetism, Dynamos*, ed. I. Tuominen, D. Moss and G. Rüdiger (Berlin: Springer), pp. 223-233.

Brants, J. J. and Zwaan, C. 1982. The structure of sunspots. IV. Magnetic field strengths in small sunspots and pores. *Solar Phys.* **80**, 251-258.

Braun, D. C., Duvall, T. L., Jr. and LaBonte, B. J. 1987. Acoustic absorption by sunspots. *Astrophys. J. (Letters)* **319**, L27-L31.

Braun, D. C., Duvall, T. L., Jr. and LaBonte, B. J. 1988. The absorption of high-degree *p*-mode oscillations in and around sunspots. *Astrophys. J.* **335**, 1015-1025.

Bray, J. R. and Loughhead, R. E. 1964. *Sunspots* (London: Chapman and Hall).

Brown, T. M., Bogdan, T. J., Lites, B. W. and Thomas, J. H. 1992. (In preparation)

Busse, F. H. 1987. A new mechanism for the Evershed effect. In *The Role of Fine-scale Magnetic Fields on the Structure of the Solar Atmosphere*, ed. E. H. Schröter, M. Vázquez and A. A. Wyller (Cambridge: Cambridge University Press), pp. 187-195.

Byrne, P. B. 1992. Starspots. (This volume.)

Cally, P. S. 1983. Umbral oscillations in the presence of a spreading magnetic field. *Solar Phys.* **88**, 77-102.

Cally, P. S. and Adam, J. A. 1983. On photospheric and chromospheric penumbral waves. *Solar Phys.* **85**, 97-111.

Cattaneo, F. 1984. Oscillatory convection in sunspots. In *The Hydrodynamics of the Sun*, (Noordwijk: ESA SP-220) pp. 47-50.

Cattaneo, F., Chiueh, T. and Hughes, D. W. 1990a. Buoyancy-driven instabilities and the nonlinear breakup of a sheared magnetic layer. *J. Fluid Mech.* **219**, 1-23.

Cattaneo, F., Chiueh, T. and Hughes, D. W. 1990b. A new twist to the solar cycle. *Mon. Not. Roy. Astron. Soc.* **247**, 6P-9P.

Cattaneo, F. and Hughes, D. W. 1988. The nonlinear breakup of a magnetic layer: instability to interchange modes. *J. Fluid Mech.* **196**, 323-344.

Cattaneo, F., Hughes, D. W. and Weiss, N. O. 1991. What is a stellar dynamo? *Mon. Not. Roy. Astron. Soc.* **253**, 479-484.

Chandrasekhar, S. 1952. On the inhibition of convection by a magnetic field. *Phil. Mag. (7th Ser.)* **43**, 501-532.

Chandrasekhar, S. 1961. *Hydrodynamic and Hydromagnetic Stability* (Oxford: Clarendon Press).

Childress, S. 1992. Fast dynamo theory. In *Topological Aspects of the Dynamics of Fluids and Plasmas*, ed. H. K. Moffatt, G. M. Zaslavsky, M. Tabor and P. Comte (Dordrecht: Kluwer), in press.

Chitre, S.M. 1963. The structure of sunspots *Mon. Not. Roy. Astron. Soc.* **126**, 431-443.

Chitre, S. M. 1968. The Evershed motion in sunspots. *Solar Phys.* **4** , 168-175.

Chitre, S. M. 1992. Theory of umbral oscillations and penumbral waves. (This volume.)

Chitre, S. M. and Davila, J. M. 1991. The resonant absorption of *p*-modes by sunspots with a twisted magnetic field. *Astrophys. J.* **371**, 785-792.

Chitre, S. M. and Shaviv, G. 1967. A model of the sunspot umbra. *Solar Phys.* **2**, 150-157.

Choudhuri, A. R. 1986. The dynamics of magnetically trapped fluids. I. Implications for umbral dots and penumbral grains. *Astrophys. J.* **302**, 809-825.

Choudhuri, A. R. 1989. The evolution of loop structures in flux rings within the solar convection zone. *Solar Phys.* **123**, 217-239.

Choudhuri, A. R. 1992. The cluster model of sunspots. (This volume.)

Choudhuri, A. R. and Gilman, P. A. 1987. The influence of the Coriolis force on flux tubes rising through the solar convection zone. *Astrophys J.* **316**, 788-800.

Clark, A., Jr. 1979. Thermal models of sunspots. *Solar Phys.* **62**, 305-330.

Cowling, T. G. 1933. The electrical conductivity of an ionized gas in the presence of a magnetic field. *Mon. Not. Roy. Astron. Soc.* **93**, 90-98.

Cowling, T. G. 1934. The magnetic field of sunspots. *Mon. Not. Roy. Astron. Soc.* **94**, 39–48

Cowling, T. G. 1946. The growth and decay of the sunspot magnetic field. *Mon. Not. Roy. Astron. Soc.* **106**, 218-224.

Cowling, T. G. 1953. Solar electrodynamics. In *The Sun*, ed. G.P. Kuiper (Chicago: University of Chicago Press), pp. 532-591.

Cowling, T. G. 1957. *Magnetohydrodynamics* (New York: Interscience).

Cowling, T. G. 1976. On the thermal structure of sunspots. *Mon. Not. Roy. Astron. Soc.* **177**, 409-414.

Cowling, T. G. 1981. The present status of dynamo theory. *Ann. Rev. Astron. Astrophys.* **19**, 115-135.

Cowling, T. G. 1985. Astronomer by accident. *Ann. Rev. Astron. Astrophys* **23**, 1-18.

Cram, L. E., Nye, A. H. and Thomas, J. H. 1981. Conjectures regarding the structure of a sunspot penumbra. In *The Physics of Sunspots*, ed. L. E. Cram and J. H. Thomas (Sunspot: Sacramento Peak Observatory), pp. 384-388.

Danielson, R. E. 1961a. The structure of sunspot penumbras. I. Observations. *Astrophys. J.* **134**, 275-288.

Danielson, R. E. 1961b. The structure of sunspot penumbras. II. Theoretical. *Astrophys. J.* **134**, 289-311.

Danielson, R. E. 1964. The structure of sunspot umbras. I. Observations. *Astrophys. J.* **139**, 45-47.

Danielson, R. E. 1965. Sunspots: theory (introductory report). In *Stellar and Solar Magnetic Fields*, IAU Symp. No. 22, ed. R Lüst (Amsterdam: North-Holland), pp. 314-329.

Danielson, R. E. 1966. Observations of sunspot fine structure from the stratosphere and theoretical interpretation. *Atti del Convegno sulle Macchie Solari*, ed. G. Righini (Florence: Barbèra), pp. 77-106.

Defouw, R. J. 1976. Wave propagation along a magnetic tube. *Astrophys. J.* **209**, 266-269.

Degenhardt, D. 1989. Stationary siphon flows in thin magnetic flux tubes. *Astron. Astrophys.* **222**, 297-306.

Degenhardt, D. 1991. Stationary siphon flows in thin magnetic flux tubes. II. Radiative heat exchange with the surroundings. *Astron. Astrophys.* **248**, 637-646.

Deinzer, W. 1965. On the magneto-hydrostatic theory of sunspots. *Astrophys. J.* **141**, 548-563.

Deinzer, W., Hensler, G., Schüssler, M. and Weisshaar, E. 1984. Model calculations of magnetic flux tubes II. Stationary results for solar magnetic elements. *Astron. Astrophys.* **139**, 435-449.

D'Silva, S. Z. and Choudhuri, A. R. 1991. The effect of Kelvin-Helmholtz instability on rising flux tubes in the convection zone. *Solar Phys.* **136**, 201-219.

Dziembowski, W. A., Goode, P. R. and Libbrecht, K.G. 1989. The radial gradient in the Sun's rotation. *Astrophys. J. (Letters)* **337**, L53-L57.

Eschrich, K.-O. and Krause, F. 1977. Uber die Möglichkeit, die Temperaturverteilung in Sonnenflecken durch einen anisotropen Wärmetransport zu erklären. *Astron. Nachr.* **298**, 1-8.

Evershed, J. 1909. Radial movement in sunspots. *Kodaikanal Obs. Bull.* No. 15, pp. 63-69.

Foukal, P. V., Fowler, L. A. and Livshits, M. 1983. A thermal model of sunspot influence on solar luminosity. *Astrophys. J.* **267**, 863-871.

Galloway, D. J. 1975. Fine structure and Evershed motions in the sunspot penumbra. *Solar Phys.* **44**, 409-415.

Galloway, D. J. 1978. The origin of running penumbral waves. *Mon. Not. Roy. Astron. Soc.* **184**, 49P-52P.

Galloway, D. J. and Moore, D. R. 1979. Axisymmetric convection in the presence of a magnetic field. *Geophys. Astrophys. Fluid Dyn.* **12**, 73-105.

Galloway, D. J. and Proctor, M. R. E. 1992. Numerical calculations of fast dynamos for smooth velocity fields with realistic diffusion. *Nature*, in press.

Galloway, D. J., Proctor, M. R. E. and Weiss, N. O. 1978. Magnetic flux ropes and convection. *J. Fluid Mech.* **87**, 243-261.

Galloway, D. J. and Weiss, N. O. 1981. Convection and magnetic felds in stars. *Astrophys. J.* **243**, 945-953.

Garcia de la Rosa, J. I. 1987. Umbral dots: a case of penetrative convection between sunspot fragments. *Solar Phys.* **112**, 49-58.

Gilman, P. A. 1970. Instability of magnetohydrostatic stellar interiors from magnetic buoyancy. I. *Astrophys. J.* **162**, 1019-1029.

Gilman, P. A. 1983. Dynamically consistent nonlinear dynamos driven by convection in a rotating shell. II. Dynamos with cycles and strong feedbacks. *Astrophys. J. Suppl. Ser.* **53**, 243-268.

Giovanelli, R. G. 1972. Oscillations and waves in a sunspot. *Solar Phys.* **27**, 71-79.

Giovanelli, R. G., Harvey, J. W. and Livingston, W. 1978. Motions in solar magnetic tubes. III. Outward wave propagation in sunspot umbras. *Solar Phys.* **58**, 347-361.

Glatzmaier, G. A. 1985. Numerical simulations of stellar convective dynamos. II. Field propagation in the convection zone. *Astrophys. J.* **291**, 300-307.

Gokhale, M. H. and Zwaan, C. 1972. The structure of sunspots. I: Observational constraints; Current sheet models. *Solar Phys.* **26**, 52-75.

Golub, L., Rosner, R., Vaiana, G. S. and Weiss, N. O. 1981. Solar magnetic fields: the generation of emerging flux. *Astrophys. J.* **243**, 309-316.

Goode, P. R. 1991. The Sun's internal differential rotation from helioseismology. In *The Sun and Cool Stars: Activity, Magnetism, Dynamos*, ed. I. Tuominen, D. Moss and G. Rüdiger (Berlin: Springer), pp. 157-171.

Goode, P. R. and Dziembowski, W. A. 1991. Solar-cycle dependence of the Sun's deep internal rotation shown by helioseismology. *Nature* **349**, 223-225.

Goode, P. R., Gough, D. O. and McIntyre, M. E. 1992. Is the solar cycle driven by a torsional oscillator? *Nature*, submitted.

Goossens, M. and Poedts, S. 1992. Linear resistive magnetohydrodynamic computations of resonant absorption of acoustic oscillations in sunspots. *Astrophys. J.* **384**, 348-360.

Grossman-Doerth, U., Knölker, M., Schüssler, M. and Weisshaar, E. 1989. Models of magnetic flux sheets. In *Solar and Stellar Granulation*, ed. R. J. Rutten and G. Severino (Dordrecht: Kluwer), pp. 481-490.

Grossman-Doerth, U., Schmidt, W. and Schröter, E. H. 1986. Size and temperature of umbral dots. *Astron. Astrophys.* **156**, 347-353.

Gurman, J.B. and Leibacher, J.W. 1984. Linear models of acoustic waves in sunspot umbrae. *Astrophys. J.* **283**, 859-869.

Hale, G. E. 1908a. Solar vortices. *Astrophys. J.* **28**, 100-116.

Hale, G. E. 1908b. On the probable existence of a magnetic field in sun-spots. *Astrophys. J.* **28**, 315-343..

Hale, G. E. and Nicholson, S. B. 1938. *Magnetic Observations of Sunspots 1917-1924* (Washington: Carnegie Institution).

Harvey, K. and Harvey, J. 1973. Observations of moving magnetic features near sunspots. *Solar Phys.* **28**, 61-71.

Hasan, S. S. 1984. Convective instability in a solar flux tube. I. Nonlinear calculations for an adiabatic inviscid fluid. *Astrophys. J.* **285**, 851-857.

Hasan, S. S. 1985. Convective instability in a solar flux tube. II. Nonlinear calculations with horizontal radiative heat transport and finite viscosity. *Astron. Astrophys.* **143**, 39-45.

Hasan, S. S. 1991. Magnetoatmospheric oscillations in sunspot umbrae. *Astrophys. J.* **366**, 328-339.

Haugen, E. 1967. The velocity field surrounding sunspots, as derived from observations of Hα. *Solar Phys.* **2**, 227-230.

Haugen, E. 1969. On the chromospheric velocity field in sunspot regions. *Solar Phys.* **9** , 88-101.

Herschel, W. 1801. Observations tending to investigate the Nature of the Sun, in order to find the Causes or Symptoms of its variable Emission of Light and Heat; with Remarks on the Use that may possibly be drawn from Solar Observations. *Phil. Trans. Roy. Soc. Lond.* **91**, 265-318.

Hevelius, J. 1647. *Selenographia* (Gdansk: published by the author).

Hollweg, J. V. 1988. Resonance absorption of solar p-modes by sunspots. *Astrophys. J.* **335**, 1005-1014.

Hoyle, F. 1949. *Some Recent Researches in Solar Physics* (Cambridge: Cambridge University Press), chapter 1.

Hoyng, P. 1990. Excitation of dynamo modes. In *Solar Photosphere: Structure, Convection and Magnetic Fields*, IAU Symp. No. 138, ed. J. O. Stenflo (Dordrecht: Kluwer), pp. 359-374.

Hughes, D. W. 1985a. Magnetic buoyancy instabilities for a static plane layer. *Geophys. Astrophys. Fluid Dyn.* **32**, 273-316.

Hughes, D. W. 1985b. Magnetic buoyancy instabilities incorporating rotation. *Geophys. Astrophys. Fluid Dyn.* **34**, 99-142.

Hughes, D. W. 1987. Finite-amplitude solutions for interchange instabilities driven by magnetic buoyancy. *Geophys. Astrophys. Fluid Dyn.* **37**, 297-343.

Hughes, D. W. 1991. Magnetic buoyancy. In *Advances in Solar System MHD*, ed. E. R. Priest and A. W. Hood (Cambridge: Cambridge University Press), pp. 77-104.

Hughes, D. W. 1992. The formation of flux tubes at the base of the convection zone. (This volume.)

Hughes, D. W. and Proctor, M. R. E. 1988. Magnetic fields in the solar convection zone: magnetoconvection and magnetic buoyancy. *Ann. Rev. Fluid. Mech.* **20**, 187-223.

Hurlburt, N. E., Matthews, P. C. and Proctor, M. R. E. 1992. Nonlinear compressible magnetoconvection in an oblique magnetic field. In preparation.

Hurlburt, N. E., Proctor, M. R. E., Weiss, N. O. and Brownjohn, D. P. 1989. Nonlinear compressible magnetoconvection. Part I. Travelling waves and oscillations. *J. Fluid Mech.* **209**, 587-628.

Hurlburt, N. E. and Toomre, J. 1988. Magnetic fields interacting with nonlinear compressible convection. *Astrophys. J.* **327**, 920-932.

Ichimoto, K. 1987. Evershed effect observed in various solar photospheric lines. I. Dependence of the velocity distribution across the penumbra on the equivalent width. *Publ. Astron. Soc. Japan* **39**, 329-342.

Jahn, K. 1989. Current sheet as a diagnostic for the subphotospheric structure of a spot. *Astron. Astrophys.* **222**, 264-292.

Jahn, K. 1992. Magnetohydrostatic equilibrium in sunspot models. (This volume.)

Jahn, K. and Schmidt, H. U. 1992. In preparation.

Jensen, E. 1955. On tubes of magnetic force embedded in stellar material. *Ann. Astrophys.* **18**, 127-140.

Kawakami, H. 1983. Polarimetric study of unipolar sunspots. *Publ. Astron. Soc. Japan* **35**, 459-489.

Kinman, T. D. 1952. Motions in the Sun at the photospheric level. II. The Evershed effect in sunspot Mt. Wilson No. 9987. *Mon. Not. Roy. Astron. Soc.* **112**, 425-444.

Knobloch, E. and Weiss, N. O. 1984. Convection in sunspots and the origin of umbral dots. *Mon. Not. Roy. Astron. Soc.* **207**, 203-214.

Knölker, M. and Schüssler, M. 1988. Model calculations of magnetic flux tubes IV. Convective energy transport and the nature of intermediate size flux concentrations. *Astron. Astrophys.* **202**, 275-283.

Knölker, M., Schüssler, M. and Weisshaar, E. 1988. Model calculations of magnetic flux tubes III. Properties of solar magnetic elements. *Astron. Astrophys.* **194**, 257-267.

Krause, F. and Rädler, K-H. 1980. *Mean-field Magnetohydrodynamics and Dynamo Theory* (Oxford: Pergamon).

Kruskal, M. and Schwarzschild, M. 1954. Some instabilities of a completely ionized plasma. *Proc. Roy. Soc. Lond. A* **223**, 348-360.

Lamb, S. A. 1975. An investigation of material motions in a sunspot.*Mon. Not. Roy. Astron. Soc.* **172**, 205-220.

Larmor, J. 1919. How could a rotating body such as the Sun become a magnet? *Brit. Assoc. Reports*, 159–160.

Lites, B. W. 1986. Photoelectric observations of chromospheric sunspot oscillations. IV. The Ca II H line and He II 0830. *Astrophys. J.* **301**, 1005-1017.

Lites, B. W. 1992. Sunspot oscillations: observations and implications. (This volume.)

Lites, B. W., Scharmer, G. and Skumanich, A. 1990. High-resolution spectra of solar magnetic features. I. Analysis of penumbral fine structure. *Astrophys. J.* **355**, 329-341.

Lites, B.W. and Skumanich, A. 1990. Stokes profile analysis and vector magnetic fields. V. The magnetic field structure of large sunspots observed with Stokes II. *Astrophys. J.* **348**, 747-760.

Lites, B. W. and Thomas, J. H. 1985. Sunspot umbral oscillations in the photosphere and low chromosphere. *Astrophys. J.* **294**, 682-688.

Livingston, W. 1991. Radial filamentary structure in a sunspot umbra. *Nature* **350**, 45-46.

Lou, Y.-Q. 1990. Viscous magnetohydrodynamic modes and *p*-mode absorption by sunspots. *Astrophys. J.* **350**, 452-462.

Makita, M. 1963. Physical states in sunspots. *Publ. Astron. Soc. Japan* **15**, 145-176.

Maltby, P. 1975. The chromospheric Evershed flow. *Solar Phys.* **43**, 91-105.

Maltby, P. 1992. Continuum observations and empirical models of the thermal structure of sunspots. (This volume.)

Maltby, P., Avrett, E. H., Carlsson, M., Kjeldseth-Moe, O., Kurucz, R. L. and Loeser, R. 1986. A new sunspot umbral model and its variations with the solar cycle. *Astrophys. J.* **306**, 284-303.

Maltby, P. and Eriksen, G. 1967. The Evershed effect as a wave phenomenon. *Solar Phys.* **2**, 249-257.

Massaglia, S., Bodo, G. and Rossi, P. 1989. Overstability of magnetic flux tubes in the Eddington approximation. *Asstron. Astrophys.* **209**, 399-405.

Matthews, P. C., Hurlburt, N. E., Proctor, M. R. E. and Brownjohn, D. P. 1992. Compressible magnetoconvection in oblique fields: linearised theory and simple nonlinear models. *J. Fluid Mech.*, in press.

Meyer, F. and Schmidt, H. U. 1968. Magnetisch ausgerichtete Strömungen zwischen Sonnenflecken. *Z. Ang. Math. Mech.* **48**, T218-T221.

Meyer, F., Schmidt, H. U. and Weiss, N. O. 1977. The stability of sunspots. *Mon. Not. Roy. Astron. Soc.* **179**, 741-761.

Meyer, F., Schmidt, H. U., Weiss, N. O., and Wilson, P.R. 1974. The growth and decay of sunspots. *Mon. Not. Roy. Astron. Soc.* **169**, 35-57.

Moffatt, H. K. 1978. *Magnetic Field Generation in Conducting Fields* (Cambridge: Cambridge University Press).

Montesinos, B. and Thomas, J. H. 1989. Siphon flows in isolated magnetic flux tubes. II. Adiabatic flows. *Astrophys. J.* **337**, 977–988.

Montesinos, B. and Thomas, J. H. 1992. Siphon flows in isolated magnetic flux tubes. V. Radiative flows. *Astrophys. J.*, submitted.

Moore, R. L. 1973. On the generation of umbral flashes and running penumbral waves. *Solar Phys.* **30**, 403-419.

Moore, R. L. 1981. Dynamic phenomena in sunspots. In *The Physics of Sunspots*, ed. L. E. Cram and J. H. Thomas (Sunspot : Sacramento Peak Observatory), pp. 259–311.

Moore, R. L. and Rabin, D. 1985. Sunspots. *Ann. Rev. Astron. Astrophys.* **23**, 239-266.

Moreno-Insertis, F. 1986. Nonlinear time-evolution of kink-unstable magnetic flux tubes in the convection zone of the Sun. *Astron. Astrophys.* **166**, 291-305.

Moreno-Insertis, F. 1987. The subsurface structure of sunspots and the origin of solar active regions. In *The Role of Fine-scale Magnetic Fields on the Stucture of the Solar Atmosphere*, ed. E. H. Schröter, M. Vázquez and A. A. Wyller (Cambridge: Cambridge University Press), pp. 167-182.

Moreno-Insertis, F. 1992. The motion of magnetic flux tubes in the convection zone and the subsurface origin of active regions. (This volume.)

Moreno-Insertis, F. and Spruit, H. C. 1989. Stability of sunspots to convective motions. I. Adiabatic instability. *Astrophys. J.* **342**, 1158-1171.

Mullan, D. J. and Yun, H. S. 1973. Can oscillations grow in a sunspot umbra? *Solar Phys.* **30**, 83-91.

Muller 1973a. Etude morphologique et cinematique des structures fines d'une tache solaire. *Solar Phys.* **29**, 55-73.

Muller 1973b. Etude photometrique des structures fines de la penombre d'une tache solaire. *Solar Phys.* **32**, 409-420.

Muller, R. 1976. Characteristics of the displacement of the penumbral bright grains of sunspots. *Solar Phys.* **48**, 101-102.

Muller, R. 1992. Fine structure of umbrae and penumbrae. (This volume.)

Musman, S. 1967. Alfvén waves in sunspots. *Astrophys. J.* **149**, 201-209.

Musman, S., Nye, A. H. and Thomas, J. H. 1976. Observations of penumbral waves in the photosphere. *Astrophys. J.* **206**, L175-L178.

Newcomb, W. A. 1961. Convective instability induced by gravity in a plasma with a frozen-in magnetic field. *Phys. Fluids* **4**, 391-396.

Nordlund, Å. 1983. Numerical 3-D simulations of the collapse of photospheric flux tubes. In *Solar and Stellar Magnetic fields: Origins and Coronal Effects*, IAU Symp. No. 102, ed. J. O. Stenflo (Dordrecht: Reidel), pp. 79-83.

Nordlund, Å. 1986. 3-D model calculations. In *Small Scale Magnetic Flux Concentrations in the Solar Photosphere*, ed. W. Deinzer, M. Knölker and H. H. Voigt (Göttingen: Vanderhoeck und Ruprecht), pp. 83-102.

Nordlund, Å., Brandenburg, A., Jennings, R. L., Rieutord, M., Ruokolainen, J., Stein, R. F. and Tuominen, I. 1992. Dynamo action in stratified convection with overshoot. *Astrophys. J.*, submitted.

Nordlund, Å. and Stein, R. F. 1989. Simulating magnetoconvection. In *Solar and stellar granulation*, ed. R.J. Rutten and G. Severino (Dordrecht: Kluwer), pp. 453-470.

Nordlund, Å. and Stein, R. F. 1990. Solar magnetoconvection. In *Solar Photosphere: Structure, Convection, and Magnetic Fields,* IAU Symp. No. 138, ed. J. O. Stenflo (Dordrecht: Kluwer), pp. 191-211.

Nye, A. H. and Thomas, J. H. 1974. The nature of running penumbral waves. *Solar Phys.* **38**, 399-413.

Nye, A. H. and Thomas, J. H. 1976. Solar magneto-atmospheric waves. II. A model for running penumbral waves. *Astrophys. J.* **204**, 582-588.

Obridko, V.N. 1974. Origin of bright umbral dots in sunspots. *Astr. Zh.* **51**, 1272-1276 (*Sov. Astron.* **18**, 758-767).

Obridko, V.N. 1979. The origin of bright umbral elements and forced oscillatory convection. *Solnechniye Danniye* No. 3, pp. 101-106.

Parker, E. N. 1955a. The formation of sunspots from the solar toroidal field. *Astrophys. J.* **121**, 491-507.

Parker, E. N. 1955b. Hydromagnetic dynamo models. *Astrophys. J.* **122**, 293-314.

Parker, E. N. 1963. Kinematical hydromagnetic theory and its application to the low solar photosphere. *Astrophys. J.* **138**, 552-575.

Parker, E. N. 1966. The dynamical state of the interstellar gas and field. *Astrophys. J.* **145**, 811-833.

Parker, E. N. 1974. The nature of the sunspot phenomenon. I: Solutions of the heat transport equation. *Solar Phys.* **36**, 249-274.

Parker, E. N. 1975 The nature of the sunspot phenomenon. IV: the intrinsic instability of the magnetic configuration. *Solar Phys.* **40**, 291-301.

Parker, E. N. 1978. Hydraulic concentration of magnetic fields in the solar photosphere. VI. Adiabatic cooling and concentration in downdrafts. *Astrophys. J.* **221**, 368-377.

Parker, E. N. 1979a. *Cosmical Magnetic Fields* (Oxford: Clarendon Press).

Parker, E. N. 1979b. Sunspots and the physics of magnetic flux tubes. I. The general nature of the sunspot. *Astrophys. J.* **230**, 905-913.

Parker, E. N. 1979c. Sunspots and the physics of magnetic flux tubes. IX. Umbral dots and longitudinal overstability. *Astrophys J.* **234**, 333-347.

Parker, E. N. 1987. The dynamical oscillation and propulsion of magnetic fields in the convective zone of a star. I. General considerations. *Astrophys J.* **312**, 868-879.

Parker, E.N. 1992. The sunspot phenomenon: a commentary. (This volume.)

Petrovay, K. 1991. On the properties of toroidal flux tubes in the solar dynamo. *Solar Phys.* **134**, 407-410.

Pikel'ner, S. B. 1961. *Osnovy Kosmicheskoy Elektrodinamiki* (Moscow: Nauka).

Pizzo, J.V. 1986. Numerical solution of the magnetostatic equations for thick flux tubes, with applications to sunspots, pores, and related structures. *Astrophys. J.* **302**, 785-808.

Priest, E. R. 1981. Theory of loop flows and instability. In *Solar Active Regions*, ed. F. Q. Orrall (Boulder: Colorado Associated University Press), pp. 213-275.

Priest, E. R. 1982. *Solar Magnetohydrodynamics* (Dordrecht: Reidel).

Proctor, M. R. E. 1992. Magnetoconvection. (This volume.)

Proctor, M.R.E. and Weiss, N.O. 1982. Magnetoconvection. *Rep. Prog. Phys.* **45**, 1317-1379.

Roberts, B. 1976. Overstability and cooling in sunspots. *Astrophys. J.* **204**, 268-280.

Roberts, B. 1986. Dynamical processes in magnetic flux tubes. In *Small Scale Magnetic Flux Concentration in the Solar Photosphere*, ed. W. Deinzer, M. Knölker and H. H. Voigt (Göttingen: Vandenhoeck und Ruprecht), pp. 169-190 .

Roberts, B. 1992. Magnetohydrodynamic waves in structured magnetic fields. (This volume.)

Rosenthal, C. S. 1991. Sunspots and sunquakes: Acoustic oscillations as a probe of magnetic structure in the Sun. Ph.D. dissertation, University of Cambridge.

Rosenthal, C. S. 1992. Absorption of acoustic waves by sunspots. II. Resonance absorption in axisymmetric fibril models. *Solar Phys.*, in press.

Rosner, R. and Weiss, N. O. 1992. The origin of the solar cycle. *Publ. Astron. Soc. Pac.* , in press.

Rosseland, S. 1926. On the transmission of radiation through an absorbing medium in motion, with applications to the theory of sun-spots and solar rotation. *Astrophys. J.* **63**, 342-367.

Rüedi, I., Solanki, S. K. and Rabin, D. 1992. Infrared lines as a probe of solar magnetic features. IV. Discovery of a siphon flow. *Astron. Astrophys.*, submitted.

Russell, H. N. 1921. Note on cooling by expansion in sun-spots. *Astrophys. J.* **54**, 293-295.

Ryutova, M. P. 1990. Waves and oscillations in magnetic fluxtubes. In *Solar Photosphere: Structure, Convection, and Magnetic Fields*, IAU Symp. No. 138, ed. J. O. Stenflo (Dordrecht: Kluwer), pp. 229-249.

St. John, C. E. 1913. Radial motions in sunspots. *Astrophys. J.* **37**, 322-353; **38**, 341-391.

Sakurai, T., Goossens, M. and Hollweg, J. V. 1991. Resonant behaviour of MHD waves on magnetic flux tubes. II. Absorption of sound waves by sunspots. *Solar Phys.* **133**, 247-262.

Sams, B. J., III, Golub, L. and Weiss, N. O. 1992. X-ray observations of sunspot penumbral structure. *Astrophys. J.* , submitted.

Savage, B. D. 1969. Thermal generation of hydromagnetic waves in sunspots. *Astrophys. J.* **156**, 707-729.

Scharmer, G. 1991. Private communication.

Scheuer, M. A. and Thomas, J. H. 1981. Umbral oscillations as resonant modes of magneto-atmospheric waves. *Solar Phys.* **71**, 21-38.

Schlüter, A. and Temesváry, St. 1958. The internal constitution of sunspots. In *Electromagnetic Phenomena in Cosmical Physics*, IAU Symp. No. 6, ed. B. Lehnert (Cambridge: Cambridge University Press), pp. 263-271.

Schmidt, H. U. 1987. Joint discussion. In *The Role of Fine-scale Magnetic Fields on the Structure of the Solar Atmosphere*, ed. E.H. Schröter, M. Vázquez and A. A. Wyller (Cambridge: Cambridge University Press), pp. 219-222.

Schmidt, H. U. 1991. Sunspots. *Geophys. Astrophys. Fluid Dyn.* **62**, 249-270.

Schmidt, H. U., Simon, G. W. and Weiss, N. O. 1985. Buoyant magnetic flux tubes II. Three-dimensional behaviour in granules and supergranules. *Astron. Astrophys* **148**, 191-206.

Schmidt, H. U., Spruit, H.C. and Weiss, N.O. 1986. Energy transport in sunspot penumbrae. *Astron. Astophys.* **158**, 351-360.

Schmidt, H. U. and Wegmann, R. 1983. A free boundary value problem for sunspots. In *Dynamical Problems in Mathematical Physics*, ed. B. Brosowski and E. Martensen (Frankfurt: Lang), pp. 137-150.

Schüssler, M. 1979. Magnetic buoyancy revisited: analytical and numerical results for rising flux tubes. *Astron. Astrophys.* **71**, 79-91.

Schüssler, M. 1980. Flux tube dynamo approach to the solar cycle. *Nature* **288**, 150-152.

Schüssler, M. 1983. Stellar dynamo theory. In *Solar and Stellar Magnetic Fields: Origins and Coronal Effects*, IAU Symp. No. 102, ed. J.O Stenflo (Dordrecht: Reidel), pp. 213-236.

Schüssler, M. 1990. Theoretical aspects of small-scale photospheric magnetic fields. In *Solar Photosphere: Structure, Convection and Magnetic Fields*, IAU Symposium No. 138, ed. J. O. Stenflo (Dordrecht: Kluwer), pp. 161–179.

Schüssler, M. 1991. Solar magnetic elements. *Geophys. Astrophys. Fluid Dyn.* **62**, 271-281.

Secchi, P. A. 1870. *Le Soleil* (Paris: Gauthier-Villars).

Sheeley, N. E. 1972. Observations of the horizontal velocity field surrounding sunspots. *Solar Phys.* **25**, 98-103.

Shine, R., Smith, K., Tarbell, T. and Title, A. 1990. Penumbral flows and magnetic fields. *Bull. Am. Astron. Soc.* **22**, 878.

Simon, G. W. and Weiss, N. O. 1970. On the magnetic field in pores. *Solar Phys.* **13**, 85-103.

Simon, G. W., Weiss, N. O. and Nye, A. H. 1983. Simple models for magnetic flux tubes. *Solar Phys.* **87**, 65-75.

Skumanich, A. 1992. Observations of the mesoscale magnetic structure of sunspots. (This volume.)

Small, L. M. and Roberts, B. 1984. On running penumbral waves. In *Hydromagnetics of the Sun*, ESA SP-220, pp. 257-259.

Solanki, S. K. 1987. Magnetic fields: observations and theory. In *Proc. Tenth European Regional Astronomy Meeting of the IAU. Vol. 1: The Sun*, ed. L. Hejna and M. Sobotka (Prague: Astron. Inst. Czech. Acad. Sci.), pp. 95-102.

Solanki, S. K. 1990. Empirical models of photospheric flux tubes. In *Solar Photosphere: Structure, Convection and Magnetic fields*, IAU Symp. No. 138, ed. J.O Stenflo (Dordrecht: Kluwer), pp. 103-120.

Spiegel, E. A. and Weiss, N. O. 1980. Magnetic activity and variations in solar luminosity. *Nature* **287**, 616-617.

Spruit, H. C. 1976. Pressure equilibrium and energy balance of small photospheric flux tubes. *Solar Phys.* **50**, 269-295.

Spruit, H. C. 1977. Heat flow near obstacles in the solar convection zone. *Solar Phys.* **55**, 3-34.

Spruit, H. C. 1979. Convective collapse of flux tubes. *Solar Phys.* **61**, 363-378.

Spruit, H.C. 1981a. A cluster model for sunspots. In *The Physics of Sunspots* ed. L. E. Cram and J. H. Thomas (Sunspot: Sacramento Peak Observatory), pp. 98-103.

Spruit, H. C. 1981b. Small scale phenomena in umbrae and penumbrae. The role of convective processes. In *The Physics of Sunspots*, ed. L. E. Cram and J. H. Thomas (Sunspot: Sacramento Peak Observatory), pp. 359–368.

Spruit, H. C. 1981c . Magnetohydrodynamics of sunspots. *Space Sci. Rev.* **28**, 435-448.

Spruit, H. C. 1981d. Magnetic flux tubes. In *The Sun as a Star*, ed. S. Jordan (Washington: NASA SP-450), pp. 385-412.

Spruit, H. C. 1982a. Effects of spots on a star's radius and luminosity. *Astron. Astrophys.* **108**, 348-355.

Spruit, H. C. 1982b. The flow of heat near a starspot. *Astron. Astrophys.* **108**, 356-360.

Spruit, H. C. 1987. How is the penumbra formed? In *The role of Fine-scale Magnetic Fields on the Structure of the Solar Atmosphere*, ed. E.H. Schröter, M. Vázquez and A. A. Wyller (Cambridge: Cambridge University Press), pp. 199-209.

Spruit, H. C. 1991. Absorption of *p*-mode waves by magnetic fields. In *Challenges to Theories of the Structure of Moderate Mass Stars*, Lecture Notes in Physics No. 388, ed. J. Toomre and D. O. Gough (Berlin: Springer Verlag), pp. 121-134.

Spruit, H. C. 1992. The fate of the heat flux blocked by sunspots. (This volume.)

Spruit, H. C. and Bogdan, T. J. 1992. The conversion of *p*-modes to slow modes and the absorption of acoustic waves by sunspots. *Astrophys. J.*, submitted.

Spruit, H. C. and Roberts, B. 1983. Magnetic flux tubes on the Sun. *Nature* **304**, 401.

Spruit, H. C. and Van Ballegooien, A. A. 1982. Stability of toroidal flux tubes in stars. *Astron. Astrophys.* **106**, 58-66.

Spruit, H. C. and Zweibel, E. G. 1979. Convective instability of thin flux tubes. *Solar Phys.* **62**, 15-22.

Stein, R. F., Brandenburg, A. and Nordlund, Å. 1992. Magneto-convection. In *Stellar Systems and the Sun*, ed. M. S. Giampapa and J. A. Bookbinder (APS Conference Series), in press.

Stenflo, J. O. 1989. Small scale magnetic structures on the Sun. *Astron. Astrophys. Rev.* **1**, 3-48.

Stix, M. 1989. *The Sun* (Berlin: Springer).

Stix, M. 1991. The solar dynamo. *Geophys. Astrophys. Fluid Dyn.* **62**, 211-228.

Sweet, P. A. 1955. The structure of sunspots. *Vistas in Astron.* **1**, 675-685.

Syrovatsky, S. I. and Zhugzhda, Y. D. 1967. Oscillatory convection of a conducting gas in a strong magnetic field. *Astron. Zh.* **44**, 1180-1190 (*Sov. Astron.* **11**, 945-952).

Thomas, J. H. 1978. The reflection of Alfvén waves and the cooling of sunspots. *Astrophys. J.* **225**, 275-280.

Thomas, J. H. 1981. Theories of dynamical phenomena in sunspots. In *The Physics of Sunspots*, ed. L. E. Cram and J. H. Thomas (Sunspot: Sacramento Peak Observatory), pp. 345–358.

Thomas, J. H. 1983. Magneto-atmospheric waves. *Ann. Rev. Fluid Mech.* **15**, 321-343.

Thomas, J. H. 1984a. Flow in an isolated magnetic flux tube. In *Small-Scale Dynamical Processes in Quiet Stellar Atmospheres*, ed. S. L. Keil (Sunspot: National Solar Observatory), pp. 276-277.

Thomas, J. H. 1984b. Umbral oscillations in sunspots. *Astron. Astrophys.* **135**, 188-189.

Thomas, J. H. 1985. Oscillations in sunspots. *Aust. J. Phys.* **38**, 811-824.

Thomas, J. H. 1988. Siphon flows in isolated magnetic flux tubes. *Astrophys. J.* **333**, 407–419.

Thomas, J. H. 1990. The structure of photospheric flux tubes. In *Physics of Magnetic Flux Ropes*, ed. C. T. Russell, E. R. Priest and L. C. Lee (Washington: American Geophysical Union), pp. 133–140.

Thomas, J. H., Cram, L. E. and Nye, A. H. 1982. Five-minute oscillations as a subsurface probe of sunspot structure. *Nature* **297**, 485-487.

Thomas, J. H., Cram, L. E. and Nye, A. H. 1984. Dynamical phenomena in sunspots. I. Observing procedures and oscillatory phenomena. *Astrophys. J.* **285**, 368-380.

Thomas, J. H., Lites, B. W., Gurman, J. B. and Ladd, E.F. 1987. Simultaneous measurements of sunspot umbral oscillations in the photosphere, chromosphere, and transition region. *Astrophys. J.* **312**, 457-461.

Thomas, J. H. and Montesinos, B. 1990. Siphon flows in isolated magnetic flux tubes. III. The equilibrium path of the flux-tube arch. *Astrophys. J.* **359**, 550–559.

Thomas, J. H. and Montesinos, B. 1991. Siphon flows in isolated magnetic flux tubes. IV. Critical flows with standing tube shocks. *Astrophys. J.* **375**, 404–413.

Thomas, J. H. and Montesinos, B. 1992. A siphon-flow model of the Evershed flow in a sunspot. (In preparation.)

Thomas, J. H. and Nye, A. H. 1975. Convective instability in the presence of a nonuniform horizontal magnetic field. *Phys. Fluids* **18**, 490-491.

Thomas, J. H. and Scheuer, M. A.. 1982. Umbral oscillations in a detailed model umbra. *Solar Phys.* **79**, 19-29.

Thompson, W. B. 1951. Thermal convection in a magnetic field. *Phil. Mag. (7th Ser.)* **42**, 1417-1432.

Title, A. M., Frank, Z. A., Shine, R. A., Tarbell, T. D., Topka, K. P., Scharmer, G. and Schmidt, W. 1991. On the magnetic and velocity field geometry of simple sunspots. *Astrophys. J.*, submitted.

Title, A. M., Frank, Z. A., Shine, R. A., Tarbell, T. D., Topka, K. P., Scharmer, G. and Schmidt, W. 1992. High resolution observations of the magnetic and velocity field of simple sunspots. (This volume.)

Tsinganos, K. C. 1980. Sunspots and the physics of magnetic flux tubes. X. On the hydrodynamic instability of buoyant fields. *Astrophys. J.* **239**, 746-760.

Uchida, Y. and Sakurai, T. 1975. Oscillations in sunspot umbras due to trapped Alfvén waves excited by overstability. *Pub. Astron. Soc. Japan* **27**, 259-274.

von Uexküll, M., Kneer, F. and Mattig, W. 1983. The chromosphere above sunspot umbrae. IV. Frequency analysis of umbral oscillations. *Astron. Astrophys.* **123**, 263-270.

Unno, W. and Ando, H. 1979. Instability of a thin magnetic tube in the solar atmosphere. *Geophys. Astrophys. Fluid Dyn.* **12**, 107-115.

Van Ballegooien, A. A. 1982. The overshoot layer at the base of the solar convective zone and the problem of magnetic flux storage. *Astron. Astophys.* **113**, 99-112.

Van Ballegooien, A. A. 1983. On the stability of toroidal flux tubes in differentially rotating stars. *Astron. Astrophys.* **118**, 275-284.

Venkatakrishnan, P. 1983. Nonlinear development of convective instability within slender flux tubes. *J. Astrophys. Astron.* **4**, 135.

Venkatakrishnan, P. 1985. Nonlinear development of convective instability within slender flux tubes. II. The effect of radiative heat transport. *J. Astrophys. Astron.* **6**, 21-34.

Vrabec, D. 1974. Streaming magnetic features near sunspots. In *Chromospheric Fine Structure*, IAU Symp. No. 56, ed. R.G. Athay (Dordrecht: Reidel), pp. 201-231.

Walén, C. 1949. *On the Vibratory Rotation of the Sun* (Stockholm: Henrik Lindstahls Bokhandel).

Webb, A. R. and Roberts, B. 1978. Vertical motions in an intense magnetic flux tube. II. Convective instability. *Solar Phys.* **59**, 249-274.

Wegmann, R. 1981. Trefftz integral equation method for free boundary problems of potential theory. In *Numerical Treatment of Free Boundary Value Problems*, ed. J. Albrecht (Stuttgart: Birkhäuser), pp. 335-349.

Weiss, N. O. 1964a. Convection in the presence of restraints. *Phil. Trans. Roy. Soc. Lond. A* **256**, 99-147.

Weiss, N. O. 1964b Magnetic flux tubes and convection in the Sun. *Mon. Not. Roy. Astron. Soc.* **128**, 225-235.

Weiss, N. O. 1966. The expulsion of magnetic flux by eddies. *Proc. Roy. Soc. Lond. A* **293**, 310-328.

Weiss, N. O. 1969. Sunspots and photospheric convection. In *Plasma Instabilities in Astrophysics*, ed. D. G. Wentzel and D. Tidman (New York: Gordon and Breach), pp. 153-169.

Weiss, N. O. 1977. Magnetic fields and convection. In *Problems of Stellar Convection*, ed. E. A. Spiegel and J.-P. Zahn (Berlin: Springer), pp. 176-187.

Weiss, N. O. 1981a. Convection in an imposed magnetic field. Part l. The development of nonlinear convection. *J. Fluid Mech.* **108**, 247-272.

Weiss, N. O. 1981b. Convection in an imposed magnetic field. Part 2. The dynamical regime. *J. Fluid Mech.* **108**, 273-289.

Weiss, N. O. 1989. Dynamo processes in stars. In *Accretion Discs and Magnetic Fields in Astrophysics*, ed. G. Belvedere (Dordrecht: Kluwer), pp. 11-29.

Weiss, N.O. 1990. Magnetohydrodynamics of sunspots. In *Basic Plasma Processes on the Sun*, IAU Symp. No. 142, ed. E. R. Priest and V. Krishan (Dordrecht: Kluwer), pp. 139-148.

Weiss, N. O. 1991. Magnetoconvection. *Geophys. Astrophys. Fluid Dyn.* **62**, 229-247.

Weiss, N. O., Brownjohn, D. P., Hurlburt, N. E. and Proctor, M. R. E. 1990. Oscillatory convection in sunspot umbrae. *Mon. Not. Roy. Astron. Soc.* **245**, 434-452.

Wentzel, D. G. 1992. Structure of sunspot penumbrae: fallen magnetic flux tubes. *Astrophys. J.* **388**, 211-217.

Wiehr, E. and Stellmacher, G. 1989. Velocity and magnetic field fluctuations in penumbral fine-structures. *Astron. Astrophys.* **225**, 528-532.

Wiehr, E., Stellmacher, G., Knölker, M. and Grosser, H. 1986. The sharp decrease of Evershed effect and magnetic field at the outer sunspot border. *Astron. Astrophys.* **155**, 402-406.

Wilson, P. 1968. The structure of a sunspot. II. The magnetohydrodynamics of the penumbra. *Solar Phys.* **3**, 454-465

Wittmann, A. 1974. Computation and observation of Zeeman multiplet polarization in Fraunhofer lines III. Magnetic structure in Spot Mt. Wilson 18488. *Solar Phys.* **36**, 29-44.

Yoshimura, H. 1983. Solar cycle dynamo wave origin of sunspot intensity and X-ray bright point number variation. *Solar Phys.* **87**, 249-274.

Yun, H. S. 1970. Theoretical models of sunspots. *Astrophys. J.* **162**, 975-986.

Zel'dovich, Y. B., Ruzmaikin, A. A. and Sokoloff, D. D. *Magnetic Fields in Astrophysics* (New York: Gordon and Breach).

Zhugzhda, Y. D. 1970. The properties of low-frequency oscillatory convection in a strong magnetic field. *Astron. Zh.* **47**, 340-350 (*Sov. Astron.* **14**, 274-282).

Zhugzhda, Y. D., Locans, V. and Staude, J. 1983. Seismology of sunspot atmospheres. *Solar Phys.* **82**, 369-378.

Zhugzhda, Y. D., Locans, V. and Staude, J. 1987. The interpretation of oscillations in sunspot umbrae. *Astron. Nachr.* **308**, 257-269.

Zhugzhda, Y. D., Staude, J. and Locans, V. 1984. A model of the oscillations in the chromosphere and transition region above sunspot umbrae. *Solar Phys.* **91**, 219-234.

Zirin, H. and Stein, A. 1972. Observations of running penumbral waves. *Astrophys. J. (Letters)* **178**, L85-L87.

Zirin, H. and Wang, H. 1991. High resolution sunspot observations. *Adv. Space. Res.* **11**, 225-231.

Zwaan, C. 1965. Sunspot models: a study of sunspot spectra. *Rech. Astron. Obs. Utrecht* **17**, no. 4, pp. 1-182.

Zwaan, C. 1968. The structure of sunspots. *Ann. Rev. Astron. Astrophys.* **6**, 135-164.

Zwaan, C. 1978. On the appearance of magnetic flux in the solar photosphere. *Solar Phys.* **60**, 213-240.

Zwaan, C. 1981. The MHD of sunspots. *Space. Sci. Rev.* **28**, 385-386.

Zwaan, C. 1992. The evolution of sunspots. (This volume.)

II. SETTING THE SCENE

This section contains two chapters that put the study of sunspots into a broader context and set the scene for more detailed discussions of their physics in the following sections. Over the last fifteen years, the solar-stellar connection has extended our understanding of magnetic activity in the Sun. Solar activity can now be related to detailed studies of activity on other cool stars with different structures, ages and rotation rates, many of which are far more active than the Sun. The most striking effect is the variation (by up to half a magnitude) in the luminosity of active stars, which is ascribed to dark spots rotating on their surfaces. The detection of starspots raises many new and interesting theoretical questions and gives new importance to the study of sunspots within stellar astrophysics. The chapter by P. B. Byrne summarizes what is known about starspots. Their properties are consistent with their being essentially scaled-up versions of sunspots but, while a large sunspot occupies less than 0.1% of the solar surface, starspots may occupy up to 60% of the surface of a highly active star. In addition, there may be small-scale fields of up to 5 kG over much of the remaining area.

The Sun, of course, remains the only star on which surface features can be observed in detail. Magnetic flux emerges at the photosphere to form active regions within which pores and sunspots can be found. The chapter by C. Zwaan describes the formation and dissolution of sunspots and their relation to other aspects of solar activity. It also raises the question of how the magnetic fields in sunspots are connected to the large-scale field associated with the solar cycle. This issue will be taken up again in Section VI, while the next three sections focus on the properties of individual sunspots.

Several of the contributed papers at the Workshop were related to the evolution of sunspots. V. Bumba presented measurements of the line-of-sight velocity in over thirty sunspot groups, revealing a general pattern of vertical motion, with downflows dominating where there are local magnetic fields. L. Strous showed high-resolution observations of the flow field in a young active region. On the theoretical side, the relationship between isolated flux tubes and turbulent convection raises many fundamental problems. J. K. Lawrence proposed that the formation of sunspots or starspots should be regarded as a magnetohydrodynamic phase transition: in his model the nonlinear back-reaction of the magnetic field on turbulent diamagnetism can lead either to a uniformly disordered state or to an ordered state with strong intermittent fields. E. N. Parker considered hydrodynamic interactions between magnetic flux tubes; he argued that convection will create downdraft vortex rings around individual flux tubes and that these vortex rings will attract nearby magnetic flux bundles, thus providing the attractive force needed to overcome magnetic stresses and to build a sunspot. K. H. Schatten discussed the ion-hurricane mechanism for sunspot formation, and M. P. Ryutova introduced the possibility that secondary plasma-streaming motions driven by small-scale currents may influence the evolution of magnetic flux tubes on the Sun.

STARSPOTS

P. B. BYRNE
Armagh Observatory,
Armagh BT61 9DG,
N. Ireland

1. INTRODUCTION

Astrophysics is perhaps unique among the physical sciences in not having direct physical access to the objects which it studies. The universe compensates for this, however, by providing a large variety of environments in which a given type of phenomenon takes place. Thus our understanding of such processes can be tested by observing a variety of objects. This has been the motivation behind the field of activity known as the *Solar-Stellar Connection.*

Since this is a workshop on the subject of Sunspots, it is therefore appropriate that we should take a little time to examine the properties of their more extreme stellar counterparts, starspots.

2. THE BASIC OBSERVATIONAL PHENOMENON

Starspots are known to be a widespread phenomenon on three classes of stars, *viz.* the dMe flare (= BY Dra spotted) stars, the RS CVn binaries and the FK Com stars. The general properties of these groups of stars are summarized in Table 1. Note that FK Com stars are single, RS CVn's are binary, while the dMe flare stars can be either binary or single.

TABLE 1

	Sp Type(s)	Mass(es) ($M_\odot$)	R ($R_\odot$)	Period (*days*)
dMe's	KVe-MVe	0.8-0.1	0.9-0.2	0.25-10
RS CVn's	KIVe/GV	1	2-4	1-50
FK Com's	GIII	?	?	1

Of course the surfaces of these stars cannot be resolved directly by present means. So the presence of starspots must be inferred. Fig. 1 shows a typical light

J. H. Thomas and N. O. Weiss (eds.), Sunspots: Theory and Observations, 63–73.

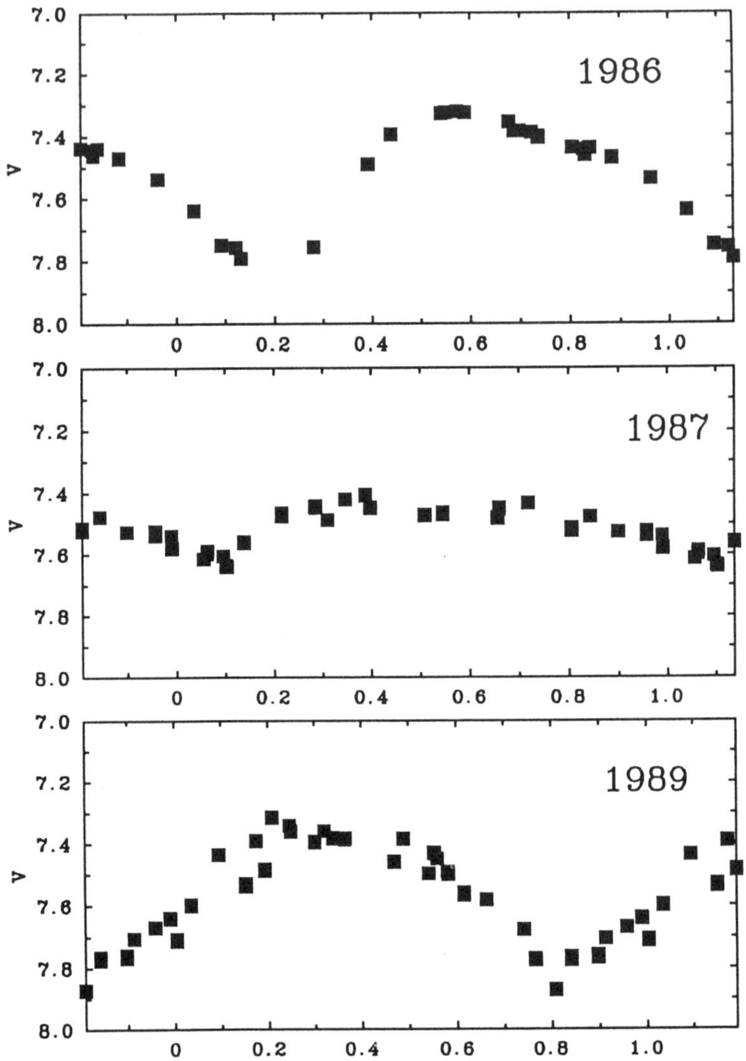

Figure 1. V light curves for the RS CVn star, II Peg in *(upper panel)* 1986, *(middle panel)* 1987 and *(lower panel)* 1989 (Byrne, private communication).

curve of a spotted star. These lightcurves repeat over intervals of 10-100 periods. While the underlying period remains constant, the form of the light curves (amplitude, phase, mean light level and detailed shape) frequently change. The period of the modulation is interpreted as the rotation period of the star, while the variation itself arises from a different total spotted area between opposite stellar hemispheres. Evolution of the light curve is then attributed to the growth or decay of individual spots or groups of spots.

3. STARSPOT AREAS

It is apparent that, in order to achieve modulation of visible light of the order of 40% as in Fig. 1, the total area coverage by starspots must be very much greater than in the solar case. A minimum figure can be calculated by assuming a perfectly black (0°K) spot which is circular in outline as seen from the centre of the star and which crosses the observer's line-of-sight. The latter restrictions arise because of the heavy limb darkening in these late-type stars, which reach values as high as 75 - 90% in V for stars of spectral type K0 - M0, respectively. A spot which is circular in outline and crosses the observers line-of-sight will occupy the least limb-darkened portion of the stellar disk on meridian transit and will thus achieve greatest darkening per unit area. Similarly a circular outline will ensure that the spot area is likewise concentrated to the brightest portions of the stellar disk.

In the case of the star whose light curves are illustrated in Fig. 1, II Peg, we need a minimum of two spots whose radii subtend angles at the star's centre of at least 37° and 20° (Byrne and Marang, 1987). This represents a minimum global area coverage by spots of 13%! II Peg does not hold the record for V amplitude however. Recently Nolthenius (1991) found an even larger amplitude modulation in the V light curve of the star HD12545. This light curve requires an area coverage by 0°K spots of 16% of the global photospheric area.

Any relaxation in the conditions of circular symmetry of the spot, a latitude which results in meridian passage across the observer's line-of-sight, or perfect blackness at all wavelengths results in larger total areas.

There is another aspect of the observations which influences the calculated area coverage by spots. The assumption that, when the star is brightest, we are looking at an unspotted surface, is inherent in our descriptions so far and in the derivation of the areas above. A star with a large spot on one hemisphere and an immaculate opposite hemisphere would result in a constant interval in the the light curve when the single spot was invisible. This is never seen. Thus it would appear from the shapes of the light curves alone that we are never viewing an entirely unspotted surface.

There is additional evidence to support this view. Fig. 2 contrasts the light curve of the star II Peg at two different epochs separated by a decade and a half. It is obvious that the least spotted hemisphere at the earlier epoch was at least 10% brighter than that observed more recently. So, unless the star has undergone some change in luminosity, the unspotted photosphere must be at least 10% brighter than indicated by the brightest part of the light curves in Fig. 1. It seems reasonable to assume that the depression of light maximum is due to a spot distribution which contributes equally at all phases of the stellar rotation. In the case of II Peg this gives rise to an additional area to be covered by 0°K spots of 10%, making a total minimum area coverage of almost one quarter of the star's photosphere occupied by starspots.

What about spot geometry? At first sight, based on the solar experience, a circular spot seems an unlikely model for stellar spots. If, however, we try to move

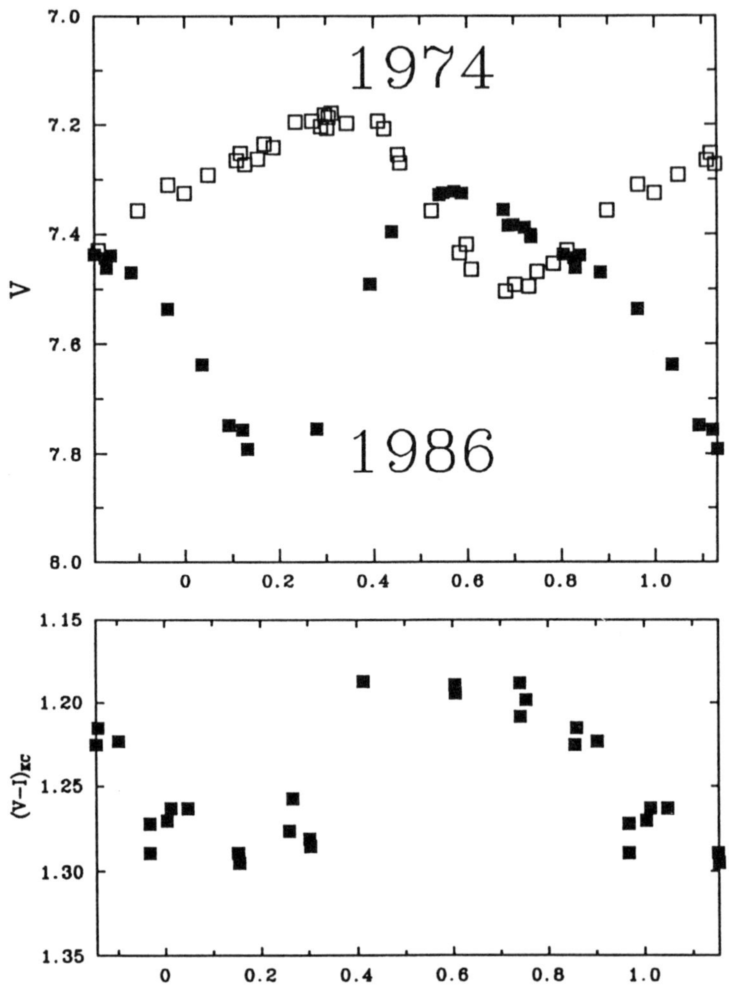

Figure 2. V light curve for the RS CVn star, II Peg in 1974 (Chugainov, 1976) *(upper panel, open symbols)* and in 1986 *(upper panel, filled symbols)*, and the $(V\text{-}I)_{KC}$ curve for 1986 *(lower panel)* (Byrne and Marang, 1987).

away from a circular outline we rapidly run into the problem that we are forcing spot area away from disk centre at meridian passage and total spot areas grow very rapidly. Not only is this true but the rapid rates of change in brightness either side of light minimum demand a compact and symmetrical geometry. So circular spots may be a better first order model for large starspots than seems likely at first sight.

4. SPOT TEMPERATURES

Thus far we have attempted to derive minimum area coverage by starspots by assuming perfectly black spots. Such an assumption is physically at odds with the solar paradigm, however. In principle, some measure of temperatures can be got from broadband photometry. Fig. 2 shows the $(V\text{-}I)_{KC}$ colour curve corresponding to one of the large amplitude V light curves of II Peg. It is clear that there is a modulation in colour accompanying that in brightness and this is in the sense that $(V\text{-}I)_{KC}$ is a maximum (reddest) at light minimum, or maximum spot visibility.

This type of observation is consistent with a variation in the mean temperature of the stellar hemisphere presented to the observer. Some estimate can, in principal, be made of the mean spot temperature from such curves. There is, of course, an interplay between the spot area and its temperature (see Byrne and Marang, 1987) such that, the higher the spot temperature, the larger the spot area needed to account for the light modulation. Many authors have derived spot temperatures using such colour curves and a thorough discussion of the method and representative results for a number of systems will be found in Poe and Eaton (1985). Temperature differences between spotted and unspotted photosphere derived in this way range from 600°K - 1500°K, with most results clustering 900 - 1200°K. Byrne and Marang (1987) have pointed out, however, that there is a selection effect which leads to temperatures precisely in this range. So the derived figures should be treated with caution.

The derived temperature differences between stellar spot and photosphere are intermediate between those for the solar photosphere and, respectively, sunspot umbrae and penumbrae. This raises the important question as to what we are measuring with these colour variations. Going by the solar example, we would expect starspots to have both umbrae and penumbrae. Because the umbra emits so much less light than the penumbra, it seems entirely possible that the light variation is dominated by the area of the umbra and the colour variation by the relatively brighter penumbra. Thus the interplay between spot area and colour temperature referred to above may lead to erroneous results.

5. DOPPLER IMAGING

There is a second, but observationally much more demanding, technique used to determine spot distributions, *viz.* doppler imaging. Doppler imaging takes advantage of two basic aspects of the spot phenomenon. First, spots and related magnetic activity are most pronounced on rapidly rotating stars. In fact many of these objects are rotating sufficiently rapidly that the rotational doppler broadening of photospheric line profiles dominates over all other sources of line broadening. Secondly, the fact that spots are dark.

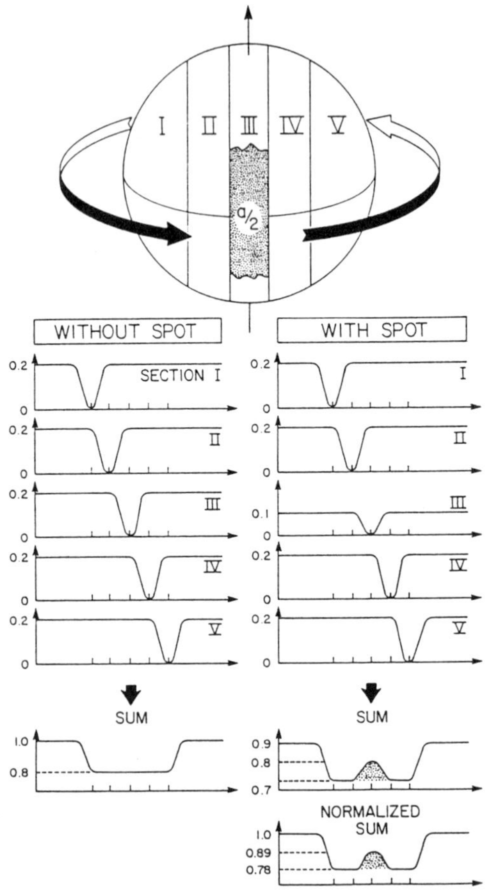

Figure 3. Schematic illustration of the formation of a photospheric absorption line on a rapidly rotating, spotted star (Reproduced from Vogt and Penrod, 1983).

The principle of doppler imaging is illustrated schematically in Fig. 3, reproduced from Vogt and Penrod (1983). Note first that, apart from some function to account for limb darkening, all velocity sectors of the stellar disk contribute equally (per unit area) to the line profile of an unspotted star. The introduction of a dark spot, assumed for the moment to possess the same intrinsic line profile as the unspotted star, results in an apparent *emission* feature at a location in the global line profile corresponding to the spot's projected radial velocity. The apparent *emission* bump arises because of the lower continuum contribution of the spotted velocity sector and, therefore, a *lack of absorption* of continuum photons at that velocity.

Provided we have a good template profile of the unspotted star, and we will return to this point below, a single spectrum of a spotted star can, in principle, provide *longitude* information on the location of the spotted region on the stellar disk. To gain *latitude* information one needs a sequence of spectra, sufficient to determine the velocity amplitude of the pseudo-emission features. Obviously, high-latitude spot bumps will exhibit a smaller velocity amplitude than those close to the equator. Inclination of the star's rotational axis helps to remove north-south hemispheric redundancy.

Since doppler imaging is primarily an effect of the lack of continuum from the cool spot, it suffers from some of the same basic uncertainties as photometric imaging in relation to the trade-off between spot area and spot temperature. A warmer spot will need greater area to produce a bump of given intensity and *vice versa.* We also note that the straight-forward procedure as described makes no mention of the variation in line formation physics between unspotted photosphere, spot penumbra and umbra.

Notwithstanding these limitations, doppler imaging is able to map the locations of starspots on a stellar disk with more accuracy than photometric techniques. Its principal weakness is its inability to make any kind of temperature determination. In this regard photometric and doppler imaging each provide information which is complementary in an important way.

6. POLAR SPOTS

One of the most interesting results to come from starspot imaging is the possibility that, unlike the Sun, stars have substantial dark spots on their rotational poles. Polar spots are very common in doppler images and, since this would be a result requiring a major modification of the solar paradigm for interpreting the stellar results, it is worth spending some time discussing the evidence.

The first point to be made is that a spot which is at the stellar pole does not produce a time-dependent signal. Thus polar spots are invisible in a single epoch photometric light curve, except in that they would depress the light level at all phases equally. Thus, if one had a long time sequence of photometric observations, a variation in the brightness of the light curve maximum would be seen as polar spots waxed and waned. However, such a variation in maximum light could equally be accounted for by a distribution of spots which were uniformly distributed in longitude at any latitude. The only requirement on their distribution being that they be uniform at the limiting resolution of the photometric image itself. Thus polar spots which have been shown in the literature on photometric images (see *e.g.* Rodonò et al. 1986) are simply a computational convenience to account for a depression of the maximum of the light curve with respect to the historical maximum.

The situation in regard to doppler imaging polar spots is more subtle. A polar spot will produce a stationary bump on the non-spot line profile. So, in principle, if one knows the non-spot profile to a high degree of accuracy a polar spot can be detected. The problem with this procedure lies in determining this non-spot profile. One of the most common ways of doing this is to take a high signal-to-noise spectrum of a slowly rotating, non-active star of the same spectral type as the star to be imaged and to spin its line profile up to match that of the rapidly rotating active star. When this is done it is very common to see, in the difference spectrum, a stationary pseudo-emission at line centre. The problem in this procedure is the assumption that the atmosphere of the slowly rotating non-active star is the same as that of the active rapidly rotating object outside of the spots themselves.

Evidence for a systematic difference in the line profiles of strong absorption lines of neutral metals between active and non-active late-type stars comes from two sources. First, there is the evidence of the solar spectrum wherein there is chromospheric filling-in of strong lines of FeI lines of the Fraunhofer spectrum, especially in the vicinity of plages and network. Second the equivalent widths of lines of strong neutral metal lines (*e.g.* CaIλ 4227Å NaI D) are systematically greater in slowly rotating dM than in fast-rotating dMe stars, an effect which has been shown to be due to chromospheric emission reversals in the line cores (see Fosbury, 1974; Doyle et al. 1990; Houdebine et al. 1990). Such emission reversals in fast rotators will mimic the flat-bottomed profiles usually interpreted as polar spots.

So, at this stage of our knowledge of spot distribution, the deduction that rapidly rotating active late-type stars possess spots at their poles should be treated with caution.

7. MAGNETIC FIELDS

Although magnetic field strengths in sunspots are high, globally the magnetic field of the Sun is weak. Therefore we would have little prospect of detecting the globally averaged field on a solar-like star. Furthermore, the dipolar nature of the solar magnetic fields make detection of polarization, the most unambiguous signature of magnetic field, extremely difficult. Nevertheless, given the greatly enhanced levels of magnetic activity on the active late-type stars, it seems worthwhile to attempt to detect and measure the fields using alternative techniques.

The most widely used technique for detection of magnetic field might be termed spectropolarimetry. Basically this relies on the fact that the Zeeman splitting of a line depends on the product of $g\,\lambda^2\,B$, where g is the Lande factor, λ the wavelength of the line and B is the field strength. The global line profile of a star will be a convolution of the many non-magnetic sources of broadening (stellar rotation, microturbulence, *etc*) and the Zeeman splitting due to many individual magnetic dipoles, each perhaps of different magnetic field strength. The result is that, even if the magnetic signature is strong enough to be detected over the other sources of broadening, it will produce an excess broadening of the line profile rather than a distinct Zeeman pattern.

The Robinson-Harvey approach has been to observe two or more lines of different Lande factors from the same atomic species and of similar excitation (to avoid the complication of the lines being formed under very different physical conditions in the stellar atmosphere) and look for differential broadening. Marcy (1984) and Saar (1987) have been particularly successful in applying this technique to active late-type stars. Their results suggest that globally averaged field can have values as high as 5 000 G in the dMe stars. Furthermore, their measured field strengths correlate impressively well with the fields calculated from equipartition with the photospheric gas pressure. This latter result suggests that the measured field strengths represent network fields rather than spots fields and would probably represent a lower bound for spot umbral fields.

What are the prospects for measuring spot fields directly? There are two factors which make this problem particularly intractable. The first is the difficulty of detecting spot signatures in the global flux of the star simply because the spots contribute so little to it. The second is that spots of detectable size occur on the most rapidly rotating of the active stars. Therefore rotational broadening tends to dominate over Zeeman splitting in optical lines for typical rotational velocities of 10 km sec^{-1} (dMes) to 40 km sec^{-1} (RS CVns).

8. SPOT LIFETIMES

Having made maps of starspot distributions at various epochs it is possible to consider the rates of growth and decay of individual large spots or spot groups. Some instances of extremely rapid spot growth and decay exists. For instance, the light curve of the RS CVn star V711 Tau (HR1099) in late 1981 showed an extremely rapid evolution over a period $\sim$2 months (Rodonó et al. 1986). Generally, however, spot light curves are stable over an observing season ($\sim$3-4 months) and only show major changes on longer timescales, 1 year $\geq \tau \geq$ 4 months. These results are not very different from those for the largest solar spot groups.

Of possibly more interest is the long-term behaviour of starspots. Several lines of evidence point to long-term trends in the occurrence and distribution of spots. Hartmann et al. (1979) examined a sequence of photographic images of the RS CVn star, II Peg, extending from $\sim$1900 - 1950. These show no evidence of variability greater than $\sim$10%. On this evidence its behaviour over the past two decades, with amplitudes rarely less than $\sim$30%, differs from that seen during the first five decades of this century. Whether there is some long-term behaviour in these stars in general in which the degree of global spottedness varies on a time scale of a century or more is therefore an unanswered question whose resolution must await the passage of time.

A second interesting observation in this regard is the long-term trend in spot coverage of the star AB Dor found by Anders et al. (1992). These data show the mean magnitude of the star declining monotonically by $\sim$20% over the past $\sim$12 yr. Similar, but less well sampled long-term trends in mean magnitude of spotted stars have been recorded by Cutispoto and Rodonò (1992). Again, whether this represents some kind of cyclical behaviour must await the collection of further data.

9. CONCLUSIONS

The study of spots on late-type stars has progressed a long way since the first recognition of the nature of the phenomenon in the early 70's. While there are many physical parameters still to be measured reliably, it is true that the solar paradigm appears to work well but on a much expanded scale. In Table 2 we suggest an "Identikit" picture of stellar spots. The reader is reminded however that, like many "Identikit" when the fine details are filled in the current sketch may only provide a very general resemblance to reality.

TABLE 2

R/R_*	A/A_*	$T_{phot} - T_{spot}$	B	Life
≤ 1	10-15%	$> 1\,000°$ K	$\geq 5\,000$ G	1 month - years?

The immediate future, however, holds out exciting possibilities for this field, perhaps the most interesting of which is the application of optical interferometry to the process of imaging stellar surfaces. Instruments presently being designed (*e.g.* SUSI (Sydney University Stellar Interferometer)) offer the promise of a few μarc sec resolution. At the distance of the brightest spotted stars of both the dMe and RS CVn class, a resolution of 10 μarc sec corresponds to the diameters of spots with radii of $\sim 2 - 3°$. Given the kinds of spot dimensions mentioned in Section 3 above we therefore have the prospect in the coming decade of imaging spot distributions and structures directly.

REFERENCES

Anders, G.J., Coates, D.W., Thompson, K., Innis, J.L.: 1992, in *Surface Inhomogeneities on Late-type Stars,* eds P.B. Byrne & D.J. Mullan, Springer-Verlag, in press

Byrne, P.B., Marang, F.: 1987, *Irish Astron. J.* **18,** 84

Chugainov, P.F.: 1976, *Izv. Krymskoi Astron. Obs.* **54,** 89

Cutispoto, G., Rodonò, M.: 1992, in *Surface Inhomogeneities on Late-type Stars,* eds P.B. Byrne & D.J. Mullan, Springer-Verlag, in press

Doyle, J.G., Mathioudakis, M., Panagi, P.M., Butler, C.J.: 1990, *Astron. Astrophys. Suppl.* **86,** 403

Fosbury, R.A.E.: 1974, *Mon. Not. Roy. Astron. Soc.* **169,** 147

Hartmann, L., Londono, C., Phillips, M.J.: 1979, *Astrophys. J.* **229,** 183

Houdebine, E.R.: 1990, PhD thesis, Universitè Paris XI - Orsay

Marcy, G.W.: 1984, *Astrophys. J.* **276,** 286
Nolthenius, R.: 1991, *Inf. Bull. Var. Stars,* No. 3589
Poe, C.H., Eaton, J.A.: 1985, *Astrophys. J.* **289,** 644
Rodonò, M., et al.: 1986, *Astron. Astrophys.* **165,** 135
Saar, S.H.: in *Cool Stars, Stellar Systems and the Sun,* eds. J.L. Linsky & R.E. Stencel, Springer-Verlag, p. 10
Vogt, S.S., Penrod, G.D.: 1983, *Publ. Astron. Soc. Pacific,* **95,** 565

THE EVOLUTION OF SUNSPOTS

CORNELIS ZWAAN

Sterrekundig Instituut Utrecht
Postbus 80 000
3508 TA Utrecht
The Netherlands

ABSTRACT. The evolution of sunspots is described in the context of the evolution of active regions. The nesting tendency in formation of sunspots is pointed out. Birth and growth of an active region are attributed to an Ω-shaped loop emerging from a deep-seated toroidal flux strand. It is argued that the field strength in the toroidal strand exceeds the equipartition value; the top of the Ω-loop, however frayed, reaches the upper boundary of the convection zone at intrinsic field strengths close to the local equipartition value. The formation of pores is preceded by the convective collapse of flux tubes. The coalescence of the buoyant tops of flux tubes leads to the growth of large sunspots. Inclusions of bright structure in umbrae, and the formation of dark cores are discussed. Penumbrae form after the spot has reached a critical flux level. Most sunspots decay fast by fragmentation. Spots decaying gradually are surrounded by moat cells, across which magnetic features of both polarities are streaming away from the spot; the decay rate appears to be approximately constant. The removal of magnetic flux from the atmosphere, including the seeming disappearance *in situ*, is addressed. Finally, the complementary roles of Ω-loops and U-loops in the magnetic structure in convection zone and atmosphere are discussed.

1. Introduction

Sunspots—blemishes on the face of the Sun—intrigued or worried the early observers. Galileo and the cardinals of his time were aware that these phenomena, if real, manifest a reality that is more complex than the idealized philosopher's world consisting of perfect spheres. More recently, it was realized that sunspots indicate some force(s) acting in the Sun, in addition to gravity and gas pressure gradients. Indeed, Hale's discovery of magnetic fields in spots introduced magnetohydrodynamics into solar physics and astrophysics. The intricate and dynamic magnetic structure in the solar atmosphere still presents a collection of hard test cases to astrophysical magnetohydrodynamics.

Sunspots are of particular interest because they are the best observable features in the intermittent solar magnetic field. It is known that the magnetic field responsible for sunspot formation originates from the solar interior. Scrutiny of the evolution of spots may provide clues to the physics of sunspots.

J. H. Thomas and N. O. Weiss (eds.), Sunspots: Theory and Observations, 75–100.

Since sunspots occur in characteristic patterns in large active regions during their first evolutionary stages, it is useful to consider the evolution of sunspots in the context of their encompassing active regions. In the decay of spots, features appear that belong to the class of intrinsically weak and turbulent magnetic fields. Hence sunspots are linked with various aspects of solar magnetism. While this paper concentrates on the evolution of sunspots, we search for more insight in solar magnetism at large, both in the atmosphere and in the solar interior.

The paper focuses on main observational features and their immediate interpretation. The selection of the "main" features is given without much discussion, so the reader should be aware: the author may be biased or otherwise short-sighted. As to interpretation, the paper refers to several theoretical and numerical studies but without going into details. What is attempted is to display the sunspot evolution as a movie picture, with the interpretation subtitled. In order to make the movie continuous, quite a few hazy frames had to be touched-up by speculation.

2. Sunspots in the Context of Active Regions

The evolution curve of sunspots is very asymmetric: the growth phase is much shorter than the decay phase. Even the largest spots reach maximum magnetic flux within some days, whereupon they live typically for some weeks, and in rare cases even for some months.

Sunspots and pores are exclusively formed within magnetic flux that emerged very recently in the atmosphere, in very young active regions. (Occasionally there are rumors that transient pores may form within "old" network flux but I have never seen a truly convincing example. Let us keep our eyes open for such a case). Hence the evolution of a spot is so intimately related with the development of the encompassing active region that it is useful to consider spots in their context.

The asymmetry characterizing sunspot evolution is even more striking in the evolution of bipolar active regions as entities. K. Harvey (1991) analysed the time scales of bipolar active regions as depending on the phase of the activity cycle and the size of the region at maximum development. The *emergence time* is defined as the time span from the first appearance of the bipolar region on Kitt Peak magnetograms up to *maximum development*, that is: the moment that the flux emergence ceases. The moment of maximum development, and hence the emergence time, can be determined with a precision of about half a day. Harvey found that the emergence time tends to increase with increasing area at maximum development, but that the distribution of emergence times is broad for any area range. All 486 active regions in Harvey's sample reached their maximum development within 5 days; for medium-large and large bipolar regions the typical emergence times range from 2 to less than 4 days. For every active region the emergence time is a small fraction of the total lifetime; this fraction is smaller than one third for small regions with lifetimes of two or three days. This fraction decreases with increasing lifetime, down to less than 6 % for regions with lifetimes larger than one solar rotation.

The clear-cut properties summarized above refer to active regions as *bipolar entities*, which evolve during one continuous process of short duration to maximum development. About one third of the active regions assume a more complex configuration by emerging within an existing active region, or by having a new active region emerging within their perimeter. This *nesting tendency* is very pronounced: Harvey (1991) found that per unit area at the solar surface the frequency is more than 25 times larger for emergence of new active regions within an already existing active region than it is for emergence elsewhere in the then available activity belt. Note that this enhancement factor does not depend on the size of the region: the distribution functions $p(A)$ for regions emerging within existing regions and for regions surfacing outside existing activity are virtually similar in shape (see Figure 1). Incidentally, the shape of the distribution function of emergence frequency

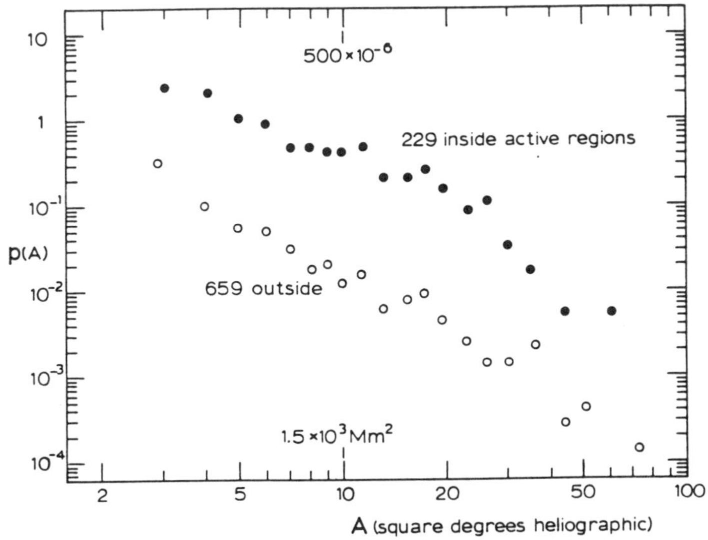

Figure 1: Distribution function $p(A)$ of emergence frequency of active regions, per day and per unit area bin, normalized per 10^6 Mm^2 of available area at the solar surface, plotted against region area A at maximum development. (The area A is counted in square degrees on the solar surface; equivalents in Mm^2 and in millionths of the visible hemisphere are given for reference). The filled symbols indicate emergences within existing regions, the open symbols stand for emergences outside existing regions. Note that within existing regions the emergence frequency is larger by more than a factor of 25; more than one third of the active regions emerge within existing regions. (Figure adapted from Harvey (1991)).

$p(A)$, if averaged over several months, does not depend on cycle phase either—it is just the amplitude that varies with the cycle phase (Harvey 1991).

The relatively short emergence times of active regions, the rapid formation process of sunspots (Section 3), and the immediate onset of the decay process (Section 4) suggest that active regions, with their sunspots, are prepared for future display deep below the photosphere, in conditions that differ drastically from those in the atmosphere and top of the convection zone. Apparently the adjustment of the magnetic structure to atmospheric conditions occurs rapidly during and shortly after the emergence process.

The observed evolutionary patterns indicate that immediately after their emergence active regions and their sunspots reveal most about the subsurface structure from which they originate. Harvey's (1991) study of properties of bipolar active regions over more than one cycle yields clear-cut relationships because of the procedure to count every active region only once during its lifetime and to determine its basic properties at *maximum development*, that is, immediately after the emergence of magnetic flux has stopped.

Fundamental properties of the magnetic configuration at the roots of active regions can be inferred from typical features of large active regions at maximum development—see Figure 2. With Zwaan (1978) let us consider such a large region, with a magnetic flux $\Phi = 3 \times 10^{22}$ Mx in one polarity. The bipole axis is slightly inclined with respect to the direction of rotation, the leading polarity being closer to the equator. The inclination angle α varies from a few degrees near the equator to about 15° at 35° latitude (Joy 1919, Brunner 1930). The distance between the centroids of opposite polarities is typically between 100 and 150 Mm. The ordered bipolar orientation of active regions, the majority following Hale's polarity law, with small inclination angles α, inspired the topology of the tightly wound toroidal flux strand under the solar surface in the descriptive dynamo models by Babcock (1961) and by Leighton (1964).

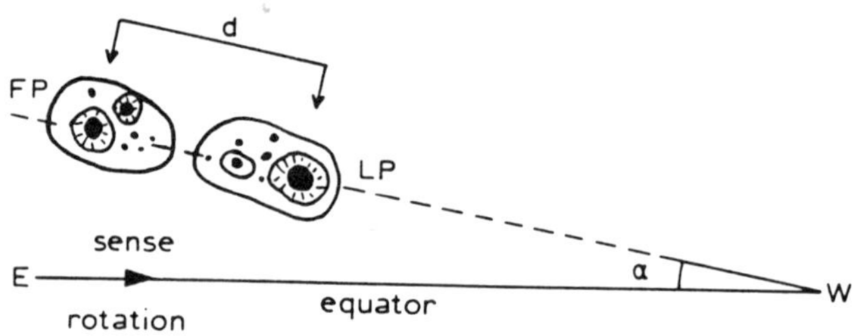

Figure 2: Sketch of large bipolar active region at maximum development. Distances d between the centroids of opposite polarity range from about 100 to about 150 Mm. The inclination angle α between the magnetic axis and the solar equator ranges from 0° to about 15°. (α depends on latitude, see text).

The good visibility of the toroidal magnetic field pattern in the atmosphere indicates that the magnetic field must be strong enough to resist distortion by the turbulent convection over the entire relevant range in the convection zone, from where the flux is rooted in the toroidal field system all the way up into the atmosphere. In other words, the field strength should exceed everywhere the local equipartition field strength B_{eq}, such that magnetic energy exceeds the kinetic energy in the convection.

The flux seen in an active region must have been stored in a toroidal flux rope with a diameter small enough to allow the top of the emerging loop to move as an entity. If the field strength $B(z)$ in that toroidal rope exceeds the local equipartition strength $B_{eq}(z)$ only moderately, then the only convenient location to store horizontal flux bundles is close to the bottom of the convection zone (Zwaan 1978). For a field strength of $B(z) = 1 \times 10^4$ gauss, slightly exceeding the local equipartition value, a flux of 3×10^{22} Mx fits into a flux bundle with a diameter of 20 Mm, which is about one third of the local pressure scale height H_p—see Figure 3. Note also that the separation between the centroids of opposite polarity $d \approx 150$ Mm compares nicely with the depth of the convection zone, and that the width of the suggested toroidal rope is smaller than the latitudinal extent of the corresponding active region. From these considerations it follows that the field strengths in the toroidal flux system must be at least as large as the equipartition field strength, and that the system must be close to the bottom of the convection zone.

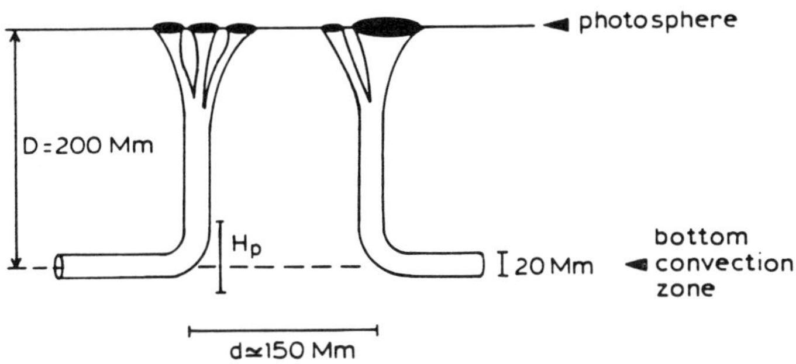

Figure 3: Schematic model for an Ω -shaped flux loop emerged from a toroidal flux rope supposedly located at the bottom of the solar convection zone, producing a large active region. The numbers are: magnetic flux $\Phi = 3 \times 10^{22}$ Mx and distance of centroids of flux $d = 150$ Mm (as in Figure 2), depth of the convection zone $D = 200$ Mm, pressure scale height near the bottom $H_p = 60$ Mm. For the field strength in the toroidal rope we assume the modest value $B = 1 \times 10^4$ gauss, which is slightly larger than the equipartition value, and which yields a diameter of 20 Mm for the toroidal rope.

A variety of arguments has been advanced why the toroidal field system must be located in the interface between the convection zone and the radiative interior, see Schüssler (1983). Gilman, Morrow and DeLuca (1989) argue that the interface is the most plausible location for the dynamo because only there may the proper gradients in the angular velocity be found.

According to van Ballegooijen (1984), toroidal flux strands field, with $B_{\rm eq} \lesssim B \lesssim 10B_{\rm eq}$, somehow held down near the bottom of the convection zone, can be connected with observed sunspot structure by vertical legs with a smoothly decreasing magnetic field strength that is nowhere far below the local equipartition value. Van Ballegooijen calculated strictly static tube interiors (no turbulence), embedded in a convective medium that departs from hydrostatic equilibrium because of the vari - ation of the vertical component of the Reynolds stress $< \rho v_z^2 >$ with depth. (The upper layers of the convection zone are slightly uplifted, causing the gas pressure to be somewhat higher than it would be in strictly hydrostatic equilibrium. The magnetic field in the tubes is confined by this increase of the external gas pressure). Van Ballegooijen found that the field strength $B(z)$ in a leg decreases smoothly with height, first from the adopted value near the bottom $B(z_0) \gtrsim B_{\rm eq}(z_0)$ to about $0.6B_{\rm eq}(z)$, and then from there the field strength follows $B(z) \simeq 0.6B_{\rm eq}(z)$ up to $z \simeq 1$ Mm below the photosphere. (The field strength $B(z)$ in the very top of the convection zone and in the photosphere exceeds the local equipartition field strength after the convective collapse, which occurs during flux emergence, as described in the next section).

The quantitative elaboration of the model of strong toroidal flux strands, from whence the active-region loops emerge, presents some substantial problems which are discussed in the contribution by Moreno-Insertis (1992) to this volume. Here I just summarize some of these problems:

(*i*) Strong flux bundles are buoyant—see Chapter 8 in Parker (1979), hence some mechanism is required to keep the toroidal flux system anchored in the solar interior for a sufficiently long period;

(*ii*) The Coriolis force deflects the rising top of a flux loop poleward, which could make active regions emerge at high latitudes, outside the activity belts (Choudhuri and Gilman 1987, Choudhuri 1989);

(*iii*) During its rise to the surface the top of the flux loop would expand enormously because the ambient gas pressure drops by many orders of magnitude over the height of the convection zone (Moreno-Insertis 1992).

As to (*i*), several mechanisms have been proposed for the suppression of the buoyancy. One is to place the toroidal flux tubes in the sub-adiabatic overshoot layer below the convection zone (see Galloway and Weiss 1981, van Ballegooijen 1982 a,b). Another possibility is an equatorward meridional flow, which induces within the toroidal tubes a longitudinal flow that creates a Coriolis force opposing the buoyancy force (van Ballegooijen 1982, van Ballegooijen and Choudhuri 1988).

Yet another mechanism has been proposed by Parker (1987): a strong magnetic field produces a cool shadow on the upper side of the flux strand, which presses the field downward. At times the accumulated heat beneath the field causes thermal plumes which push flux loops upward—Parker envisages a relaxation oscillation that sends thermal plumes, with entrained field, to the surface at irregular intervals of a week or so. Such an oscillation might account for the sequence of emergences of active regions in one nest.

As to (*ii*), the poleward deflection of the emerging top of the flux loop by the Coriolis force is sufficiently small if the risetime is sufficiently short or, in other words, if the initial field strength B_0 is sufficiently high. Computational results to date suggest that the existence of the activity belts requires a field strength B_0 of the order of 10^5 G rather then the equipartition value around 10^4 (Moreno-Insertis 1992).

As to (*iii*), observations of the flux emergence (see Section 3.1) prove beyond doubt that the expected ballooning of the rising flux loops does not happen. Apparently some essential physical aspect is missing in the calculations to date, and in Section 3.1 it is suggested that it is the departure from hydrostatic equilibrium caused by the convective turbulence that leads to a slight overpressure in the ambient convection zone which confines the magnetic field in the rising loop to strengths close to the local equipartition values. This is the mechanism invoked to confine the static vertical legs of flux tubes in van Ballegooijen's (1984) models, as mentioned above.

Admittedly the present model of toroidal flux strands and emerged loops sustaining active regions and sunspots is still sketchy—some of its conceptual ideas and several of its quantitative aspects require further study. For instance, the field strength in the toroidal strands in poorly known: probably it is at least as large as the local equipartion strength, but it may be larger than that by an order of magnitude. However sketchy, the model is strongly supported by observational indications, hence it is used as a working hypothesis in the rest of this paper.

3. Emergence of Active Regions and Growth of Sunspots

3.1. EMERGING FLUX REGIONS AND PORES

The phenomena of the birth of an active region are readily pictured by the emergence of the top of an Ω-shaped flux loop (Figure 4). Some of the chromospheric features have been known for quite some time: the first manifestation of an active region is the appearance of a small, compact and very bright (bipolar) plage (Fox 1908, Waldmeier 1937, Sheeley 1969). Soon the *arch filament system* (AFS) appears in the Hα line core: a set of roughly parallel dark fibrils connecting faculae of opposite polarity (Bruzek 1967, 1969). Every fibril is a loop, whose top ascends with a speed of several km s^{-1}, up to 10 km s^{-1}. In both legs, the matter flows down at speeds of some tens of km s^{-1}, and locally up to 50 km s^{-1}, as measured in Hα.

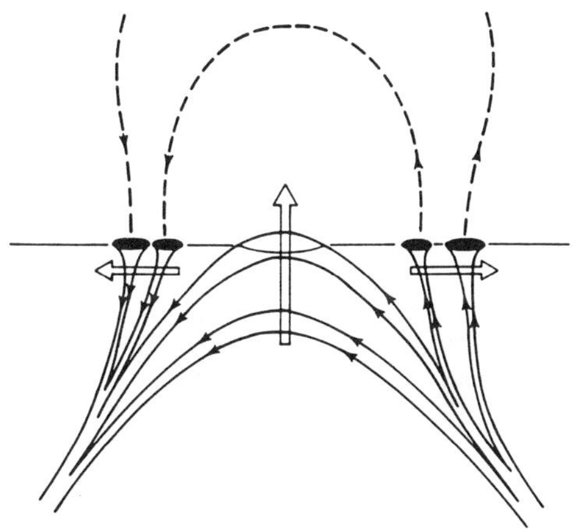

Figure 4: Top of an Ω-shaped loop as a model for the emergence of magnetic flux, separation of the polarities, and coalescence of sunspots. Broad open arrows indicate local displacements of flux tubes. Note that real emerging flux regions consist of many more separate flux tubes.

The development of an Emerging Flux Region (EFR, this term was coined by Zirin (1972)) follows distinct patterns. It appears as an intruding bipolar magnetic feature that, at first, does not disturb the ambient "old" magnetic structure. The faculae of opposite polarity move apart. During the first half hour the rate of expansion may exceed 2 km s^{-1}, then it drops to a value between 1.3 and 0.7 km s^{-1} during the next 6 hours (Harvey and Martin 1973). The expansion continues at a gradually slower pace for several days. New magnetic flux emerges within the polarity inversion zone, which separates the diverging opposite polarities made up of flux that emerged earlier. In other words, an EFR grows from inside outward. The arch filament system lasts as long as flux is emerging, but the individual dark Hα fibrils live only for about 20 minutes. Hence after some time the AFS connects only faculae in the inner part of the growing bipolar region.

Initially, the emerging flux regions are oriented almost at random (Weart 1970), but they rotate and after about one day most EFRs have assumed the "correct" orientation of a mature active region, nearly parallel to the equator, as sketched in Figure 2 (Weart 1972,

Frazier 1972).

At closer examination, large active regions appear to be formed by contributions from several emerging flux regions that emerge separately but in close proximity, simultaneously or in rapid succession, within a few days (Weart 1970, 1972; Schoolman 1973). In many cases the distribution of the polarities is somewhat messy because the contributing EFRs may emerge at different orientations, and at somewhat different times. Usually, however, the various EFRs align and merge in such a way that within a few days of flux emergence one single, simply bipolar, active region results. This process is like a bird preening its ruffled feathers after a rough experience.

The photospheric features of flux emergence are particularly telling. In the central part of an emerging flux region the granulation is disturbed: it looks fuzzy, there are remarkable alignments in the dark intergranular lanes, and there is no "filigree" as yet. The alignments in the intergranular lanes have been noticed before—see Section 3.3.4 in Bray and Loughhead (1964). Fascinating new observations of emerging flux regions have been collected at the Swedish Solar Telescope on La Palma, with Lockheed instrumentation—see Figure 5. During the workshop some of these data were shown by Strous; this type of data will deepen our insight in the emergence process and the formation stage of active regions. Here I summarize the present knowledge, largely based on observations in 1979 with the

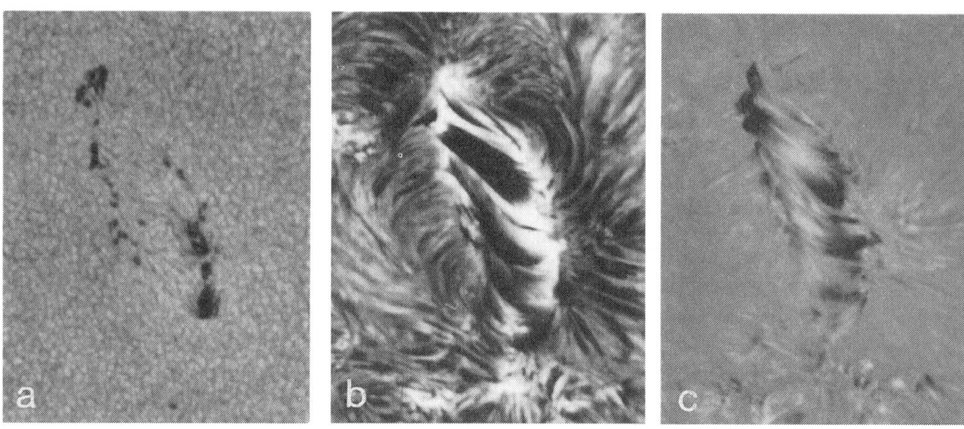

Figure 5: Three nearly simultaneous aspects of an emerging active region (AR 5617 on 29 July 1989). Panel a. is the image in continuum intensity: dark pores are located in the outer strip, the internal granulation looks fuzzy, with alignments in the dark intergranular lanes. Panel b. is recorded in the core of Hα: it shows the Arch Filament System. Panel c. shows the Doppler shift in Hα, white marks upward velocity, and black downward. The tops of the Hα loops are rising, at the edges there are strong downward flows which are sharply bounded near the inner edge of the strip with pores. The combination of these three aspects illustrates emergent magnetic flux as a sharply outlined intrusion from below. (Images prepared by L. Strous from data collected at the Swedish Solar Observatory at Roque de Los Muchachos, La Palma by the Lockheed group).

Sacramento Peak Vacuum Tower Telescope and on supporting data from Kitt Peak and Ottawa River Solar Observatory (Zwaan, Brants and Cram 1985); the results are presented in Brants's (1985 a) thesis and in a review paper by Zwaan (1985).

The dark alignments in the fuzzy granulation (Figure 5, panel a.) are compatible with tops of magnetic loops passing through the photosphere. These alignments are parallel to the fibrils in the overlying arch filament system (panel b.). The following estimates apply to a well-defined alignment: width $\simeq$ 1.5 Mm, visible in the photosphere for about 10 minutes. The speed of rise is estimated at $v \simeq 3$ km s^{-1}—here *dark* matter is rising! If this speed equals the Alfvén speed then the field strength is about 600 gauss and the flux about 1×10^{19} Mx.

The combination of the observations of the arch filament system and the system of dark alignments in the photosphere indicates that an emerging flux region consists of a bundle of many more or less discrete flux loops, which arrive at the photosphere with intrinsic field strengths of at least a few hundred gauss (in order to be so well visible as faculae on magnetograms), up to about 600 gauss, which roughly equals the equipartition field strength at the top of the convection zone.

The high concentration of magnetic field in an Emerging Flux Region, which follows from intrinsic field strengths in the individual flux tubes and the rather dense packing of such tubes, is not explained by the present theory. As shown in this volume by Moreno-Insertis (1992), a flux loop rising from the bottom of the convection zone is expected to expand enormously, particularly so in the upper part of the convection zone where the gas pressure drops rapidly with height. Hence the comparison with the observations indicates that an essential physical aspect is missing in the models used in the calculations. I wonder whether allowance for "turbulent pressure" in the convection zone outside the rising flux tubes could remove the discrepancy. As summarized in Section 2, "puffing up" the upper part of the convection zone by turbulence provides an efficient confinement of the static isothermal flux tubes in van Ballegooijen's (1984) calculations, hence this mechanism may restrict the ballooning of the adiabatically rising tops of flux loops.

At the lower tips of the strong chromospheric downflows (Figure 5, panel c.) there are downflows in the photosphere with velocities up to about 2 km s^{-1} (Brants 1985 a,b). There is a well-documented case of a small downdraft area, which developed from a darkish "protopore", with a diameter barely larger than one arcsec, into a dark sunspot pore, which later coalesced with a larger pre-existing pore.

Zwaan (1985) interprets this process as the *convective collapse* that transforms the emergent field of intermediate strength ($B \lesssim$ 500 gauss) into a strong field ($B \gtrsim$ 1500 gauss), referring to Spruit's (1979) calculations of the convective collapse of thin flux tubes. The physics is that any initial downdraft is enhanced by the convective instability in the top of the convection zone. As a result, the top of the flux tube becomes nearly evacuated, so it is compressed and hence the magnetic field is strongly enhanced until the magnetic pressure balances the external gas pressure. (Note that an initial upward flow would work in the opposite way: the updraft would be enhanced, the flux tube would expand, and the magnetic field would become weaker). In emerging flux regions all processes point downward: the draining of the chromospheric parts of the tubes does, the radiative cooling causes the photospheric plasma to sag, and the convective instability in the top of the convection zone amplifies the downdraft and carries the cooling and the partial evacuation deeper down.

The reduction of the turbulent heat exchange in magnetic fields deeply affects the structure of magnetic field in the top of the convection zone and in the photosphere. Not only is this effect responsible for the darkness of sunspots, as pointed out by Biermann and Cowling long ago, it also leads to the convective collapse in the flux emergence process, which causes all the strong and vertical magnetic structure in the solar photosphere, consisting of nearly evacuated flux tubes. In addition, the darkness of the alignments in the granulation indicates that turbulent heat transport is impeded already in the tops of the emergent flux loops, apparently at field strengths not far from the local equipartition strength $B_{\rm eq} \simeq$ 500 gauss.

Sunspot pores grow by increase of area and by decrease of continuum brightness. Both phenomena happen simultaneously in the immediate vicinity of downdrafts indicating convective collapse. In fact, wherever sufficient magnetic flux becomes available fast enough, pores are formed in one continuous process with the convective collapse. The local downward flow stops, however, once a truly dark pore mass, with a field strength larger than about 2000 gauss, has formed.

There is some converging motion in the rapidly darkening matter forming pores but the main impression is that of *in situ* darkening matter, as cases displayed in Brants and Steenbeek (1985) and Brants (1985a) demonstrate. In white-light pictures the (lateral) motions are small—it takes Doppler measurements and chromospheric observations (Figure 5, panels b. and c.) to appreciate the dynamics in the formation process of pores.

3.2. FORMATION OF SUNSPOTS

Larger sunspots grow by advection of small umbrae, pores and faculae, at typical velocities of about 0.25 km s^{-1}, and in some cases at speeds up to about 1.0 km s^{-1} (Vrabec 1974). Because there are small spots and pores among the inflowing magnetic patches, the process of growth of a sunspot by coalescence can be seen quite well on any white-light movie of good resolution.

In large active regions the magnetic flux emerges in more than one bipolar Emerging Flux Region, which can lead to a seemingly irregular mix of polarities within the confines of the growing active region. Magnetographic movies, such as Vrabec's, show that the opposite polarities disentangle while separating, and in this way a neatly bipolar active region grows out of an initially messy mix of polarities.

A substantial fraction of the moving magnetic patches of one polarity ends up in one or more sunspots. The movies indicate that the magnetic fragments ending in one sunspot travel along a few well-defined tracks, which may be markedly curved. Do these tracks correspond to lanes between convective cells of some intermediate scale, between granules and supergranules? The large-scale pattern in these motions of the magnetic patches must be controlled by forces other than those in the observable convective cells.

Excellent illustrations of very elongated strings of small spots, which rapidly coalesce into single roundish sunspots, are given by McIntosh (1981, his Figures 13 and 15).

The pictures published by García de la Rosa (1987) show how some of the larger fragments in the coalescence of a sunspot can remain visible in later evolutionary stages, including the final dissolution phase. These fragments remain indicated, or reappear, because they are outlined by (partial) light bridges in the umbra. Conversely, probably every light bridge in an umbra is the relict of a boundary between fragments that formed the spot.

More in general, the coalescence of spots from many magnetic patches agrees with the cluster model for sunspot structure. Note that every pore, before it gets incorporated in a larger sunspot, probably was formed by a fusion of smaller magnetic elements. Yet some sunspot umbrae develop very dark cores, virtually without fine structure visible on the best high-resolution pictures that have been obtained to date. Apparently in parts of some umbrae the constituent flux tubes fuse together without visible joints. These dark cores (or voids, in Livingston's (1991) terminology) are particular to long-lived spots.

In large active regions the coalescence of small spots into the large sunspots often continues for several days after all magnetic flux has emerged. Hence the growth of a leading sunspot may last for a period up to about 10 days. This is a net growth: while the coalescence proceeds in a small sector in the periphery, the spot is already losing magnetic flux across the rest of its boundary (see Section 4.2).

The growth of the leading sunspot and the increase in the distance between the centroids of opposite polarity appear to be related. The expansion of an active region is caused mainly by the leading polarity moving faster than the average photospheric rotation. This proper motion of the leading sunspot decreases with time, and it drops to zero at the same time when the growth of the spot stops. In the mean time, the following sunspot moves backward in longitude at a much slower pace, or it stays put. The typical dynamics in young active regions, in relation to the growth of the leading spots, has been known for a long time, see Maunder (1919), Astronomer Royal (1925), and Waldmeier (1955, his section 45). (Note that in recent literature pathological patterns of motions in complex active nests have received particular attention.) As mentioned in Section 2, the eventual separation between the centroids of the polarities in a fully developed active region is between 8° and 12° in longitude (100–150 Mm).

In the evolution of nearly every medium-large or large active region one dominant sunspot is formed, namely in the leading polarity (i.e., the polarity preceding in the solar rotation). Generally this leading spot is quite close to the leading (Western) edge of the region. Not rarely a conspicuous secondary spot appears in the following polarity; the position of such a follower tends to be close to the trailing (Eastern) edge of the region.

During and shortly after the formation of the major spot(s), in a large region many small short-lived spots and pores are formed, mainly at positions between the dominant spots. (The observers compiling the white-light data in the *Greenwich Photoheliographic Results* often used the term "stream of sunspots" instead of "sunpot group").

Sunspot pores with diameters exceeding 5″ (or 3.5 Mm) show a strong tendency to develop at least some penumbral structure (see Sections 3.4 and 3.5 in Bray and Loughhead (1964)). If much flux becomes available rapidly then the growing spot may overshoot the mark for penumbra formation, and form a large and ofter irregular pore mass, initially without penumbral structure. Apparently this is a transient situation that does not last longer than half a day.

There are only a few, brief descriptions of the formation process of penumbrae. Bray and Loughhead (1964, their Section 3.5.2) describe the formation of rudimentary and transitory penumbrae. In their study of the development of an active region Bumba and Suda (1984) show the development of penumbrae around the large leading sunspot and the main following spot. From this study, from Bumba (1965), and from miscellaneous data on large spots it is gathered that the penumbra develops in spurts, sector by sector. One such penumbral sector is completed within an hour or so, and the penumbra is completed around (most of) the umbral periphery in a couple of hours, within half a day. Often the first development of the penumbra in a large spot is toward the nearby edge of the region. Thereupon the penumbra nearly closes, leaving a gap toward the inside of the active region—in this gap more magnetic flux parcels march in and coalesce with the umbra. After the entry of more flux has stopped, the penumbra closes completely, and gradually the sunspot assumes a more roundish shape.

As to the formation of penumbral filaments, Bumba (1965) notes that the dark filaments come about by intergranular lanes in the photosphere lining up, usually in the direction radially away from the umbra. The chain of granules between these dark alignments rapidly change into the bright penumbral filaments.

Apparently the formation of the penumbra marks a sudden transition in the structure of the flux bundle near the photosphere, from a nearly vertical system with little fanning, to a widely fanning sheaf, with nearly vertical field in the umbra and nearly horizontal field in the outer penumbra. The formation of a penumbra must be preceded by reconnection of magnetic field. The field lines responsible for the first penumbral structure, pointing then toward the nearest outer edge of the region, were connected originally with the opposite end of the bipolar region. It is not difficult to picture how the required reconnection happens when the expanding coronal field of the emerging flux region runs into the overlying "old" coronal field. (More demanding on the imagination are the magnetic events introducing the formation of penumbral sectors pointing toward the interior of the active region).

There are no observational indications that the formation of the penumbra requires a sudden increase of the total magnetic flux in the spot—the penumbra forms at the expense of the umbra. Often the total magnetic flux in the spot keeps gradually increasing while the penumbra is formed, and for several days after that event. It has been mentioned already that in such a case magnetic patches keep streaming into the umbra across a small sector without penumbral structure. There are beautiful examples in the pictures of Vrabec (1974), and in the movie supporting that paper.

There is a crude linear relation between the umbral and the penumbral areas for spots larger than 50×10^{-6} of the visible hemisphere, or total area A larger than 150 Mm^2: the umbral area is 17 ± 3 % of the total area A (Tandberg-Hanssen 1956). There are suggestions that the area ratio depend on parameters such as the relative age of the spot, or the phase of the solar cycle (Ringnes 1964). It is doubtful, however, whether the quality of the routine photographs warrants such detailed conclusions.

3.3. THE PLAGE STATE

Even at maximum development of an active region, only part of the magnetic flux is incorporated in sunspots—only up to about 60 % of the leading polarity is in one or more sunspots (N.R. Sheeley, private communication; Schrijver 1986 (Ch. 11), 1987). Apparently the flux bundle is not quite coherent since some flux escapes being incorporated in spots.

Outside the sunspots the magnetic flux is in the small faculae, which are seen to be arranged in more or less coherent sheets, which are called *plages*, where the supergranular convection is suppressed but convection on smaller scales is visible. There appears to be a pattern hidden in these plages. In his study of the relations between the chromospheric and coronal emissions from entire solar active regions and their magnetic fluxes, Schrijver (1987) discovered the well-defined *plage state*: within some convenient isogauss contour (e.g., 50 gauss in Kitt Peak magnetograms) the average flux densities of plages are remarkably constant: $\Phi/A = 100 \pm 20$ gauss on Kitt Peak magnetograms, independent of the size or the evolutionary stage of the plage. In other words, the magnetic filling factor of any plage is somewhere between 5 and 10 %. Zwaan (1987) concluded that there are *two* magnetically controlled quasi-equilibrium states in the solar convective mantle and photosphere: (*i*) in umbrae, where the magnetic filling factor is nearly 100 %, and (*ii*) in plages.

3.4. SYNOPSIS OF FLUX EMERGENCE AND SUNSPOT FORMATION

Observational features of flux emergence and sunspot formation may be brought into perspective by the model of an emerging top of an Ω-shaped loop bundle (Figure 4) whose ends are anchored near the bottom of the convection zone (Figure 3).

Apparently the top part of the Ω-loop bundle is frayed into many distinct flux loops. In this frayed system two scales can be distinguished. The thinnest loops are responsible for the individual dark alignments in the photosphere and the fibrils in the Arch Filament Systems in the chromosphere. Somewhat deeper down these many thin loops combine into a few bundles, each one of which is responsible for one of the Emerging Flux Regions that make up for one active region. This ordering in EFR-bundles is seen in the parallel ordering of the dark photospheric alignments, in the systems of parallel fibrils in the Arch Filament Systems, and the brief duration of the emergence process of any one EFR.

The various flux bundles defining the individual Emerging Flux Regions combine in the grand loop bundle defining the entire bipolar active region. This rooting explains why the initially somewhat messy mixed-polarity distribution, produced by the variety of emerging flux regions, is subsequently ordered into a nearly bipolar active region; we return to the ordering mechanism presently.

The distribution of features and the dynamics within a growing active region is qualitatively explained by the emergent flux loop bundle. (For an early comprehensive description of the phenomena see Zirin (1974)). The magnetic flux which emerged earliest is found at the periphery of the EFR (see Figure 5 panel a.), in particular at the leading and following edges, and that is why the earliest pores are found there. More recently surfaced flux, still in the state of convective collapse, borders the edges of pores toward the interior of the EFR, and that is why pores grow toward the interior (while the EFR keeps expanding).

After the convective collapse the tops of the flux tubes are nearly evacuated, particularly those in pores and spots, and hence they are buoyant. Consequently tops of tubes drift together, if their roots are held together somewhere deeper in the convection zone—like a bunch of balloons whose strings are clenched in a fist. So it is the buoyancy that drives the magnetic fragments to coalesce into sunspots. On the way to their destination the fragments percolate along lanes in between convective cells—that is why the fragments are seen marching in single files along a few specific tracks which are often curved. From the coalescence of magnetic fragments originating from different Emerging Flux Regions it is concluded that the flux bundles at the roots of the Emerging Flux Regions must be held firmly together deeper down in the grand loop bundle defining the bipolar active region as an entity. The growth of sunspots ends within several days after all flux has emerged.

The buoyancy of the tops of the flux tubes also explains why the polarities in a young active region keep separating after all the flux has emerged, be it at a steadily reducing rate. Eventually the separation comes to a halt, presumably because the legs of the Ω-loop have reached the equilibrium position. Often the largest leading and following spots are found close to the leading and following edges of the region.

As to the loop bundle as a whole, after its release from the bottom of the convection zone, its top rises up into the atmosphere within 3 to 4 weeks if it rises at the Alfvén velocity at the local equipartition field strength $B_{\rm eq}(z)$. If the actual field strength in the bottom half of the convection zone is appreciably higher than $B_{\rm eq}(z)$, the real rise time is more likely near two weeks, or even as small as one week. These time spans compare favorably with the emergence times of active regions of a few days, as found by Harvey (1991) and discussed in Section 2. (The emergence time is the *spread* in the rise times of the tubes constituting the loop bundle).

It has been suggested that flux loops creating active regions are brought to the atmosphere by rising convective bubbles. At the present time it seems more likely that the top of a loop completes its rise by magnetohydrodynamic forces, and then a convective upwelling in the wake of the rising loop top might be expected. In any case, the vertical and the horizontal components of the velocity field within and around emerging flux regions call for scrutiny.

Two facts are particularly important in controlling the orientation and the compact structure of active regions: the strong toroidal field near the bottom of the convection zone, which apparently is tightly held in place for a couple of years (see Section 2), and the buoyancy of the collapsed flux tubes in the top of the convection zone. These principles make up for a surprisingly robust magnetic fabric of the Ω-loop bundle throughout the convection zone, in which the sunspots are embedded. Indeed the *cluster model* describes the formation of sunspots well, but this term should not be replaced by "spaghetti model" because this terminology suggests a soft and messy mass underneath the spots, and that view is not supported by the observations. If a term more picturesque than "cluster model" were called for, then I suggest "composite model", with a composite flower in mind as an analogy for the clear-cut structure of sunspots.

The bipolar active region has reached its maximum development shortly after all flux has emerged. Then the full hierarchy of magnetic elements is there: from the small faculae arranged in plages, up to the large sunspots, with umbrae and penumbrae. All that splendor springs up well within a week. During that emergence time the magnetic flux remains contained within the growing region (see Figure 5)—there is little interaction with the surrounding photospheric field. But thereupon the active region is left to the decay processes, which may entail interactions with its surroundings.

4. The Decay of Sunspots

4.1. DECAY OF ACTIVE REGIONS

Before turning to the decay phenomena of sunspots let us consider the main decay properties of active regions as entities (see Zwaan (1987) for references and details).

Once all the available flux has surfaced and the active region has reached its maximum development, obvious decay phenomena are observed. The sunspots diminish in size and in magnetic flux, only one or two dominant spots may grow by further coalescence for some more days. The spectrum of magnetic elements becomes simpler: the pores and the "magnetic knots" (or "micropores") disappear, and so do most of the sunspots. Often one leading sunspot remains for some time, slowly shrinking during a period of a few weeks—very rarely the leading sunspot lives for a few months.

The total area of the active region increases, while maintaining the average flux density $\Phi/A = 100\pm20$ gauss over all of its plages. Particularly at the periphery of the active region, large roundish holes ("cells") appear in the plages, with sizes typical of supergranules. These empty cells are resposible for the gradual decrease of the magnetic filling factor over the whole active region. Gradually the plages are transformed into the "enhanced network". This enhanced network keeps expanding and it becomes fainter, until it is lost in the "quiet" network. This picture of a gradually expanding region, with an ever decreasing average filling factor, applies to relatively isolated active regions. Active regions appearing in rapid succession in active nests are not given the opportunity of a long and undisturbed old age (see Section 4.3).

Although the large-scale magnetic structure changes appreciably, from plages without holes to thin network, the nature of the individual elements making up these fabrics does not change: the flux tubes consist of intrinsically strong field, of kilogauss order. Despite recent statements to the contrary (e.g., Stenflo 1991), I still believe that all the magnetic flux in large-scale magnetic structure, also that at higher latitudes and polar caps, originates exclusively in the sunspot belts (Howard and LaBonte (1981)). It is known that so-called "ephemeral active regions" may emerge at high latitudes, but these short-lived regions do not contribute effectively to the large-scale magnetic field. In this picture, magnetic flux that stays for a while in the solar atmosphere emerges in an active region, where it is transformed into strong-field flux tubes by the convective-collapse process. (This process probably occurs in ephemeral regions as well.) In that strong-field state a magnetic network patch is carried over the solar surface until it disappears in an encounter with a magnetic patch of opposite polarity.

In many active regions, particularly in those occurring in active nests, the magnetic field disappears largely *within* the periphery of such a region. Zirin (1984) shows a striking example in which, after all flux had emerged, the faculae of opposite polarities returned to come back together and to vanish completely. Martin *et al.* (1985) describe the flux disappearance during the decay of an active region of modest size. The plages are seen to fragment; small fragments of opposite polarity move toward the polarity dividing line, where they cancel.

The lifetime of an active region is roughly proportional to the magnetic flux at maximum development. Golub (1980) found for the constant of proportionality about 1×10^{20} Mx day^{-1}.

Many decaying sunspots are surrounded by so-called "moat cells"; the "moving magnetic features" across these cells resemble features of the "intranetwork field". Therefore the main features of these dynamic phenomena are sketched in the next section, before returning to sunspot decay.

4.2. MOATS, MOVING MAGNETIC FEATURES, AND INTRANETWORK FIELD

A *sunspot moat*, or *moat cell*, is an annular zone without stationary magnetic field which surrounds (part of) the periphery of a sunspot (Sheeley 1969, 1972, 1981; Harvey and Harvey 1973, Vrabec 1974). These moats are 10 to 20 Mm wide; they persist with little change for at least several days. They are found around many sunspots: around virtually all leading spots, except the smallest, and around many of the follower spots. A penumbra is not a prerequisite: moats are found around some large pores as well.

Note that still growing sunspots may show moats already: in Vrabec's (1974) Figure 7 there is a beautiful example, with a well developed penumbra and moat around most of the periphery, except for the "back gate" through which new fragments keep marching into the umbra along three tracks.

The shape of the moat depends on the magnetic environment, and a moat affects its environment. Completely circular moats are only found around sunspots that are well isolated; moats show breaks where there is substantial magnetic structure nearby. During its formation, however, a moat cell sweeps the faculae aside into the wreath around the moat; Sheeley (1981, his Section 2.4) noticed that during the westward motion of a leader sunspot its moat "almost like a snowplow" pushed and compressed scattered patches of old flux into a compact plage.

Across moats magnetic elements are seen to drift radially away from the spot with velocities of several tenths up to 2.0 km s^{-1}. These outward moving features differ in several aspects from the inflowing magnetic fragments during the coalescence of spots. First: all outward moving features are small: no more than about 1 arcsec in diameter. They are not be seen on movies in the continuum, but they are clearly visible on magnetograms and on CN or Ca II K spectroheliograms. Second: whereas the coalescing fragments move along a few distinct tracks, the outward moving features are seen to stream all over the moat. Third: whereas all coalescing fragments are of the same polarity as the target-umbra, the outstreaming features are of both polarities. The pattern is "pepper and salt": mixed polarities, with only locally some indication of opposite-polarity pairs. The range of estimated fluxes of the individual features is large: $6 \times 10^{17} - 8 \times 10^{19}$ Mx (Harvey and Harvey 1973).

Many authors have noticed that the shape, size and horizontal velocity pattern suggest that moat cells are similar to supergranulation cells. There is no intuitive difficulty in accepting for moat cells the longer lifetimes, and somewhat larger sizes and velocities.

The puzzling aspect is the mixed-polarity character of the outflowing features—the magnetic structure and dynamics differ drastically from magnetic structure elsewhere in the photosphere, except in the so-called *intranetwork field.*

Two types of magnetic fields in the solar photosphere are ubiquitous, i.e. present with some filling factor everywhere outside active regions: the network field and the intranetwork field.

The network field consists of isolated magnetic patches, with magnetic fluxes from about

2×10^{18} up to about 3×10^{19} Mx. Over several hours some of the patches change markedly, by splitting, by merging with a fragment of the same polarity, or by a complete or partial cancellation in an encounter with a patch of opposite polarity (see Martin (1984) and Livi *et al.* (1985)). Over one day most of the patches remain quite steady in position and appearance, however, so the pattern retains a recognizable coherence. The crude network suggested by the magnetic patches corresponds to the boundaries and, in particular, to the vertices of the supergranular cells (Simon and Leighton 1964).

The network patches are bundles of intrinsically strong flux tubes, with field strengths between 1 and 2 kG. According to the interpretation given in Section 4.1, these tubes are made up of flux that emerged in some active region, contracted during the convective collapse, and maintained its strength during the transport from its emergence site.

The *intranetwork field* (INF) is only detected during fair seeing conditions by sensitive magnetographs. Livingston and Harvey (1975, see also Harvey 1977) found magnetic flux between the network patches. The polarities are mixed on a scale of a few arcseconds, and the polarity distribution is independent of the dominant polarity of the surrounding network field. The existence of the INF has been confirmed and more properties have been found with the videomagnetograph at Big Bear Solar Observatory (see Martin 1984). The magnetic fluxes in the patches of the INF typically range from a few times 10^{16} Mx (the detection limit, see Zirin 1985) to a few times 10^{17} Mx—so they are one to two orders of magnitude smaller than the fluxes in the network patches.

The fragments of the INF are in motion, with typical velocities of 0.3 km s^{-1}. In many network cells INF patches are seen to appear near the center, from where they are swept to the boundaries. These displacements are readily explained as transport by supergranular flow. Locally, however, nonradial displacements are seen as well. Patches of the same polarity may coalesce, and it is reported that patches with fluxes up to a few times 10^{18} Mx can be built up.

Because of these motions, the pattern of the intranetwork field changes markedly within one hour. If seeing permits, however, relatively large fragments can be traced for some hours. Eventually the fragments disappear by merging or cancelling with network patches, or with other INF fragments (Martin 1974, Livi *et al.* 1985).

The description above indicates that the intranetwork field differs radically from the network field and the field in ephemeral active regions. From the absence of any intranetwork field on their magnetograms of high angular resolution (0.″5) but limited sensitivity (50 gauss), Tarbell *et al.* (1979) conclude that the INF must be rather diffuse and *intrinsically* weak. There is independent proof for this conclusion (see Zwaan 1987): While the magnetograph signal from network field and plages decreases rapidly toward the solar limb (as is to be expected of thin vertical flux tubes with a "Wilson depression"), the contrast in the intranetwork field in magnetograms hardly decreases toward the limb (S.F. Martin has shown this at several places). This is nicely compatible with a diffuse and intrinsically weak field which, unlike strong flux tubes, are not set vertical by buoyancy, hence the weak INF intersects the photosphere at all angles.

The intranetwork field and the magnetic field in the features streaming across sunspot moats share some characteristics: The polarities are mixed on scales of a few arcseconds. The magnetic features move in the same direction as the dominant velocity field in the photosphere and the top of the convection zone; many patches appear to originate in the sources of the flows: in the centers of supergranules, or in the source region of sunspot moats. There is a quantitative difference, however: the magnetic patches in moats carry much more flux per patch than do the INF patches.

4.3. CHARACTERISTICS OF SUNSPOT DECAY

There are two modes of sunspot decay: *fragmentation* and *gradual decay*, so termed in this paper for easy reference.

The great majority of sunspots break up into several parts within one day or some days after their maximum development, the fragments disappearing soon after that. This fragmentation also befalls the majority of dominant following spots. Consequently of all the recurrent groups 95 % return with only the leader spot, without a trace of a follower (Astronomer Royal 1925). In many small groups also leaders may disappear by fragmentation (one very clear, slowly progressing case is shown in Figure 20 of McIntosh (1981)).

The fragmentation starts with the sudden appearance of very bright umbral dots (Zwaan 1968, 1987)—or is it a rapid brightening of existing dots? The umbra crumbles away, and the spot disappears so rapidly (typically overnight) that the process is hard to follow.

Spots dying by fragmentation show one, if not all of the following properties: irregular shape, conspicuous bright structure in the umbra well before the fragmentation starts, absence of a moat cell.

During the *gradual decay* the spot just *shrinks*: the initial ratio between the umbral area to total area is maintained, and the umbral and penumbral brightnesses do not change substantially. This means that magnetic flux tubes initially embodied in the umbra must be transferred to the penumbra, and pass through there, before they are severed from the spot altogether (as discussed in Section 4.4 below). The loss rate, $\mathrm{d}\Phi/\mathrm{d}t$, or $\mathrm{d}A/\mathrm{d}t$, is nearly constant throughout the gradual decay phase. Also the magnetic field strength in the umbra remains constant (Cowling 1946).

The gradual decay by shrinkage may continue until the spot is reduced to a tiny pore. In other cases the spot subsides to a fraction of its size, whereupon that remnant suddenly fragments.

During the gradual decay there is a moat around (most of) the spot's periphery, carrying the outflow of magnetic features. The loss of flux goes disguised under the cover of mixed polarities—Harvey and Harvey (1973) verified that the polarity of the net flux carried across the moat agrees with the polarity of the decaying spot. This decay may start already during the coalescence of the spot: as mentioned above, while magnetic fragments are still marching in over a small umbral sector, a moat may have developed already around a large part of the periphery, across which magnetic flux is streaming out.

There is a great variety in flux loss rates, and it is my impression that the loss rate is indicated indeed by the density of the magnetic features streaming out across the moat.

The fastest gradual decay is found for spots in active nests, where even large spots may dwindle within a few days. There are many outstreaming magnetic features, which look "fat", and there seems to be a strong excess of patches with the same polarity as that of the spot. The flux disappears *within* the periphery of the nest; in some cases this disappearance has been seen as a cancellation of flux against flux of opposite polarity within the nest. The details of the process are not yet known for certain—probably loops of flux are retracted from the atmosphere into the convection zone, with or without preceding reconnection in the corona. In any case, in view of the huge amount of flux processed within a large active nest during its lifetime (Gaizauskas *et al.* 1983, see also Parker 1987), it seems necessary that the flux is recycled such that parts of the same flux bundle emerge and submerge several times. For the active nest to survive for many months within its relatively very small area, with virtually no photospheric magnetic flux escaping across the boundary, such an up-and-down shuttle should proceed in quite an orderly manner. The thermal relaxation oscillation proposed by Parker (1987) is an interesting mechanism for repeated flux emergence in active nests.

Many leading sunspots live for a period between one week and a few weeks after maximum development. Such spots develop completely circular moats, with the typical mixed-polarity magnetic outflow.

During a sunspot cycle only a few sunspots are observed that live for more than two solar rotations. Even though these spots are exceptional, they present the intriguing problems of how sunspots can be so stable, and what the stringent conditions for such extreme stability are. However slow, the decay proceeds steadily, at a surprisingly constant rate. Presumably also these spots display a moat with the mixed-polarity outflow—with a thinly spread magnetic outflow?

In summary, a long-lived sunspot showing gradual decay has the following properties: (*i*) it is a leading spot, (*ii*) it has one or two very dark umbral cores, (*iii*) its umbra/penumbra structure is nearly circularly symmetric (often this circular shape is acquired during a relatively long process which may last for several days after the spot had reached its maximum area), (*iv*) it is encircled by a moat.

The phenomena during gradual decay suggest that the flux tube bundle remains well contained, while thin flux tubes are peeled off from the circumference (as discussed in Section 4.4 below). This process may continue until only a small fraction of the initial flux is left. Then the containment lets go, and the remnant of the flux bundle desintegrates by fragmentation.

It is not quite clear why flux tubes being peeled off the circumference lead to an approximately constant decay rate, as seems to be the case for slowly decaying spots. Gokhale and Zwaan (1972) remarked that a constant rate is compatible with a model of a flux tube wrapped in a thin current sheet whose thickness decreases in proportion with the radius of the tube. Such a radial similarity in the deep sunspot structure, maintained during the gradual decay, is in harmony with the observed overall radial similarity of sunspots, as, e.g., in the roughly constant ratio between the umbral and penumbral areas (see Section 3.2). This radial and evolutionary similarity does not explain the spread in the rates of gradual decay, however.

Most pores have a short lifetime, from less than one day to some days. They either coalesce into a larger sunspot, or they disappear. (It is conceivable that bunches of pores may live longer than some days in particular pockets within active nests). Even during the first hours of a growing active region, some of its pores are already decaying (Brants and Steenbeek 1985, see also Zwaan 1985).

There are few descriptions of the decay of pores. In their best case, Brants and Steenbeek

(1985) noticed a decrease in area and a marked loss in contrast, but the pore did not disappear altogether during the observing run of a few hours. There was no opposite-polarity flux in the immediate vicinity. Such a case may serve as a reminder that the convective instability, which leads to the convective collapse in case of an initial downdraft, could work the other way: an initial upward flow would make the flux tube expand and the field become weaker (see Spruit 1979).

At least some of the larger and relatively long-lived pores do show a well-defined moat. Presumably such pores pass through a gradual decay phase.

There are several reports of disappearance of magnetic flux without visible transport of flux. Wallenhorst and Howard (1982) and Wallenhorst and Topka (1982) found that magnetic flux disappeared from decaying sunspots without an indication as to where it went. A way out of these problems is to invoke flux retraction from the atmosphere that had escaped observation. However, there are reports of flux disappearance from isolated magnetic features, without a trace where that flux went (Wilson and Simon 1983, Simon and Wilson 1985, Topka *et al.* 1986), so there is good reason to search for covert ways of flux transport in order to explain the observed "disappearance *in situ*", while satisfying Maxwell's law $\nabla \cdot \mathbf{B} = 0$.

4.4. U-LOOPS AND THEIR EFFECTS

Whereas the Ω-loop is an indispensable concept in the explanation of the emergence and the features of an active region at maximum development, the complementary U-*loop* is an important device in the interpretation of the decay phenomena of sunspots. Spruit, Title and van Ballegooijen (1987) discussed these U-loops in some detail; Parker (1984) had investigated these in connection with the problem of the escape of magnetic fields from stars.

In the solar convection zone U-loops may be created in various ways. In the present context the formation by two adjacent Ω-loops that have emerged from the same toroidal flux strand is of most interest (Figure 6). Note that one of the two active regions AR1 and

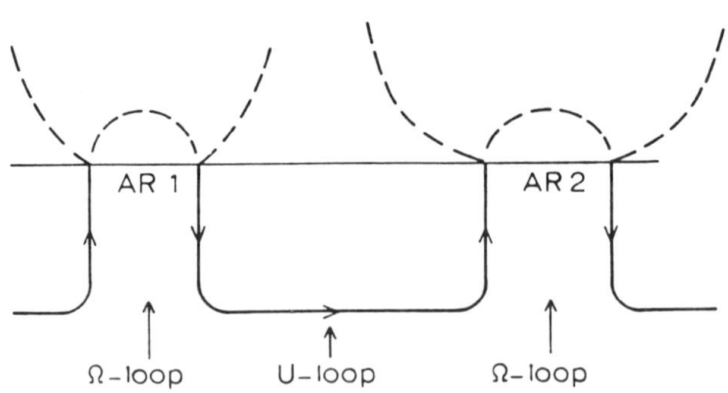

Figure 6: The formation of a U-loop in between two Ω-loops forming the active regions AR1 and AR2 from about the same toroidal flux strand.

AR2 may have decayed and been dispersed beyond recognition.

An important property of a U-loop is that the matter inside is trapped on the field lines. Whereas an Ω-loop can rise rapidly from the bottom of the convection zone into the atmosphere by draining mass down the legs, U-loops cannot leave the Sun until some subtle process disengages the gas from the field lines (Parker 1984).

Spruit *et al.* (1987) argue that the bottom part of the U-loop must be (or become) in thermal equilibrium with its surroundings, so it is buoyant and must start rising. The loop expands strongly, and when its size becomes comparable to the scale height, it cannot continue to rise as a unit. The field strength drops by some orders of magnitude, so the field becomes weak, undulatory, and it is at the mercy of the convection in the top of the convection zone. As a result of this "*sea-serpent*" process, an irregular and weak field is formed in the photosphere, and this field is identified as the intranetwork field by Spruit and coworkers.

The crests of the undulatory field expand above the photosphere into the rarefied outer atmosphere, where they run into the "canopy" of the overlying chromospheric field, which is rooted in the intrinsically strong field in the legs of the two original Ω-loops. The crests expand laterally as well, bumping into their neighbors. On both sides of the contact zones the magnetic fields are opposed, hence there are current sheets. Reconnection occurs, with the result that antiparallel fields in two different atmospheric layers have been replaced by a set of closed loops near the photosphere (see Figure 5 in Spruit *et al.* 1987)—these loops disappear eventually by Ohmic decay. Then new sea serpents of the U-loop system will appear, and the process can carry on.

The net result is that horizontal field disappears from the U-loop system, and that flux in two areas of opposite polarity (in the legs) is reduced. Spruit and coworkers pointed out that those two areas may be far apart so that they are not recognized as being related. The underlying sea-serpent process does not catch the eye in a photosphere full of sea serpents, so flux in a leg is seen to "disappear in place".

The efficiency of this flux disappearance at a distance is hard to estimate because it depends on the decay rate of the small loops that are formed during the process and milled down by turbulence at a variety of scales. On the basis of their estimates, Spruit *et al.* (1987) do suggest that the process may explain observations of flux disappearance *in situ*, and contribute significantly to the decay of active regions.

The sea-serpent process in rising U-loops presents a nice explanation for the mixed-polarity outflow of magnetic patches across the moat—see Figure 7. The flow in the moat cell grabs the top part of the rising U-loop system and carries it radially outward, with "sea serpents" and all. This model provides a natural explanation for the mixed-polarity outflow across the moat. Sunspots decay by flux tubes that are peeled off, from the bottom upward, by the combined action of the rising U-loops and the moat cell. This explains why a sunspot gradually shrinks: apparently the remaining flux tube bundle keeps adjusting such that radial similarity, including the roughly constant ratio between umbral and penumbral areas, is maintained.

Spruit and coworkers remark that the effects of U-loops and the sea-serpent process are seen best near sunspots because these strong flux concentrations contain a large number of end points of U-loops. The explanations for the intranetwork field and for the moving magnetic features in sunspot moats by the model of rising U-loops are similar, but with differences in the convective cells propelling the magnetic patches, and the high flux density

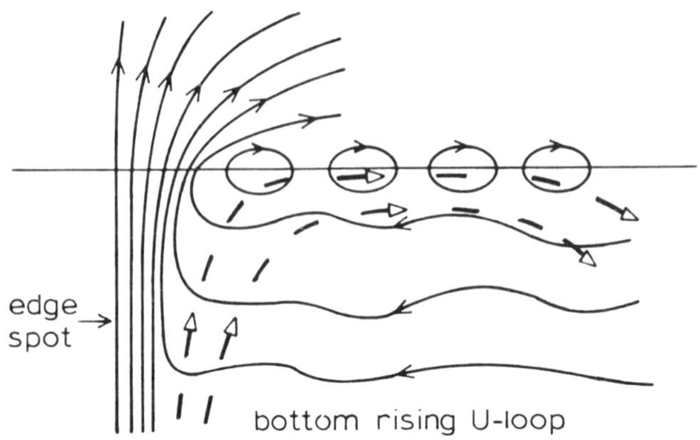

Figure 7: Moving magnetic features across a sunspot moat as explained by a rising U-loop, with one end in the spot. The moat flow, indicated by broken stream lines and open arrowheads, grabs the top part of the U-loop system, and so lets patches of mixed polarities parade across the moat. (Figure adapted from Spruit *et al.* (1987)).

near spots.

There is observational evidence for an important role of U-loops in solar magnetism. All sunspots that go through the gradual decay phase of steady flux loss, do develop moats with magnetic outflow immediately after, or even during, the growth phase. This suggests that U-loops are already rising in the earliest phase of the active region's Ω-loop (which poses an interesting synchronization problem).

Sunspots going through the gradual decay phase may shrink to a small fraction of their size at maximum development. This indicates that the remnant of the local stretch of the matrix toroidal flux strand survives the loss of U-loops until only a small fraction of its original flux is left. Sunspot nests indicate that some stretches of toroidal flux strands must survive for many months, probably by the trick of letting roughly the same Ω-loops retract and emerge again.

5. Concluding Remarks: Magnetic Patterns and Convection

Here follows in broad outline the sequence of magnetic patterns in the convection zone that are responsible for the time-dependent magnetic structure in the solar atmosphere. The emphasis is on the evolving sunspots, but these are treated here as embedded in the overall magnetic field.

The differential rotation must be responsible for the ordered and strong toroidal fields, with the field strength at least of the order of the equipartition strength: $B_{\rm eq} \lesssim B \lesssim 10B_{\rm eq}$. Probably this toroidal system is produced in the thin interface layer at the bottom of the convection zone, because the proper gradients in the angular rotation rate are expected only in that layer.

The differential rotation is maintained by the interaction between rotation and convection; the differential rotation, winding up the toroidal field, is powered by convection.

In other words, in the toroidal system the magnetic field harnesses convection. In these toroidal strands lies all the potential splendor of activity ready for display in the solar atmosphere.Because the field is buoyant, some mechanism is required to keep the toroidal strands pressed down at the bottom of the convection zone.

An active region results from the emergence of an Ω-loop. The top of such a loop is estimated to rise from the bottom within a few weeks. The top is frayed. The legs of the loop bundle must be firmly contained near its roots in the toroidal system. The convective collapse occurring in the atmosphere and the top of the convection zone transforms the magnetic flux into its strong quasi-equilibrium state, and so controls the magnetic structure in the atmosphere. The coalescence of the spots and the separation of the polarities is caused by the buoyancy of the flux tubes in the top of the convection zone.

Apparently tops of Ω-loops may show up in two ways: as one single active region (as from a "Jack-in-the-box"), or as an appearance in a series, occuring within an "active nest" of small dimensions. The observations suggest that in such a nest large parts of the same flux strand emerge and are retracted in succession, so that flux loops shuttle up and down (as in a "puppet show").

The Ω-loops create all the intrinsically strong magnetic structure in the solar atmosphere by emergence in active regions and subsequent convective collapse: upon flux emergence, the sunspots and the faculae in the plage are formed, and later, after the dispersal of remnants of active regions, the patches in the network field. The precise nature of the ephemeral active regions is not yet established beyond doubt—my guess is that these are normal active regions, be it of the smallest size.

The U-loops are created as inevitable by-products of Ω-loops, for instance in between two Ω-loops formed at some distance in largely the same toroidal flux strand. U-loops expand enormously when rising in the convection zone, and so they create magnetic phenomena of their own. One ubiquitous feature is the intrinsically weak and turbulent intranetwork field. Two features bear on sunspot decay: conspicuous are the mixed-polarity patches moving outward across the sunspot moat, and covert is the process leading to the seeming "disappearance *in situ*".

With the introduction of the rising U-loops there are two utterly different types of magnetic structure in the Sun. There is the strong toroidal field in the thin interface between the convection zone and the radiative interior. The whole convection zone is filled with rising and tumbling U-loops; at least in the top part of the convection zone this field is intrinsically weak and turbulent. Locally Ω-loops, reaching up from the toroidal strands, poke through the weak mess.

Convection reclaims its freedom in the decay processes of sunspots and active regions. The combined efforts of the moat cell and rising U-loops cause the gradual decay of sunspots. The supergranular holes developing in the plage indicate that the flux bundles cannot keep out the large-scale convection for long, which leads to a gradual dispersal of the flux tubes.

Various modes of flux removal from the atmosphere are operating in the Sun. In large and long-lived active nests retraction of flux loops must be the main process, because this seems the only efficient step in the rapid reprocessing of the magnetic flux in a nest. (Note that I do not speculate whether it is a direct retraction of emerged loops, or a retraction after some reconnection of field in the corona). Active nests appear to survive for a long time by retraction and re-emergence of Ω-loops, so the underlying toroidal flux bundle must maintain its coherence for the lifetime of the nest.

Part of the flux in an active region may seemingly "disappear in place" by the elimination of U-loops against field in the chromospheric "canopy", which is sustained by intrinsically strong field in active regions and in the network. It is not yet known what fraction of the magnetic flux is removed from the photosphere by this process.

Finally, part of the flux leaves the site of an active region by gradual dispersal. This process is not important for active regions in nests—in the horizontal plane most of their magnetic flux remains within the periphery of the nest. Flux dispersal is significant from isolated active regions, and from an active region that marks the end of a nest's lifetime. But probably even in these cases only a fraction of the magnetic flux leaves the site by dispersal. Yet it is probable that the magnetic patches dispersed from active regions make up all the large-scale magnetic field at higher latitudes, including the polar caps.

The dispersed magnetic patches constitute the "quiet" network; its patches are clusters of intrinsically strong flux tubes, but of low filling factor. It is an interesting question whether these tubes remain rooted in the toroidal flux system, and if so, how. And if these tubes do not remain rooted in some deep and strong flux system, how can they stay intrinsically strong?

The toroidal flux system, the Ω-loops and the U-loops are related. Only after the toroidal system has been wound up enough, can the Ω-loops form. In the present scenario, U-loops can form readily only after Ω-loops have formed.

Since neighboring Ω-loops and ambient U-loops contribute to the destruction of active regions and sunspots, there is a chance that observational studies may relate the lifetimes of spots and nests to previous activity (or lack thereof) along the presumed toroidal strand. Studies along such lines may explain the large range of sunspot decay rates, and reveal more of the magnetic structure near the bottom of the convection zone.

In the end the key question is: how does the movie picture presented in this paper fit into a viable model for the solar dynamo?

Acknowledgements. This paper was prompted by the invitation from the Directors of Workshop, J.H. Thomas and N.O. Weiss, to present a talk on "The evolution of sunspots and pores: observations and theory". This task was broadened by placing the sunspot evolution in the context of the evolution of active regions, and then made performable by leaving the interpretation of the title addition: "observations and theory" up to the speaker.

Thanks to reactions by several colleagues, the presentation has evolved markedly from the original talk to the present paper. That process started with lively discussions during the workshop, and it continued by oral and written comments and discussions. I wish to mention explicitly the input of the following colleagues: K.L. Harvey allowed me to quote from unpublished work, which added substance to the talk and this paper. Pointed questions by the referee P. Maltby, the editor N.O. Weiss, and by F. Moreno-Insertis and A.M. Title made me rethink and reformulate some of the issues. V. Gaizauskas, K.L. Harvey and L. van Driel-Gesztelyi indicated some aspects to be worked out and many improvements in presentation and text. After close reading of the first draft, the Ph.D. students J.H.G.M. van Geffen and L.H. Strous recommended clarifications and text improvements.

C.J. Schrijver read the final draft.

I thank all these colleagues for their constructive responses, which which have added greatly to the pleasure in working on this paper.

References

Astronomer Royal: 1925, *Mon. Not. R. Astron. Soc.* **85**, 553
Babcock, H.W.: 1961, *Astrophys. J.* **133**, 572
Brants, J.J.: 1985a, *Observational Study of the Birth of a Solar Active Region*, Thesis, University of Utrecht
Brants, J.J.: 1985b, *Solar Phys.* **98**, 197
Brants, J.J., Steenbeek, J.C.M.: 1985, *Solar Phys.* **96**, 229
Bray, R.J., Loughhead, R.E.: 1964, *Sunspots*, Chapman and Hall, London
Brunner, W.: 1930, *Astron. Mitt. Zürich* Nr. 124
Bruzek, A.: 1967, *Solar Phys.* **2**, 451
Bruzek, A.: 1969, *Solar Phys.* **8**, 29
Bumba, V., Suda, J.: 1984, *Bull. Astron. Inst. Czechosl.* **35**, 28
Choudhuri, A.R.: 1989, *Solar Phys.* **123**, 217
Choudhuri, A.R., Gilman, P.A.: 1987, *Astrophys.J.* **316**, 788
Cowling, T.G.: 1946, *Mon. Not. R. Astron. Soc.* **106**, 218
Frazier, E.N.: 1972, *Solar Phys.* **26**, 130
Fox, Ph.: 1908, *Astrophys. J.* **28**, 253
Gaizauskas, V., Harvey, K.L., Harvey, J.W., Zwaan, C.: 1983, *Astrophys. J.* **265**, 1065
Galloway, D.J., Weiss, N.O.: 1981, *Astrophys. J.* **243**, 945
García de la Rosa, J.I.: 1987, *Solar Phys.* **112**, 49
Gilman, P.A., Morrow, C.A., DeLuca, E.E.: 1989, *Astrophys. J.* **338**, 528
Gokhale, M.H., Zwaan, C.: 1972, *Solar Phys.* **26**, 52
Golub, L.: 1980, *Philos. Trans. R. Soc. London Ser. A* **207**, 595
Harvey, J.: 1977, in E.A. Müller, *Hightlights of Astronomy* **4(2)**, 223
Harvey, K.L.: 1991, in preparation
Harvey, K.L., Harvey, J.W.: 1973, *Solar Phys.* **28**, 61
Harvey, K.L., Martin, S.F.: 1973, *Solar Phys.* **32**, 389
Howard, R., LaBonte, B.J.: 1981, *Solar Phys.* **74**, 131
Joy, A.H.: 1919, *Astrophys. J.* **49**, 167
Leighton, R.B.: 1964, *Astrophys. J.* **140**, 1547
Livi, S.H.B., Wang, J., Martin, S.F.: 1985, *Australian J. Phys.* **38**, 855
Livingston, W.C.: 1991, *Nature* **350**, 45
Livingston, W.C., Harvey, J.: 1975, *Bull. Am. Astron. Soc.* **7**, 346
Martin, S.F.: 1984, in S.L. Keil (ed.), *Small-Scale Dynamical Processes in Quiet Stellar Atmospheres*, NSO-Sacramento Peak, p. 30
Martin, S.F., Livi, S.H.B., Wang, J.: 1985, *Australian J. Phys.* **38**, 929
Maunder, E.W.: 1919, *Mon. Not. R. Astron. Soc.* **79**, 451
McIntosh, P.S.: 1981, in L.E. Cram and J.H. Thomas (eds.), *The Physics of Sunspots*, NSO-Sacramento Peak, p.7
Moreno-Insertis, F.: 1992, in these proceedings
Parker, E.N.: 1979, *Cosmical Magnetic Fields*, Clarendon Press, Oxford
Parker, E.N.: 1984, *Astrophys. J.* **281**, 839
Parker, E.N.: 1987, *Astrophys. J.* **312**, 868
Ringnes, T.S.: 1964, *Astrophys. Norvegica* **8**, 303
Schoolman, S.A.: 1973, *Solar Phys.* **32**, 379

Schrijver, C.J.: 1986, *Stellar Magnetic Activity*, Thesis, University of Utrecht
Schrijver, C.J.: 1987, *Astron. Astrophys.* **180**, 241
Schüssler, M.: 1983, in J.O. Stenflo (ed.), *Solar and Stellar Magnetic Fields: Origins and Coronal Effects*, IAU Symposium **102**, Reidel, Dordrecht, p.213
Sheeley, N.R.: 1969, *Solar Phys.* **9**, 347
Sheeley, N.R.: 1972, *Solar Phys.* **25**, 98
Sheeley, N.R.: 1981, in F.Q. Orrall (ed.), *Solar Active Regions*, Colorado Assoc. Univ. Press, p.17
Simon, G.W., Leighton, R.B.: 1964, *Astrophys. J.* **140**, 1120
Simon, G.W., Wilson, P.R.: 1985, *Astrophys. J.* **295**, 241
Spruit, H.C.: 1979, *Solar Phys.* **61**, 363
Spruit, H.C., Title, A.M., van Ballegooijen, A.A.: 1987, *Solar Phys.* **110**, 115
Stenflo, J.O.: 1991, in A. von Alvensleben (ed.), *JOSO Annual Report 1990*, p.49
Tandberg-Hanssen, E.: 1956, *Astrophys. Norvegica* **5**, 207
Tarbell, T.D., Title, A.M., Schoolman, S.A.: 1979, *Astrophys. J.* **229**, 387
Topka, K.P., Tarbell, T.D., Title, A.M.: 1986, *Astrophys. J.* **306**, 304
van Ballegooijen, A.A.: 1982 a, *Astron. Astrophys.* **113**, 99
van Ballegooijen, A.A.: 1982 b, *Sunspots and the Physics of Magnetic Flux Tubes in the Sun*, Thesis, University of Utrecht
van Ballegooijen, A.A.: 1984, in S.L. Keil (ed.), *Small-Scale Dynamical Processes in Quiet Stellar Atmospheres*, NSO-Sacramento Peak, p.260
van Ballegooijen, A.A., Choudhuri, A.R.: 1988, *Astrophys. J.* **333**, 965
Vrabec, D.: 1974, in R.G. Athay (ed.), *Chromospheric Fine Structure*, IAU Symposium **56**, Reidel, Dordrecht, p. 201
Waldmeier, M.: 1937, *Z. Astrophys.* **14**, 91
Waldmeier, M.: 1955, *Ergebnisse und Probleme der Sonnenforschung*, Geest & Portig K.-G., Leipzig
Wallenhorst, S.G., Howard, R.: 1982, *Solar Phys.* **76**, 203
Wallenhorst, S.G., Topka, K.P.: 1982, *Solar Phys.* **81**, 33
Weart, S.R.: 1970, *Astrophys.J.* **162**, 987
Weart, S.R.: 1972, *Astrophys.J.* **177**, 271
Wilson, P.R, Simon, G.W.: 1983, *Astrophys.J.* **273**, 805
Zirin, H.: 1972, *Solar Phys.* **22**, 34
Zirin, H.: 1974, in R.G. Athay (ed.), *Chromospheric Fine Structure*, IAU Symposium **56**, 161
Zirin, H.: 1984, *Astrophys. J.* **291**, 858
Zirin, H.: 1985, *Australian J. Phys.* **38**, 961
Zwaan, C.: 1968, *Ann. Rev. Astron. Astrophys.* **6**, 135
Zwaan, C.: 1978, *Solar Phys.* **60**, 213
Zwaan, C.: 1985, *Solar Phys.* **100**, 397
Zwaan, C.: 1987, *Ann. Rev. Astron. Astrophys.* **25**, 83
Zwaan, C., Brants, J.J., Cram, L.E.: 1985, *Solar Phys.* **95**, 3

III. OVERALL STRUCTURE OF SUNSPOTS

This section is concerned with azimuthally averaged properties of ideal sunspots, leaving fine structure to be discussed in section IV. Although a sunspot owes its existence to the presence of a strong magnetic field, it is only recently that the form of this field has been unequivocally established. At the photosphere the field strength approaches 3000 G in the centre of the umbra and declines gradually to a value around 1000 G at the outer edge of the penumbra, where it drops to zero over a distance too small to be resolved. The mean field is vertical at the centre and at the edge it is still inclined at 20° to the horizontal. It follows that the penumbra must be deep and not superficial. This field hinders convection but semi-empirical models show that convective transport is essential in order to supply the energy radiated from the umbra. Moreover, the umbral intensity varies by about 15% during the solar cycle – a remarkable effect that has never been satisfactorily explained. It seems at first sight that the missing energy that would have been radiated had the sunspot not been present ought to emerge elsewhere and be detectable. In fact, however, this energy can readily be absorbed within the Sun without appearing at the surface.

Elementary models of sunspots in magnetohydrostatic equilibrium consist of an axisymmetric flux tube with a meridional field enclosed by a current sheet at the magnetopause, where **B** drops discontinuously to zero. In the simplest models (corresponding to a pore without a penumbra) the field within the flux tube is current–free. Magnetic buoyancy can stabilize such a configuration against fluting (or interchange) instabilities. More elaborate models allow distributed volume currents in the penumbra or introduce a second current sheet at the umbra–penumbra boundary. The convective processes responsible for heat transport in the umbra and penumbra seem to be very different: in the umbra energy is transported along the field lines, whereas in the penumbra there is an interchange of flux tube or sheets that gives rise to filamentary structure.

The first two chapters in this section are concerned with observations: P. Maltby reviews observations in the continuum and the thermal structure that is deduced from them, while A. Skumanich describes the magnetic field configuration in a sunspot (which is also discussed by Title *et al.* in the next section). Then K. Jahn surveys models of sunspots in magnetohydrostatic equilibrium, from the first attempts 25 years ago to his own most recent calculations. Finally, H.C. Spruit provides an elegant account of the fate of the missing energy.

At the Workshop there were also five contributed papers on this topic. D. O. Gough pointed out that the Wilson depression depends in part on the decrease in opacity with decreasing temperature. S. K. Solanki described measurements of the magnetic flux through the penumbra: the observations show that less than half the total flux emerges through the umbra, confirming that there must be a deep penumbra. V. J. Pizzo explored the effects of radiative transfer on two–dimensional magnetohydrostatic equilibria, while B. C. Low showed how to construct magnetostatic structures that model the fields above complex spot groups. Lastly, M. Bünte analysed the stability of realistic flux tube models to interchange modes and found that they are more easily stabilized than thin tubes.

CONTINUUM OBSERVATIONS AND EMPIRICAL MODELS OF THE THERMAL STRUCTURE OF SUNSPOTS

P. Maltby
Institute of Theoretical Astrophysics,
University of Oslo
P.O. Box 1029
0315 Oslo

ABSTRACT. Intensity observations of sunspot umbrae and penumbrae in the visible and infrared are reviewed. The continuum intensity in the darkest part of the umbra and the average penumbral intensity are known with relatively high accuracy in large sunspots. With the aim of understanding conflicting observations of umbral dots it is suggested that the umbral core is nearly free of umbral dots, and that an umbral core only occurs in sunspots above a certain size.

The present status of semi-empirical models of the temperature stratification in umbrae and penumbrae is discussed and the need for extending the model calculations to more than one dimension is pointed out.

1. Introduction

Our empirical knowledge of sunspots is considerable (e.g. Bray and Loughhead 1964; Cram and Thomas 1981; Stix 1989; Cox et al. 1992). As we intend to focus on the thermal structure of sunspots it is appropriate to draw attention to a comprehensive review by Obridko (1985) that includes evaluations of the intensity observations and discussions of semi-empirical sunspot models, i.e. models that are in hydrostatic equilibrium and are adjusted to fit the observations.

In the 1960s (e.g. Bray and Loughhead 1964) it was generally accepted that large sunspots were darker than small sunspots. However, this view was based on observations that were insufficiently corrected for scattered light originating in the Earth's atmosphere or in the instrument. Zwaan (1965) examined earlier observations and pointed out the sharp increase in the amount of stray light with decreasing sunspot size. One may state that the corrections for stay light are so critical that one is tempted to disregard intensity observations that are published without them. Observers who carefully correct their observations for scattered light have hitherto not found any evidence for a decrease in umbral brightness with increasing umbral diameter (see Rossbach and Schröter 1970; Mykland 1973; Albregtsen and Maltby 1981). This does not exclude the possibility that a systematic change in umbral brightness with umbral size may exist.

J. H. Thomas and N. O. Weiss (eds.), Sunspots: Theory and Observations, 103–120.

For sunspots with umbral radii larger than 5" differences between different sunspots have been detected. However, these differences are found to have other causes (see below) and may have prevented the detection of a possible small change in brightness with umbral size for larger sunspots. For smaller sunspots we have few reliable observations because extremely good observing conditions and space observations have not been used often enough to determine with accuracy the intensity and other physical parameters of smaller sunspots.

In textbooks on stellar atmospheres it is usual to discuss the determination of the temperature stratification based on center-limb intensity observations in detail (e.g. Mihalas 1978). Less attention is given to the determination of the temperature stratification from observations of the spectral energy distribution in the continuum. We note, however, that with an extended coverage in wavelength the latter method gives higher accuracy in the temperature stratification determination. Semi-empirical models of solar features should preferably account for both sets of observations.

The observed fine structures in the form of bright umbral dots (e.g. Adjabshirzadeh and Koutchmy 1983) and penumbral filaments (e.g. Muller 1973; Collados et al. 1987) show dimensions comparable to the scale height, and knowledge about their thermal stratification may be important for understanding the physics of sunspot atmospheres. Such observations are important for constructing two-component models with one hot and one cold component.

In order to study layers in the sunspot atmosphere formed at greater heights than the continuum, observations of spectral lines in the visible, infrared as well as the ultraviolet are used (e.g. Avrett 1981). The non-LTE effects involved are similar to those of the undisturbed atmosphere, but are complicated by the fact that the sunspot atmosphere interchanges radiation laterally with the surrounding atmosphere (e.g. Lites et al. 1987). In the following we shall concentrate on the continuum observations and the deepest observable layers, as the information from these layers puts stringent limitations on theoretical models of sunspots.

The sunspot temperature stratification is determined by the energy transport processes taking place. We will show that in the deepest observable layers the radiative flux decreases with depth and other processes control the energy transport. Semi-empirical models of the layers close to the solar "surface" may also be used to study the difference in gas pressure between the photosphere and the umbra. The lateral difference in gas pressure may be determined provided we know the height difference, termed the Wilson depression, between continuum optical depths of unity in the umbra and the photosphere. In order to maintain a stable sunspot the magnetic field configuration must compensate for the lateral difference in gas pressure.

1.1. NOMENCLATURE AND SUNSPOT STRUCTURE

Solar physics is a branch of science where the same phenomenon may have different names and it is therefore useful to make clear the nomenclature that will be used. The darkest part of the sunspot will be referred to as the umbral core.

Sunspot umbrae are not uniformly dark but contain extensions of penumbral filaments and small, bright regions called umbral dots. For larger sunspots the bright umbral dots are usually located close to the rim of the umbra and may be called peripheral umbral dots (Krat et al. 1972). In umbral cores the umbral dots show less contrast (Loughhead et al. 1979) and to some observers the core gives the impression of being a featureless void (Livingston 1991; Ewell 1991). One should be aware, however, that examples of sunspots with complex structure showing umbral dots distributed over nearly the whole umbra have been presented (Lites et al. 1991).

The penumbral structure has been studied in detail; see Shine et al. (1987) and the review by Muller (this volume). Here we shall limit the discussion to the thermal stratification of the penumbra and regard the penumbra as consisting of dark and bright filaments.

2. Continuum observations

Sunspot intensity observations were pioneered by Henry and Alexander (1846). The development of the field has been reviewed by several authors (e.g., Bray and Loughhead 1964; Zwaan 1965; Obridko 1985). We emphasize the importance of photoelectric photometry (Makita and Morimoto 1960) and the introduction of a proper correction method for stay light (Stumpff 1961; Birkle and Mattig 1965; Zwaan 1965; Iuell and Staveland 1975). The stray light originates in the Earth's atmosphere and/or in the instrument and causes small amounts of light to be scattered from each position in the solar image to other positions. Since the sunspot is darker than the surroundings, the corrections for stay light become critical. In order to measure the amount of stray light, observations outside the solar disk have to be carried out. In passing we note that a test of the correction method is possible, using the transit of Mercury as an artificial spot on the solar disk (Maltby 1971; Mattig 1971).

One should be aware of the fact that this correction method will not produce a complete restoration of the sunspot image. The method gives the corrected intensity in the center of the umbra as well as the corrected penumbral intensity as averaged over penumbral filaments. In observations of penumbral filaments image restoration methods have been applied by Grossmann-Doerth and Schmidt (1981) and by Collados et al. (1987). It seems likely that new algorithms like those for maximum entropy image restoration (Skilling 1984; Myrheim and Rue 1991) will be used in routine sunspot observations in the near future.

A two-dimensional study of the sunspot atmosphere will require simultaneous observations with a series of two-dimensional detectors covering the wavelength region from 0.4 to 2.5 μm, properly corrected for stray light through the use of image restoration techniques.

2.1. UMBRAL CONTINUUM OBSERVATIONS

Two methods in photoelectric photometry have been used to measure sunspot continuum intensities: 1) observations in "continuum windows" between spectral lines, and 2) observations in broad spectral bands, corrected for spectral lines. One problem with the first approach is that continuum windows are scarce, especially in the visible umbral spectrum where the high spectral line density depresses the continuum. Furthermore, since high spectral resolution and relatively long exposure times are required, extremely good and stable observing conditions are necessary for the method to work. With broad

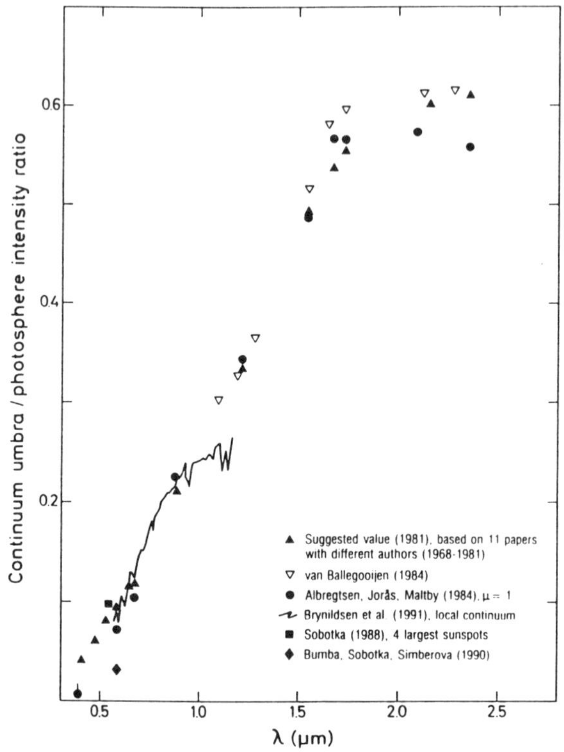

Fig. 1. The spectral distribution of the umbra/photosphere continuum intensity ratio as published by different observers.

band observations good signal to noise ratios are obtained with short integration times. The observations are corrected for spectral lines by using the best available knowledge of the sunspot spectrum within each spectral band. The problem with this approach is that possible differences in spectral characteristics between different sunspots cannot be taken into account without supplementary observations with high spectral resolution.

As relatively few intensity observations have been published recently, we have to rely on earlier observations, such as the sunspot intensity observations carried out for 10 months each year at the Oslo Solar Observatory between 1967 and 1986. The broad band pinhole photometers recorded intensities in a total of 11 wavelength regions in the spectral range 0.387 - 3.8 μm. Only measurements with an accuracy of 0.015 in the umbra/photosphere intensity ratio were retained. The observations allowed studies of spectral energy distribution, center-limb variation, solar cycle variation, as well as comparison with other sunspot parameters, both for the umbrae and the penumbrae.

2.1.1. *Spectral Energy Distribution of the Dark Umbral Component.* Let us consider the continuum emission in the darkest part of the sunspot umbra, called the umbral core. Figure 1 gives the spectral distribution of the umbra/photosphere continuum intensity ratio in umbral cores for sunspots with umbral radii larger than 4". For comparison Figure 1 includes the intensity values suggested by Albregtsen and Maltby (1981) after an evaluation of eleven papers published by different observers during the period 1968 to 1981. The results for the four largest sunspots observed by Sobotka (1988) appear to be in general agreement with the results of Albregtsen et al. (1984), van Ballegooijen (1984), as well as the previously suggested values. The extreme darkness observed in one sunspot by Bumba et al. (1990) is probably related to the fact that no correction for spectral line haze has been applied to this result. Brynildsen et al. (1991) have recently completed the reduction of an umbral spectrum with high spectral resolution and the results for the *local* continuum within each 10 cm^{-1} are shown in Figure 1; the shape of the curve indicates deviations between the local and the true continuum in some wavelength regions.

Real differences in radiation temperature between different sunspots have been detected for large sunspots. Accordingly, Figure 1 only gives an indication of the run of umbra/photosphere intensity ratio versus wavelength since the results for different sunspots and different heliocentric positions are included in the same diagram. Part of the scatter among the results in Figure 1 may be explained by the center-limb variation and by a solar cycle variation, as described below.

Although the data do not show any systematic change of brightness temperature with sunspot size for large sunspots, such a relation could exist for smaller sunspots and has been suggested (see Sobotka 1988 and section 2.1.4). One way to study this matter is by selecting spectral lines that are known to behave differently in the umbra and the photosphere (see Stellmacher and Wiehr 1988). However, a better way would be to observe the smaller sunspots with high spatial resolution during excellent seeing conditions.

2.1.2. *Center-Limb Variation.* The importance of the limb-darkening observations for the study of the temperature stratification in the umbra was emphasized by Minnaert and Wanders (1932). Older data, not properly corrected for stay light, showed the umbral limb-darkening to be less than that of the photosphere, whereas more recent observers were unable to detect any center-limb variation in the umbra/photosphere intensity ratio. By observing the very same sunspot on several days during its passage over the solar disk Albregtsen et al. (1984) found that all sunspots studied showed a real and significant decrease in the umbra/photosphere intensity ratio towards the limb. We note that the umbral limb darkening is easiest to detect in the infrared, a wavelength region not frequently studied by earlier observers. Based on these observations the limb-darkening observations of the umbra/photosphere intensity ratio may, for a sunspot with index i, be written with sufficient accuracy as,

$$\phi_{u,i}(\mu,\lambda) = a_i(\lambda) + b_i(\lambda)\mu$$

for $1 \geq \mu \geq 0.3$. Here $\mu = cos\theta$ and θ is the heliocentric angle. Hitherto differences in limb-darkening between sunspots have not been detected.

2.1.3. *Solar Cycle Intensity Variation.* A solar cycle variation in the umbra/photosphere intensity ratio was detected by Albregtsen and Maltby (1978). The umbra/photosphere intensity ratio appears to be a linear function of the phase in the solar cycle. Figure 2 shows the umbra/photospere intensity ratio as function of the phase in the solar cycle, after correction for limb-darkening. The phase is measured as the time t elapsed since the

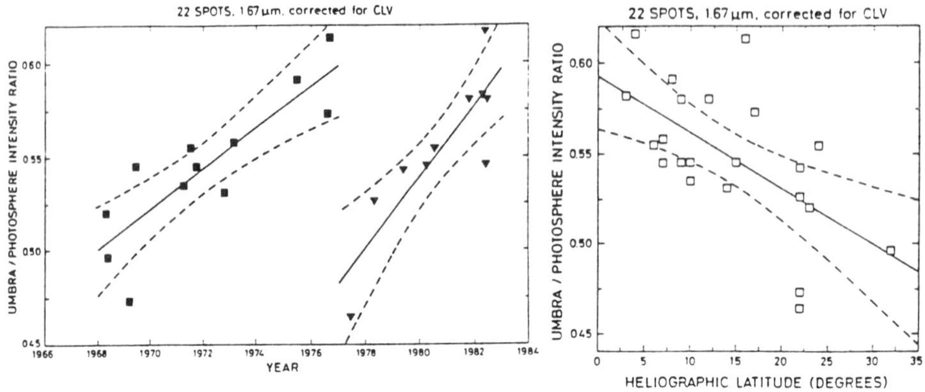

Fig. 2. Umbra/photosphere intensity ratio at 1.67 μm , corrected for the center-limb variation, is given as function of time for cycle 20 (squares) and cycle 21 (triangles). The ratio is also shown as function of sunspot heliographic latitude. All large sunspots observed during good seeing conditions are included. Least squares regression lines as well as 95 per cent confidence limits are given.

last minimum in the cycle, the duration of which is t_0. This corresponds to expressing the umbra/photosphere intensity ratio as

$$\phi_u(\mu = 1, \lambda, t/t_0) = c(\lambda) + d(\lambda)t/t_0$$

At 1.67 μm the correlation coefficient between the umbra/photosphere intensity ratio and t/t_0 is 0.86 (+0.07, 0.12). The variation is caused by differences in umbral intensity; the observed change in the intensity ratio is too large to be explained by a variation in the photospheric intensity.

It is well known that the average heliographic latitude of sunspots varies throughout the solar cycle. Figure 2 also shows the connection between the umbra/photosphere intensity ratio and the heliographic latitude, with a correlation coefficient of - 0.63 (- 0.12, + 0.16) at 1.67 μm. Figure 2 and the corresponding correlation coefficients show that the umbra intensity is better correlated with t/t_0 than with the sunspot heliographic latitude.

The hypothesis (e.g. Adjabshirzadeh and Koutchmy 1983) that a variation in the relative number of umbral dots from one sunspot to another may account for the observed difference in umbra/photosphere intensity ratio has been investigated. A change in the number of bright umbral dots corresponds to a change in umbral intensity from one sunspot to another that is wavelength dependent. Whereas the hypothesis would give the largest change in the visible, the observations show the largest change in the infrared. This does not exclude a real variation in the number or in the temperature of umbral dots from one sunspot to another, but makes the hypothesis unlikely as an explanation for the observed solar cycle variation.

Suggestions as to the nature of the observed relation between the umbral temperature and the solar cycle have been presented. Since the sunspot temperature appears to be a function of the phase in the solar cycle, the responsible subphotospheric flux tubes probably have a lifetime comparable to the solar cycle. The umbral brightness may be determined by the age of the subphotospheric flux tube (Schüssler 1980). An alternative interpretation, where the temperature depends on the height of the flux tube within the convection zone at the time of formation, was presented by Yoshimura (1983). Nordlund and Stein (1990) have recently performed a preliminary simulation of the central part of a sunspot umbra. Considering the total mass in the flux tube as approximately conserved, they argue that the relative pressure difference might control the "looseness" of a sunspot, and thus determines the umbra brightness dependence on cycle phase.

The data presented above were examined for other connections between umbral temperature and sunspot parameters like the magnetic field induction, size, age and type of sunspot. No connection was apparent.

2.1.4. *Umbral Dot Intensities.* According to Kusoffsky and Lundstedt (1986) the half-life time of "normal" umbral dots is about 60 min. Since umbral dots have sizes comparable to the resolution of the telescope, considerable uncertainty as to their temperatures exist. Different methods of observing the temperatures of umbral dots have been tried (e.g. Wiehr 1987). Studies of magnetically insensitive spectral lines with different excitation potentials may give information regarding both the umbral dots and the dark background.

It is often assumed that the umbral dots have almost photospheric temperature and sizes of 100-200 km. Since considerably lower temperatures of the umbral dots have been observed in the center of the umbra (Grossmann-Doerth et al. 1986), Garcia de la Rosa (1987) remarked that temperatures close to that of the photosphere very probably apply to the peripheral umbral dots, whereas umbral dots in the central part of the umbra are cooler than the photosphere.

Recently umbral dots have been observed in the infrared at 8500 Å (Ewell 1991) with brightness contrasts between 2 and 15 per cent. One type of dot seems to be associated with bright penumbral grains. The other type tends to occur near the center of the umbra and with lower contrast. We note that the darkest part of this sunspot did not contain any umbral dots, in agreement with the observations by Livingston (1991). The observations by Ewell (1991) and Livingston (1991) contrast other high resolution observations (e.g. Lites et al. 1991).

The following hypothesis may possibly unite the conflicting results. Let us assume:

- The umbral core is nearly free of umbral dots; and
- Only umbrae with radius larger than a certain value, say 4", have an umbral core.

This hypothesis is compatable with low-contrast dot intensity (Loughhead et al. 1979; Grossmann-Doerth et al. 1986) or lack of umbral dots (Livingston 1991; Ewell (1991) in the darkest part of the umbra. For smaller sunspots a larger fraction of the umbra is covered with umbral dots. Based on intensity observations, Sobotka (1988) suggested that the fraction β of the umbra covered with dots increases from 0.05 for large sunspots to 0.30 for a sunspot with umbral radius equal to 2". Since the sunspot umbral areas are distributed lognormally (Bogdan et al. 1988), one would, on the basis of the assumptions made above, expect most sunspot umbrae to show umbral dots in agreement with observations (e.g. Lites et al. 1991).

In order to construct two-component models for the umbra we need to study the fraction of the umbra covered with umbral dots as a function of height. Often a constant

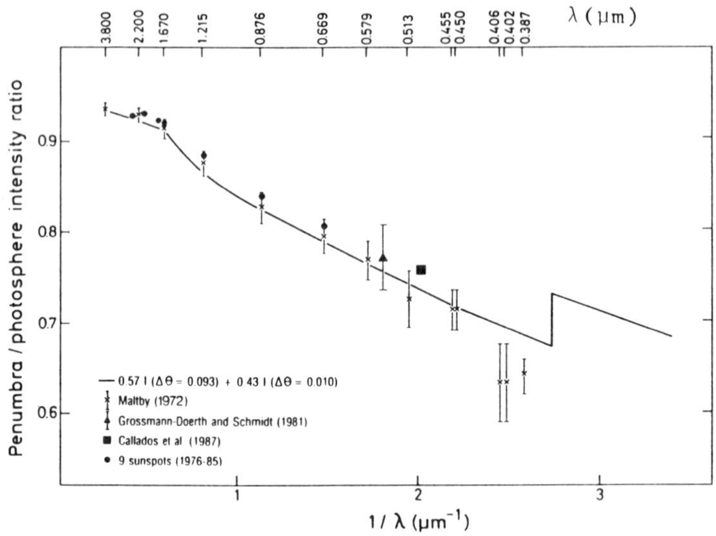

Fig. 3. The observed spectral distribution of the penumbra/photosphere intensity ratio as averaged over penumbral filaments. For comparison the results for a two-component penumbral model is shown.

fraction has been assumed. This may be a reasonable assumption within a limited height range. However, comparing the deepest observable layers with the transition region of the sunspot atmosphere we know that the sunspot appearance changes with height.

2.2. PENUMBRAL CONTINUUM OBSERVATIONS

A description of the morphology of the penumbra is given by Muller (this volume). Here it is sufficient to regard the penumbra as consisting of bright and dark filaments. Before we discuss individual filaments it is of interest to study the average intensity, as its value may be determined with less uncertainty than that of the filaments.

2.2.1.*Spectral Energy Distribution.* Figure 3 shows the wavelength dependence of the average continuum penumbra/photosphere intensity ratio, i.e. as averaged over penumbral filaments. No corrections for spectral line differences between the penumbra and the photosphere have been applied to these observations. The observations by Grossmann-Doerth and Schmidt (1981) and Callados et al. (1987) agree well with the results given by Maltby (1972) based on observations of 54 sunspots. Lites (1984) pointed out that his observations of penumbral intensities are influenced by scattered light. Hence, we have not included his results in the diagram. Figure 3 includes also the results from 9 sunspots, selected from the Oslo Solar Observatory material, as particularly well suited for penumbral intensity measurements.

The temperature of the penumbra may, in a first approximation, be described by a single parameter value of $\Delta\theta$ = constant, where $\Delta\theta = 5040\ (1/T_p - 1/T_{ph})$ and T_p and

T_{ph} are respectively the temperature of the penumbra and the photosphere, as averaged over penumbral filaments and granules. The average value for penumbrae is $\Delta\theta = 0.055$.

Although the average penumbral intensity only changes slightly from one sunspot to another, differences in penumbral intensities between different sunspots are observed and found to be real (Ekmann 1974). Sunspots with a darker than average umbra usually have a relatively dark penumbra. The correlation coefficient between penumbral and umbral intensities is 0.90 at a wavelength of 1.67 μm.

Within the accuracy of the observations it has not been possible to detect any center-limb variation in the penumbra/photosphere intensity ratio.

2.2.2. *Temperature of Penumbral Filaments.* Figure 3 gives for comparison the results for a two-component model with $\Delta\theta = 0.010$ and $\Delta\theta = 0.093$ and with the ratio of the areas covered by bright and dark filaments equal to 0.75 (Kjeldseth-Moe and Maltby 1974). This model is based on observations of penumbral filaments by Muller (1973).

Grossmann-Doerth and Schmidt (1981) and Callados et al. (1987) have studied the intensities in bright and dark penumbral filaments. It is apparent that the bright (dark) filaments within the very same sunspot are not equally bright (dark). This led Grossmann-Doerth and Schmidt (1981) to argue that their results were in conflict with those of Muller (1973). Callados et al. (1987) confirm the spread in brightness between the filaments, but find that their results are in general agreement with those of Muller (1973).

3. Semi-empirical Models

In semi-empirical models the atmosphere is assumed to be either in hydrostatic or in hydrodynamical equilibrium and the temperature-optical depth relation is adjusted until the calculated intensities account for the observations.

Consider first one limitation learned from one-component model calculations for the photosphere. The calculated intensities should ideally agree with both the observed spectral intensity distribution and the observed limb darkening. However, if the observations are of sufficiently high quality, one is be able to detect inconsistencies. A one-component model adjusted to fit the limb-darkening observations differs slightly from a model adjusted to fit the spectral energy distribution. The reason is that the photosphere consists of granules and intergranular lanes and we "see" down to different depths in bright and dark regions, a fact that will influence the limb-darkening observations. This means that a first approximation to the temperature stratification may be derived from a one-component model, but a two-component model including a height difference between dark and bright columns is necessary to account for both sets of observations in detail.

Semi-empirical models of umbrae and penumbrae are usually constructed by comparing theoretical umbra/photosphere intensity ratios with the corresponding ratios derived from observations. The alternative approach with a direct comparison of umbral intensities on an absolute scale with calculated spectra requires extremely good knowledge of the spectral absorption in the Earth's atmosphere.

We shall first discuss a one-component model for the umbral core. The observations suggest that the umbral core temperature may differ from one sunspot to another. Hence, a complete discussion will require more than one core model. In this paper the emphasis will be on the core model that fits the observations for an average large sunspot.

Next, we shall take into account the presence of umbral dots and discuss a few aspects of two-component umbral models, like the area covered with umbral dots at different heights.

In penumbral semi-empirical models one has to consider that a non-vertical magnetic field may introduce a lifting force. Furthermore, the marked filamentary structure of the penumbra calls for two or more components in the horizontal direction.

3.1. UMBRAL CORE MODEL

In semi-empirical umbral core models the atmosphere is assumed to be in hydrostatic equilibrium along a vertical magnetic field such that the field does not influence the stratification.

The easiest way to construct a model for the umbral core is by scaling the quiet Sun model to another effective temperature. This approach has been used by several authors, most recently by Sobotka (1988) who sets $\Delta\theta$ = constant, where $\Delta\theta = 5040$ $(1/T_u - 1/T_{ph})$ and T_u and T_{ph} are respectively the temperature of the umbra and the photosphere. The advantage of this method is that the $\Delta\theta$ value characterizes the entire model. The disadvantage is that the model does not fit the observations (see below). Other scaling methods like T_u/T_{ph} = constant have been tried, but unfortunately not found to be satisfactory.

Figure 4 shows the temperature-optical depth relations presented by van Ballegooijen (1984), Maltby et al. (1986), Obridko and Staude (1988) and Sobotka (1988) for the umbral core. Since these models have been calculated with different computer codes and input parameters, differences other than those caused by the different temperature-optical depth relations may occur. In order to minimize these other effects the models have been recalculated using the same computer code.

A comparison between calculated continuum intensities and the observed continuum intensities at the center of the solar disk is presented in Figure 5. It is apparent that all four models may account for the observations in the visible and near infrared part of the spectrum up to approximately 1 μm. However, the model by Sobotka (1988) with $\Delta\theta = 0.45$ shows considerably lower intensities than those observed in the far infrared. The model by Obridko and Staude (1988), which is based on the model by Staude et al. (1983), predicts intensities that are too low at wavelengths above 1.7 μm. Both the models by van Ballegooijen (1984) and Maltby et al. (1986) may account for the observations shown in Figure 5. A recent model by Ming-de and Cheng (1990) was not included in Figures 4 and 5; the model is quite similar to the model by Sobotka (1988).

Figure 6 shows the difference in continuum umbra/photosphere intensity ratios betweeen disk center and heliocentric angle of 60 degrees. A reasonable fit to the observations is obtained with the models by van Ballegooijen (1984) and by Maltby et al. (1986). The deviations between the calculated and observed values are significant for the two other models, most apparently for the model by Sobotka (1988).

For three out of the four models we show in Figure 7 the radiative flux density, integrated over wavelength, as a function of optical depth at the reference wavelength λ

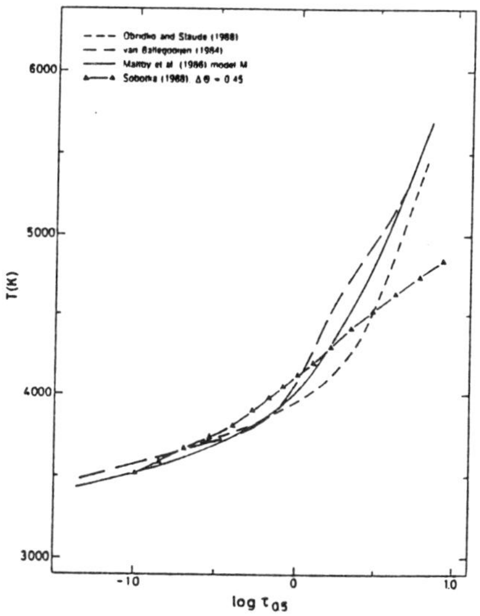

Fig. 4. Relations between umbral core temperature T and optical depth $\tau_{0.5}$ for the models by van Ballegooijen (1984), Maltby et al. (1986), Obridko and Staude (1988), and Sobotka (1988).

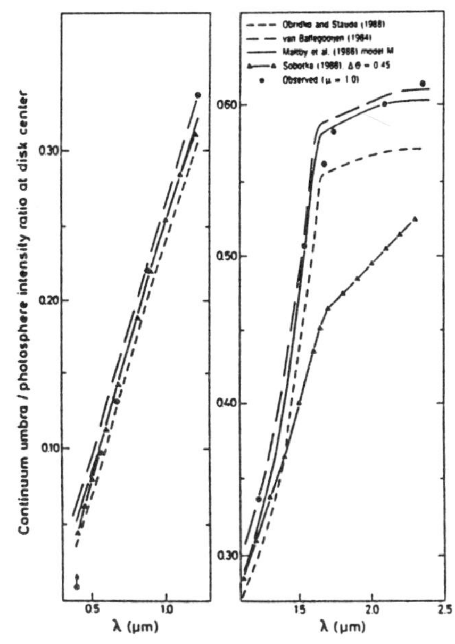

Fig. 5. Observed continuum umbra/photosphere intensity ratios at disk center compared with the predicted intensity ratios for four different umbral core models. The same computer code was used to obtain the four sets of intensity ratios.

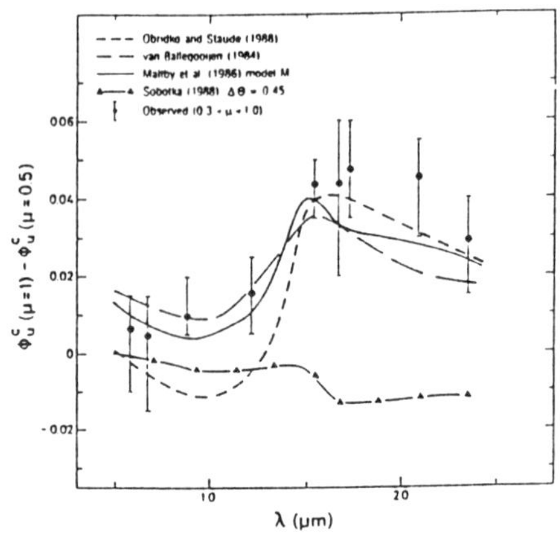

Fig. 6. Differences in continuum umbra/photosphere intensity ratio, $\phi_u(\mu = 1) - \phi_u(\mu = 0.5)$, between the disk center ($\mu = 1$) and heliocentric angle $\theta = 60^o$ ($\mu = 0.5$) vs. wavelength for four models. The observed average values are plotted for comparison, with error bars that indicate the total range of measured values.

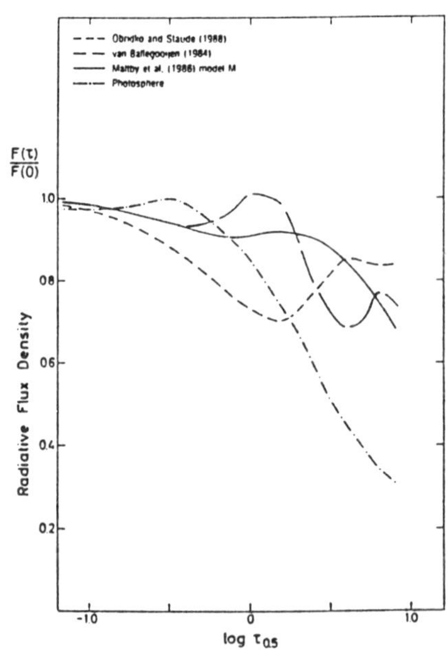

Fig. 7. Depth dependence of the radiative flux density for three umbral core models. The result for the quiet Sun is shown for comparison.

$= 0.5\ \mu m$. For comparison the radiative flux in the photosphere is shown. Whereas both the photospheric model and the umbral model by Maltby et al. (1986) vary relatively smoothly with depth, the models by van Ballegooijen (1984) and Obridko and Staude (1988) show more complicated variations. For the models that are within the uncertainties introduced by the observational limits, a transition from radiative to another form of energy transport occur close to the solar surface. The transition appears to occur at a larger optical depth in the umbral core than in the photosphere.

The semiempirical umbral and photospheric models give the stratification in the umbra and the photosphere, respectively. In order to make a comparison of theoretical and empirical sunspot models meaningful, a value for the Wilson depression is needed.

3.2. TWO-COMPONENT UMBRAL MODEL

The presence of umbral dots have initiated investigations of two-components models (e.g. Adjabshirzadeh and Koutchmy, 1983; Obridko and Staude, 1988). Each umbra is regarded as consisting of a dark main component and a bright component that originates in umbral dots. In the two-component working model of Obridko and Staude (1988) their dark main component is identical to their model discussed above and the bright component is given by $\Delta\theta = 0.10$, covering a fraction $\beta = 0.05$ to 0.10 of the umbral area. The fraction β changes only slightly with height. Obridko and Staude (1988) assumed a Wilson depression of 500 km (200 km) between the optical depths $\tau_{0.5} = 0.1$ in the umbral main component (bright component) and the undisturbed photosphere.

In other two-component models by Pahlke and Wiehr (1990) and Sobotka (1988) the bright component has a range in temperature between that of the photosphere and photospheric temperature minus 500 K and between $\Delta\theta = 0.03$ and 0.11, respectively. Pahlke and Wiehr (1990) set $\beta = 0.05$, whereas Sobotka (1988) increases β with decreasing sunspot size from 0.05 up to 0.30 for the smallest sunspots considered.

Several questions arise in connection with the two-component umbral models. The temperature of the dark component may change from one sunspot to another (Sobotka 1988). This finding agrees with the detected solar cycle variation.

The intensity of the bright component is uncertain. We know that umbral dots in the central part of large sunspots show less contrast than peripheral umbral dots. It is possible that the distribution of umbral dot intensities is such that the bright component changes in temperature from one umbral dot to another and from one sunspot to another.

The number of umbral dots is observed to change from one sunspot to another. This will change the value of β from 0.05 for large sunspots to 0.30 for small sunspots according to Sobotka (1988). Furthermore, it is unlikely that the fraction covered with bright emission remains constant with height since the cooling time is short close to the temperature minimum region.

3.3. PENUMBRAL MODELS

The penumbra is much closer in temperature to the photosphere than is the umbra. Accordingly, a first approximation to a penumbral model may be derived by scaling the quiet Sun model. With $\Delta\theta = 0.055$ a good fit of the calculated to the observed penumbra/photosphere intensity ratios is obtained. Such a model is also able to account for the spectral lines formed at low heights (Kjeldseth-Moe and Maltby 1969). The model shows that the penumbra, like the photosphere, is not in radiative equilibrium in the deepest observable layers.

The more recent one-component penumbral models of Yun et al. (1984) and Ding and Fang (1989) are primarly aimed at modelling the penumbral chromosphere. In passing we note that these models are too cold in deeper layers to account for the continuum observations.

In the penumbra the filament structure is so evident that it is tempting to construct a two-component model. The model should agree with the observed intensities in dark and bright filaments. As a check on the model the calculated intensity as averaged over penumbral filaments may be compared with observations. In Figure 3 the observed intensities may be compared with the results for a two-component model with $\Delta\theta_b = 0.010$, $\Delta\theta_d = 0.093$ and the ratio of the areas covered by bright and dark filaments equal to 0.75 (Kjeldseth-Moe and Maltby 1974). This model is based on observations of penumbral filament intensities by Muller (1973).

Evidently the intensity may change from one bright penumbral filament to another within the same penumbra so that the two-component model must be used with care. The construction of a two-component model also brings up the question of keeping the two atmospheres in balance in the horizontal direction over an extended height range. This implies that the sum of the gas pressure and the magnetic pressure must be the same in bright and dark filaments. As an additional effect, the gas pressure stratification may be influenced by a variation in curvature of the magnetic field lines with depth that differs in the two components. It is interesting that large fluctuations in inclination angle for the penumbral magnetic field have been reported (see paper by Title in this volume).

4. Concluding remarks

The present observations of umbral dots show conflicting results. With the aim of understanding the observations the hypothesis has been presented that the umbral core is nearly free of umbral dots and that umbral cores only occur in sunspots above a certain size.

Semi-empirical models give us knowledge about the temperature stratification and the radiative flux density as a function of depth. The models indicate that the sunspot regions are in radiative equilibrium in the upper layers, but other energy transport processes take over close to the solar surface.

The sunspot brightness is primarily determined by the energy transport processes, but the energy equation is mathematically coupled to the momentum equation. Along the magnetic field lines the gas is nearly in hydrostatic balance, but not quite since flows are observed. The largest contribution to the lateral gas pressure difference between the photosphere and the umbra comes from the magnetic field, that acts both through the magnetic pressure $B^2/2\mu$ and through the magnetic tension. The importance of this relation was first pointed out by Alfvén (1943), who found that static equilibrium in a vertical magnetic field is only possible if the temperature in the field is lower than in the environment. Using semi-empirical models one finds that the lateral pressure difference reaches a maximum that exceeds the magnetic pressure at heights relatively close to an optical depth of unity in the umbra. The exact value depends on the adopted value for the Wilson depression.

A future goal could be to extend the model calculations to more than one dimension and to compare computer simulations of sunspot spectra with observations with high spatial and spectral resolution.

References

Adjabshirzadeh, A. and Koutchmy, S. (1983) 'Photometric analysis of sunspot umbral dots', Astron. Astrophys. 122, 1-8.

Albregtsen, F., Jorås, P.B., and Maltby, P. (1984) 'Limb-darkening and solar cycle variations of sunspot intensities', Solar Phys. 90, 17-30.

Albregtsen, F., and Maltby, P. (1978) 'New light on sunspot darkness and the solar cycle', Nature, 274, 41-42.

Albregtsen, F. and Maltby, P. (1981) 'Constraints on umbral core models as derived from broad-band intensity observations' in L.E. Cram and J.H. Thomas (eds.), The Physics of Sunspots, Sacramento Peak Obs. Conf., Sunspot, N.M., pp. 127-139.

Alfvén, H. (1943) 'On the effect of a vertical magnetic field in a conducting atmosphere', Arkiv Mat. Astr. Fysik 29, No.11, 1-6.

Avrett, E.H. (1981) 'Reference model atmosphere calculation: the Sunspot sunspot model', in L.E. Cram and J.H. Thomas (eds.), The Physics of Sunspots, Sacramento Peak Obs. Confer., Sunspot, N.M., pp 235-255.

Birkle, K., and Mattig, W. (1965) 'Über den einfluss der bildunruhe auf die intensitätsverteilung in sonnenflecken', Zeitsch. Astrophys. 60, 264-272.

Bogdan, T.J., Gilman, P.A., Lerche, I., and Howard, R. (1988) 'Distribution of sunspot umbral areas: 1917-1982', Astrophys. J. 327, 451-456.

Bray, R.J. and Loughhead, R.E. (1964) Sunspots, Chapman and Hall, London.

Brynildsen, N., Engvold, O., Hansteen, V., and Brault, J.W. (1991) 'An umbral spectrum of a large sunspot - $\sigma\sigma = 8000 - 1800cm^{-1}$', Inst. Theoret. Astrophys., Oslo, (in press).

Bumba, V., Sobotka, M., and Simberová, S. (1990) 'Photometry of the internal structure of a large sunspot', in L. Dezsø (ed.), 'The Dynamic Sun', Publ. Debrecen Heliophys. Obs. 7, 84-87.

Collados, M., del Toro, J.C., and Vázquez, M. (1987) 'The intensity distribution in sunspot penumbras' in in E.-H. Schröter, M. Vázquez, and A.A. Wyller (eds.),

Proc. Inaugural Workshop D-E-S Telescope Install. Canary Islands, Cambridge Univ. Press, Cambridge, pp. 214-218.

Cram, L.E. and Thomas, J.H. (eds.) (1981) 'The Physics of Sunspots', Sacramento Peak Obs. Conf., Sunspot, N.M.

Cox, A.N., Livingston, W.C., and Matthews, M. (1992) The Solar Interior and Atmosphere, Univ. Arizona Press., Tucson (in press).

Ding, M.D. and Fang, C. (1989) 'A semi-empirical model of sunspot penumbra', Astr. Astrophys. 225, 204-212.

Ekmann, G. (1974) 'The relation between intensities of umbra and penumbra of large sunspots', Solar Phys. 38, 73-75.

Ewell, M.W. (1991) 'Infrared CCD observations of umbral dots', Bull. Amer. Astr. Soc. 23, 938-938.

Grossmann-Doerth, U. and Schmidt, W. (1981) 'The brightness distribution in sunspot penumbrae', Astron. Astrophys. 95, 366-372.

Grossmann-Doerth, U., Schmidt, W. and Schröter, E.H. (1986) 'Size and temperature of umbral dots', Astron. Astrophys. 156, 347-353.

Garcia de la Rosa, J.I. (1987) 'Established and non-established properties of umbral and penumbral fine structures' in E.-H. Schröter, M. Vázquez, and A.A. Wyller (eds.), Proc. Inaugural Workshop D-E-S Telescope Install. Canary Islands, Cambridge Univ. Press, Cambridge, pp. 140-155.

Henry, J., and Alexander, S. (1846) 'Experiments on the spots on the Sun', London, Edinburgh, and Dublin Phil. Mag. J. Sci. 28, 230-232.

Iuell, P. and Staveland, L. (1975) 'Correction of an irregular sunspot or facula for stray light', Inst. Theor. Ap. Oslo Rept. No. 43, pp.1-32.

Kjeldseth Moe, O., and Maltby, P. (1969) 'A model for the penumbra of sunspots', Solar Phys. 8, 275-283.

Kjeldseth-Moe, O., and Maltby, P. (1974) 'The temperature of penumbral filaments', Solar Phys. 36, 101-108.

Krat, V.A., Karpinsky, V.N., and Pravdjuk, L.M. (1972) 'On the sunspot structure', Solar Phys. 26, 305-317.

Kusoffsky, U. and Lundstedt, H. (1986) 'Lifetimes of umbral dots in sunspots', Astron. Astrophys. 160, 51-55.

Lites, B.W. (1984) 'The color temperature of a sunspot penumbra' Solar Phys. 90, 1-12.

Lites, B.W., Bida, T.A., Johannesson, A.,and Scharmer, G.B. (1991) 'High-resolution spectra of solar mahnetic features. II. Magnetic fields of umbral brightenings', Astrophys. J. 373, 683-694.

Lites, B.W., Skumanich, A., Rees,D.E., Murphy, G.A., and Carlsson, M. (1987) 'Stokes profile analysis and vector magnetic fields. III. Extended temperature minimum of sunspot umbrae as infered from Stokes profiles of MgI λ4571', Astrophys. J. 318, 930-939.

Livingston, W. (1991) 'Radial filamentary structure in a sunspot umbra', Nature 350, 45-46.

Loughhead, R.E., Bray, R.J., and Tappere, E.J. (1979), 'Improved observations of sunspot umbral dots', Astron. Astrophys. 79, 128-131.

Makita, M., and Morimoto, M. (1960) 'Photoelectric study of sunspots', Publ. Astr. Soc. Japan. 12, 63-76. **12**, 63.

Maltby, P. (1971) 'The effect of scattered light on solar intensity observations as derived from 9 May, 1970 Mercury transit', Solar Phys. 18, 3-21.

Maltby, P. (1972) 'Observations of the intensity of the penumbra of sunspots', Solar Phys. 26, 76-82.

Maltby, P., Avrett, E.H. Carlsson, M., Kjeldseth-Moe, O., Kurucz, R.L., and Loeser, R. (1986) 'A new sunspot umbral model and its variation with the solar cycle', Astrophys. J. 306, 284-303.

Mattig, W. (1971) 'Observations of stray-light and sunspot intensities during the Mercury transit of 1970 May 9', Solar Phys. 18, 434-442.

Mihalas, D. (1978) Stellar Atmospheres, Freeman, San Francisco.

Ming-de, D. and Cheng, F. (1990) 'A semi-empirical model of sunspot umbra', Acta Astron. Sinica 31, 276-283.

Minnaert, M. and Wanders, A.J.M. (1932) 'Zur theorie der sonnenflecke', Zeitschr. Astrophys. 5, 297-320.

Muller, R. (1973) 'Étude photométrique des structures fines de la pńombre d'une tache solaire', Solar Phys. 32, 409-420.

Mykland, N. (1973) 'Relative umbral intensity of a large sunspot', Solar Phys. 28, 49-60.

Myrheim, J., and Rue, H. (1991) 'New algorithms for maximum entropy image restoration', Computer Vision, Graphics and Image Processing (in press).

Nordlund, Å., and Stein, R.F. (1990) 'Solar magnetoconvection' in J.O. Stenflo (ed.) Solar Photosphere, Convection and Magnetic Fields, IAU Symp. 138, pp. 191-210.

Obridko, V.N. (1985) Sunspots and Activity Complexes (in Russian), Nauka, Moscow.

Obridko, V.N. and Staude, J. (1988) 'A two-component working model for the atmosphere of a large sunspot', Astron. Astrophys. 189, 232-242.

Pahlke, K.-D. and Wiehr, E. (1990) 'Magnetic field, relative Doppler shift and temperature for an inhomogeneous model of sunspot umbrae' Astron. Astrophys. 228, 246-252.

Rossbach, N., and Schröter, E.H. (1970) 'On the dependence of sunspot minimum intensity on area', Solar Phys. 12, 95-100.

Schüssler, M. (1980) 'Flux tube dynamo approach to the solar cycle', Nature 288, 150-152.

Shine, R.A., Title, A., Tarbell, T. and the SOUP Team (1987) 'Sunspot observations from the SOUP instrument on Spacelab 2' in G. Athay and D.S. Spicer (eds.) Theoretical Problems in High Resolution Solar Physics II, NASA Conf. Publ. 2483, pp. 133-141.

Skilling, J. (1984) 'The Cambridge maximum entropy algorithm' in J.H. Justice (ed.), Maximum Entropy and Bayesian Methods in Applied Statistics, Cambridge University Press, pp. 179-193.

Sobotka, M. (1988) 'Two-component models of individual sunspot umbrae', Bull. Astr. Inst. Czechosl. 39, 236-242.

Stellmacher, G. and Wiehr, E. (1988) 'Spectrophotometric analysis of small sunspots', Astron. Astrophys. 191, 149-153.

Stix, M. (1989) The Sun, Springer Verlag, Berlin.

Stumpff, P. (1961) 'Photometrie des kontinuierlichen spektrums von sonnenflecken im spektralbereich 4000Å-8600Å', Zeitschr. Astrophys. 52, 73-109.

Van Ballegooijen, A.A. (1984) 'On the temperature structure of sunspot umbrae', Solar Phys. 91, 195-218.

Wiehr, E. (1987) 'Modelling active region finestructures from observations' in L. Hejna and M. Sobotka (eds.), The Sun, Publ. Astro. Czech. Acad. Sci., 66, pp. 29-32.

Yoshimura, H. (1983) 'Solar cycle dynamo wave origin of sunspot intensity and X-ray bright point number variation', Solar Phys. 87, 251-259.

Yun, H.S., Beebe, H.A., and Baggett, W.E. (1984) 'A model of a penumbral chromosphere', Solar Phys. 92, 145-151.

Zwaan, C. (1965) 'Sunspot models', Rech. Astr. Obs. Utrecht 17, Part 4, 1-182.

OBSERVATIONS OF THE MESOSCALE MAGNETIC STRUCTURE OF SUNSPOTS

A. SKUMANICH
High Altitude Observatory
National Center for Atmospheric Research[1]
P.O. Box 3000
Boulder, Colorado 80307, U.S.A.

ABSTRACT. Recent inferences of the mesoscale structure of spot magnetic fields based on high spectral resolution Stokes vector polarization profiles for both visual and infrared lines are reviewed and partly considered from the point of view of self-similarity. The Evershed flow is briefly examined with regard to the non-uniqueness of the inferred intrinsic flow field and its relation to the inclination of the penumbral magnetic field.

1. Introduction

As is well known, the concept of self-similarity has been usefully exploited in the theoretical analysis of sunspot structure, e.g., Yun (1968). In this brief review, I reconsider this concept with regard to recent determinations of the lateral magnetic vector structure of fairly isolated single spots as inferred from the inversion of spectrally well-resolved Stokes-vector polarization profiles. The data has a spatial resolution of $\sim$ 3.6 Mm (5") and thus are averages over the finer structures present in spots.

The merit of using spectrally resolved Stokes-vector polarization profiles to infer the vector magnetic field is that profile asymmetries and scattered light can be explicitly accounted for and that no profile shape assumptions are imposed by the calibration procedure, refer Lites et al. (1988) and references therein. In addition, it is possible to determine an absolute velocity zero point using telluric absorption lines. The analytic solution used in the least squares fit to the Stokes profiles assumes that the field is continuous through the line-forming region. This assumption may not hold in the outer penumbral regions where the field may have a canopy-like structure because of the 'passage' of a current sheet through the layers of formation. However, recent synthetic studies by Skumanich, Grossman-Doerth, and Lites (1992) have shown that reasonable estimates of the field magnitude occurs in such 'canopy' cases if one restricts the fit to the three polarization parameters (refer Skumanich and Lites 1987) by excluding or reducing the role of the intensity profile. The use of the fill factor parameter allows for such a procedure, but only approximately. Lateral gradients are necessarily averaged over the resolution element. The exact effect of such a non-linear averaging on the field magnitude and geometry are currently unknown. Thus, the nature of the outer penumbra requires a more careful analysis than currently available.

---

[1]The National Center for Atmospheric Research is sponsored by the National Science Foundation.

J. H. Thomas and N. O. Weiss (eds.), Sunspots: Theory and Observations, 121–137.

2. Scaling of Umbral Field

Nicholson's umbral field magnitude, |B|, measurements (Nicholson 1933), using the Zeeman splitting of Fe I λ6173, showed an essentially constant strength for large spots, but a drop of field magnitude with decreasing spot size for umbral radii R_u < 5 Mm (or penumbral radii R_p < 12Mm). Brants and Zwaan (1982) showed that Nicholson's measurements for small spots were affected by systematic instrumental effects. The points labeled by Z in Figure 1 show the Brants and Zwaan measurements using a purely umbral absorption line, Ti λ606.46 nm. Nicholson's measurements for spots greater than 5 Mm are shown by the symbol N with the range in size and field as indicated. The range in field values is believed to be correlated with a range in umbral brightness (Zwaan 1968) and is not a fine structure effect. Also indicated in Figure 1 are recent Stokes profile inferences (except for Adam – see below) by other observers (identified by letter symbols).

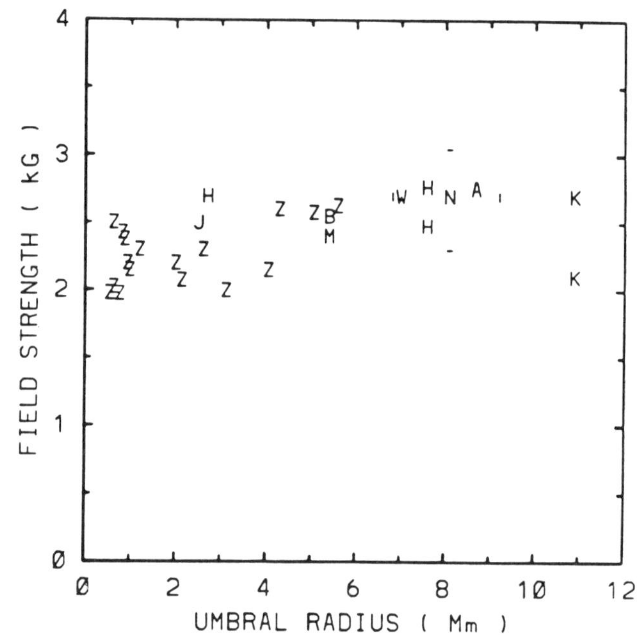

Figure 1. Measured maximum umbral field strength (kG) vs. umbral radius (Mm). The letters refer to measurements by: Adam (1990) — A (R_p = 22Mm); Beckers and Schröter (1969) —B (R_p = 17 Mm); Lites and Skumanich (1990) — H (R_p = 19Mm); Lites et al. (1991) — H (R_p = 6.7Mm); Johannessan et al. (1991) — J (R_p = 6.8Mm); Kawakami (1983) — K (R_p = 19Mm); Mickey (1991) — M (R_p = 13Mm); Nicholson (1933) — N (R_p = 2.4R_u); Wittmann (1974) — W (R_p = 17Mm), and Brandts and Zwaan (1982) —Z (only R_u given). The horizontal and vertical bars show the range in both variables for Nicholson's data. $R_p(R_u)$ is the penumbral (umbral) radius.

It is apparent that for quasi-steady spots there is an approximate invariance of umbral field strength with spot size. This is consistent with photometric observations of a similar invariance, at least for spots with R_u > 3.6 Mm where scattered light effects are minimal, of umbral brightness with size (Maltby et al. 1986). These observations support the use of the concept of umbral invariance in describing the structure of developed spots.

3. Scaling of Lateral Structure

3.1. FIELD MAGNITUDE AND ZENITH ANGLE

Representative of the Stokes vector profile technique for visual spectral lines is the analysis by Lites and Skumanich (1990, hereafter LS) and Arena et al. (1990). LS present data on a large, fairly symmetric spot of 09 March 1980 with penumbral radius $R_p = 19$ Mm and with a spatial resolution of $\simeq 3.6$ Mm (5″) obtained with the High Altitude Observatory (HAO) Stokes II polarimeter. The inferred magnetic field distribution, $|\mathrm{B}(R)|$, with cylindrical radius from spot center is indicated by the open circles in Figure 2 with different azimuths representing the fluctuations ($\simeq \pm$ 200G rms) at a given radius. The various curves indicate the (interpolative mean) field distribution found by other observers, after scaling to the 09 March '80 central value of 2480 G, as identified in the legend. It is evident that the recent observations fall significantly below the solid line of Beckers and Schröter (1969, hereafter BS), as was the case for most of the earlier measurements discussed by BS[2] The 2.2 Mm (3″) wide 'drop' in the field strength inferred by BS at the penumbral edge was not detected by the two data sets which extend past $R/R_p = 1.0$, viz. Kawakami (1983) (long dash curve) and LS (points). This may be due to the lack of sufficient spatial resolution and the effects of seeing. Except for this issue of the penumbral 'edge' the recent observations do not confirm the high penumbral fields of BS.

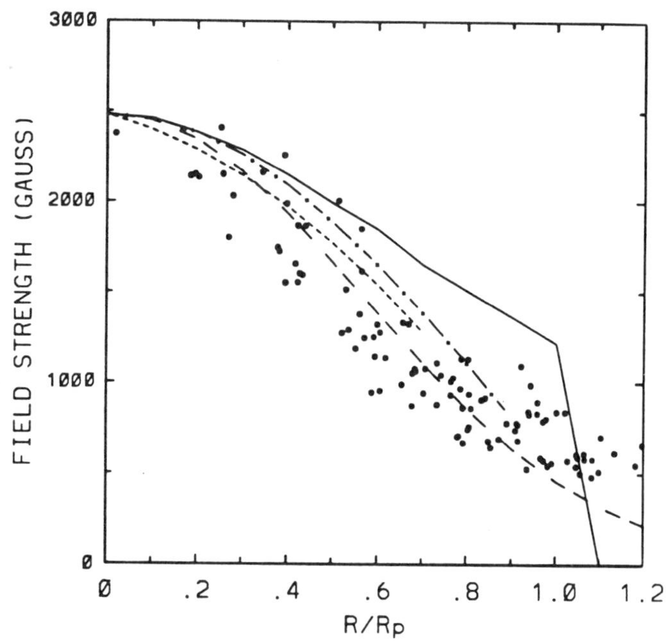

Figure 2. Inferred magnetic field strengths (G) (scaled to 2480G central value) vs. scaled radial distance from spot center for different spots. Dots are results for the LS data at different azimuths; solid curve — B; short dash — A; long dash — K; dot-dash — W. For identification of letter code see Figure 1.

The radial distribution of the field zenith angle or inclination, ζ, to the local solar *vertical* and azimuth, relative to the local cylindrically radial (from spot center) direction, for the 09 March '80 spot are plotted as circles and stars, respectively, in Figure 3. The curves are as identified in Figure 2. Here the Wittman (1974) inversion, uppermost curve, appears to deviate the most.

[2]It is to be noted that the BS results are based on an inversion of only two Stokes parameter profiles (I, V) while the field angles for Adam are based on a direct geometric determination of the polarization ellipse, spectrally averaged over the separate σ profiles, which in turn is a non-linear function of the Stokes parameters.

The two BS (solid) curves refer to two different inversion techniques. It would appear that the geometric 'neutral-line' method, which is model free and gives the lower solid line, compares more favorably with the results of Stokes vector inversion of LS. Thus, the more recent inversions do not yield fields as horizontal throughout the penumbra as suggested by Wittman. It should be noted that both scattered light *and* fill-factor effects have been *included in* the LS inversion, but *not in* the Kawakami inversion (dashed curves). This would account for the higher LS field strengths evident in the outer penumbral regions. For $R/R_p \gtrsim 0.7$, the putative (Chapman 1943) relation cited by BS as representing their data, namely, $\zeta = (R/R_p)90°$, does not approximate the LS values, nor for that matter, the Nicholson data as presented by Chapman. Note that here, seeing and resolution effects yield only average values. A consideration of the relative azimuth (stars) shows that the horizontal field component for this spot appears to be essentially radial although there may be a weak azimuthal twist of $\sim 3°$ in the penumbra.

3.2. FLUX DISTRIBUTION: POTENTIAL AND OTHER REPRESENTATIONS

Any invariance of the lateral distribution of the scaled field magnitude, $|\mathbf{B}(R/R_p)|/|\mathbf{B}(0)|$, for different size spots can be confused by the existence of different degrees of azimuthal twist in spots. An alternate comparison is to consider the distribution of magnetic flux, i.e., of the vertical field component, $B_z(R/R_p)$. In Figure 4 we consider the flux-distribution for two spots of significantly different sizes.

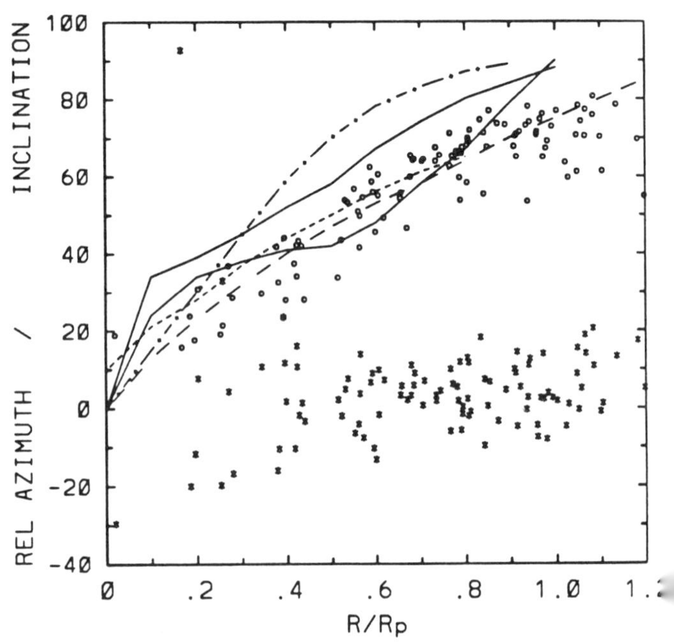

Figure 3. Magnetic field zenith angle or inclination from local vertical vs. scaled radius for the same spots as in Figure 2. The azimuth relative to the local radius for the LS spot is indicated by the asterisks.

The development of a new HAO Stokes vector polarimeter (the Advanced Stokes Polarimeter, hereafter ASP), which has a spatial resolution of $\simeq 0.7$ Mm (1″) at the Sacramento Peak Vacuum

Tower Telescope, and records wavelength and slit position profiles simultaneously, allows one to study smaller spots than was possible with the previous HAO Stokes II vector polarimeter. An engineering prototype was used to observe a spot on 04 June '90 with $R_p = 6.7$ Mm (Lites et al. 1991). The B_z distribution inferred for this spot is shown by the open (limb side) and filled (center side) circles in Figure 4. For comparison, the solid line indicates the B_z-distribution for the 15 September '80 spot (LS), as represented by a dipole potential fit, scaled only by R_p (= 19 Mm) and not in magnitude. Except for a magnetic feature with an associated 15° smaller zenith angle at $R/R_p \cong 0.6$ (center side), the agreement is quite good. The spatial resolution , in units of R_p, is comparable for both spots so that the penumbral 'edge' of BS, if it has the same relative scale (i.e., $\Delta R = 0.15R_p$) in both spots, would not be resolved in this spot either. However, the effects of seeing were reduced for the 04 June '90 spot by means of an active mirror. Thus, it would appear that the flux-distribution is essentially the same for these two significantly different size spots, and supports the concept of a radial self-similarity of $B_z(R/R_p)$ as proposed by Gokhale and Zwaan (1972).

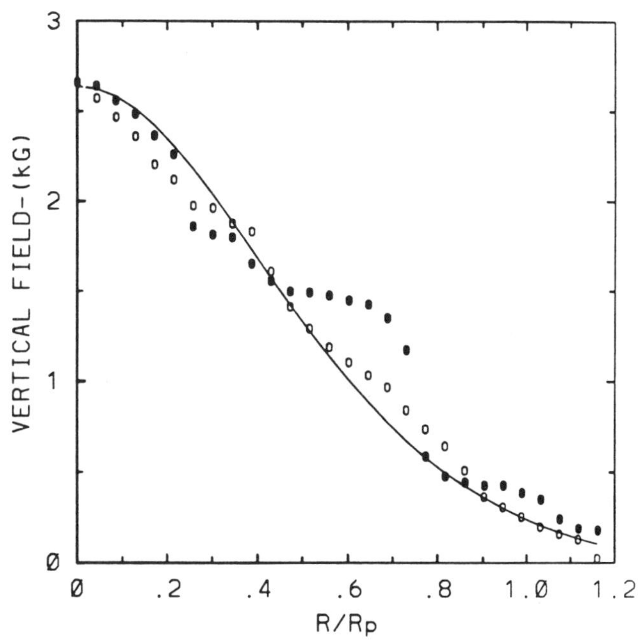

Figure 4. Comparison of inferred vertical field from ASP data ($R_p = 6.7$Mm) with that from Stokes II data ($R_p = 19$Mm).

The magnetic 'field line' function for the dipole representation used in Figure 4 is given by

$$u(R,z) = (1/2)B_{00}z_0^2\left(\frac{|z_0|}{z+|z_0|}\right)\left(\frac{\alpha^2}{(1+\alpha^2)^{3/2}}\right)$$

where $\alpha = R/(z+|z_0|)$, $z = 0$ is the spot photosphere, $z = -|z_0|$ is the depth of the dipole, B_{00} is the axial ($R = 0$) photospheric ($z = 0$) field and $B_z = (1/R)\partial u/\partial R$, $B_R = -(1/R)\partial u/\partial z$. For the 15 September '80 spot LS found ($B_{00} = 2740G, |z_0| = 0.965R_p$) for $0 \leq R \leq R_p$. I note in passing that the use of a buried dipole to represent spot fields was first proposed by Kiepenheuer (1938), see also Broxon (1942, 1949). A potential fit to $B_z(R/R_p)$ allows one to assess the degree to which the horizontal field deviates from that given by the potential representation. Figure 5 presents the dipole fit (solid curve) to the photospheric B_z data (open circles) over the range $0 \leq R \leq R_p$ for the 9 March '80 spot. Here the values are ($B_{00} = -2480\text{G} \pm 90\text{G}, |z_0| = (0.90 \pm .03)R_p$). The cited parameter errors are based on a fit to normal points which reduce the azimuthal fluctuations. The resulting potential radial field, B_R, (dashed curve) can be compared with the observed horizontal field B_h (stars). As LS have noted, this particular spot appears to be close to a potential state. The slight excess of the observed horizontal field in the outer penumbra may be interpreted, using $B_R = B_h$ cos (rel. az.), as the presence of a twist of $\simeq 20°$. This is larger than that shown in Figure 3 and may indicate the inapplicability of the potential fit. Nonetheless, the dipole field yields a local representation that is useful for various purposes. For example, as an interpolatory representation of B_R one can use the dipole fit to calculate $(1/R)\partial RB_R/\partial R$ which, with $\nabla\cdot \mathbf{B}= 0$, yields the result that $dB \equiv -R_p(dlnB_z(R = 0, z)/dz)_{z=0} = 3R_p/|z_0| = 3.3$. Since $|z_0| = O(R_p)$ one has that the magnetic flux $\Phi_p = 2\pi u(R_p, 0) \simeq (1/2\sqrt{2})B_{00}\pi R_p^2$ which yields the flux scaling coefficient 0.35 compared to 0.39 as given in Allen (1973).

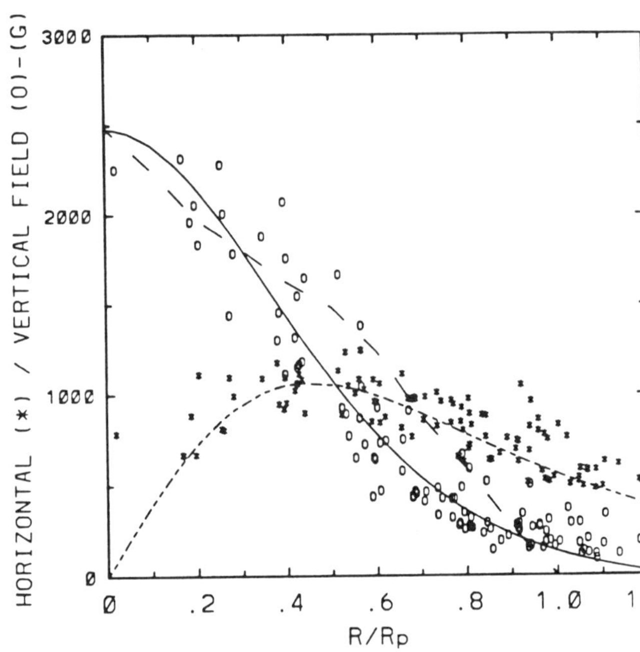

Figure 5. Comparison of LS data (circles and asterisks) with dipole fit (solid) to vertical field and with predicted radial field (dash). Beckers and Schröter B_z field is indicated by the long dash curve.

I note that if one wants to assign physical significance to the dipole field one can 'cut' it along the field line passing through the penumbral radius. With the field truncated by a current sheet supported in static equilibrium by a gas pressure jump, one finds a spot pressure deficiency distribution, $\Delta P(z) = P_{\text{external}}(z) - P_{\text{spot}}(z)$, with depth that differs by a factor of 2 or less from extant models as calculated by Jahn (1989), refer Skumanich (1991). It would appear that more accurate determinations of the spot field are needed if one is to use them to specify the nature of the subsurface current sheet.

In spite of the significant differences in field magnitude and zenith angle, the BS flux distribution is similar to other observed B_z distributions except in the penumbra. This can be seen in Figure 5 where their flux distribution is indicated by the dashed curve. I have used the BS inclinations based on the neutral-line method so that B_z should be less affected by model dependent effects. Except for the Wittman data, the other observations in Figures 2 and 3 yield similar B_z values as for the 09 March '80 spot and are not plotted.

Since the Schlüter-Temesváry (1958, hereafter ST) exponential flux distribution representing vertical self-similarity has had much currency in theoretical discussions of spot structure, we consider it as a fit to the 09 March spot. The ST field line function is given by

$$u(R,z) = (1/2)B_{00}\xi_0^{-2}\ \{1 - \exp\ [-R^2\xi^2(z)]\}\ .$$

Here $\xi(z)$ is the ST similarity variable (the structure function) and $\xi_0 = \xi(0)$. This field function yields the photospheric distributions at $z = 0$,

$$B_z(R, z=0) = B_{00}\ \exp[-R^2\xi_0^2]$$
$$B_R(R, z=0) = 0.5RB_z(R, z=0)(-2\xi_0'/\xi_0)$$

with $\xi_0' = (d\xi/dz)_{z=0}$. One also has that $dB = (-2\xi_0' R_p/\xi_0)$.

By fixing B_{00} to the observed maximum value for the 09 March '80 spot, viz $B_{00} = 2480$G one finds a least-squares value for $\xi_0 R_p = 1.80 \pm 0.07$ (for $0 \leq R \leq R_p$) which is reasonably consistent with the value of 1.63 based on imposing the flux scaling relation discussed above, Yun (1968). This fit is illustrated in Figure 6 by the solid curve. Solving for dB pointwise by equating $B_R(R,0)$ to the observed $B_R(R)$ and using the B_z (fit), one finds dB varies from 3.3 in the inner umbra (in agreement with the dipole fit) to 5.0 at $R = .65R_p$ with a mean of 4.0 $\pm$ 0.7. This shows that ξ_0' varies by $\sim$ 35% in this range. By choosing $dB = 4.0$ one finds a resulting B_R distribution as indicated by the dashed curve. The outer penumbra, $R \geq 0.75R_p$, is poorly fit with the chosen dB, at $.85R_p, dB = 7.8$, which implies a significant 'break down' of vertical self-similarity in these regions. The umbral value of $dB(= 3.3)$ is in agreement with the value, $dB = 3.3$, calculated by Flå et al. (1982, hereafter FOS) for an unstressed spot using the ST representation. For a laterally stressed spot with a Wilson depression of 0.5 Mm, FOS found a lower eigenvalue of $dB = 2.2$, indicating that stressed fields extend higher into the atmosphere. Thus, here also one would conclude that a current-free description of the spot field is a reasonable approximation. Note that for the ST representation given above, the flux scaling relation $\Phi(R_p) = 2\pi u(R_p, 0) = 0.30 B_{00}\pi R_p^2$, yields a scaling coefficient 14% lower than with the dipole representation. The canonical value of $\xi_0 R_p = 1.63$ is based on the scaling coefficient of 0.35.

There have been a variety of efforts led by Osherovich, refer to FOS for references, to 'fix' the ST representation in the outer penumbra. We postpone consideration of these efforts to the appendix; suffice to say here that they have not been overly successful.

4. Twisted Spots

The issue of the nature of the twist, or azimuthal field, in spots has been most difficult to address due to the sensitivity of the Stokes linear polarization parameters (Q, U), to magneto-optic, instrumental cross-talk, scattered light, and other noise effects. Beckers and Schröter used the geometry of the observer's magnetic neutral line, i.e., where V_λ passes through zero, and found a possible weak twist (or magnetic "vorticity") of $\sim 6°$ in the horizontal field. Wittman also found "vorticity" variations but these were not described in the solar frame. Inversion of the Stokes-vector profiles, taking into full account: (1) cross-talk, i.e., a full 4x4 calibration matrix, (2) magneto-optics (Macaluso-Corbino) effect, and (3) scattered light and fill-factor effects for the 09 March '80 spot also showed the possible presence of a weak "vorticity," i.e., a non-zero relative azimuth of the horizontal field as is evident in Figure 3. Such a "vorticity" was more evident in the observations of the 15 September '80 spot (refer to LS). Interpreting the "vorticity" or azimuthal field as due to a field aligned Ampere current, *i*, one can derive the lateral variation of the current for comparison with theoretical assumptions such as that of Yun (1971). To facilitate such a comparison one can express the Ampere current in terms of the dipole field function, u, derived above from the B_z-distribution. In Figure 7, the data points show the dependence of the normalized Ampere current, i/i_p, on the normalized potential u/u_p, where $i_p = 8.48TA = u(R_p)/R_p$.

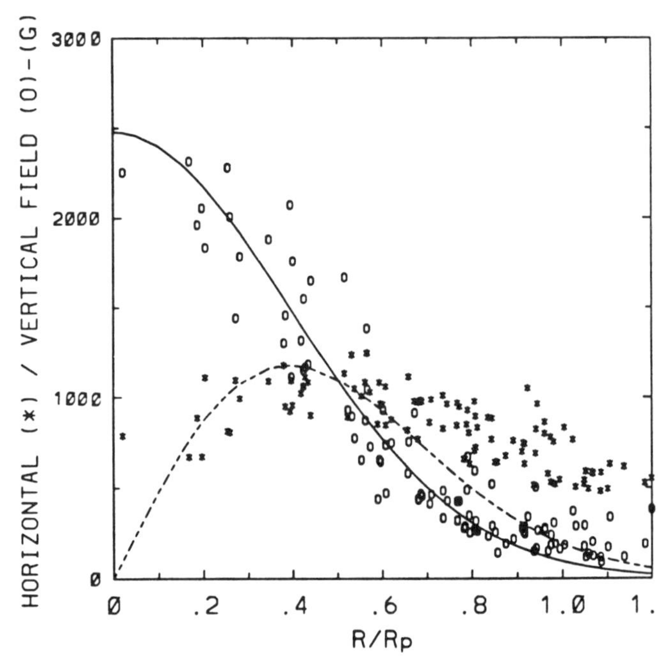

Figure 6. Comparison of LS data with exponential ST fit (solid) to vertical field and with predicted radial field (dash) using a scale factor $dB = 4.0$.

The solid curve represents the distribution assumed by Yun scaled downward by 6.4. It appears that the quantity $A = R_p di/du \approx$ constant within the spot, i.e., for $R \lesssim 0.7R_p$, and is a good approximation to the various assumed theoretical forms for $i(u)$. In the case of Yun one has that $A = 2.0$, a factor of 6.4 larger than the value of 0.3 indicated for the 15 September '80 spot. Thus the effects derived by Yun, which depend quadratically on A, *q.v.* Skumanich (1991) for a rediscussion of these, are unlikely to occur.

5. Comparison of Vector Fields Derived from 12μm and Visual Stokes Vector Profiles: Height Variation

The discovery of a narrow non-LTE emission reversal in the core of a broad but weak absorption line of Mg I at 12.32μm which is Zeeman sensitive has been a boon to the determination of essentially error-free magnetic field *magnitudes* in the penumbra of sunspots. The reversal is absent in the inner umbra but this is not a great loss as visual Stokes observations are most reliable there. Deming and coworkers, (D. E. Jennings, T. Hewagama, G. Wiedemann, P. Zipoy, among others) have developed a Stokes-vector polarimeter for the $\lambda 12\mu$m line and observed (Hewagama et al. 1992) a visually symmetric sunspot on 28 October 1989. The nominal spatial resolution was 3.3 Mm (4.″5), while the penumbral (umbral) radius was 13 (5.9) Mm. Seeing was below the diffraction limit of 1.6 Mm (2.″2) and image excursions were within $\pm$ 0.7 Mm (1″). Visual (λ6302) Stokes-vector observations of the same spot were made by D. Mickey, with the Haleakala vector polarimeter on 27 October '89 with a similar spatial resolution. Hewagama (1991) inverted the $\lambda 12\mu$ Stokes-vector profiles using an optically thin homogeneous emitting slab model with the inclusion of a fill factor derived from the percentage polarization. I consider his results here (kindly made available by Hewagama) and, following Hewagama, compare them with Mickey's inversion which used the Lites and Skumanich least squares algorithm but without any correction for scattered light or fill factor (D. Mickey, private communication). A re-reduction of this data by Mickey with the scattered light and fill factor option is now available but yet to be analyzed.

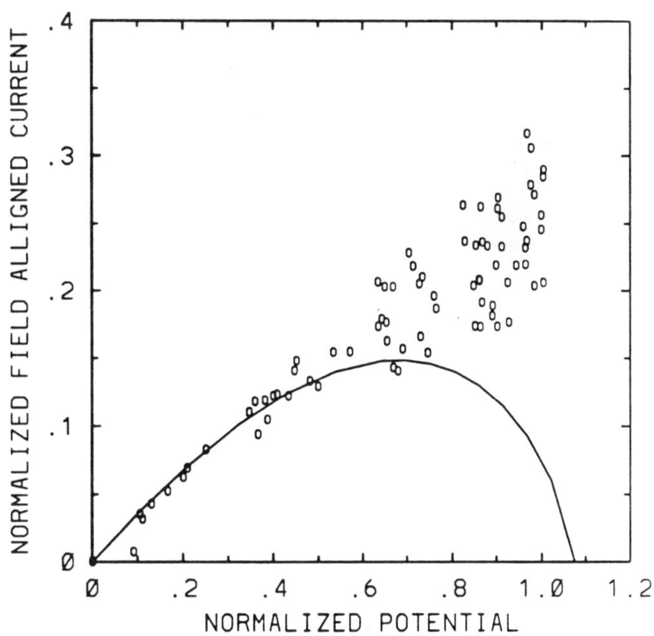

Figure 7. Comparison of field aligned current derived by LS (open circles) with theoretical form of Yun.

The Mickey λ6302 results for the magnetic field magnitude for the southern segment of the spot (azimuth position angles from Solar west of $-46°$ to $-116°$, i.e., clockwise) which contains the $\lambda 12\mu$m linear raster scan (azimuth $\simeq -90°$) is shown in Figure 8 by the open circles. The $\lambda 12\mu$m results are shown by the filled circles. Also shown, scaled only by R_p (= 19 Mm) are the λ6302 results for the 09 March '80 spot from Figure 2 as represented by the interpolation fit (dashed curve) given by Lites and Skumanich. The corresponding zenith angles are shown in Figure 9.

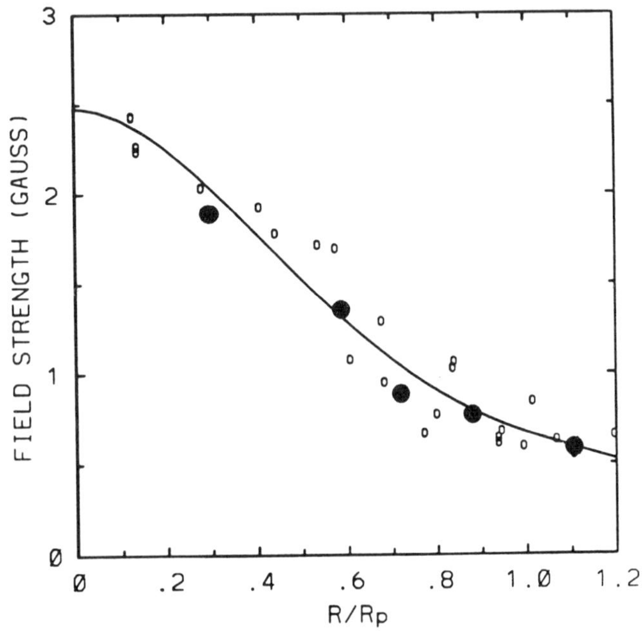

Figure 8. Comparison of field strengths derived from $\lambda 12\mu$m data (filled circles) with those from λ630 nm data (open circles) for 27/28 October 1989 spot (R_p = 13 Mm). The scaled curve is the LS result for 09 March 1980 spot (R_p = 19 Mm).

The close similarity of the $\lambda 12\mu$m data to both the 09 March '80 spot and the Mickey data is evident. Note that both λ6302 derived fields are quite similar, giving further evidence to lateral self-similarity. The tendency of the $\lambda 12\mu$m data to be at the lower envelope of the λ6302 data is, Hewagama believes, real, and implies that the field is somewhat weaker, especially in the umbra. Using various interpolation functions to fit the $\lambda 12\mu m$ $B(R)$ data shown in Figure 8 one finds an extrapolated central value $B_{00}(\lambda 12\mu m) = 2200\text{G} \pm 200\text{G}$. The error is that derived for a non-linear least-squares procedure. The λ630 nm value derived from the LS data, reduced to normal points, is $B_{00}(\lambda 630\text{nm}) = 2480\text{G} \pm 90\text{G}$. This yields $\Delta B_{00} = 280\text{G} \pm 290\text{G}$ which, with $dB = 3.3$ derived above and $R_p = 13$ Mm yields $\Delta z = 440\text{km} \pm 440$ km for the relative height difference between the two lines. These values differ somewhat but not significantly from those of Hewagama et al (1991). This Δz value is in good agreement with the difference in the height of formation, Δh, of the 12μm line relative to the continuum in the quiet Sun as derived by Chang et al. (1991) and Carlsson et al. (1992).

6. Non-Uniqueness of Penumbral Flow and Velocity Zero Point

It is common knowledge that the reconstruction of the intrinsic penumbral vector velocity field from its projection onto the line of sight is non-unique and requires additional information, e.g., a sequence of projections obtained by a center-to-limb sequence, Schröter, 1965, and the assumption of temporal stability. Lacking such a sequence one has to make certain assumptions about the geometry of the intrinsic field. In addition, for observations without an absolute wavelength reference, an assumption must be made regarding the velocity zero point, i.e., the standard of rest. Presumably, one wants a rest frame in which the spot has no mean peculiar translational (or rotational) motion. Beckers (1962) called attention to the fact that the surrounding photosphere is not a suitable rest standard since it is not at rest relative to a coordinate frame rotating with the angular speed appropriate to the latitude in question. This is due to a spatial average over the granulation which produces a blue shift of $\simeq 0.3$ km/s. By applying this correction, Schröter (1965) removed a limb vs. disk-center asymmetry in his penumbral velocities.

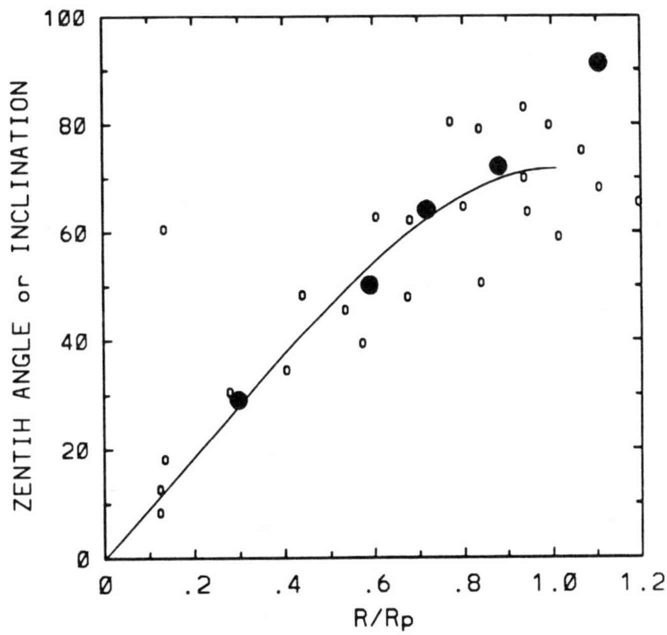

Figure 9. Comparison of zenith angle or inclination for same data as in Figure 8.

The Stokes inversion method also yields the relative, to some arbitrary fiducial, wavelength position (of the mean bisector) of the line profile *if* the terrestrial fiducial lines are ignored. Note that the average over the spatial resolution implies an average over any primary and satellite lines (i.e., two or more velocity components). In the following, I present a study by Sánchez Almeida of HAO of the Stokes derived relative velocity shift for the 09 March '80 spot. If one presumes that the radial velocity at the spot center is zero and that the intrinsic flow refers to locally horizontal filaments, then one finds that the limb (squares) and center-side (stars) penumbra have an intrinsic outflow speed vs. radius dependence as indicated in Figure 10a. A significant limb-center asymmetry is found. If one assumes the intrinsic flow is field aligned and uses the field geometry shown in Figure 3, then Sánchez finds, Figure 10b, that the limb side shows an outflow while the center side a weak inflow, i.e., the asymmetry remains. However, if one assumes that at the spot center an outflow (blue shift) of 1/2 km/s is present then a field-aligned flow shows little, if any, asymmetry, Figure 10d. The asymmetry remains if the spot center has an inflow (red shift) of 1/2 km/s, Figure 10c.

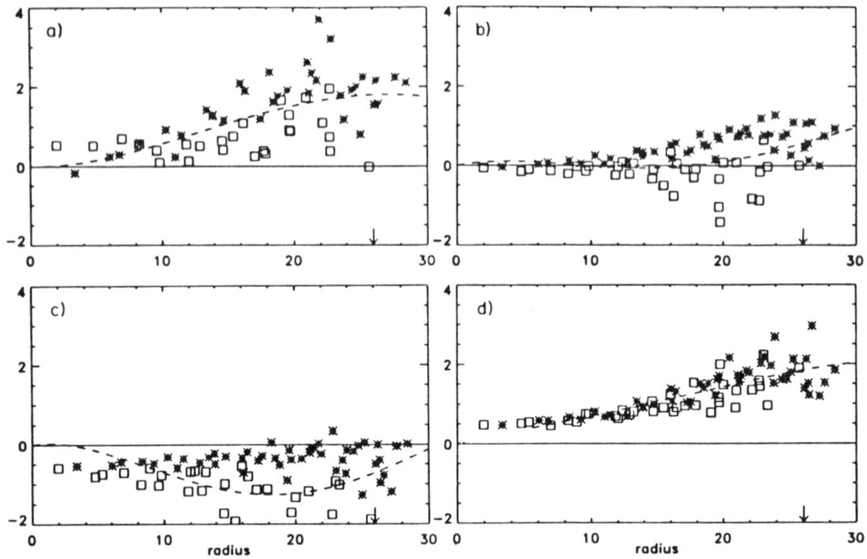

Figure 10. Inferred Evershed flow (km/s) vs. radius (arcseconds) at different azimuths for 09 March '80 spot under different assumptions, limb side data are indicated by squares, center-side by asterisks. The arrow indicates penumbral radius. Case a: Horizontal flow (v $\perp$ to g) with spot center at rest. Case b: Field aligned flow (v $\|$ to **B**) with center at rest. Case c: Field aligned with center redshifted by $-1/2$ km/s. Case d: Field aligned with center blueshifted by $+1/2$ km/s.

We have demonstrated here the role of the adopted rested frame in the interpretation of the inferred Evershed flow. These results compliment the discussion of 15 September '81 spot by Arena et al. (1990) who noted similar rest frame difficulties in the *observers frame* description of the Evershed flow. Note that Title et al. (1992) *assume* an azimuthal symmetry to derive a unique Evershed flow.

7. Summary

It is found that recent determinations of the magnetic field structure of developed spots confirms the extant concepts of umbral field invariance to spot size as well as lateral structure self-similarity. Vertical self-similarity as represented by the Schlüter-Temesváry model is found to hold approximately but only for regions inside $0.7R_p$. Within the current accuracy of the data, spots appear to have a potential or force-free field distribution suggesting that the interaction between the spot and its surroundings takes place in a boundary current sheet. If such a current sheet were to pass as a canopy through the line-forming layers with a zenith angle of $\zeta_{cs} = tan^{-1}(\Delta x/\Delta h) \approx tan^{-1}$ (2.2Mm / .4Mm) $\simeq$ 80° one would be able to represent the Beckers and Schröter penumbral 'edge'. Such an angle is consistent with the recent observations reviewed here keeping in mind the fact that the observations are averages over the azimuthal structure reported by Title et al. at this conference. Further discussion of this issue requires a more careful treatment of the inversion of canopy-like field geometries, refer Skumaich et al. (1992).

The field magnitude distribution derived from the well-split $\lambda 12\mu$m Mg I line confirm the general lateral shape derived from model dependent λ630.2nm Fe I line although in both cases the spatial resolution is insufficient to see the penumbral 'edge' detected by Beckers and Schröter. In detail, the $\lambda 12\mu$m derived central field appears to be 10% ($\pm 10\%$) lower than the visually derived field while the lateral width of the distribution is 3% ($\pm 7\%$) larger. The central field difference suggests a height difference of 0.4Mm ($\pm$0.4Mm) between the two lines. The larger lateral size is consistent with the prediction of a 5% size increase by the ST solutions of FOS for the derived height difference. It is shown that care must be exercized in assigning a rest frame to non-absolute Evershed velocities if one is to avoid a limb-disk center, i.e., earth direction dependent, asymmetry to the spatial distribution of such flows.

8. Appendix

In an effort to improve the ST model in the outer penumbra, Osherovich (1982) introduced a return-flux term into the ST field-line function. If $-X(\geq 0)$ is a measure of the fractional amount of return-flux, then the return-flux function can be written as

$$u(R,z) = (1/2)B_{00}\xi_0^{-2}(1-X)^{-1}\{1-[1+XR^2\xi^2(z)\hat{f}(z)]\,\exp[-R^2\xi^2(z)]\}$$

Here $\hat{f}(z)$ is the Osherovich function such that $d\hat{f}(z)/dz = -2\sqrt{2}\hat{f}(z)\xi(z)$ with $\hat{f}(0)=1$. This representation yields the photospheric distributions

$$B_z(R, z=0) = B_{00}Q(R^2\xi_0^2, X)\,\exp(-R^2\xi_0^2)$$
$$B_R(R, z=0) = 0.5RB_z(R, z=0)\{(-2\xi_0'/\xi_0) + 2\sqrt{2}[-X]\xi_0[1-X(1-R^2\xi_0^2)]^{-1}\}$$

Here $Q(x,y) = 1 + xy(1-y)^{-1}$ is a slowly varying bounded function of (x,y) in the domain of interest. One also has that $dB = (-2\xi_0' R_p/\xi_0) + 2\sqrt{2}\xi_0 R_p[-X(1-X)^{-1}]$. Thus, $(-2\xi_0'/\xi_0)$ is augmented in B_R by the return-flux term which has its maximum contribution at $R \sim R_p$.

A least-squares fit to $B_z(R)$ of the 09 March '80 spot for the two parameters $(\xi_0 R_p, X)$, keeping B_{00} fixed at the maximum value as before, converged very slowly to the solution with $X=0$, i.e., the ST fit had the minimum χ^2. The problem is the increasing departure of B_z(fit) from the data in the outer penumbra with increasing $-X$. In view of this problem we consider X at a fixed value and determine $\xi_0 R_p$ from B_z by least-squares, keeping B_{00} fixed, and solve the B_R equation for $-2\xi_0' R_p/\xi_0$ at each R. One finds, as in the ST case above, that no constant value is found. By considering the largest value of $-X$ that allows a minimum variation in $-2\xi_0' R_p/\xi_0$ maximally over R one finds that $-X \simeq 1.0$. The case $-X = 0.62$, which yielded $\xi_0 R_p = 1.47$ from B_z, with a mean value $-2\xi_0' R_p/\xi_0 = 1.9 \pm 0.3$ is illustrated in Figure 11. In the inner umbra $dB = 3.2$ as expected. There is some improvement in the representation of B_R compared to that of ST in Figure 6 but it is not much and a strongly varying ξ_0' in the outer penumbra is still found, $-2\xi_0' R_p/\xi_0 = 3.0$ at $.75R_p$. One must note that ξ_0' is an eigenvalue of the lateral force balance equation and depends both on the Wilson depression and the value of X. The value of $-X = 5.56$ as used by FOS, which is based on the identification of the umbra-penumbra boundary as the transition surface from open to closed (return) field lines, appears to be inadmissible by the data. Indeed, Livingston's (1991) recent demonstration that penumbral structure can be seen to extend deep into the umbra, if the exposure is suitable, shows that the umbra-penumbra boundary is not an invariant feature in visual images. In addition, soft X-ray observations show inner penumbral flux tubes extending very large distances from the spot to other active regions (Sams et al.1991).

Figure 11. Comparison of LS data with fit represented by return-flux potential. Solid curve is fit to vertical field with fixed central field ($B_{00} = 2480$G) and fixed return-flux $X = -0.62$. Dash curve is predicted radial field with (mean) $-\xi_0' R_p/\xi_0 = 1.9$.

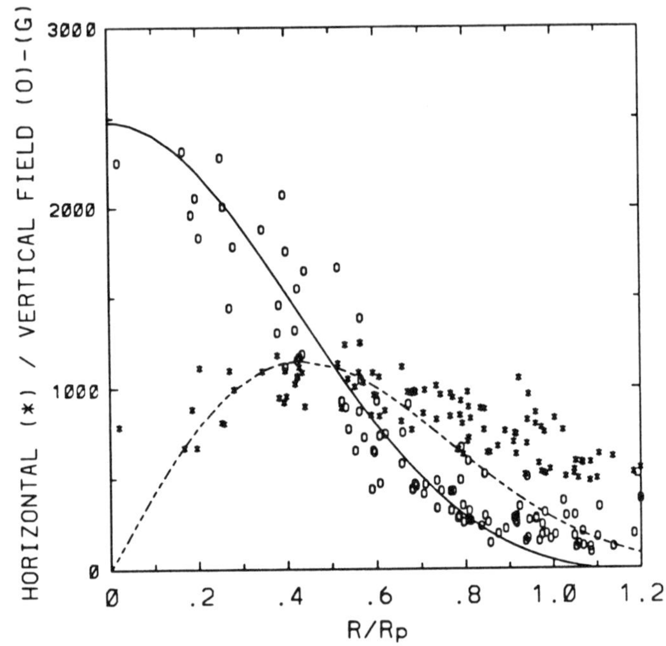

In a recent paper, Osherovich and Garcia (1990, hereafter OG), present a further claimed improvement to the return-flux representation of spot structure. By allowing the X parameter, the return-flux, to be a function of z which is determined from a second non-linear differential equation in addition to that for $\xi(z)$, a third parameter appears in the expression for B_R. In particular, the *numerator* of the second term (the return flux term) changes from $2\sqrt{2}[-X]\xi_0$ to $2\sqrt{2}[-X(0) - X'(0)(2\sqrt{2}\xi_0)^{-1}]\xi_0$. This does not change the *shape* of this term only its *amplitude*. This forces ξ_0' to continue to change both in amplitude and shape with R. OG present values that represent a fit to the Kawakami data. Except near $1.0R_p$ this data, as can be seen from Figures 2 and 3, is essentially the same as our 09 March '80 data which we have considered here. Note that the OG value for $\xi_0 R_p$ must be misprinted; I believe their intended value is 1.37, although I find a larger value of 1.47, if I fix $B_{00} = 2480$G and $X = -0.62$ (essentially their values). Their parameters do not improve on the constancy of ξ_0' for the data considered here. Note that the Kawakami data differs from the HAO data in the outer penumbra in the direction to improve the applicability of the vertical self-similarity. However, lack of fill factor corrections, we believe, account for the difference.

Of course, three parameters should allow better curve fitting. It is difficult to see what has been gained further. A new function[3] $X(z)$ is to be determined from the pressure distribution in the outer penumbra which is not easy to derive from observations. A more natural recourse is to solve 2D-lateral force balance pointwise in (R, z) as illustrated by Pizzo (1986) or Jahn (1989).

The ST or return flux (with $X = -1$) models are useful in that they provide an initialization or benchmark structure (or topology) for 2D numerical force balance solutions. They could also prove useful in the inversion of the transfer equation in the penumbra, see Skumanich (1991). Certainly they provide an *approximate* representation of the mesoscale structure of spots but fine tuning the model, as some propose, is a fruitless exercise.

[3] Actually this redefines the $\hat{f}(z)$ function.

Acknowledgments

The author is grateful to B. C. Low, D. Degenhardt, J. Sánchez Almedia, and two anonymous referees for their careful reading of and comments on this review. He is also indebted to L. Croom for her careful and competent transcription to camera-ready form.

References

Adam, M. G. 1990, "An observational examination of models for sunspot magnetic field configurations," *Solar Phys.*, **125**, 37-44.

Allen, C. W. 1973, *Astrophysical Quantities*, 3rd ed., University of London, The Athlone Press, London.

Arena, P., Landi Degl'Innocenti, E., and Noci, G. 1990, "Velocities and magnetic fields observed in a sunspot," *Solar Phys.*, **129**, 259-276.

Beckers, J. M., and Schröter, E. H., 1969, "The intensity, velocity and magnetic structure of a sunspot region. IV. Properties of unipolar sunspot," *Solar Phys.*, **10**, 384-403.

Brants, J. J. and Zwaan, C. 1982, "The structure of sunspots. IV. Magnetic field strengths in small sunspots and pores," *Solar Phys.*, **80**, 251-258.

Broxon, J. W. 1942, "Relation of the cosmic radiation to geomagnetic and heliophysical activities," *Phys. Rev.*, **62**, 508-522.

__________ 1949, "Magnetic moments of sunspots," *Phys. Rev.*, **78**, 606-611.

Carlsson , M., Rutten, R. J., and Shchukina, N. G. 1992, "The formation of the Mg I emissin features near 12μm," *Astron. Astrophys.*, in press.

Chapman, S. 1943, "Magnetism in the Sun's atmosphere," *Mon. Not. R. Astr. Soc.*, **103**, 117-130.

Chang, E. S., Avrett, E. H., Mauas, P. J., Noyes, R. W., and Loeser, R. 1991, "Formation of the infrared emission lines of Mg I in the solar atmosphere," *Ap. J. Lett.*, **379**, L79-82.

Flå, T., Osherovich, V. A., and Skumanich, A. 1982, "On the magnetic and thermodynamic consequences of the return-flux sunspot model," *Ap. J.*, **261**, 700-709.

Gokhale, M. H. and Zwaan, C. 1972, "The structure of sunspots. I: Observational constraints: current sheet models, " *Solar Phys.*, **26**, 52-75.

Hewagama, T. 1991, Ph.D. thesis, "An infrared polarimetric study of sunspots," University of Maryland.

Hewagama, T., Deming, D., Jennings, D. E., Osherovich, V., Wiedemann, G., Zipoy, D., Mickey, D., and Garcia, H. 1991, "Solar magnetic field studies using the 12μm emission lines. II. Stokes profiles and vector field samples in sunspots," preprint.

Jahn, K. 1989, "Current sheet as a diagnostic for the subphotospheric structure of a spot," *Astron. Astrophys.*, **222**, 264-292.

Johannesson, A., Bida, T., Lites, B., and Scharmer, G. B. 1991, "Very high spatial resolution two-dimensional solar spectrocopy with video," submitted to *Astron Astrophys.*

Kawakami, H. 1983, "Polarimetric study of unipolar sunspots," *Publ. Astron. Soc. Japan*, **35**, 459-489.

Kiepenheuer, K. O. 1938, "Zur dynamik der Sonnenprotuberanzen," *Zeitschrift für Astrophysik*, **15**, 53-68.

Lites, B. W. and Skumanich, A. 1990, "Stokes profile analysis and vector magnetic fields. V. The magnetic field structure of large sunspots observed with Stokes II," *Ap. J.*, **348**, 747-760.

Lites, B. L., Skumanich, A., Rees, D. E., and Murphy, G. A. 1988, "Stokes profile analysis and vector magnetic fields. IV. Synthesis and inversion of the chromospheric Mg I *b* lines," *Ap. J.*, **330**, 493-512.

Lites, B. W., Elmore, D., Murphy, G., Skumanich, A., Tomczyk, S., and Dunn, R. B. 1991, "Preliminary results from the HAO/NSO advanced Stokes polarimeter prototype observing run," in *Solar Polarimetry*, proc. of 11th National Solar Observatory/Sacramento Peak Summer Workshop, ed. L. J. November, Sunspot, N.M., p. 3-15.

Livingston, W.C. 1991, "Radial filamentary structure in a sunspot umbra," *Nature*, **350**, 45-47.

Maltby, P., Avrett, E. H., Carlsson, M., Kjeldseth-Moe, O., Kurucz, R. L., and Loeser, R. 1986, "A new sunspot umbral model and its variation with the solar cycle," *Ap. J.*, **289**, 284-303.

Mickey, D. 1991, private communication.

Nicholson, S. B. 1931, "The area of a sunspot and the intensity of its magnetic field," *P.A.S.P.*, **45**, 51-52.

Osherovich, V. A. 1982, "A new magneto-hydrostatic theory of sunspots," *Solar Phys.*, **77**, 63-68.

Osherovich, V. A. and Garcia, H. A. 1990, "Electric current in a unipolar sunspot with an untwisted field," *Geophys. Res. Let.*, **17**, 2273-2276.

Pizzo, V. J. 1986, "Numerical solution of the magnetostatic equations for thick flux tubes, with application to sunspots, pores, and related structures," *Astrophys. J.*, **302**, 785-808.

Sams, B. J., Golub, L., and Weiss, N. O., "Coronal structure above sunspots," submitted to *Astrophys. J.*

Schlüter, A. and Temesváry, S. 1958, in *IAU Symposium 6, Electromagnetic Phenomena in Cosmical Physics*, ed. B. Lehnert (Cambridge: Cambridge University Press), p. 263.

Schröter, E. H. 1965, "Der Evershed-effekt in einem inaktiven symmetrischen Einzelfleck," *Zeitschrift für Astrophysik*, **62**, 228-255.

Skumanich, A. 1991, "Physical models of solar and stellar spots," in *Lecture Notes in Physics*, proc. of Armagh Bicentenneary Colloquium, Springer-Verlag, Berlin, in press.

Skumanich, A., Grossman-Doerth, U., and Lites, B. W. 1992, "Stokes inversion in the presence of canopy-like structures and unresolved flux tubes," in *Methodes de Determination des Champs Magnetiques Solaires et Stellaires*, eds. M. Faurobert, H. Frisch, and N. Mein (Obs. de Paris-Meudon), in press.

Skumanich, A., and Lites, B. W. 1987, "Stokes profile analysis and vector magnetic fields. I. Inversion of photospheric lines," *Ap. J.*, **322**, 473-482.

Title, A. M., Frank, Z. A., Shine, R. A., and Tarbell, T. D. 1992, "Fine structure of sunspot magnetic fields and velocities," in *Sunspots: Theory and Observations*, proc. of NATO Advanced Research Workshop, submitted.

Wittmann, A. 1974, "Computation and observation of zeeman multiplet polarization in Fraunhofer lines. III: Magnetic field structure of Spot Mt. Wilson 18488, *Solar Phys.*, **36**, 29-44.

Yun, H. S. 1970, "Theoretical sunspot models," *Astrophys. J.*, **162**, 975-986.

Yun, H. S. 1971, "A magnetostatic sunspot model with 'twisted' field," *Solar Phys.*, **16**, 398-403.

Zwaan, C. 1968, "The structure of sunspots," *Ann. Rev. Astr. & Astrophy.*, **6**, 135-164.

MAGNETOHYDROSTATIC EQUILIBRIUM IN SUNSPOT MODELS

K. JAHN
Max Planck Institut für Astrophysik
Karl Schwarzschild Str. 1, 8046 Garching, FRG
and
Warsaw University Observatory
Al. Ujazdowskie 4, 00-478 Warsaw, POLAND

ABSTRACT. This review describes global models of sunspots in overall magnetostatic equilibrium, extending over many scale heights into the solar convection zone. Different explanations of the low temperature of spots are illustrated by quantitative models, and new methods of solving the equation of magnetostatic equilibrium for an isolated flux tube are presented. These global models make it possible to probe the deep structure of a spot. Particular attention is paid to the structure of a deep penumbra. The hypothesis that energy transport within the penumbra occurs through convection by an exchange of magnetic flux tubes is used in order to construct a magnetostatic approximation for the penumbra. We describe the principal features of a global sunspot model in which this approximation is used. Finally, the controversial topology of the magnetic field below the photosphere is discussed.

1. Introduction

Magnetic flux ropes, emerging through the solar photosphere and forming active regions with bipolar sunspot groups, have a very complex structure. Despite this complexity, sunspots, the largest concentrations of magnetic flux, have recurrent features. They consist of two distinct regions that have different photometric and magnetic characteristics. An umbra, the darkest and relatively most homogeneous part of a spot, is permeated by a strong and almost vertical magnetic field. It is cooler than the solar photosphere by about 2000 K and forms a photometric plateau with a sharp edge independently of its shape. A penumbra contains a weaker and more horizontal magnetic field. It has an intermediate brightness and usually surrounds the umbra. Although it can often be very irregular or even partly missing, it always has a filamentary structure, usually with a radial orientation.

These features are common to all spots, although the interaction of coagulated or neighbouring magnetic flux in sunspot groups may deform a configuration and obscure some typical properties. However, occasional long–lived spots, sufficiently separated from other large flux concentrations, tend to be round and to have regular, ring–like penumbrae. Therefore an isolated, axially symmetric spot with a unipolar magnetic field, often considered by theoreticians as a model prototype, does not seem to be an unrealistic oversimplification. The surface characteristics of such a model should comply with the observations. The magnetic field strength should amount to 2500–3000 gauss in the centre of the umbra, decrease to one half or one third of that value at the outer edge of the penumbra, and drop abruptly to zero in the quiet Sun. The distribution of the heat flux should reflect a rapid transition from the value of 20% of the solar heat flux in the uniform umbra to about 75% in the

J. H. Thomas and N. O. Weiss (eds.), Sunspots: Theory and Observations, 139–162.

penumbra.

The lifetime of sunspots is much longer than any dynamical time scale in the solar photosphere. Some sunspots persist for several rotations of the Sun in essentially unchanged form, whereas perturbations that propagate with the Alfvén speed would need about an hour to cross a large sunspot at its photosphere. That justifies the assumption of global magnetohydrostatic equilibrium in a spot, at least for the upper part of the emerging flux tube. The assumption becomes questionable deep below the photosphere, near the base of the convection zone of the Sun, because any information about perturbations in those layers would need 30–40 days to reach the photosphere, if the gas density at the base is equal to 0.2 g/cm^3 and a field strength of ten kilogauss is assumed. Most quasi-static models do not, however, extend to such deep layers of spots.

This review concentrates on models describing the overall magnetostatic structure of a spot extending many scale heights below the photosphere. Any evolutionary effects are neglected. The reader is referred to the review by Wilson (1981), who discussed several models taking into account the growth and the decay of sunspots, as well as models that include a general twist of the magnetic field, which is neglected here. A summary of theories of dynamical phenomena in sunspots has been presented by Thomas (1981). Spruit (1981a) presented a review of basic theoretical problems posed by observed properties of spots. More recently, the structure of sunspots has been discussed by Moreno-Insertis (1987) in the context of the origin of solar active regions.

In the next section general approaches to the construction of magnetostatic equilibria in the convection zone are briefly recalled. Basic concepts concerning the coolness of spots and their utilization in early models are discussed. Section 3 presents new methods for numerical treatment of the equation of magnetohydrostatic equilibrium, developed in the past decade. Some of the results are recalled and possibilities of diagnosing the subphotospheric layers of a spot are discussed. Section 4 is devoted to the particular problem of penumbral structure. It presents a magnetostatic approximation for this region, based on the concept of convection by an exchange of flux tubes. Finally, controversies concerning the morphology of the subphotospheric magnetic field in spots and some implications for heat flow are discussed in Section 5.

2. The Magnetostatic Approximation for a Sunspot

The term magnetostatic equilibrium, as utilized in existing sunspot models, refers to the overall structure of a spot, averaged horizontally over inhomogeneities of dynamic origin which are much smaller than the size of a spot. Most of these are due to magnetoconvective processes that are responsible for the supply of a large part of the energy emitted from a spot. They are clearly visible in the penumbral photosphere, which is composed of alternating bright and dark filaments. This time-dependent filamentation shows an upward motion of bright matter with respect to dark (Beckers, 1981), and a correlation between the photometric and magnetic signals (Title, 1989; Title *et al.*, 1992) which indicates the convective nature of the penumbra. Convection in the uniform umbra is less evident. An upper limit for the convective blueshift of spectral lines, which is only about 25 m/s (Beckers, 1977), may indicate that the convective heat flux is negligible, at least in the upper photosphere where the lines are formed. On the other hand, empirical models of the umbral photosphere suggest that radiation alone cannot transport as much as the 20% of the average solar heat flux that is actually emitted from the umbra. It is impossible to obtain agreement with observations, if the radiative temperature gradient is assumed below an optical depth $\tau_{cont} \approx 1.5$; the gradients derived in empirical models are much smaller (Zwaan 1968, 1974, 1975; van Ballegooijen, 1981; Maltby, 1992). That may imply some form of convection just below the visible surface.

In the magnetostatic description of the balance of forces in a sunspot all fluctuations related to convective motions are ignored. The Evershed motion is also neglected in such models, though the velocity of the gas may become nearly sonic (6 km/s) in the outermost penumbra. This field–aligned flow does not affect the overall magnetostatic equilibrium, however. The outward motion may occur in a subset of almost horizontal flux tubes as a siphon flow (Meyer and Schmidt, 1968; Thomas, 1988; Montesinos and Thomas, 1989; Thomas, 1990), in which the forces due to velocity gradients are negligible in the upward branch of the flow. Thus it is assumed that the matter is stratified hydrostatically along the magnetic field lines, and that the gas pressure gradient across the field is balanced by the magnetic and gravitational forces.

Although the convective mass motion and heat transport do not directly contribute to the overall force balance they do affect the temperature gradient, and hence the pressure distribution. Therefore the magnetostatic equilibrium is strongly coupled to the thermodynamic equilibrium; for a discussion of the mathematical treatment of these equilibria and their coupling see Low (1985).

The wide range of mutual interaction between convection and the magnetic field (see *e.g.* Proctor, 1992) applicable to sunspots is poorly understood. Furthermore, there is no rigorous macroscopic theory of heat transport by magnetoconvection that could be applied quantitatively over many scale heights in global models. Therefore some additional parametrized assumptions concerning energy transport are required in order to close the mathematical description of the overall configuration. Although the question: 'how is the energy transported in the spot?' cannot be addressed directly within such an approach, the assumptions made prescribe the temperature gradient in the model. Due to the coupling of the magnetostatic and the thermodynamic equilibria the temperature gradient affects not only the photometric, but also the magnetic properties of the global configuration. Features at the surface of the model may then be tested against the observed properties of spots, so that some implications for energy transport in deep layers may be derived.

2.1. BASIC CONCEPTS

The low temperature of large concentrations of magnetic flux in the photosphere is related to the properties of strongly magnetized plasma and its ability to transport energy in optically thick regions. Three general concepts proposed to explain the coolness of sunspots point to different possible mechanisms that might provide the observed heat flux.

Biermann (1941) suggested that a strong magnetic field suppresses convective energy transport. The inhibition cannot be complete, however, because even in the coolest part, *i.e.* in the umbra, spots radiate only 80% less energy than the solar photosphere – they are too hot.

Hoyle (1949) argued that convective motions in a spot are channelled by the magnetic field, while the heat exchange across the field is inefficient. According to this concept the spot is a tube thermally isolated from its surroundings. The tube fans out with height, so that the energy flux should be diluted as the surface is approached.

Parker (1974, 1976) proposed an explanation of the spot's coolness by conversion of the thermal energy into the energy of mechanical waves. He argued that the inhibition of energy transport increases the temperature beneath the inhibition level and causes a pressure build-up at depths where the efficiency decreases. That would lead to a fast destruction of a spot unless the heat excess is efficiently removed from the region. Cooling by hydromagnetic waves does not seem to give a satisfactory explanation. As Cowling (1976) argued, the generation of the Alfvén waves cannot be so efficient as to achieve a conversion of 80% of the energy into waves.

2.2. QUANTITATIVE MODELS

The discussion below refers to a particular set of mathematically self-consistent models describing an entire spot. Most of these investigations were based on Biermann's concept and were aimed at its quantitative evaluation.

2.2.1. The similarity law assumption. Schlüter and Temesváry (1958) proposed an approach which significantly simplifies the mathematical treatment of the equation of magnetohydrostatic equilibrium. They assumed that the distribution of magnetic flux on horizontal planes is geometrically similar at each depth. That makes it possible to reduce the partial differential equation to a second-order ordinary differential equation for the field strength at the axis of a spot. To solve this equation the difference between the gas pressures in the solar convection zone and at the axis of the spot must be known. Thus the stratification in the central parts of a configuration must be determined by some auxiliary assumptions concerning the energy transport. Comparison of a model with the field strength and the temperature observed in the central parts of the umbra yields a test of the validity of the assumptions made. Schlüter and Temesváry were the first to note that complete suppression of convection is rather unlikely. No regular solution existed in their models for more than a few scale heights below the photosphere, if they assumed that energy is carried by radiation only. They also argued that the relative gas pressure difference between the solar envelope and the spot decreases with depth and stays low far below the photosphere, so that the stratification inside is almost the same as the stratification in the quiet Sun.

Deinzer (1965) presented a generalization of the model by Schlüter and Temesváry. In his model the similarity law for the magnetic field is coupled with the thermodynamic structure along the axis of a spot, as described by mixing length theory. The *partial* inhibition of convection is parametrized by a reduced mixing length parameter which is kept constant with depth. The model yields a relation between the magnetic field strength at the surface and the heat flux (or mixing length parameter). Comparison with the umbral heat flux indicates that the efficiency of convection, as measured by the mixing length parameter, is reduced to about 20% of its efficiency in the undisturbed solar convection zone.

Deinzer also checked the effect of channelling of the energy flux along the field lines in his family of models and argued that the funnelling was too small to account for the values of the heat flux in spots. This test may not be conclusive, however, as it was only made for magnetic configurations implied by the stratifications computed with a heat flux independent of depth, *i.e.* ignoring the effects of funnelling on the pressure distribution along the axis. Variations of the energy flux with depth do affect the stratification, so that the rate of divergence of the field lines depends on the heat funnelling itself.

The concept of similarity of the magnetic field has been used by many authors. Yun

(1970) improved Deinzer's model by the introduction of an "effective surface monopole", which controlled the inclination of the particular field lines identified with the outer edge of a spot at the surface. In this way he improved the upper boundary conditions taking into account the fact that the gas pressure difference at photospheric levels is not negligible, as had been assumed in Deinzer's model.

The lower boundary conditions are also slightly modified in Yun's model: he included the effects of partial ionization on the relation between the internal and external pressures and temperatures. Some inconsistency in the lower boundary conditions remains, however; namely, a discontinuity of the temperature gradient across a spot, which Yun ascribed to different mean molecular weights inside and outside. In fact, it is rather the result of a not quite consistent application of the assumed condition that the magnetic field is vertical at the lower boundary (see 2.3).

Yun (1971) also demonstrated that the introduction of a moderately twisted field only slightly changes the main characteristics of the model. Such parameters as the mixing length parameter, the effective temperature, the Wilson depression, or the central field strength are practically the same in models with and without a twist.

A detailed discussion of models utilizing the similarity assumption has been presented by Staude (1978). Such assumptions, offering a great mathematical simplification, represent an attractive approach. They do, however, have some drawbacks. For instance, the Gaussian distribution of the magnetic flux ignores the fact that sunspots have sharp edges. Moreover, the emerging intensity calculated in similarity models has a maximum not far from the centre of a spot (see *e.g.* Jakimiec and Zabża, 1966). Landman and Finn (1979), who found such a bright ring inside their models, suggested that the assumption of magnetohydrostatic equilibrium was inadequate. However, this is rather a consequence of the similarity law, which enforces a somewhat arbitrary distribution of electric current. Those currents determine to some extent the horizontal temperature variations, which in general need not comply with the observed photometric profile (see 3.1).

2.2.2. Channelling of the heat flux. Hoyle's proposal, that the low temperature of sunspots could be explained by a dilution of the energy flux due to the divergence with height of the magnetic field lines in the spot, found a quantitative evaluation in the model constructed by Chitre (1963) and refined by Chitre and Shaviv (1967). These models apply again to the central parts of the umbra. Both effects, the suppression of convection and the channelling of the convective heat flux along the magnetic field, were taken into account. The inhibition of convection was parametrized by an empirical factor (Chitre, 1963) or by a function of depth (Chitre and Shaviv, 1967), while the funnelling effect was incorporated in the model by mapping the heat flux vector on the magnetic field vector.

The stratification in the spot was computed along the axis down to a depth at which the energy flux reached the value of the solar heat flux. This level was defined as the base of a spot model. The energy flux varied with depth proportionally to the field strength, which had been determined approximately from the condition that the horizontal gas pressure difference must be equal to the magnetic pressure at the axis, *i.e.* ignoring the tension force due to curvature of the field lines.

The authors concluded that neither complete suppression of convection nor channelling of the heat flux alone had yielded a plausible model. In both those extreme cases the temperature gradient was so large that the gas pressure in the spot already exceeded the gas pressure in the undisturbed convection zone well above the base of a spot. They found consistent models when the convective flux increased with depth – at the surface less than 1% of the heat flux was carried by convection, and near the base almost all.

The model discussed by Chitre and Shaviv was fitted to slightly too high a heat flux at the surface ($0.33F_\odot$), and too high a field strength (3550 gauss). Further, the curvature force of the magnetic field is not negligible in models in which an arbitrary distribution of electric currents is introduced. For instance, in Deinzer's models at the level of the Wilson depression the radially integrated curvature force reaches half the value of the total gas pressure difference between the exterior and interior. In Chitre's (1963) model the currents are implicitly predefined by the assumed distribution of the heat flux and magnetic field at the surface. Those currents are ignored in the balance of forces. Thereby the gradient of the field strength at the axis is overestimated, so that the compressed heat flux already reaches the average solar flux $F_\odot$ at depths 2000–3000 km below the photosphere.

2.2.3. The cluster model. All conventional models which assume a continuous distribution of the magnetic field in a spot require some heat transport by magnetoconvection below the photosphere of the umbra. An alternative explanation of the surplus energy flux in the umbra is offered by the cluster model discussed by Parker (1979a,b) and Spruit (1981b). This model represents a spot as a dynamic clustering of small magnetic tubes which preserve their identity down to large depths. Inside the tubes the energy is carried by radiation only, so that most of the heat flux emitted in the umbra has to be supplied by field free columns of hot gas penetrating the spot immediately below the photosphere. Spruit (1981a) argued that even if the magnetized component of the umbra, *i.e.* the flux tubes, did not carry any energy from below, the observed heat flux could still be provided by a cluster of very small unresolved hot columns covering about 20% of the area of the umbra. Unfortunately, the cluster model has not yet been worked out in detail.

2.3. A COMMENT ON BOUNDARY CONDITIONS

The very low densities of chromospheric and coronal gas allow a free expansion of the magnetic field above the photosphere. Then the field of a single isolated spot may assume a monopole–like configuration. Such an approximation for the magnetic field at the upper boundary seems to be appropriate provided the formulation of the model allows the boundary conditions to be placed high enough in the atmosphere. More caution is required if the boundary conditions are specified close to or in the photosphere, as in the models of Deinzer (1965) or Yun (1970).

The behaviour of the sunspot magnetic field deep below the surface is not known. The lower boundary condition usually utilized is based on the conjecture that the field forms a vertical column below some depth (however, see van Ballegooijen, 1982). This is quite a severe condition, which strongly constrains the thermal structure of those deep layers in a spot. The magnetostatic equilibrium implies that the gas density at the axis of a vertical column is the same as in the undisturbed convection zone, since the horizontal difference of gas pressure, ΔP, is then constant with depth. Thus, with the stratification of the quiet Sun given and the value of ΔP assumed, all thermodynamic parameters in the column, including variations of the temperature gradient, can be calculated. For a given heat flux one can formally define a mixing length parameter required by this gradient, and compare that with the mixing length in similarity models, where it is an eigenvalue of the problem (see *e.g.* Deinzer, 1965). The value implied by imposing a vertical field is larger by a factor of more than two. The concurrent difference of the temperature gradients is very small but noticeable (Yun, 1970).

It can be argued that this inconsistency results from the assumption that both the heat flux and the mixing length parameter are independent of depth. The mixing length formalism (with a constant parameter) does not yield a proper approximation for the temperature gradient in a vertical magnetic column with the density distribution the same as in the quiet Sun, even in very deep layers where the stratification becomes practically adiabatic.

3. New Methods

The past decade has brought progress in the numerical treatment of magnetostatic equilibria. There are two general approaches to solving the elliptic problem posed by the partial differential equation describing the magnetohydrostatic equilibrium of a conventional model of a spot with a continuous magnetic field. First, a direct solution of the partial differential equation and, second, the solution of a free surface problem formulated within the scope of potential theory.

3.1. DIRECT SOLUTION OF THE PARTIAL DIFFERENTIAL EQUATION (PDE)

Pizzo (1986) developed a numerical technique for constructing two–dimensional equilibria in the solar photosphere. His method is based on the formalism of Low (1975), who proposed transforming the thermodynamic parameters (the pressure and temperature of the gas) into functions of the magnetic vector potential and depth. In this approach a functional form is prescribed for the gas pressure, but not for the magnetic field as in the similarity models (see *2.2.1*). The elliptic equation for magnetostatic equilibrium is then reduced to a simplified form that is more convenient for numerical analysis.

In this paper Pizzo does not address the question of the spot structure in deeper layers. The gas pressure difference between the normal photosphere and the axis of the spot, needed for the computation, is determined from semi–empirical models of the umbral atmosphere. Therefore in the basic parameter study the configuration considered by Pizzo has its base placed only 120 km below the visible surface of the umbra.

For comparison with previous models based on the similarity law Pizzo assumed a Gaussian profile of the magnetic field across the base. Of course, that ignores the existence of a discontinuous transition from magnetized to field free plasma. Also, a local maximum of the emerging intensity appears in some of models. However, Pizzo has demonstrated that this bright ring inside a spot model is an artifact of the similarity assumption, which does not allow for a self–consistent treatment of the energy balance. By a minor rearrangement of the field lines Pizzo could remove this feature without affecting the magnetostatic balance (see Figs. 7 and 9 of Pizzo, 1986).

The magnetic field in the central parts of Pizzo's models for larger spots assumes a configuration that is very close to being current free. But near the periphery the field lines in these models become more and more constricted and the hypothetical magnetic geometry for a large spot (Fig.21 of Pizzo, 1986) with an hour glass shape seems to be exaggerated, as it would artificially enforce fluting instabilities at the level of the Wilson depression.

Recently, Pizzo (1990) has extended his method by incorporating the free surface problem in the solution of the equation of magnetostatic equilibrium for a flux tube surrounded by a infinitely thin current sheet. He utilizes a body-fitted coordinate system which allows an efficient and accurate discretization of the domain considered. Preliminary models, presented to illustrate the numerical technique, extend many scale heights below the surface, so that the method might be used to diagnose the deep structure of a spot.

3.2. THE FREE BOUNDARY PROBLEM FOR A SUNSPOT

An alternative approach used in modelling the overall magnetostatic equilibrium of spots utilizes potential theory. The direct solution of the PDE is avoided. Instead, an integral equation is solved to determine the position of the magnetopause, which is the free surface separating the magnetized interior of the spot from the surrounding field-free plasma. A method of solving such a free boundary problem for a potential field has been developed by Wegmann (1981). The linearization of the integral equation allows an iterative improvement of the trial position of the free surface, with corrections that are a function of the geometry of the surface itself, so that calculation of the potential throughout space can be avoided.

Schmidt and Wegmann (1983) used this method to construct models for magnetic flux tubes embedded in the solar convection zone. The spot is represented by a potential magnetic field separated from the quiet Sun by a magnetopause. This is schematically illustrated in Fig.1a. The boundary condition requires that the total pressure (magnetic plus gas pressure) across the magnetopause is continuous. The jump in the gas pressure associated with the jump in the magnetic field defines the electric surface currents which generate the magnetic field of the whole configuration. In this way the profile of the magnetic field calculated at the surface of the model is related to pressure variations with depth along the flux tube. Such an approach yields a handy tool for probing the stratification in deeper layers of spots.

Schmidt and Wegmann identify the pressure stratification and the shape of the magnetopause that yields a magnetic field profile that fits the observations of Beckers and Schröter (1969). It appears that the magnetopause near the surface of the model is not flat but significantly inclined to the horizontal. As Schmidt (1987, 1992) has pointed out, the high values of the magnetic field observed at the outer edge of a spot are not consistent with the concept of a shallow penumbra with an almost horizontal magnetopause, as considered *e.g.* by Schmidt *et al.* (1986). The value of the field strength at the outer edge (which amounts to 0.3–0.5 of the central field strength), combined with a force-free expansion of the magnetic field in the corona, implies that a significant fraction of the total magnetic flux crosses the surface in the penumbra. Consequently the penumbra should be an extended thick layer between the umbra and the quiet Sun, similar to that proposed *e.g.* by Gokhale and Zwaan (1972).

An extension of the approach by Schmidt and Wegmann (1983) has been used to incorporate such a thick penumbra into a global model of a spot (Jahn, 1989). The free boundary problem for the magnetopause and its surface current was supplemented by trial volume currents distributed in the outer parts of a flux tube underneath the photosphere of the penumbra, where the gas has a thermal structure different from that in the umbra and in the quiet Sun. Fig.1b shows a part of the flux tube with an extended penumbra in the region enclosed by a dashed rectangle in Fig.1a. The flux tube is still surrounded by a magnetopause carrying surface currents related to the jump in the magnetic field.

The volume currents contribute to the lateral balance of forces and affect the pressure stratification, so that gas in the penumbra is hotter; hence layers of equal gas pressure assume higher levels than in an umbra where the field is current free. Both currents are related to the magnetic and photometric profiles at the surface. The specific parameters of the global current system for which the magnetic and the photometric signatures of the model agree with photospheric observations were determined by trial and error.

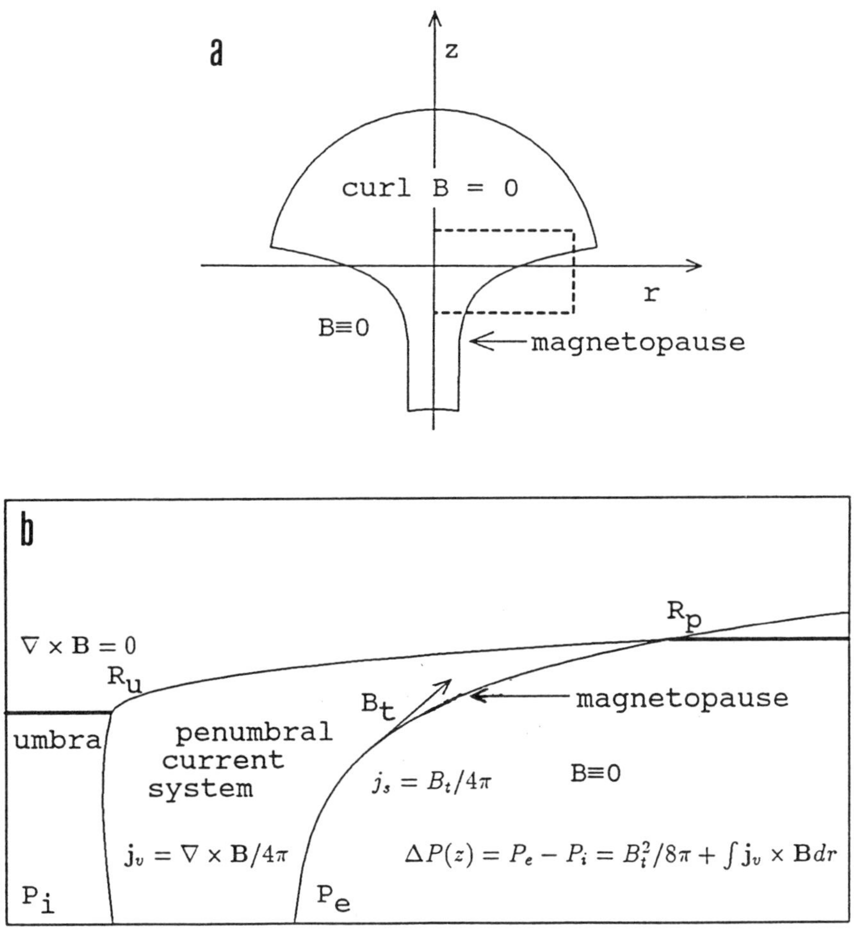

Fig. 1. The magnetic configuration of a sunspot as a free boundary problem for the magnetopause: (a) The potential model – the domain considered is bounded by an infinitely thin magnetopause, and by two spherical cups; the interior is permeated by a current–free magnetic field.
(b) The volume current model with a thick penumbra – the structure in the region bounded by a dashed rectangle in (a). R_u and R_p indicate radii of an umbra and of a spot. The gas pressure difference between the quiet Sun and the umbra ($\Delta P(z)$) is balanced by Lorentz forces due to surface (j_s) and volume ($\mathbf{j}_v$) currents.

In this way the model has been used to diagnose the self-consistent structure of deeper layers in the spot. The resulting stratification in the umbra is well constrained down to depths comparable with the radius of the spot. Due to the inverse-square dependence on distance in the Biot-Savart law the distribution of currents in still deeper layers has a smaller effect on the surface characteristics of the model. The pressure stratification in the umbra derived in the model implies a variation of temperature with depth, so that the temperature gradient can be determined. Assuming that the heat flux does not change with depth some implications for the efficiency of energy transport, as parametrized by mixing length formalism, can then be derived. Variations of the ratio of the mixing length to the pressure scale height as implied by the temperature gradient have been calculated along the tube. Down to a depth of about 1500 km the resulting mixing length parameter, 0.18, is small. Then it increases abruptly in a shallow layer and below a depth of about 2000 km reaches the value 0.45. Such an increase of efficiency concurs with the results of Meyer *et al.* (1974) who pointed to the effect of the magnetic diffusivity on the efficiency of possible modes of convection in a spot. A local analysis of convective stability applied in their model shows that below a depth of about 2000 km, where the magnetic diffusivity η becomes larger than the thermal diffusivity κ, the convection becomes more efficient than in the layers above, where the ratio η/κ is less than unity.

Diagnostics for the thermal structure in deep layers of the penumbra are more difficult and less reliable. The trial function describing the distribution of volume currents is a function of both the vertical and the horizontal coordinate, and so its prescription is difficult. Moreover, it is impossible to find an unambiguous distribution of currents in the whole penumbra, such that the thermodynamic properties of the gas can be determined reliably at all depths. It has been demonstrated that the surface properties of the model do not depend on how the volume currents are distributed at depths larger than about 2000 km below the photosphere (compare models TE and HP of Jahn, 1989). Apparently, the average brightness of the penumbra can be produced in the model by means of a relatively shallow layer permeated by continuously distributed currents.

The ambiguity mentioned above is a consequence of the extreme sensitivity of the temperature at the visible surface to the distribution of electric currents in static models of the atmosphere. Although a small change in the current distribution does not change the overall field configuration it must locally affect the gradient of the gas pressure, and hence the density distribution. A small change in the density causes a large change in the opacity, which varies in the photosphere by orders of magnitude. As a result, the optical depth $\tau \approx 1$ may be shifted to a slightly different geometrical level. Even for very small shifts the associated change in the temperature at $\tau \approx 1$ will be large, since the temperature gradient in the photosphere is very steep. A similar problem was encountered by Jakimiec (1970) who modelled the current distribution at the surface of a spot, and also by Pizzo (1986) when removing the unwanted "bright ring" appearing in his models.

3.3. THE PRESSURE DEFICIT IN THE UMBRA

The variation of the gas pressure difference between the solar convection zone and the umbra, $\Delta P(z)$, exhibits a peculiar feature. This function has a local maximum at the level of the Wilson depression in the umbra. Such a maximum is present in the similarity models considered by Deinzer (1965; see his Fig.1) and in the models by Pizzo (1986), in which it leads to an hour-glass shape for the spots. A very pronounced maximum of the pressure difference is found in the volume current model (see *e.g.* Fig.28 of Jahn, 1989). It is difficult to ascribe this feature to inadequate assumptions made in the magnetostatic models, since it also appears when $\Delta P(z)$ is determined by other means (for instance, if the pressure difference is calculated from a comparison between semi-empirical models of the stratification in the umbral and solar atmospheres, when reasonable values of the Wilson depression are assumed – see Fig.2 of Jahn, 1989). Such a maximum also appears in the numerical simulations of umbral structure discussed by Nordlund and Stein (1990; see their Fig.6).

Thus the local peak in the pressure difference is likely to be real; it is probably caused by the Wilson depression and variations of the umbral temperature gradient with depth. These variations decide how fast does the gas pressure increase with depth as compared with the increase of the external pressure. The gas pressure is negligibly small above the photosphere in the umbra, so that $\Delta P(z)$ is practically equal to the pressure in the quiet Sun. Below the umbral photosphere, the pressure in the interior increases much more rapidly than in the surrounding plasma, since the temperature (and hence the pressure scale height) is much lower than in the normal Sun at the same depth. As a result, the pressure difference may decrease with depth just below the umbral photosphere, provided that the value of the Wilson depression is not too large. After reaching a local minimum, $\Delta P(z)$ may increase again in still deeper layers as the difference between the temperatures inside and outside diminishes. The shape of $\Delta P(z)$ near its minimum depends on the variations of the temperature gradient, and thus on the efficiency of energy transport processes. In particular, for a purely radiative gradient, $\Delta P(z)$ would become negative a few hundred kilometers below the photosphere. If the umbral photosphere were to lie significantly deeper than about 500 km below the solar surface, then the internal pressure would still be a small fraction of the external pressure and the maximum of $\Delta P(z)$ would be absent. But then the field strength at the spot's surface would have to be unrealistically large.

Realistic global models of the structure of a sunspot must account for this local maximum in the pressure difference in such a way that the magnetostatic configuration of the field is free of strong constrictions which would lead to instability. A thick penumbra, forming an extended buffer separating the innermost part of a spot from the solar convection zone, can apparently accommodate such an excess.

4. The Penumbra

In the penumbra we witness a peculiar magnetoconvective process that transports about 75% of the solar heat flux with a radial variation of less than 5%. The concept of convection in horizontal field-aligned rolls (Danielson, 1961), although attractive because of this low radial variation, can no longer be upheld in view of recent observations of the field inclination. The magnetic field at the outer edge of the penumbra is only two to three times smaller than the central field strength, indicating a steep magnetopause, and the average inclination of the field is about 20° to the horizontal (Title, 1989; Adam, 1990; Lites and Skumanich, 1990). High resolution observations by Title and coworkers (Title *et al.* 1992) are particularly interesting. Their measurements of the magnetic field in a nearly circular spot show that the penumbra actually contains a mixture of field directions. The mean inclination to the horizontal is found to vary from 45–50° at the umbra–penumbra boundary to 15–20° at the outer edge of the spot. The inclination fluctuates by about 15° over small distances in the azimuthal direction and is correlated with brightness: it is larger in bright and smaller in dark filaments. The fluctuations show a radial coherence, *i.e.* they are nearly independent of radial distance in the penumbra. The azimuthal variations of the magnetic signal are mostly due to fluctuations in the angle of inclination.

Observations of motions in the penumbra show systematic flows (Muller, 1976, 1981; Title, 1989; Zirin and Wang, 1990; Title *et al.* 1992): in the inner half of the penumbra an inflow of bright elements is visible, while the outward Evershed flow, occurring predominantly in the dark filaments, is seen in the outer penumbra.

4.1. INTERCHANGE CONVECTION

On the basis of these observations we propose that heat transport in the penumbra takes the form of convection by an exchange of entire flux tubes between the magnetopause and the photosphere (Schmidt, 1992). We explore the consequences of this hypothesis in a new magnetostatic model of a sunspot with a thick penumbra (Jahn and Schmidt, 1992). In this concept the energy flux in the penumbra is carried by convective elements identified with magnetic flux tubes or sheets. Their azimuthal thickness is very small compared with their extent along the field. The flux elements are heated at the magnetopause, gain buoyancy and start to move towards the photosphere. They replace the overlying magnetic flux elements containing cooler gas. As a result of this interchange cool matter sinks toward the magnetopause. Such a motion should be rather coherent along the field, because the Lorentz forces tend to minimize the curvature of the field and smooth out any sharp bends in the field lines. The flux sheets somehow avoid crossing each other – otherwise convection of this type would be stopped. It is conceivable therefore that very long magnetic flux units participate in the convective process, and that the motion preserves coherence over many scale heights. That is, long sections of the flux units are either in the phase of "inward/upward" or in that of "outward/downward" movement. Such lateral coherence is clearly visible at the surface in the filamentary form of the penumbra. Emerging flux units are brighter and have a larger inclination of the magnetic field to the horizontal. Cooler elements have a more nearly horizontal field which may form a subset of flux tubes in which the Evershed flow is allowed. Eventually they sink or are expelled towards the magnetopause by the newly emerging hot flux units.

This interchange convection can only start if the buoyancy of the heated elements provides a transverse force large enough to overcome magnetic forces and to pull the field lines inwards. That depends on the inclination of the magnetopause. The inclination to the vertical increases with the total magnetic flux of a tube. Apparently, for fluxes large enough the inclination angle exceeds the critical value at which this form of convection sets in, so that a sunspot with a penumbra, and not a pore, is formed.

In sunspot groups the dependence of the inclination on the total magnetic flux can be strongly perturbed by neighbouring flux concentrations and the magnetopause may have substantially different inclinations on different sides. Then, for instance, parts of the penumbrae of large spots may be missing, or rudimentary penumbrae may form adjacent to very small flux concentrations.

The inclination of the magnetic field also affects the range of inward penetration of the interchanging flux sheets. When the field becomes too close to the vertical the transverse component of the buoyancy force cannot compete with the magnetic forces. Then lateral motion ceases rather abruptly as indicated by the sharp transition from the penumbra to the umbra.

The global nature of this process implies that at any depth in a sunspot roughly the same fraction of the total magnetic flux should participate in this type of convection: this fraction constitutes the penumbra. If so, one can expect that the concurrent lateral mixing of gas in the penumbra efficiently smooths out the horizontal gradients of thermodynamic parameters when averaged over a suitable azimuthal range. Thus the whole penumbra assumes a strongly vertical stratification, although its photosphere seems to be slightly inclined (by 3 to 6 degrees).

4.2. A MAGNETOSTATIC APPROXIMATION FOR THE PENUMBRA

The hypothesis of global exchange allows a crude description of the penumbra as a one–dimensional, vertically stratified medium in which all variables except a current–free magnetic field depend on depth only. We simply replace the volume currents of our previous model by a second current sheet, which is also a free surface like the magnetopause. It represents the location at which the interchanging flux elements turn back in their peripatetic movement across the penumbra. So we call it the peripatopause. This interface between the umbra and the penumbra is a magnetic surface, so that the magnetic flux contained in either region is constant with depth.

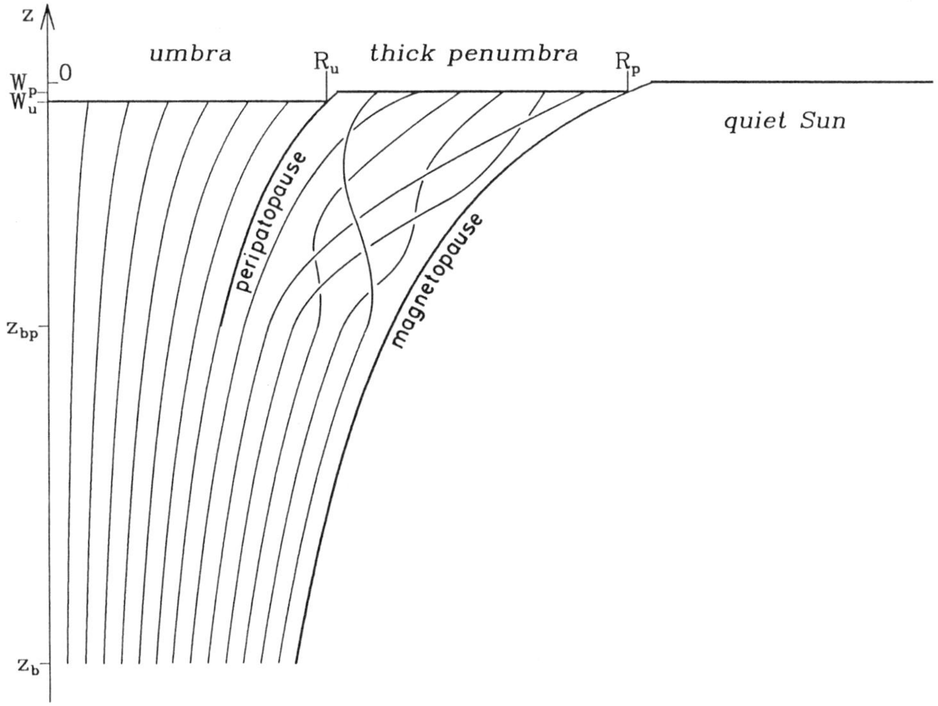

Fig. 2. The magnetic configuration of a spot model without volume currents: the magnetopause and the peripatopause are treated as two free surfaces. The bottom of the model, z_b, and the depth at which the penumbra merges with the umbra, z_{bp}, are indicated as well as the depths of the photosphere in the three stratifications.

We now construct a model of the whole spot by solving the free surface problem for both the magnetopause and the peripatopause with boundary conditions that imply exact magnetostatic equilibrium at each depth (that is, continuity of the total pressure across both interfaces). The local jump in the gas pressure, and so the value of the surface currents, is obtained by comparing the local pressures in three one–dimensional stratifications representing the umbra, the penumbra, and the quiet Sun. A schematic illustration of the magnetic field in this model is given in Fig.2.

The three stratifications have been computed with several assumptions. The most important of these concerns the heat flow in the model. Fig.3 illustrates these assumptions.

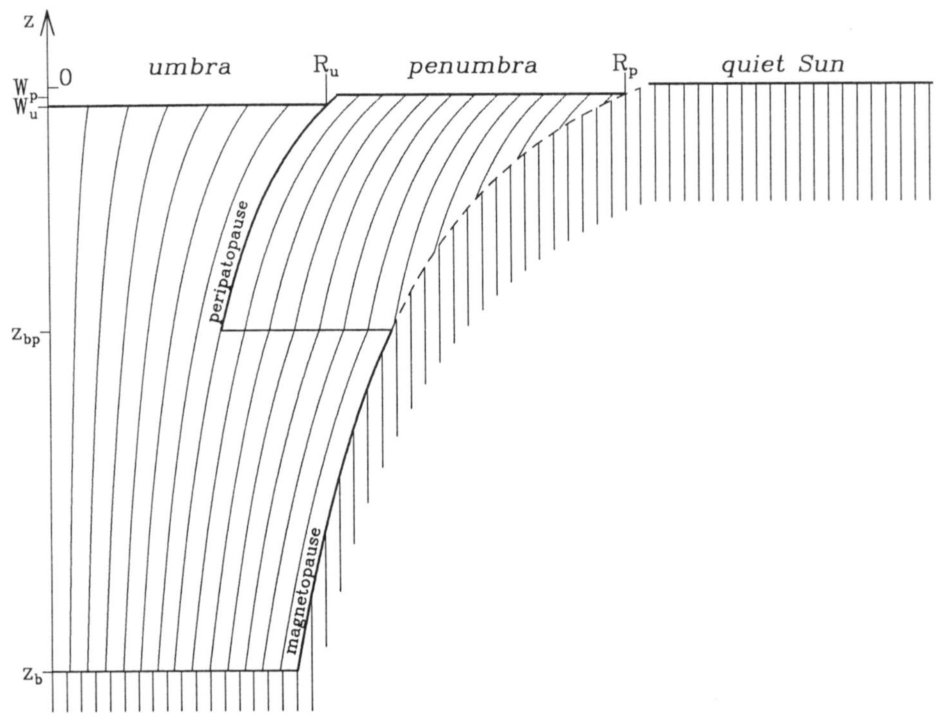

Fig. 3. The heat flow in the model with two free surfaces. Description is the same as in Fig.2. The dashed line represents the part of the magnetopause that transmits heat flux.

We conjecture that only above a certain level z_{bp}, representing the bottom of the penumbra, is the magnetopause able to transmit a (fixed) fraction of the impinging solar heat flux into the penumbra. The transmissivity of the magnetopause is one of the free parameters of the model. The fraction of the heat flux that enters the penumbra at some level above z_{bp} is instantly distributed in the horizontal plane between the magnetopause and the peripatopause. The remainder of the impinging flux is diverted and distributed over a large area in the quiet solar photosphere as shown in Spruit's (1977) calculations. Below the level z_{bp} there is no heat exchange at all across the magnetopause and all impinging flux is diverted. The penumbra merges here with the umbra and both regions have the same thermal structure. The peripatopause does not allow any efficient heat exchange between the umbra and the penumbra. Consequently, all the energy in the umbra is supplied by the underlying layers, so that the variation of the heat flux with depth is inversely proportional to the area of the umbra, or proportional to the vertical component of the magnetic field.

Energy transport is described by the mixing length formalism in all three stratifications. In the umbra changes of the mixing length parameter with depth have been introduced in order to get appropriate variations of the temperature gradient. At the surface we constrain the heat flux of the umbra ($0.23F_{\odot}$, $F_{\odot}$ being the solar flux) and the penumbra ($0.75F_{\odot}$), the central field strength (3000 gauss), and the fractional radius of the umbra (0.5). The model is governed by three free parameters: the total magnetic flux Φ, the transmissivity of the magnetopause ϵ, and the position z_b of the bottom of the model, which is the level where the continuing flux tube becomes a vertical column. z_{bp} is not a free parameter, but a part of the solution.

In our models the penumbra must contain about 50–60% of the total magnetic flux of a spot in order to achieve the proper radial extent of the visible penumbra. The resulting inclination of the peripatopause to the vertical never exceeds about 20° . This may perhaps be interpreted as a value at which the interchange of magnetic elements driven by buoyancy against counteracting Lorentz forces becomes inefficient.

Although the transmissivity of the magnetopause is a free parameter in the model it seems to have a limited range of feasible values such that $\epsilon \approx 0.6 - 0.7$. Fig.4 presents two illustrative models with different transmissivities of the magnetopause. If it is assumed that $\epsilon = 0.65$ the penumbra merges with the umbra at depths 5–6 Mm below the photosphere, where the inclination of the magnetopause goes down to 20-30° (see Fig.4a). Note a fair agreement of the magnetic profile with observations by Beckers and Schröter (1969). For larger values of ϵ the penumbra is overfed with energy and, as a result, the merging level z_{bp} moves to a very shallow depth, where the inclination of the magnetopause is much larger (nearly 70° for a fully free inflow of heat flux, see Fig.4b). Below that depth the inclination is still large, and it is conceivable that convection by exchange of long flux elements would in reality penetrate to deeper layers of the penumbra. On the other side, for $\epsilon \leq 0.5$ the energy supply is evidently too small and we cannot obtain a model that has a properly extended penumbra with the prescribed heat flux of $0.75F_{\odot}$. These results indicate that an angle of about 25° may be the critical value of the inclination of the magnetic field above which an efficient mode of convection in the penumbra can be maintained (actually convection may start at larger inclinations and propagate downwards until the critical angle is reached).

An interesting feature of our model is connected with the assumption that the umbra is thermally isolated from its surroundings. Due to the funnelling effect, the heat flux in the umbra increases with depth. We even obtain models with the heat flux at the base (F_{ub}) equal to the solar heat flux $F_{\odot}$. Thus it is possible to construct models exclusively within the scope of Hoyle's (1949) concept. Such models do not encounter the problem of a heat pile–up at the base (Parker 1974, 1976) that might lead to a fast destruction of a spot. All the blocking of the heat flux occurs at the magnetopause and there is no need (at least not for this reason) to invoke an additional force near the base to keep the tube together.

The geometry of magnetostatic flux tubes and so the funnelling effect depends on the value of the total magnetic flux. Consequently, the heat flux at the base increases as the magnetic flux is decreased in our models. The four spots with different total magnetic fluxes shown in Fig.5 illustrate this dependence. For very small spot models the flux at the base even exceeds the value of the solar heat flux. Due to their geometry such small spots seem to require more energy to form a penumbra than actually is available. For a conservative choice of other free parameters the cirtical value of the magnetic flux for which $F_{ub} = F_{\odot}$ is equal to 3×10^{21} Mx, and we cannot construct any reasonable model of a spot *with a penumbra* for a magnetic flux smaller than this value. This is again consistent with a conjecture that the penumbra may be formed if the magnetic field fans out rapidly enough, so that a large inclination of the magnetopause allows the interchange.

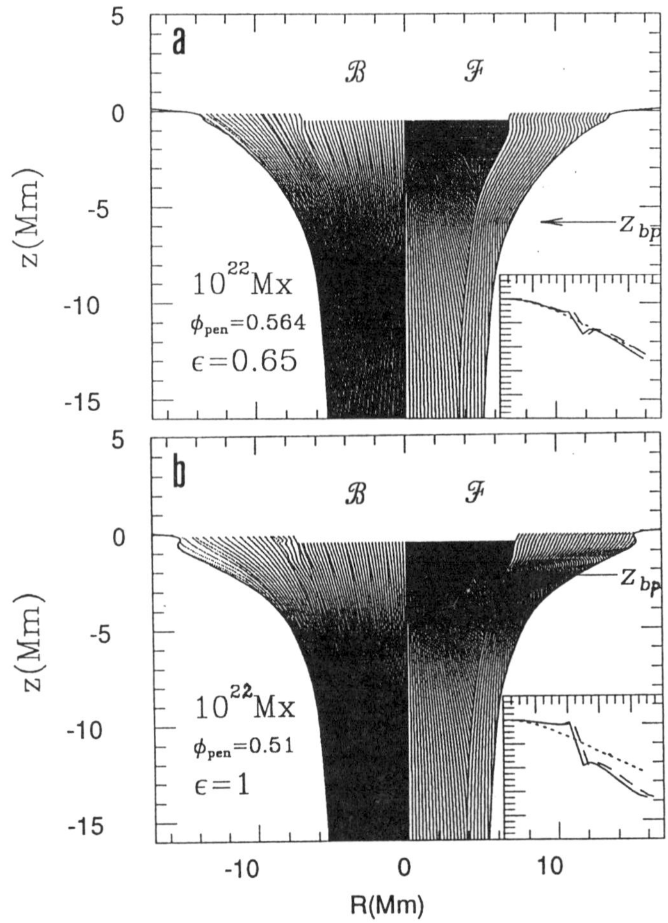

Fig. 4. An analogue representation of the magnetic field strength $\mathcal{B}$ (on the left), and the heat flux $\mathcal{F}$ (on the right) in models with different transmissivities of the magnetopause: (a) $\epsilon = 0.65$, and (b) $\epsilon = 1$. The intensity of shading is proportional to the heat flux and inversely proportional to the field strength. The normalized magnetic profile at the surface is shown in the lower corner; the dotted line represents the profile deduced from observations by Beckers and Schröter (1969).

It is tempting to suppose that this feature of the model is connected with a transition from a pore to a sunspot. Such a transition was considered by Simon and Weiss (1970) in discussing a model of a pore. The field lines in their model bend completely over near the

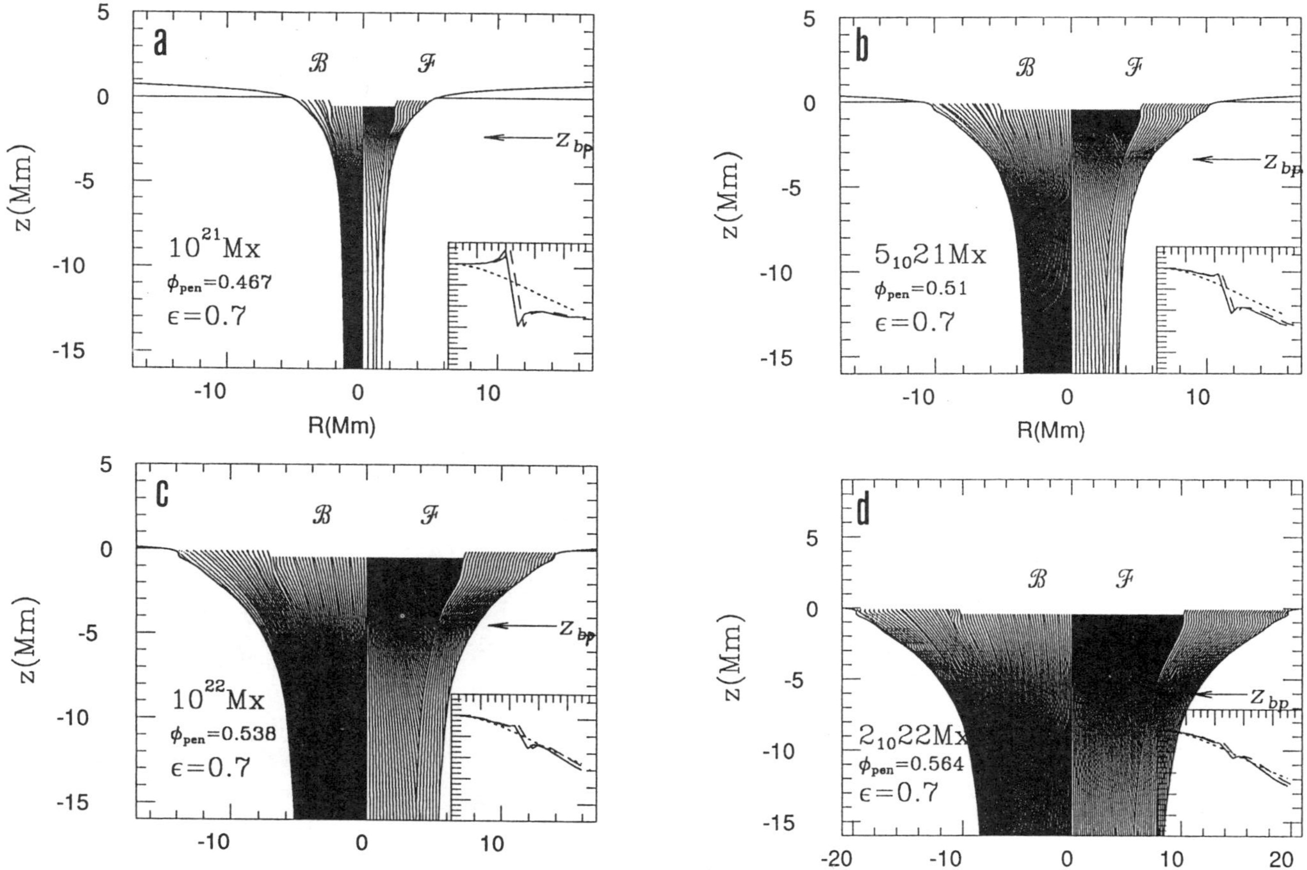

Fig. 5. Models of spots with different values of the total magnetic flux in the same representation as in Fig.4. The relative heat flux at the bottom $F_{ub}/F_{\odot}$ is equal to: 1.28 (a), 0.95 (b), 0.83 (c), and 0.71 (d).

surface for magnetic fluxes that are too large and the external pressure cannot be balanced by a field strength less than 3500 gauss. They concluded that the pore model then breaks down, because the finite thickness of a boundary layer surrounding the flux tube, *i.e.* the penumbra, has been neglected. Our present spot models, which always include a penumbra, break down for magnetic fluxes that are too small, if we require that this boundary layer be as thick as one half of the radius of the tube.

4.3. STABILITY AGAINST FLUTING

The magnetopause is concave toward the field–free plasma, and might therefore be subject to the flute instability (Parker, 1975). Meyer *et al.* (1977) investigated the stability of a flux tube in a gravitationally stratified plasma. By using the energy principle, formulated for a magnetic interface by Bernstein *et al.* (1958), they found that the magnetopause can be stabilized by the buoyancy of the magnetized matter. The condition for stability requires that the radial component of the field strength on the magnetopause decreases with height, so that the magnetic field fans out rapidly enough. In a flux tube the magnetized plasma is less dense than in the external medium, and its buoyancy counteracts the fluting tendency induced by curvature forces.

This instability has been used as an argument in favour of a cluster model. Parker (1979a) and Spruit (1981b) pointed out that as the magnetic field becomes more vertical below the photosphere the destabilization due to curvature wins over buoyancy and consequently the magnetic flux should split into individual tubes. It is not clear, however, how far and how fast this instability can proceed, and whether the nonlinear effects may not lead to a saturation of the process of splitting, so that a kind of diffuse magnetopause may be formed.

Applying Meyer *et al.'s* condition to our model we find that the magnetopause is stable against fluting down to depths of 5–6 Mm. At those levels the density inside the tube becomes comparable to that outside, and buoyancy becomes indeed slightly too weak to compensate for the remaining curvature forces. The layer at which the instability appears coincides with the depth at which the penumbra merges with the umbra in our model. It is conceivable that the transmissivity of the magnetopause, which is parametrized in our model, might somehow be related to the fluting instability. The fastest growing modes are those with the shortest wavelengths, according to a rough estimate by Bernstein *et al.* (1958). Perhaps they could make the magnetopause diffuse and so also increase its transmissivity for heat flow to 60–70% which can hardly be achieved by radiative exchange.

5. Morphology of a sunspot below the photosphere

The process that supplies 20% of the normal solar heat flux to the surface of the umbra depends crucially on the structure of the magnetic field beneath the photosphere. The two competing pictures of the topology of a spot, the conventional monolithic plug and the dynamic clustering of magnetic strands, imply qualitatively different answers. In the monolithic plug a magnetoconvective process must operate in which the magnetic field, though inhibiting the heat transport, still allows some convection which is capable of transporting energy more efficiently than radiation. In the cluster model most of the heat is carried by efficient overturning convective motions in the field–free space between magnetic tubes. The magnetic regions may allow only radiative transport, so that the required amount of energy is supplied to the photosphere by the unmagnetized plasma. It is important to establish the efficiency of the proposed mechanisms quantitatively for both pictures, so that they can be tested in global sunspot models.

The stability of a magnetized medium with a superadiabatic temperature gradient has been the subject of many investigations. The review by Proctor and Weiss (1982) summarizes the studies of the interaction of the magnetic field with convection in the Boussinesq approximation. The effects of compressibility on magnetoconvection have been discussed by Hughes and Proctor (1988). Such investigations deal with idealized models, in which the complexity of the solar atmosphere is to a large degree neglected. That makes it possible to pose well defined problems and to study them in the detail. The results yield a valuable guide to the nature of the convective processes that may occur in sunspots. For instance, different types of instabilities can appear at different depths in the umbra, depending on the ratio of the magnetic to the thermal diffusivity η/κ. When $\eta/\kappa < 1$ overstable oscillations are preferred, while for $\eta/\kappa > 1$ plasma may diffuse across the magnetic field, and steady convection sets in. Meyer *et al.* (1974) used this feature to describe the energy transport in their sunspot model.

Near the photosphere of the umbra $\eta/\kappa < 1$ and the oscillatory convection may manifest itself in the form of umbral dots. Knobloch and Weiss (1984) provided a consistent description of the umbral dots as a result of nonlinear oscillatory convection within the monolithic plug. More recently, Weiss *et al.* (1990) included the effects of stratification and variation of the diffusivity ratio in a study of compressible nonlinear magnetoconvection. Very interesting results were obtained for η/κ increasing with depth ($\eta/\kappa < 1$ at the top and $\eta/\kappa > 1$ at the base of the domain). They point strongly to the importance of nonlocal effects on magnetoconvection. The two types of instability are coupled with each other, so that there is a transition from nonlinear oscillatory convection at the top to more efficient overturning at the base. This results in time dependent convection taking the form of plumes penetrating to the upper boundary. The work of Weiss *et al.* indicates that oscillatory convection may account for the observed heat flux in the umbra as well as for the appearance of umbral dots.

The results of studies of magnetoconvection allow at present only a somewhat limited application to global models of sunspots. The high Rayleigh numbers and the large density contrasts of the actual stratification in and around spots are inaccessible in numerical experiments on idealized models. A more realistic stratification in the umbra and its convective stability have been investigated by Moreno–Insertis and Spruit (1989). A similar problem was studied also by Lou (1989). Linear stability analysis shows that the stratification is indeed unstable to overturning convection a few thousand kilometers below the surface. In neither case were overstable oscillations discussed, since the analysis was performed for adiabatic perturbations only.

Moreno–Insertis and Spruit calculate the eigenfunctions of the unstable modes, which assume a maximum 2300 km below the surface, and determine their e-folding times of about 18 minutes, in reasonable agreement with the lifetime of umbral dots. The further evolution of the unstable modes into the nonlinear regime is uncertain. In particular, it is not quite clear whether they can produce field–free gaps, as argued by the authors, and whether the process may ultimately lead to a formation of a field–free background with a congregation of individual magnetic tubes. Therefore it is difficult to decide whether the instability discussed by Moreno–Insertis and Spruit is consistent either with the cluster model or with a monolithic plug.

Energy transport in the cluster topology is explained by unimpeded convection in field free strands. Unresolved hot columns of unmagnetized plasma penetrate to the photosphere, and irradiate cold magnetic tubes, so that the apparent heat flux in the umbra is increased. Longitudinal overstability of the gas between flux tubes may form larger intrusions, visible at the surface as umbral dots (Parker, 1979b). In the cluster configuration the convective heat flux from below is blocked rather abruptly at the level of merging of the individual magnetic flux tubes. An efficient cooling must be postulated in order to avoid a pressure build-up (Parker, 1974). A systematic heat flow around the flux tubes rather than hydromagnetic waves might remove an energy excess of 80% of the solar flux.

It seems that the purely theoretical arguments presented so far cannot settle the question of the subphotospheric morphology of a spot. The existence of umbral dots is related to this issue, but seems to find an explanation in both the cluster and the conventional picture of a spot. Also the amount of energy emitted by the umbra can be explained within the framework of both approaches. Therefore it is interesting to consider whether global models which fit the observational constraints can distinguish between the two configurations. Although lack of a detailed cluster model makes it impossible to give a conclusive answer, a qualitative comparison can be made and some consequences implied by the different topologies can be indicated.

It can be argued, for instance, that the magnetic field lines beneath the photosphere should not diverge at depths less than or comparable with the spot radius. Otherwise, the field strength at the surface would not decrease monotonically with increasing radius across the spot (Jahn, 1989). It is rather unlikely that local electric currents in the force-free umbra could significantly modify the magnetic profile implied by the global distribution of currents. Therefore the observed distribution of the magnetic flux in a spot seems to exclude a loose cluster of flux tubes diverging just below the surface. Perhaps a deeply rooted cluster could fit the observations, but no detailed model allowing such a test has been proposed.

The comparison between the thermal structures of the two types of configuration is very uncertain. The monolithic model yields a temperature gradient that is consistent with the equilibrium of the overall configuration. Perhaps an equally consistent gradient could be constructed for the cluster by averaging over the thermal structure of magnetic tubes and field-free strands. An appropriate adjustment of the filling factor of the magnetic tubes at each depth might be possible for the umbra. More problematic is the structure of the penumbra, which forms a fairly flat photometric plateau, contains a significant amount of magnetic flux, and has a typical width of one half of the spot's radius. An explanation of all those features has not yet been proposed for a cluster topology.

There is one important difference between the cluster and the monolithic plug concepts. It is the efficiency of lateral heat exchange between the innermost regions of the spot and the surrounding convection zone. The individual flux tubes should preserve their identity to large depths and form a magnetic forest through which the heat may flow relatively unimpeded. Indeed, such a systematic heat transfer is needed to avoid a heat pile-up just below the photosphere of the umbra. The thermal contact of the monolithic plug with the surroundings should be less effective, and probably limited to the outermost regions of a single tube, at least on short time scales comparable with the lifetime of a spot.

A flux tube submerged in the solar convection zone for a long time may be subject to slow heating. The inefficient contact of the tube with its surroundings may account for the variations of the brightness of the umbra with the phase of the solar cycle, observed by

Albregtsen and Maltby (1978, 1981), and discussed in more detail by Maltby *et al.* (1986) who constructed semiempirical models for the atmosphere of umbrae in different phases. The temperature of the umbra at the beginning of the cycle is about 300 K lower than the temperature in a very late phase (Maltby *et al.*, 1986; Maltby, 1992). As Albregtsen, Jorås and Maltby (1984) note, the high correlation of the intensities in the umbra and in the penumbra of the same spot suggests that it is the brightness of the whole spot that varies with the solar cycle, and not only the temperature of the umbra. In any case, it is quite probable that such variations reflect different thermal histories of entirely 'young' and 'old' flux tubes that were heated for different times. Thus, the observations of Albregtsen and coworkers seem to yield an argument in favour of a single column as the representation for a spot. The lognormal distribution of umbral areas, found by Bogdan *et al.* (1988), might also support this picture. As they argue, the lognormal distribution, as a typical feature of fragmentation or comminution processes, may indicate that spots are fragments of larger flux tubes, although the coalescence of small flux units cannot be rigorously excluded.

Perhaps the question of the topology of the field beneath the surface could be settled by observations of periodic phenomena in spots. Thomas, Cram and Nye (1982) suggested that observations of oscillations in spots can be used to probe their deep structure. The discrete modes observed in the spectrum might indicate the coherence of the magnetic field down to several thousand km below the surface. However, Bogdan and Fox (1991) note that in order to discriminate between the two possible structures, one has to know whether the interaction of acoustic waves in the monolith is different from that in the cluster. Lack of a more detailed description of the cluster model precludes such a comparison. Their current investigation of the interaction of acoustic modes with a pair of flux tubes (Bogdan and Fox, 1991), and also with individual tubes (Bogdan, Poedts, and Goossens, 1992) will perhaps help to establish the possible differences. Thus sunspot seismology could be the most promising means of providing an answer to the long–standing question of the spot's topology.

Acknowledgments. I am deeply indebted to Hermann Schmidt for the opportunity to participate in the investigations of sunspot models in his group. Hospitality and the support at the Max Planck Institut für Astrophysik in Garching are gratefully acknowledged. I wish to thank Hermann Schmidt, Jack Thomas, and Nigel Weiss for many helpful comments and suggestions that greatly improved the manuscript. It is a pleasure to acknowledge an assistance of NATO Scientific Affairs Division during the workshop in Cambridge. This work was partly supported also by the grant 2 1177 91 01.

References

Adam, M.G. 1990. An observational examination of models for sunspot magnetic field configuration. *Solar Phys.*, **125**, 37–44.

Albregtsen, F. and Maltby, P. 1978. New light on sunspot darkness and the solar cycle. *Nature*, **274**, 41–42.

Albregtsen, F. and Maltby, P. 1981. Solar cycle variation of sunspot intensity. *Solar Phys.*, **71**, 269–283.

Albregtsen, F., Joras, P.B. and Maltby, P. 1984. Limb–darkening and solar cycle variation of sunspot intensities. *Solar Phys.*, **90**, 17–30.

Beckers, J.M. 1977. Material motions in sunspot umbrae. *Astrophys. J.*, **213**, 900–905.

Beckers, J.M. 1981. Dynamics of the solar photosphere. In S. Jordan (ed.), *The Sun as a Star*, NASA SP–450, Washington, pp. 11–64.

Beckers, J.M. and Schröter, E.M. 1969. The intensity, velocity and magnetic structure of a sunspot region. IV: properties of a unipolar sunspot. *Solar Phys.*, **10**, 384–403.

Bernstein, I.B., Frieman, E.A., Kruskal, M.D. and Kulsrud, R.M. 1958. An energy principle for hydromagnetic stability problems. *Proc. Royal Soc.*, *A*, **244**, 17–40.

Biermann, L. 1941. Der gegenwärtige Stand der Theorie konvektiver Sonnenmodelle. *Vierteljahrschr. Astron. Ges.*, **76**, 194–200.

Bogdan, T.J., Gilman, P.A., Lerche, I. and Howard, R. 1988. Distribution of sunspot umbral areas: 1917–1982. *Astrophys. J.*, **327**,451–456.

Bogdan, T.J. and Fox, D.C. 1991. Multiple scattering of acoustic waves by a pair of uniformly magnetized flux tubes. *Astrophys. J.*, **379**, 758.

Bogdan, T.J., Poedts, S. and Goossens, M. 1992. In preparation.

Chitre, S. M. 1963. The structure of sunspots. *Monthly Notices Roy. Astron. Soc.*, **126**, 431–443.

Chitre, S. M. and Shaviv, G. 1967. A model of the sunspot umbra. *Solar Phys.*, **2**, 150–157.

Cowling, T.G. 1976. On the thermal structure of sunspots. *Monthly Notices Roy. Astron. Soc.*, **177**, 409–414.

Danielson, R.E. 1961. The structure of sunspot penumbras. II: Theoretical. *Astrophys. J.*, **134**, 289–311.

Deinzer, W. 1965. On the magneto–hydrostatic theory of sunspots. *Astrophys. J.*, **141**, 548–563.

Gokhale, M.H. and Zwaan, C. 1972. The structure of sunspots. I: Observational constraints; Current sheet models. *Solar Phys.*, **26**, 52–75.

Hoyle, F. 1949. *Some Recent Researches in Solar Physics.* Cambridge.

Hughes, D.W. and Proctor, M.R.E. 1988. Magnetic fields in the solar convection zone: magnetoconvection and magnetic buoyancy. *Ann. Rev. Fluid Mech.*, **20**, 187–223.

Jahn, K. 1989. Current sheet as a diagnostic for the subphotospheric structure of a spot. *Astron. Astrophys.*, **222**, 264–292.

Jahn, K. and Schmidt, H.U. 1992. In preparation.

Jakimiec, J. 1970. Determination of the azimuthal electric currents in the surface layers of sunspots. *Sov. Astron.*, **14**, 416–425.

Jakimiec, J. and Zabża 1966. Magnetohydrostatic models of sunspots. II. *Acta Astronomica*, **16**, 73–79.

Knobloch, E. and Weiss, N.O. 1984. Convection in sunspots and the origin of umbral dots. *Monthly Notices Roy. Astron. Soc.*, **207**, 203–214.

Landman, D.A. and Finn, G.D. 1979. Radiation transfer through a model sunspot. *Solar Phys.*, **63**, 221.

Lites, B. and Skumanich, A. 1990. Stokes profile analysis and vector magnetic fields. V. The magnetic field structure of large sunspots observed with Stokes II. *Astrophys. J.*, **348**, 747–760.

Lou, Y.Q. 1989. The onset of magnetohydrodynamic convective instability. I. Ideal limit. *Astrophys. J.*, **347**, 520–528.

Low, B.C. 1975. Nonisothermal magnetostatic equilibria in a uniform gravity field. I Mathematical formulation. *Astrophys. J.*, **197**, 251–255.

Low, B.C. 1985. Some recent developments in the theoretical dynamics of magnetic fields. *Solar Phys.*, **100**, 309–331.

Maltby, P. 1992. This volume.

Maltby, P., Avrett, E.H., Carlson, M., Kjeldseth-Moe, O., Kurucz, R.L. and Loeser, R. 1986. A new sunspot umbral model and its variation with the solar cycle. *Astrophys. J.*, **306**, 284–303.

Meyer, F., and Schmidt, H.U. 1968. Magnetisch ausgerichtete Strömungen zwischen Sonnenflecken. *Zeitschrift f. Angew. Mathematik u. Mechanik*, **48**, 218–221.

Meyer, F., Schmidt, H.U., Weiss, N.O., and Wilson, P.R. 1974. The growth and decay of sunspots. *Monthly Notices Roy. Astron. Soc.*, **169**, 35–57.

Meyer, F., Schmidt, H.U., and Weiss, N.O. 1977. The stability of sunspots. *Monthly Notices Roy. Astron. Soc.*, **179**, 741–761.

Montesinos, B. and Thomas, J.H. 1989. Siphon flows in isolated magnetic flux tubes. II. Adiabatic flows. *Astrophys. J.*, **337**, 977–988.

Moreno–Insertis, F. 1987. The subsurface structure of sunspots and the origin of solar active regions. In E.H Schröter, M. Vazquez and A.A. Wyller (eds.), *The role of fine-scale magnetic fields on the structure of the solar atmosphere*, Cambridge University Press, pp. 167–182.

Moreno–Insertis, F. and Spruit H.C. 1989. Stability of sunspots to convective motions. I. Adiabatic instability. *Astrophys. J.*, **342**, 1158–1171.

Muller, R. 1976. Characteristics of the displacement of the penumbral bright grains of sunspots. *Solar Phys.*, **48**, 101–102.

Muller, R. 1981. The fine structure of sunspot penumbras. In L.E. Cram and J.H. Thomas (eds.), *The physics of sunspots*, Sacramento Peak Observatory, Sunspot, pp. 340–342.

Nordlund A. and Stein, R.F. 1990. Solar magnetoconvection. In J.O. Stenflo (ed.) *Solar Photosphere: Structure, Convection, and Magnetic Fields*, IAU Symp. No. 138, Kluwer, Dordrecht, pp. 191–211.

Parker, E.N. 1974. The nature of sunspot phenomenon: I. Solutions of the heat transport equation. *Solar Phys.*, **36**, 249.

Parker, E.N. 1975. The nature of the sunspot phenomenon. IV: The intrinsic instability of the magnetic configuration. *Solar Phys.*, **40**, 291–301.

Parker, E.N. 1976. Hydraulic concentration of magnetic fields in the solar photosphere. III. Field of one or two kilogauss. *Astrophys. J.*, **204**, 259–267.

Parker, E.N. 1979a. Sunspots and the physics of magnetic flux tubes. I. The general nature of the sunspot. *Astrophys. J.*, **230**, 905–913.

Parker, E.N. 1979b. Sunspots and the physics of magnetic flux tubes. IX. Umbral dots and longitudinal overstability. *Astrophys. J.*, **234**, 333–347.

Pizzo, J.V. 1986. Numerical solution of the magnetostatic equations for thick flux tubes, with applications to sunspots, pores, and related structures. *Astrophys. J.*, **302**, 785–808.

Pizzo, J.V. 1990. Numerical modeling of solar magnetostatic structures bounded by current sheets. *Astrophys. J.*, **365**, 764–777.

Proctor, M.R.E. 1992. This volume.

Proctor, M.R.E. and Weiss, N.O. 1982. Magnetoconvection. *Rep. Prog. Phys.* **45**, 1317–1379.

Schlüter, A. and Temesváry, St. 1958. The internal constitution of sunspots. In B. Lehnert (ed.), *Electromagnetic phenomena in cosmical physics*, IAU Symp. No. 6, Cambridge University Press, pp. 263–271.

Schmidt, H. U. 1987. Joint discussion. In E.H. Schröter, M. Vazquez and A.A. Wyller (eds.), *The role of fine-scale magnetic fields on the structure of the solar atmosphere*, Cambridge University Press, pp. 219–222.

Schmidt, H.U. 1992. Sunspots. In press.

Schmidt, H.U., and Wegmann, R. 1983. A free boundary value problem for sunspots. In B. Brosowski and E. Martensen (eds.), *Dynamical problems in mathematical physics*, Lang, Frankfurt, pp. 137–150.

Schmidt, H.U., Spruit, H.C., and Weiss, N.O. 1986. Energy transport in sunspots penumbrae. *Astron. Astrophys.*, **158**, 351–360.

Simon, G.W. and Weiss, N.O. 1970. On the magnetic field in pores. *Solar Phys.*, **13**, 85–103.

Spruit, H.C. 1977. Heat flow near obstacles in the solar convection zone. *Solar Phys.*, **558**, 3–34.

Spruit, H.C. 1981a. Magnetohydrodynamics of sunspots. *Space Sci. Rev.*, **28**, 435–448.

Spruit, H.C. 1981b. A cluster model for sunspots. In L.E. Cram and J.H. Thomas (eds.), *The physics of sunspots*, Sacramento Peak Observatory, Sunspot, pp. 98–103.

Staude, J. 1978. Models of heat flux in the subphotospheric layers of sunspots and the interpretation of the umbral granulation. *Bull. Astron. Inst. Czechosl.* **29**, 71–79.

Thomas, J.H. 1981. Theories of dynamical phenomena in sunspots. In L.E. Cram and J.H. Thomas (eds.), *The physics of sunspots*, Sacramento Peak Observatory, Sunspot, pp. 345–358.

Thomas, J.H. 1988. Siphon flows in isolated magnetic flux tubes. *Astrophys. J.*, **333**, 407–419.

Thomas, J.H. 1990. The structure of photospheric flux tubes. In C.T. Russel, E.R. Priest and L.C. Lee (eds.), *Physics of magnetic flux ropes*, AGU, Washington, pp. 133–140.

Thomas, J.H., Cram, L.E. and Nye, A.H. 1982. Five–minute oscillations as a subsurface probe of sunspot structure. *Nature*, **297**, 485–487.

Title, A.M. 1989. Personal communication.

Title, A.M., Frank, Z.A., Shine R.A., Tarbell, T.D., Topka, K.P., Scharmer, G. and Schmidt, W. 1992. This volume.

Van Ballegooijen, A.A. 1981. Energy transport in deep umbral layers. In L.E. Cram and J.H. Thomas (eds.), *The physics of sunspots*, Sacramento Peak Observatory, Sunspot, p. 140.

Van Ballegooijen, A.A. 1982. The structure of the solar magnetic field below the photosphere I. Adiabatic flux tube models. *Astron. Astrophys.*, **106**, 43–52.

Wegmann, R. 1981. Trefftz integral equation method for free boundary problems of potential theory. In J. Albrecht (ed.), *Numerical treatment of free boundary value problems*, Birkhäuser, Stuttgart, pp. 335–349.

Weiss, N.O., Brownjohn, D.P., Hurlburt, N.E. and Proctor, M.R.E. 1990. Oscillatory convection in sunspot umbrae. *Monthly Notices Roy. Astron. Soc.*, **245**, 434–452.

Wilson, P.R. 1981. Theories of sunspot structure and evolution. In L.E. Cram and J.H. Thomas (eds.), *The Physics of Sunspots*, Sacramento Peak Observatory, Sunspot, pp. 83–97.

Yun, H.S. 1970. Theoretical models of sunspots. *Astrophys. J.*, **162**, 975–986.

Yun, H.S. 1971. A magnetostatic sunspot model with 'twisted' field. *Solar Phys.*, **16**, 398–403.

Zirin, H. and Wang, H. 1990. High resolution sunspot observations. BBSO preprint No. 0318.

Zwaan, C. 1968. The structure of sunspot. *Ann. Rev. Astron. Astrophys.*, **6**, pp. 135–165.

Zwaan, C. 1974. Structure of sunspots. II: A continuum model atmosphere for dark umbral cores. *Solar Phys.*, **37**, 99–111.

Zwaan, C. 1975. *Solar Phys.*, **45**, 115.

THE FATE OF THE HEAT FLUX BLOCKED BY SUNSPOTS

H. C. Spruit
Max-Planck-Institut für Astrophysik
Karl-Schwarzschildstr. 1, 8046 Garching, Germany

ABSTRACT. A sunspot blocks a part of the heat flux from the solar surface, causing a thermal adjustment in the convection zone. Two very different time scales are involved, and the adjustments are different on the two time scales. On time scales longer than the thermal time scale τ_t of the convection zone (10^5 yr), a significant fraction of the heat flux blocked by spots can reappear elsewhere at the surface. The average spot coverage over this time scale leads to a small change in position of the Sun in the HRD. On time scales short compared with τ_t, almost all changes in the heat flux are absorbed by the convection zone. The surface temperature outside spots, and the solar radius *do not change* on such time scales. Thus for all time scales of observational interest, the irradiance variations directly reflect the instantaneous surface coverage by spots (and faculae if these are present).

1. Introduction

The absence of a bright ring around a sunspot implies that the heat flux that would normally emerge from the surface area of the spot has been diverted to somewhere else. To interpret irradiance variations such as those seen by the Nimbus and Solar Max satellites, one would like to know where this flux goes. How much of the irradiance change, for example, is due to a change of the temperature in the unspotted parts of the surface? How much does a possible radius change contribute?

I will argue here that our knowledge of structure of and energy transport in stars is sufficient to give satisfactory theoretical answers to these questions. The reason is that the answers depend only mimimally on uncertainties like the precise value of the convective efficiency (l/H ratio). They are even independent of the structure of the spots, as long as the thermal blocking effect of the spots extends to a depth that exceeds a certain (rather mimimal) value. This contrasts sharply with our theoretical understanding of the spots themselves. For a more detailed account of the effect of spots on internal structure, see Spruit (1991).

An issue of central importance is that of thermal time scales. When a dark spot area

J. H. Thomas and N. O. Weiss (eds.), Sunspots: Theory and Observations, 163–171.

is suddenly 'switched on', the Sun is no longer in equilibrium because it emits less energy than it generates in the core. It starts a thermal adjustment process. Part of this change takes place on a time scale on the order of the Kelvin contraction time of the Sun and therefore affects the position of the Sun in the HRD, but does not lead to changes that can be observed on historical time scales. Other time scales are much shorter (days to years). The reason for the existence of these widely differing time scales lies in the very high thermal conductivity of a convective envelope (more precisely: conductivity for changes in entropy, since the convective heat flux is a strongly nonlinear function of the gradient), combined with the very high heat capacity of a convective envelope. A useful analogy for a stellar convection zone that I will exploit is that of an electric heater plate, imagined to be made of copper (high heat conductivity), and being thick (high heat capacity).

But before discussing these time-dependent phonomena, consider the changes in structure on a very *long* time scale; long compared with the thermal time scale of the Sun as a whole (Kelvin-Helmholtz time scale).

2. Long term changes

On time scales long compared with the Kelvin-Helmholtz time scale, the sun is in thermal equilibrium, emitting as much energy as is produced in the core (ignoring the slower change in structure due to nuclear conversion). The questions one is interested are:

- -Do spots affect the luminosity generated in the core?
- -How large are the radius changes?
- -How large are the color changes?

2.1. Long-term luminosity and radius changes

To answer these questions one can construct solar models in the usual way, with changes in the description of the energy transport process in the convection zone to take into account the blocking effect of the spots, and a corresponding change in the outer boundary condition. Before proceeding, we note that in the presence of spots the effective temperature (defined in the usual way) is no longer equal to the surface temperature, and is not a good measure of the spectral type. Instead we will use T_s, the *surface temperature of the unspotted surface*. If f is the effective fraction of the surface covered by spots (assuming them to be completely dark), we have

$$L = 4\pi R^2(1-f)\sigma T_s^4, \tag{1}$$

Thermally relaxed models of spotted main-sequence stars have been made by Spruit and Weiss (1986). The results show that in stars with thin convection zones the luminosity is not affected by spots, while there is a significant luminosity reduction in fully and nearly fully convective stars. This can be understood qualitatively by appealing to some well-known properties of radiative cores of stars (e.g. Kippenhahn and Weigert, 1990). Integrating inward into the core, solutions with different starting conditions rapidly converge to a single track. The characteristic distance for this convergence is of the order of the pressure scale height. If the radiative core spans a large number of pressure scaleheights (as in F and early G stars), the dependence of the central conditions on changes taking place in the convection zone is small, so that the presence of spots does not influence the luminosity:

$$\Delta L/L \ll \Delta T_s/T_s. \qquad (F \text{ and early } G \text{ stars}, \quad t \gg \tau_t)$$

The heat flux blocked by spots, in such stars, is all radiated by the non-spot area of the star, which therefore has a higher temperature. The higher temperature in the convective envelope leads, by hydrostatic equilibrium, to a larger radius. Since the temperature change is limited to the convective envelope (and maybe one pressure scale height below it), the radius change is of the order

$$\Delta R/D \sim \Delta \ln T_s|_p,$$

where D is the depth of the convection zone and the index $_p$ indicates a change at constant pressure. The radius change is therefore small in the F stars, which have thin convection zones.

The situation is different if the core is small (in the sense defined). A change in the depth of the convection zone due to spots now has a direct effect on the conditions in the core. This is illustrated best by the extreme case of fully convective stars (Spruit and Ritter, 1983). For the main sequence these are stars later than about M5 according to present models. In such stars the luminosity is rather sensitive to the central temperature (index around 5). The central temperature therefore changes little compared with the luminosity. But all fully convective stars are similar: they are (nearly) polytropes of index 1.5. Their central temperature is a known function of stellar mass and radius ($T_c \propto M/R$). Thus the radius change is also small compared with the luminosity change. The entire internal structure is therefore unchanged, except for a different heat flux. In particular the entropy of the convection zone is unchanged. Since the structure of the surface layers is uniquely determined by the surface temperature T_s (normally equated to the effective temperature, but not in the presence of spots!) and surface gravity g, this means that the change in temperature of the unspotted surface is small compared with the luminosity change. With (1) we have therefore

$$\frac{\Delta L}{L} \approx -f \qquad (\textit{fully convective stars}, t \gg \tau_t),$$

$$\Delta R/R \ll \Delta L/L; \qquad \Delta T_s/T_s \ll \Delta L/L. \tag{2}$$

That is, the 'missing flux' of spots in these stars is truly missing even on the Kelvin-Helmholtz time scale. In reality the energy production is not an infinitely steep function of temperature, so that the central 'thermostat' is not perfect, and small changes in radius and surface temperature occur.

The behavior of stars of intermediate mass is between these extremes. Their surface temperature, radius, and luminosity are all affected by spots (see Spruit and Weiss, figs. 1, 2).

2.2. Bright rings

If part of the blocked heat flux reappears at the surface in the vicinity of the spot, as one might expect, there would be a 'bright ring'. The Sun does not show much evidence of such rings around spots[1]. The explanation for this has been found in the very high thermal conductivity of the convection zone due to turbulent convection (Spruit, 1977). In the mixing length model of convection, a change in heat flux $\mathbf{\Delta F}$ is related to the temperature by

$$\mathbf{\Delta F} = \kappa \rho c_p T \nabla (\Delta T/T), \tag{3}$$

where the turbulent diffusivity $\kappa \approx lv$ is found from the mixing length estimates of the eddy size l and convective velocity v. Due to the rapid increase of ρ with depth, only a small temperature difference is needed to carry a large heat flux. The importance of this fact was first discussed in the present context by Sweet (1955). For example, to carry a horizontal heat flux of the order $F_\odot = 6\,10^{10}$ over a length of 10 000 km at a depth of 10 000 km a temperature difference $\Delta T/T = 2\,10^{-4}$ is needed. This is the expected temperature enhancement below such a spot. Since it is the quantity $\Delta T/T$ (proportional to the entropy fluctuation) that is carried by convection, the surface temperature fluctuation $\Delta T_s/T_s$ cannot be larger than this. Thus we would expect a sunspot of diameter 10 000 km to show a bright ring with an intensity contrast of maybe 0.1% with respect to the average non-spot surface, if its blocking effect extends over 10 000 km.

In making this estimate, however, we have assumed that *all* the blocked heat flux is reradiated elsewhere at the surface. This assumption is justfied only if the spot lasts as long as it takes the Sun to reach thermal balance, which is hardly likely (since spots appear and disappear on much shorter time scales). The more realistic question therefore is how strong a bright ring would result around a spot that lasts for, say, ten years. This turns out to be smaller, as discussed in the next section.

The absence of bright rings can be made more understandable with a kitchen analogy. Consider an electric heater plate made of a nice thick piece of copper (rather than iron). I put a small piece of thermally insulating material on the plate, for example a ceramic tile (see figure), switch on the electricity, and wait until the heat lost from the plate equals the power dumped into the heater coil. Because of the poor thermal conductivity of the ceramic tile, its radiating top surface is cooler than the surrounding surface of the plate. Below the tile, the copper surface is a bit hotter than elsewhere because of the heat blocked by the ceramic. How hot is the resulting 'bright ring' around the ceramic tile? There is not much of a bright ring, because copper is a very good heat conductor. The heat flux blocked by the tile distributes quickly throughout the copper keeping the temperature inhomogeneity small. The surface is therefore also at a nearly uniform temperature, somewhat higher than it would have been without the tile.

[1] Most of the brightening around spots can be traced to faculae and has the characteristic patchy and irregular distribution of faculae, though a faint, regular, ring is definitely seen at just outside the penumbra in some cases (Livingston, 1991, private communication)

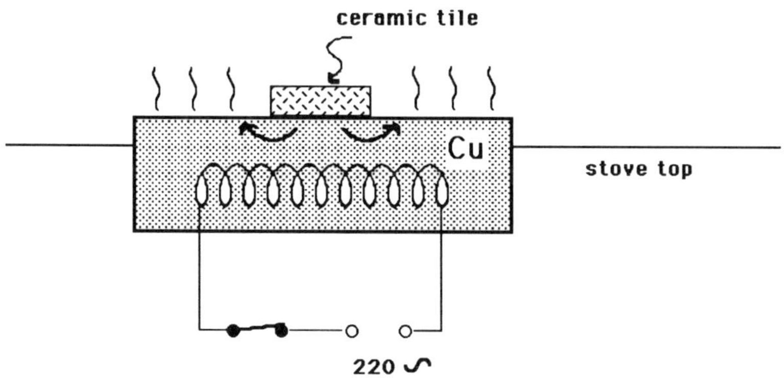

Fig.1 Heater plate analogy for the thermal adjustments in the convection zone due to heat flux blocking near the surface

3. Time dependence

Suppose that at $t = 0$ a sunspot of radius r whose blocking effect on the heat flux extends to a depth d below the surface is suddenly 'switched on'. How does the temperature in the Sun respond to this, as a function of time? A first time scale of importance for a convective envelope is the *diffusive time scale* t_d:

$$t_d = D^2/\kappa_t, \tag{4}$$

where D is the depth of the convection zone and κ_t the turbulent diffusivity (see eq. (3)). This is the time it takes a disturbance in the entropy to 'smear out', through convective turbulence, over the entire depth of the convection zone. Since this process takes place by macroscopic motions, it is much faster than the time it takes a disturbance to diffuse through a *radiative* envelope. For the Sun, it is on the order of 1 yr. Because this time scale is so short, the temperature distribution in a convective envelope is kept close to a state of marginal instability, that is, it is nearly isentropically stratified ('adiabatically' stratified in stellar evolution nomenclature). An exception to this is the thin superadiabatic surface layer (a thousand or so km deep), where convection is not as efficient at smearing out entropy disturbances.

Because turbulence dilutes heat through the entire convection zone, small temperature changes require large amounts of energy. Associated with this is a long time scale, the *thermal time* scale of the envelope:

$$t_t = U/L, \tag{5}$$

where U is the total thermal energy content of the convective envelope. This is the time it would take the temperature in the convection zone to change if the heat supply at its base were switched off. It is about $2\,10^5$ yr for the sun, and more generally it is of the order (M_e/M) $(T_e/T_c)t_{KH}$ where M_e is the mass of the stellar envelope, T_e the temperature at the base of the envelope, T_c the central temperature, and t_{KH} the Kelvin-Helmholtz time scale. In terms of the heater plate analogy of Fig.1, t_d is the time it takes the available thermal energy in the plate to spread out by conduction (on the order of maybe 10 s), and t_t is the time it takes this thermal energy content to change when the electricity is switched on or off (on the order of minutes). The ratio of these time scales is much larger in the case of a star.

According to these time scales, we have to divide the spottedness of the stellar surface into its long term ($\gg t_t$) average and the short term fluctuations around this mean. In this sense, all observationally accessible time scales are 'short', including the 'activity cycle' time scales. On these time scales, the entropy of the convective envelope is practically *unchangeable*. When a spot appears on the surface, a part of the heat flux it blocks may appear around it as a 'bright ring', but the energy emitted in this ring will usually be only a small fraction of the blocked heat flux, the remainder being lost in the envelope, slowly (on the time scale t_t) heating it up. That is, there is a very efficient *storage* of heat in the envelope, and the divergence of the heat flux is nonzero. Though this storage is temporary, the time scale involved is infinite from an observational point of view. In terms of the heater plate analogy, the appearance of a spot is analogous to putting the ceramic tile on the surface, and removing it a few seconds later. A faint 'hot ring' does form around the tile, but during its presence it emits only a small fraction of the blocked heat flux, because we have not given the plate the time it needs to reach equilibrium with the new boundary condition.

How much of the blocked heat flux returns to the surface in the form of a bright ring, and how large its maximum brightness is depends on the depth to which the blocking effect of the spot extends into the envelope. This is because the heat conductivity in a convective envelope (unlike the case of the heater plate) is a steeply increasing function of depth. Close to the surface it is not very large, due to the low gas density there. In the deeper layers, where convection is very efficient, the convective turbulence creates a very high thermal conductivity, compared with the normal radiative diffusivity. The transition takes place exactly in the superadiabatic layer at the top af the convection zone. In main sequence stars (but not in supergiants), this layer is very thin, a few hundred km for the Sun.

The storage of heat that occurs on time scales $\ll t_t$ takes place throughout the convection zone. But since the heat capacity is largest in the deepest parts of the convection zone, an accurate approximate answer is obtained by ignoring energy storage effects everywhere except in the deepest layers, and setting the entropy constant (in time) there. Since some storage takes place in other places as well, this overestimates the bright ring effects a bit. In this approximation, the bright ring effect can be calculated form a *steady* heat flow calculation. Such calculations were done in Spruit (1977) and Spruit (1982a,b). Foukal et al. (1983) and Chiang and Foukal (1984) did fully time dependent calculations which show how the ring effect develops in time. The latter calculations demonstrate the storage of heat explicitly, and the important role played by the thermal time scale (5) in this context (though for numerical reasons only the topmost layers of the convection zone were included, so that the thermal time scale was much less than the Sun's $2\,10^5$ yr).

For spots whose radius r is equal to or less than their depth d, one finds that the fraction α of the blocked heat flux that reappears at the surface in the form of a bright ring is of the order (Spruit, 1982a,b):

$$\alpha \approx 0.5[1 + d/(3H_0)]^{-2}, \qquad (t \ll t_t) \tag{6}$$

where H_0 is the pressure scale height at the surface. A fraction $1-\alpha$ of the blocked heat flux is stored in the convection zone. For spots with a radius larger than their depth, the value of α is expected to be less than given by (6) by a factor of the order d/r, so that (6) can be regarded as an upper limit. As an example, using $H_0 = 150$ km, spots with a depth of only 1000 km yield $\alpha = 0.05$: 95% of the blocked flux would be stored even for such shallow spots. Note that the bright ring effect decreases with the assumed depth of the spot. This is because a diversion of the heat flux is associated with a temperature change that decreases with this depth (c.f. the discussion near eq. (3)). For example, for $d = 150$ km (one scale height), (6) predicts that 28% of the blocked heat flux reappears at the surface, which is not a negligible fraction any more. We know however that the blocking effect of the umbra must extend to a depth at least equal to the observed Wilson depression of the spot (about 600 km), so that blocking depths as small as a scale height are not realistic, at least not for the umbra.

Though we do not know enough about sunspots to tell at which depth d their blocking effect on the heat flux disappears, these estimates show that it is highly unlikely that a significant amount of heat is radiated in bright rings for anything that looks like a sunspot.

3.1. Effect of ordered flows

The temperature enhancement below the spot generates a circulation cell with the spot at its center. This circulation is observed as the *moat flow* (for a compilation of observational results, see Brickhouse and LaBonte, 1988). It carries some of the excess heat to the surface, and one may wonder how this would modify the above results based on the turbulent diffusion model. The observed moat velocities (v) and sizes (l) are comparable to those of the supergranulation. An entropy contrast ΔS advected directly by this flow yields an energy flux proportional to $v\Delta S$, whereas in the diffusion model, with the turbulent diffusivity $\kappa_t = \frac{1}{3}lv$ based on the supergranulation velocity and length scale, would, over a distance L, yield a flux $\Delta S\kappa_t/L$. Taking for L the size of a spot, which is typically also of the order of supergranule size, we have $L \approx l$, and hence the flux carried by the circulation is of the same order as that given by the diffusion model. Fox et al. (1991) have simulated the convective flow around a small object, 1200 km wide and 132 km deep, that blocks 18% of the normal heat flux. The diffusive model (eq. 5) predicts for such an object that, after the initial transients, about 1/3 of the blocked heat flux reappears at the surface. Averaged horizontally over the 24000 km width of the Fox et al. simulations this yields a reduction of the net heat flux by 0.6%. An effect of this order of magnitude can be seen in the results of Fox et al. (see their figures 12,14), though it can not be reliably measured due to the large random fluctuations (accompanied by a curious 20 min oscillation). To measure the blocking effect of sunspots by numerical simulations of convection, more extensive computations will be needed. Such computations should also use a proper radiative boundary condition at the top surface, since the constant-entropy condition of Fox et al. overestimates the reappearance of blocked heat flux at the surface.

3.2. Variations in color and solar radius

The absence of bright rings implies that on the short term $t \ll t_t$ the temperature of the unspotted surface is unchanged:

$$\Delta T_s/T_s \ll \Delta L/L \qquad (t \ll t_t). \tag{7}$$

Since the temperature structure of the envelope does not change either, hydrostatic equilibrium implies that the solar radius is unchanged:

$$\Delta R/R \ll \Delta L/L \qquad (t \ll t_t). \tag{8}$$

Irradiance variations, as well as monochromatic variations, are therefore directly due to the spots (and faculae, see below)

3.3. Faculae

Faculae are areas of enhanced emission. Their effect on the internal structure is in principle identical to that of spots, but with the opposite sign. The net effect on internal structure is just the sum of the two, and hence is smaller than that of spots alone.

The main difference is that the facular excess emission probably derives from a depth that is not as large as the blocking depth of spots, so that they may be surrounded by 'dark rings' (the equivalent of the spot's bright ring). Calculations (Spruit, 1976; Grossmann-Doerth et al. 1989) indicate that these bright rings are narrow, and compensate for only a fraction of the excess emission in the facular structure. This agrees with the finding (Foukal and Lean, 1990; Lean 1991) that faculae and the active network account for an important part of the solar luminosity variation. They even dominate the variation on time scales of years, so that the Sun is brighter at maximum activity (it is not clear if this is also the case for other stars, especially the more active ones).

Facular areas on the Sun are bright only as long as the filling factor of the magnetic field is less than about 30% (Frazier, 1971). In areas where more magnetic field erupts the faculae are replaced by dark pores. If this is also the case in active stars, one would expect faculae to be less important than they are for the Sun.

4. Conclusions

The convective envelope of the Sun is deep enough that its thermal time scale is much longer than any observable time scale. As a result, variations in spot coverage on the surface have no effect on the internal structure of the Sun. Only the average over very long time scales has an influence on the internal temperature, on the temperature of the unspotted part of the surface, and on the radius of the Sun. Under minimal assumptions about the depth of the heat-blocking effect of sunspots one expects only very weak bright rings around spots. On observable time scales all photometric variations can be interpreted directly in terms of the fraction of the visible hemisphere covered by spots at each instant. The effects of faculae are of opposite sign to those of spots, but again are entirely local.

The changes in radius, luminosity, and color on *long* time scales ($\gg 10^5$ yr) due to spots cause a minute change in the sun's color and position in the HRD, compared with otherwise identical but less active stars.

References

Brickhouse, N.S. and LaBonte, B.J.: 1988, *Solar Phys.,* **115**, 43
Chiang, W.H. and Foukal, P.: 1984, *Solar Phys.,* **97**, 9
Foukal, P.V., Fowler, P. and Livshits, M.:1983, *Astrophys., J.,* **267**, 863
Foukal, P.V. and Lean, J.: 1990, *Science* **247**, 556
Fox, P., Sofia, S. and Chan, K.L.: 1991, *Solar Phys.,* **135**, 15
Frazier, E.N.: 1971, *Solar Phys.,* **21**, 42
Grossmann-Doerth, U., Knölker, M., Schüssler, M., Weisshaar, E.: 1989, in R. Rutten and G. Severino (eds.) *Solar and stellar Granulation*, Kluwer, Dordrecht, p.481
Kippenhahn, R., and Weigert A.: 1990, *Stellar Structure and Evolution*, Springer, Berlin, Ch. 10.3
Lean, J.: 1991, in P.B. Byrne (ed.) *Effect of starspots on internal stellar structure*, Lecture Notes in Physics, Springer, Berlin, in press
Spruit, H.C.: 1976, *Solar Phys.,* **50**, 269
Spruit, H.C.: 1977, *Solar Phys.,* **55**, 3
Spruit, H.C.: 1982a, *Astron. Astrophys.,* **108**, 348
Spruit, H.C.: 1982b, *Astron. Astrophys.,* **108**, 356
Spruit, H.C.:1991, in C. Sonnett and M. Matthews (eds.) *The Sun in Time*, University of Arizona, Tucson, in press
Spruit, H.C. and Ritter, H.: 1983, *Astron. Astrophys.,* **124**, 267
Spruit, H.C. and Weiss, A.: 1986, *Astron. Astrophys.,* **166**, 167
Sweet, P.A.: 1955, *Vistas in Astronomy* **1**, 679

IV. FINE STRUCTURE OF SUNSPOTS

The remarkable advances in high-resolution observations of sunspots within the last few years have changed our understanding of fine structure and raised many challenging theoretical problems. The most striking effects are in the filamentary penumbra. Close study shows that bright filaments are composed of separate grains which migrate inwards and may penetrate into the umbra, while the dark filaments are associated with a horizontal outflow at the outer edge of a spot (the Evershed effect). The field geometry has at long last been established: in the outer penumbra the magnetic field is nearly horizontal in dark filaments but inclined at almost 40° to the horizontal in bright filaments. Thus the Evershed flow is parallel to the local field. These structures are produced by the magnetoconvective processes that are responsible for energy transport immediately below the photosphere. Numerical studies of nonlinear convection in a vertical field reveal a variety of steady, oscillatory and travelling–wave solutions, depending critically on the ratio of the effective magnetic and thermal diffusivities. As yet, however, there is no agreement on the form of convection in the umbra. Some believe that the subphotospheric magnetic field splits into a tight cluster of isolated tubes, separated by field–free plasma, while others envisage small–scale convection in a coherent but inhomogeneous flux tube. Penumbral convection seems to involve interchanges between the filaments, a process which still needs to be modelled quantitatively.

This section begins with two chapters on observations. First, R. Muller gives an overview of the observations of fine structure in sunspot umbrae and penumbrae, emphasizing the features seen in white light but relating them to spectroscopic measurements of velocities and magnetic fields. Then A. M. Title *et al.* present new observational results concerning the fine structure of the magnetic and velocity fields in a sunspot. They show that the inclination of the magnetic field to the horizontal varies substantially (by 35–40°) in the azimuthal direction around the penumbra, on a scale comparable to the width of the penumbral filaments, while the field strength remains approximately constant. The Evershed flow is largely confined to regions where the magnetic field is horizontal.

At the Workshop R. A. Shine showed high-resolution video images of time-dependent, cloudlike structures in the Evershed flow. L. Golub described high-resolution X–ray observations of magnetic loops, showing that some flux tubes emanating from the inner penumbra are connected to regions far away from the spot. G. B. Scharmer presented important new observations of the behaviour of fine-scale intensity and magnetic features in the penumbra. He found that the bright penumbral filaments tend to move inward across the umbra-penumbra boundary and into the umbra, where they break up into several smaller features which are hard to distinguish from isolated umbral dots. There was prolonged discussion on the nature of the umbral dots: M. A. Aballe Villero presented the results of observations at two different wavelengths which show that the dots have a variety of diameters and temperatures but are cooler than the normal photosphere; W. Schmidt reported that the field strength in bright dots is reduced. It was agreed that the distinction between dots near the periphery of the umbra and those in the centre is important, but W.

C. Livingston argued that dots are actually fibril structures, while P. Maltby insisted that the umbrae of large spots have dark cores in which no fine structure can be discerned.

On the theoretical side, the chapter by M. R. E. Proctor reviews the analytical and numerical work on magnetoconvection, with emphasis on the effects of compressibility and stratification, and discusses the relationship of time–dependent model calculations to umbral convection. The chapter by A. Rai Choudhuri is concerned with the overall structure of an isolated umbra; he assesses the current status of the cluster model of a sunspot and compares it with the monolithic flux–tube model in the light of recent observations and theoretical considerations.

There were three more contributed papers on theoretical problems. P. A. Fox presented a progress report on techniques for representing subgridscale behaviour and their implementation in numerical models of magnetoconvection. P. C. Matthews described linear and weakly nonlinear studies of compressible magnetoconvection, focusing on the symmetry–breaking effects of an inclination in the overall magnetic field. B. Montesinos discussed a siphon-flow model for the photospheric Evershed flow that takes into account recent observations indicating a deep penumbra. In the model the arched flux tubes carrying the Evershed flow are embedded in an ambient magnetic field, which reduces the magnetic buoyancy of the tube and allows for long arches consistent with the lengths of observed penumbral filaments.

FINE STRUCTURE OF UMBRAE AND PENUMBRAE

R. MULLER
Observatoire du Pic du Midi
65200 Bagnères de Bigorre
France

1. Introduction

In this review paper, I describe the fine structure of a well-developed sunspot, as it appears when observed with a spatial resolution of 0"3 or 0"25. I do not restrict myself to describing morphological properties, because I also want to show how the small scale features are associated with the velocity and magnetic field patterns. This is now possible thanks to recent spectrographic and magnetographic observations obtained at the Swedish Telescope at La Palma (Canary Islands) with a resolution close to 0"3. These high-resolution observations of the intensity, velocity and magnetic structure of sunspots are very helpful in discussing the validity of more elaborate models.

A well-developed sunspot has a dark umbra surrounded by a filamentary penumbra. The boundary between the umbra and the penumbra is very sharp, in terms of brightness, so they are very easily distinguished from each other. This review is organized as follows : the umbra is described (section 2), separately from the penumbra (section 3). For each of these two distinct parts of a sunspot, I describe their morphological aspect at high resolution, and some properties of the small scale features such as their size, brightness and temperature, their lifetime and proper motion ; then the velocity and the magnetic structure ; and finally I discuss the main classes of models. I conclude the paper with a few suggestions for future observations which should be useful for a better understanding of sunspots (section 4). The review is restricted to the structure at the photospheric level.

2. Fine Structure of Sunspot Umbrae

2.1. MORPHOLOGY OF SUNSPOT UMBRAE

The structure of a sunspot umbra can be very complex. The degree of complexity depends on the stage of development of the sunspot and on its size. In general, a young sunspot (Figure 1) is more irregular than an old one (Figure 2), which may even became nearly radially symmetric. When one observes a sunspot umbra with a resolution better than 0"5, most striking is the presence of many small, bright, starlike points, for which the name "umbral dots" has been adopted, following Danielson (1964). Some of them can be aligned and bright enough to form one or several light bridges crossing

J. H. Thomas and N. O. Weiss (eds.), Sunspots: Theory and Observations, 175–193.

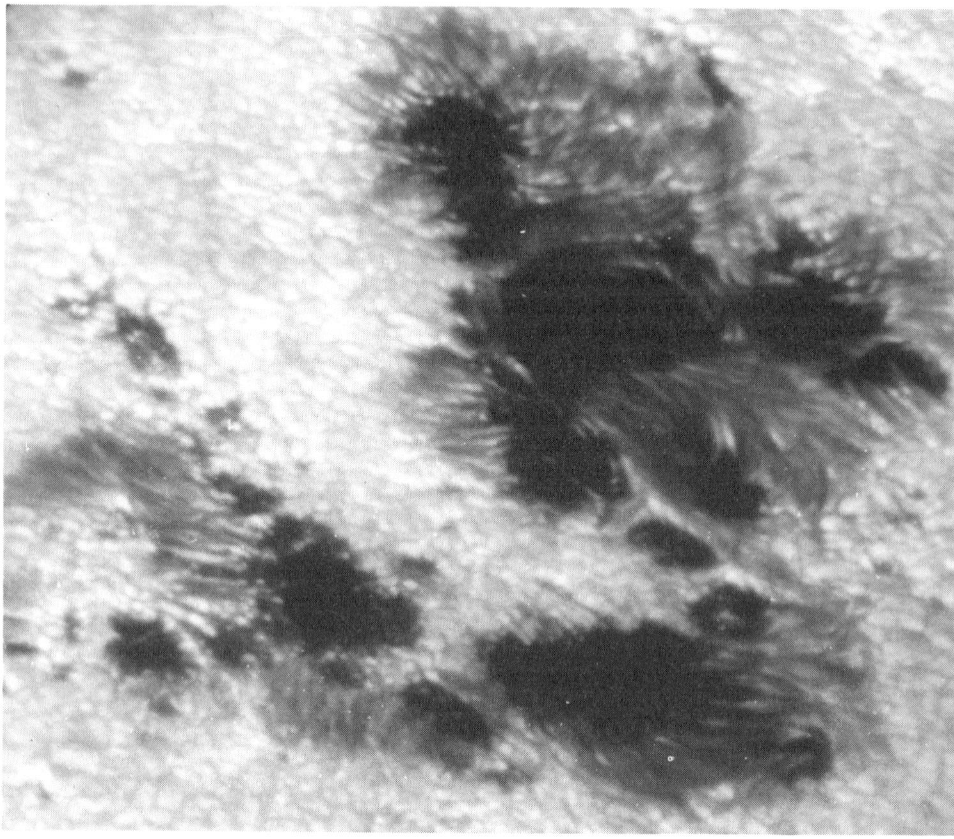

Figure 1a : A complex group of sunspots. Filamentary penumbrae have developped only at the outer boundary of the spots group. In the inner boundary the granulation is directly adjacent to sunspot umbrae. Note the existence of several light bridge, made with penumbral material or with bright umbral dots.

the umbra. They show up over a background whose brightness changes markedly throughout the umbra. In the darkest part of a large sunspot, the intensity does not exceed 5% of the photosphere or so (values as small as 2% at $\lambda 550$ nm have been reported in the literature), while at the penumbra boundary it can be as high as 50%. The background has a fine structure too, except possibly in the darkest parts, which are called "dark cores" (Zwaan and Buurman 1971) or "voids" (Livingston 1991). This fine structure has the same size as umbral dots, but its contrast is much lower. Most authors do not consider these weak features to be umbral dots, and hence they are excluded from their measurements. Are umbral dots and background fluctuations of the same origin ? This is an interesting question, which should have important physical implications. Several authors distinguish peripheral umbra dots, located at the umbra-penumbra boundary where the intensity gradient is strong, from the central umbral dots, because of differences in brightness and shape and because peripheral dots are brighter, often comet-like elongated (Soltau 1982 ; Grossmann-Doerth, Schmidt and Schröter

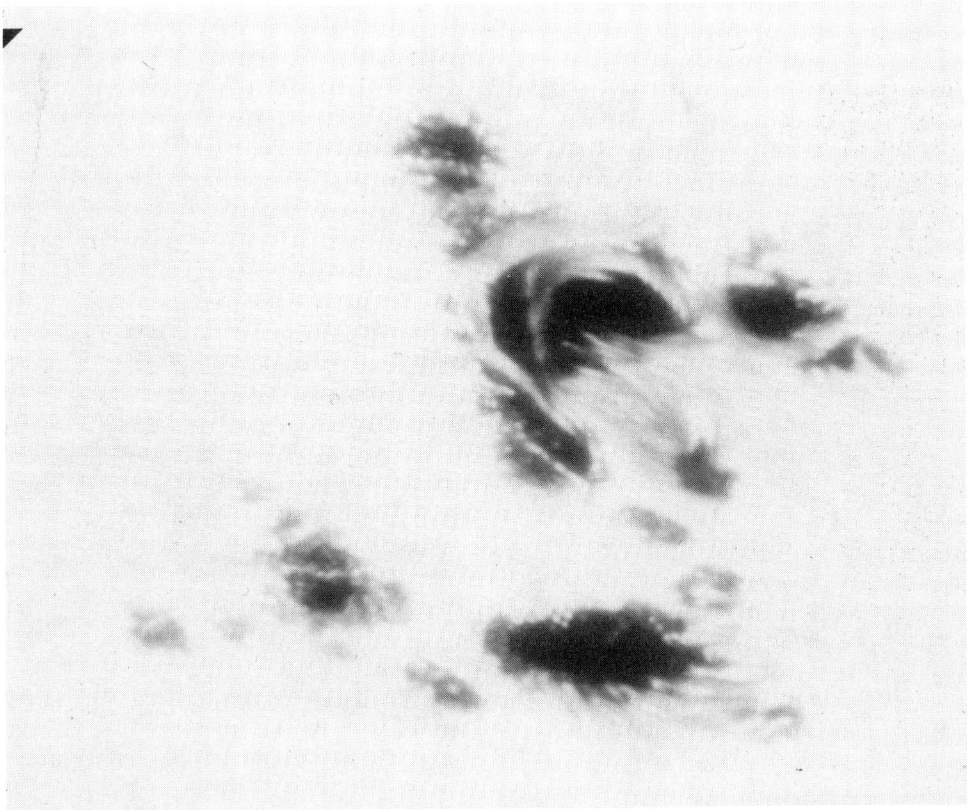

Figure 1b : Details in the umbra of the sunspots group of Figure 1a, photographed at the same time. Many umbral dots are visible in small umbrae. The central part of the main umbra near the center of the figure is free of any feature, even in strongly overexposed frames.

1986 ; Kitai 1986; Ewell 1991 ; Aballe-Villero 1991). Krat, Karpinski and Pravdjuk (1972) and Soltau (1982) reported that umbral dots are often situated on diffuse patches of size 1"-3". Inspecting some Pic du Midi high resolution photographs, I do not find this to be a general case ; rather I see umbral dots showing up over the background of weak intensity fluctuations.

When they are observed with a lower resolution (>0"5), sunspot umbrae show a cellular pattern, fairly similar to the granular pattern outside sunspots. This similarity gave some support to a convective interpretation of the so-called "umbral granules" (Loughhead and Bray 1960 ; Bray and Loughhead 1964 ; Bumba, Hejna and Suda 1975 ; Hejna 1977 ; Bumba and Suda 1980). However, high resolution observations show an umbral structure so different from the photospheric granulation that the term granulation is certainly inappropriate.

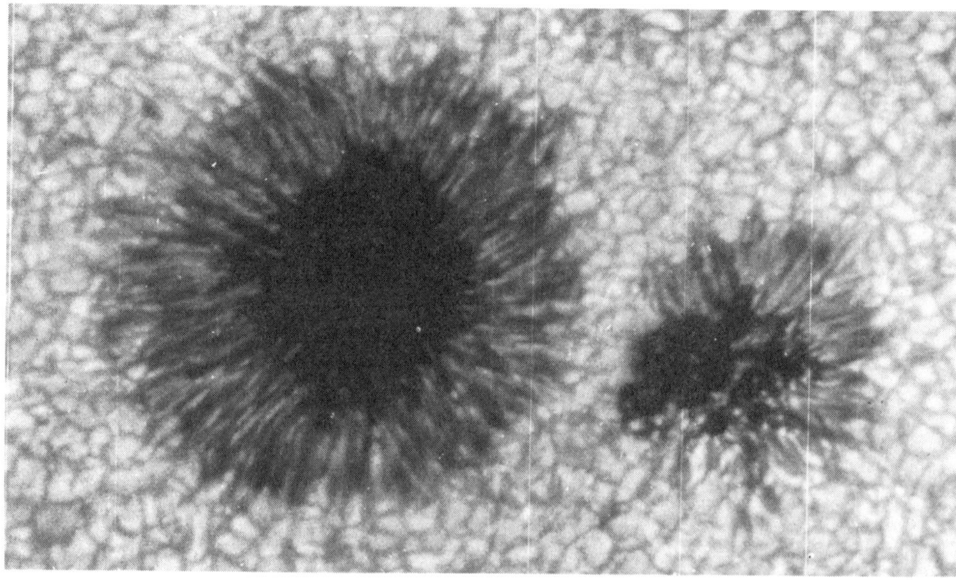

Figure 2a : A small regular sunspot showing a symmetrical filamentary penumbral. The small spot on the right contains many umbral dots.

Umbral dots often appear aligned (Adjabshirzadeh and Koutchmy 1980 ; Kusoffsky and Lindstedt 1986), sometimes with penumbral filaments (Livingston 1991). In some extreme cases, alignments of particularly bright dots form light bridges, as mentioned above.

The existence of a dark core, void of any bright features, in the largest sunspot umbrae is a matter of dispute. Loughhead, Bray and Tappere (1979) and the Ondrejov group, Bumba, Hejna and Suda (1975), Bumba (1991), claim that they have observed umbral dots everywhere in every sunspot. On the other hand, Zwaan and Buurman (1971) and more recently Ewel (1991) and Livingston (1991) have reported that some sunspots, in general of large size, can have dark cores ; Ewel used a sensitive, near-infra-red, low-noise CCD camera and did not detect any feature with a contrast higher than 2% in the darkest part of a large sunspot, covering 10% of the surface of its umbra. I have been unable to detect any trace of brightness fluctuations in the main umbra in Figure 1, even in the over-exposed photographs. On the other hand, some traces of fluctuations are present in the large sunspot umbra in Figure 2, but in the very central part, which covers only a few percent of the surface. The existence of a dark core may just be a matter of technical definition, depending on the threshold of the intensity fluctuations (2% in Ewel's work) and on the relative area free of fluctuations (2% of the umbra in Bumba, Sobotka and Simberova, 1990). We may wonder whether very low contrast fluctuations have a physical significance. These are points to be clarified, because dark cores can be places where the sunspot magnetic field is strong enough to completely inhibit the solar convection and thus tell us about the value of the critical magnetic field strength.

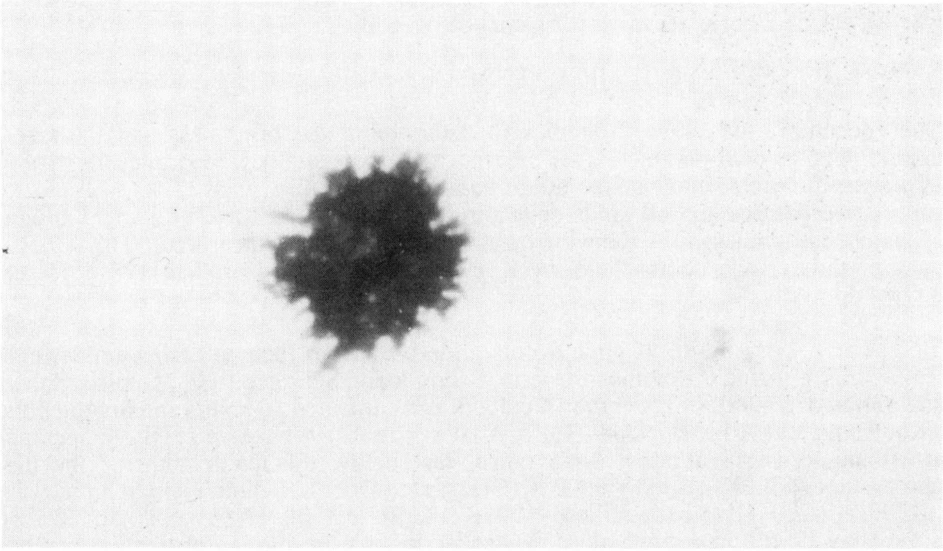

Figure 2b : Details in the umbra of the main sunspot of Figure 2a. A few conspicous umbral dots show up. The dark background shows faint brightness fluctuations, even in the central part.

2.2. SIZE OF UMBRAL DOTS

Measuring the size of umbral dots is not an easy task, as it is close to the telescopic resolution, or even smaller, and the intensity profile of the dots is much smeared. In addition, they are embedded in a non-uniform dark background, many of them are of low contrast, and the amount of scattered light from the surrounding penumbra and photosphere is high. In general, the measured sizes lie in a rather narrow range, 0"3 – 1"0, just above the telescopic resolution (Adjabshirzadeh and Koutchmy 1978, 1980 ; Loughhead, Bray and Tappere 1979 ; Koutchmy and Adjabshirzadeh 1981 ; Aballe-Villero 1991). However, Grossmann-Doerth, Schmidt and Schröter (1986) found much larger umbral dots, of size up to 1"4. These discrepant results reflect the difficulties mentioned above, in identifying and measuring umbral dots. Retrieving the real size from the observed intensity profiles is another difficult exercise. Restored values are very inaccurate, because of poor knowledge of the blurring function and uncertainties introduced by the restoration process (Adjabshirzadeh and Koutchmy, 1980 ; Grossmann-Doerth, Schmidt and Schröter, 1986). Alternatively, one can derive the true size from the profiles of umbral dots observed in two different wavelengths ; this method has the advantage of being independent of blurring, but leads again to incorrect values, because one must assume, for simplicity, that umbral dots radiate like black bodies. This is probably wrong for such small features embedded in a cool atmosphere. A size of 200-250 km has been derived, either by two-colour analysis (Beckers and Schröter 1968 ; Koutchmy and Adjabshirzadeh 1981 ; Aballe-Villero 1991) or by blurring corrections (Adjabshirzadeh and Koutchmy 1978). However, the significance of a characteristic size for umbral dots is questionable, owing to the wide range of values obtained by Aballe-Villero (0"1 – 0"6) and Grossmann-Doerth, Schmidt and Schröter (0"3-1"3). Observations of higher resolution and lower noise are required, in order to see whether the largest observed umbral dots are actually single features, and whether the dark cores are free of any convective eddies.

2.3. BRIGHTNESS AND TEMPERATURE OF UMBRAL DOTS

One encounters the same difficulties in determining the brightness and the size of umbral dots, as both parameters are derived from the same degraded intensity profiles. In a sunspot umbra, one can see bright and sharp point-like features, as well as weak and diffuse patches and all kinds of features of intermediate size. This morphological appearance is confirmed by the wide range of contrast values measured : 0.02 – 0.30, in units of the average photosphere, by Koutchmy and Adjabshirzadeh (1981), for core umbral dots only.

The real temperature of umbral dots is derived from their intensity profile, either restored for blurring (Grossmann Doerth, Schmidt and Schröter 1986 ; Bumba, Sobotka and Simbarova 1990) or taken in two colours and compared (Beckers and Schröter 1968 ; Koutchmy and Adjabshirzadeh 1981; Aballe-Villero 1991). As a result, temperatures are found to be in a rather wide range, just below the temperature of the mean photosphere (6300K) : 5300 – 6000 K (Grossmann Doerth, Schmidt and Schröter 1986) ; 4700 – 6700 K (Aballe-Villero 1991) ; 4250 – 4700 K (Bumba, Sobotka and Simbarova 1990 ; these low values may reflect a lower resolution of the observations). An average temperature close to that of the mean photosphere has been reported by Beckers and Schröter (1968) and Koutchmy and Adjabshirzadeh (1981). A two-component model, derived to fit unresolved umbra photospheric lines by Pahlke and Wiehr (1990), indirectly confirms that the temperature in umbral dots must be much lower than the photosphere (4200 – 5600 K in the central part of the umbra, 4600 – 6300 K at the periphery).

It thus appears that umbral dots can have any temperature between the cool umbral dark background, 4200 K, and a little more than the mean photosphere, 6300 K. It must be noted, however, that the temperatures, as well as the sizes, are derived assuming that umbral dots radiate like a black body, which is probably incorrect for such small, high contrast features. A proper 3-dimensional NLTE treatment is needed to compute correct temperatures.

The simultaneous presence of umbral dots with temperatures ranging between umbral (4000K) and photospheric (6500K) temperatures, in a sunspot umbra is not surprising, because they are observed at various stages of development. Therefore it would be interesting to follow the time evolution of their contrast, brightness, and temperature (as was done by Adjabshirzadeh and Koutchmy 1980, for the radiative flux), and then to analyse the dispersion of the maximum values measured in a time series, versus distance from the umbra periphery.

2.4. LIFETIME OF UMBRAL DOTS

The mean lifetime of umbral dots was first found to be relatively short, 20 to 30 min on average (Danielson 1964 ; Beckers and Schröter 1968 ; Ikhsanov 1972 ; Adjabshirzadeh and Koutchmy 1980). It was found to be larger (40 to 60 min) when more recent, higher quality photographic material was used (Kitai 1986 ; Kusoffsky and Lindstedt 1986). CCD cameras are best suitable for umbral dot observations, because of their low noise and large sensitivity range. Unfortunately, the results from two recent CCD camera data sets are in strong disagreement : according to Schmidt (1991) the lifetime of all identified umbral dots is longer than 40 min (the movie, obtained at Izaña Observatory lasts only 40 minutes) ; on the other hand, according to Ewel (1991), the lifetime is only 15 min in average, although a few umbral dots can remain visible for more than 2 hours (Kusoffsky and Lindstedt, 1986, have also seen a few umbral dots last for more than two hours). The short average lifetime reported by Ewel can be explained if all features of contrast higher than 2% have been taken into account, and if low contrast umbral dots are numerous and of short lifetime (or are detectable above the 2% threshold only for a short time). It thus seems that bright umbral dots have lifetimes in the range 30 to 60 minutes, sometimes longer, while the fainter ones have much shorter lifetimes. A possible relationship between contrast and lifetime must be checked on observational material of higher quality (in spatial resolution, duration and noise).

2.5. PROPER MOTION OF UMBRAL DOTS

Peripheral umbral dots, near the umbra-penumbra boundary, have been observed to move inward, with an average speed of 0.3-0.5 kms^{-1}, like penumbral grains (see section 3.4 and Muller 1973a ; Kitai 1986 ; Ewel 1991). In fact they often originate as penumbral grains. In a small spot, Kitai (1986) has observed that the central umbral dots were also moving toward the centre, but at a smaller speed than at the periphery : 0.1 – 0.2 kms^{-1}. Ewel (1991) has suggested that the primary distinguishing property between central and peripheral dots is their proper motion, rather than brightness or shape. A distinction between outer and inner umbral dots was also made by Kitai (1986) on the basis of their proper motion.

2.6. MAGNETIC AND VELOCITY STRUCTURE OF SUNSPOT UMBRAE

Measurements of spectra at moderate spatial resolution (not better than 1") have indicated that the magnetic field is weaker and more inclined in umbral dots than in the dark surrounding. In general, the reduction of strength is found to be a few hundred gauss (Beckers and Schröter 1969 ; Zwaan and Buurman 1971 ; Buurman 1973 ; Adjabshirzadeh and Koutchmy 1983 ; Zwaan, Brandts and Cram 1985 ; Pahlke and Wiehr 1990). However Kneer (1973) reported a much stronger reduction, about 1500G, which represents half the strength measured in the dark regions of the umbra. Similar observations of moderate resolution revealed the presence of upflows in umbral dots, relative to the dark material, with speeds less than 1 kms^{-1} (Beckers and Schröter 1969 ; Pahlke and Wiehr 1990). It is noteworthy that Kneer (1973) again reported a much larger value of 3 kms^{-1}. Was he observing a very peculiar sunspot, where an unusually strong reduction of field strength in umbral dots produced strong upflows ?

Surprisingly, the correlation between umbral bright structures and magnetic and velocity fluctuations vanished when high spatial resolution spectra became available. Such spectra have been obtained by Zwaan, Brandts and Cram (1985) at the Vacuum Tower Telescope of Sacramento Peak Observatory (resolution slightly better than 1") and Lites et al. (1991) at the Swedish Solar Telescope at La Palma (resolution : 0"3). Both groups of investigators did not detect any significant fluctuation of the magnetic field and velocity throughout sunspot umbrae. However, they have different explanations for the lack of correlation between bright features and velocity and magnetic fluctuations. According to Zwaan (1985), there are no magnetic and velocity structures at the level where the lines are formed, about one scale height above the continuum, where umbral dots are visible. Lites et al. disagree with the level of formation adopted by Zwaan ; they found it to be much lower in the photosphere and representive of the continuum. Therefore, they conclude that the energy in umbral dots is carried by radiation rather by a mass flow at $\tau=1$, and that the magnetic field is not highly structured at this level.

2.7. DISCUSSION OF UMBRAL MODELS

There are two main classes of models that try to explain the presence of bright dots in sunspot umbrae : the "cluster model" of Parker (1979) and Choudhuri (1986) and the "monolithic model". In the cluster model, a sunspot is made of a bundle of thin flux tubes, separated by field-free material. As the flux tubes rapidly expand at the top, the geometry of the field-free fluid has the form of a tapering column ending at an apex. Hot columns of gas can rise between the flux tubes and leak vertically toward the photosphere when the apex opens under the pressure of the rising gas, like a magnetic valve. In the monolithic model, which consists of a monolithic flux tube, overstable convection permits the vertical transport of thermal energy in a non-linear regime (Knobloch and Weiss 1984 ; Weiss et al. 1990).

According to Lites et al. (1991), the absence of a significant vertical flow and magnetic field weakening does not rule out the Parker/Choudhuri cluster model. It simply means that the transport of energy must be quenched below the surface, at a distance short enough so that the radiation can still heat the surface locally and cause umbral dots of small horizontal size. For the same observational reasons, in a monolithic magnetic field oscillatory convection should not reach the surface ; the long umbral-dot lifetime implies that the convection must be nonlinear (Knobloch and Weiss 1984). Such convection should produce fluctuations of the magnetic field ; this is not observed. However, the treatment of nonlinear convection in the presence of a strong magnetic field is still not developed and understood enough to draw definite conclusions.

3. Fine Structure of Sunspot Penumbrae

3.1. MORPHOLOGICAL DESCRIPTION

Examining Stratoscope photographs, Danielson (1961) described the penumbra of sunspots as being made up of bright filaments separated by narrow dark fibrils. Penumbrae are of a varying degree of complexity, depending on the sunspot size and stage of evolution. In a regular sunspot (Figure 2), or in the regular part of a complex sunspot (Figure 1), the filaments cross the penumbra almost radially. In complex parts of penumbrae, the filamentary structure is much less ordered, having filaments which are not necessarily oriented perpendicular to the umbra. A good example of a sunspot with both regular and complex penumbral parts is shown in Figure 2 of the chapter by Thomas and Weiss (1992, these proceedings). In the following, we mainly describe the properties of regular sunspot penumbrae.

By analysing an outstanding time series of high-resolution photographs of a large sunspot taken at the Pic du Midi Observatory, Muller (1973b) realised that the penumbral bright filaments are, in fact, chains of much shorter aligned grains. They have elongated, cometlike photometric contours (Muller, 1973a ; Soltau, 1982), especially when they are near the umbra boundary ; the maximum brightness of the coma is located on the umbral side of the grain. It is easy to identify a bright grain and follow its change of shape and brightness on successive frames, because it forms a well-defined entity surrounded by dark material. Conversely, it is impossible to identify dark elements, because the dark material surrounding the bright features forms a more or less coherent background. The situation is very similar in the granulation pattern, where the bright granules, surrounded by interconnected intergranular lanes, are easy to identify. This is the reason why in the literature there are no estimates of lifetimes, proper motions, etc., for penumbral dark fibrils and intergranular lanes.

In a high-resolution photograph of a sunspot obtained at Big Bear Observatory, the presence of narrow, elongated dark fibrils visible at the outer edge of the penumbra give the impression of elevated filaments over normal granules (Moore 1981a, b). This finding raises the problem how cool and dense material is suspended above granules. It is possible, however, that the observed fibril elongation was artificially created by some atmospheric astigmatism during the exposure (Muller 1985).

Sunspots have a sharp outer boundary between the photospheric granulation and the penumbra. Many dark spaces, 1 to 3 arcseconds in size and void of any feature, are visible at this boundary. Photospheric granules are directly adjacent either to these dark spaces or to elongated bright grains. It is easy to distinguish a granule from a penumbral grain, which is smaller, fainter, narrower and more elongated. Near the outer boundary, penumbral grains are usually longer than near the inner boundary and do not appear well isolated and aligned to form long filaments. Some dark spaces prolongate into the granular pattern as intergranular lanes. I do not believe they are extensions of dark penumbral fibrils overlying granules (as suggested by Moore 1981), because granules adjacent to sunspots as well as intergranules behave normally, as in the quiet Sun.

3.2. BRIGHT GRAINS : WIDTH AND SHAPE

After correction for blurring, typical grains have a width of 0"35 = 270km (Krat, Karpinski and Pravdjuk 1972 ; Muller 1973b ; Bonet, Ponz and Vazquez 1982). Danielson, in 1961, estimated the width to be 300 km. Although speckle interferometry shows power down to 0"11 (Stachnik, Nisenson and Noyes 1983), the dispersion is not large. As reported in the previous section, penumbral grains are shaped like comets, with their heads pointing toward the umbra. The length can be several times larger than the width, especially near the outer penumbra boundary.

3.3. BRIGHTNESS DISTRIBUTION

The average brightness of the penumbra is 0.75 that of the surrounding photosphere. In the radial direction, the brightness decreases sharply at the penumbra-photosphere boundary ; then it declines gently from 0.80 to 0.70 through what is called the "penumbral plateau", which occupies a large part of the penumbra ; at the umbra – penumbra boundary, it decreases again with a strong gradient, down to umbral intensities. The detailed brightness distribution is usually described as consisting of a bright and a dark component. On average the brightness, relative to the photosphere and corrected for blurring, is, at λ550 nm and in the penumbral plateau (Muller 1973b ; Collados, del Toro Iniesta and Vazquez 1988) :

$I_b = 0.95$ for the bright component (bright filaments)
$I_d = 0.60$ for the dark component (dark fibrils).

The brightness of both components slowly decreases from the photosphere to the umbra boundary. The average brightness of the penumbra, computed with the intensity and relative area of the two components, is 0.72, in very good agreement with low-resolution photometric measurements. However, despite the two-component filamentary appearance of the penumbra, the brightness distribution is singly peaked, not showing any photometric evidence of two components, not even after corrections for blurring (Grossmann-Doerth and Schmidt 1981). This can be explained by the fact that there are large variations through the penumbra, in the sense that a local maximum of brightness can be less bright than a minimum elsewhere. A maximum (or a minimum) has only a local meaning : it is brighter (or darker) than its surroundings. The single-peaked observed distribution reflects the important contribution of pixels of intermediate intensity between local maxima and minima. As discussed in the next paragraphs, bright filaments and dark fibrils have different properties and are of different origin. Consequently a 2-component description of the penumbra, in terms of brightness, is not only justified by its filamentary appearance, but, more important, by the different physical origin of the two components ; it is thus useful for modelers.

3.4. PROPER MOTIONS IN THE PENUMBRA

Penumbral bright grains move toward the umbra with a speed that increases with decreasing distance to the umbra. A maximum of 0.35 to 0.55 km s^{-1} is reached at the umbra – penumbra boundary (Muller 1973a, 1976 ; Tönjes and Wöhl 1982). Bright grains enter into the umbra, where on single pictures they cannot be distinguished from umbral dots, and then disappear (Muller 1973a, Soltau 1982, Kitai 1986, Zirin and Wang 1989). The motion of penumbral bright grains can be seen spectacularly on video movies, destretched for atmospheric distorsion and taken with the help of CCD cameras, at Big Bear (Zirin and Wang 1990) and La Palma (Scharmer 1991). In addition, the movies allow us to see some motions of the dark fibrils : in the outer penumbra, dark material is moving outward in the form of discrete features, at speeds exceeding 1 kms^{-1} at the photosphere edge (Zirin and Wang 1989, 1990 ; Scharmer 1991, shown during the Workshop ; Shine et al. 1991). It is interesting to note that I was unable to detect the motions of dark fibrils in my earlier investigation (Muller 1973a) of a series of sunspot photographs, which have not been put together to make a movie. This demonstrates the usefulness of movies to study solar dynamic phenomena. In the inner penumbra there is no clear evidence of motions in dark fibrils ; Zirin and Wang (1990) reported an inward motion, Shine et al. (1991), reported an outward motion, while Scharmer (1991) was unable to see any organised flow.

3.5. LIFETIME AND LOCATION OF BRIGHT GRAINS

A detailed description of the evolution of bright grains was given by Muller (1973a). In summary, bright grains may form anywhere in the penumbra. Their lifetime depends on the position in the penumbra : on average, the shortest lifetimes (40 min) are found at the inner and outer boundaries of the penumbra, and the longer lifetimes (3 to 4 hours) in the middle. Bright grains are not found in the photosphere or in the umbra ; once formed they migrate further into the penumbra, where most of them disappear, after having travelled a typical distance of 3" to 5". If they form near enough to the umbra, they can enter the umbra where their appearance becomes similar to that of umbral dots. Muller (1976), analysing the asymmetry of the proper motion of bright grains in limb-side and center-side penumbrae, inferred an inclination of 5° to 20° to the horizontal.

3.6. EVERSHED EFFECT

The Evershed effect consists of a shift of the line core and a line asymmetry, in the same direction, resulting in a net shift of the center of gravity of a spectral line. At the photospheric level, the shift is explained as an almost horizontal outward flow of several kilometers per second (see Bumba 1960, Schröter 1965, Wiehr et al. 1984, among several other authors). A maximum velocity, up to 6 km s^{-1}, is reached at the photosphere-penumbra boundary. The question of the extension of the Evershed effect into the photosphere, well beyond the outer edge of the penumbra, was a point of contention. It is now widely accepted that the effect extends very little less than 2" beyond the outer sunspot boundary (Wiehr et al. 1986, Wiehr and Balthasar 1989). A few authors, like Küveler and Wiehr 1985, Dialetis et al. 1985, reported extended penumbrae, probably because they confused the Evershed effect with Doppler shifts due to photospheric motions outside sunspots (Sheeley and Bhatnagar 1971). It is interesting to note that "Evershed effects" outside sunspots, were always derived from filtergrams, not from spectrograms, demonstrating how spurious results can be obtained, when the observations have a poor spectral resòlution. The Evershed effect stops abruptly at the outer sunspot boundary, probably because the flowing gas becomes transparent to the radiation in the spectral line (see Thomas 1981 ; Wiehr et al. 1986).

In general, the shift and asymmetry of photospheric lines in spectra of moderate resolution (not better than 1") can be explained if the Evershed flow is concentrated in dark fibrils (Beckers and Schröter 1969, Abdussamatov and Krat 1970, Stellmacher and Wiehr 1971, 1980). High spatial resolution movies taken in the continuum and in photospheric lines, with narrow bandpass tunable filters, shown in this workshop by Alan Title and his colleagues, confirm spectacularly that the Evershed flow coïncides with dark fibrils (Zirin and Wang 1989, 1990, Shine et al. 1991). These movies demonstrate in detail what actually happens : the Evershed effect is dominated by discrete, dark cloud-like features which move outward in the penumbra. These clouds can be several arcseconds long ; moving clouds are visible in the outer penumbra, where proper motions have been measured in the range 0.5 – 5 km s^{-1}. According to Shine et al. (1991), in the Dopplergrams the boundary between the penumbra and the photosphere is sharp. A few clouds penetrate into the photosphere, where they dissolve. This may explain why the Evershed effect extends a little (1 or 2 second of arcs) beyond the outer penumbral edge. As the boundary between the photosphere and the penumbra is ragged, it is not surprising that the coïncidence between the continuum and the Evershed boundary is not perfect. In the inner penumbra, there is no clear outward motion of the dark material which would correspond to the Evershed flow.

Surprisingly, the correlation between the Evershed effect and dark fibrils is not confirmed by spectrograms of high spatial resolution (<1", Wiehr and Stellmacher 1989) and of very high resolution (≈0"3, Lites, Scharmer and Skumanich 1990). In these two cases, only smooth Evershed fluctuations have been detected, which are not correlated with intensity fluctuations. Again, it is interesting to note that, even in high spatial resolution observations, results from spectrograms and filtergrams disagree; in this case, however, it seems that the results from filtergrams are more reliable, owing to the good coïncidence between Doppler and continuum intensity features, especially in the outer penumbra, which makes sense.

3.7. VERTICAL VELOCITIES IN THE PENUMBRA

The detection of vertical velocities is important for the interpretation of penumbral fibrils in terms of some kind of magneto-convection process. Beckers and Schröter (1969) were the first to report a relative upward motion in bright filaments. But this correlation has not been confirmed by recent spectrograms of higher spatial resolution (Wiehr and Stellmacher 1989, Lites, Scharmer and Skumanich 1990).

3.8. MAGNETIC STRUCTURE

At a large scale, the magnetic field is more or less vertical in the central part of the umbra. Away from the center, the inclination with respect to the vertical increases with increasing radius. It is about 45° at the umbra-penumbra boundary, and 70° to 75° near the photosphere (Kawakami 1983 ; Lites and Skumanich 1990). The inclination may have been underestimated, especially near the outer penumbral boundary, because of the contribution of the vertical magnetic field in the photosphere outside sunspots (Lites and Skumanich 1990). Yet, Beckers and Schröter (1969) derived an almost horizontal magnetic field from their observations. Higher resolution magnetic measurements would provide more reliable inclination values.

The first results derived from observations obtained with a relatively high resolution (≈ 1"), have shown a tendency for the magnetic field to be weaker (by a few hundred gauss) and less inclined in bright filaments (Beckers and Schröter, 1969 ; Abdussamatov 1976). High resolution (0"5, Zirin and Wang 1990) and very high resolution (0"3, Title et al. 1992) magnetograms confirm that the magnetic field is more inclined in dark fibrils. Analysing in detail an outstanding movie, Title (1991) found that the line of sight magnetic field in the "Evershed clouds" described in the section 3.6 is weaker. They interpreted this weakining as due to a higher inclination. As they measured an average inclination of 70° to 75°, with excursions of ± 18°, they come to the conclusion that the magnetic field is nearly horizontal in dark fibrils.

As in the case of the Evershed effect, in spectra of high spatial resolution no clear relation could be found between the magnetic field and the penumbra brightness (Wiehr and Stellmacher, 1989 ; Lites, Scharmer and Skumanich, 1990) : only weak fluctuations of the magnetic field strength have been detected, of spatial scale 2", much larger than and not correlated with brightness fluctuations.

3.9. DISCUSSION OF PENUMBRAL MODELS

The physical situation in sunspot penumbrae is certainly more complicated than in the umbra, mainly because the magnetic field is strongly inclined, and also because there is a difference of temperature and pressure between the photosphere and the umbra. Any realistic model should explain the filamentary structure of the penumbra, the sharp outer boundary, the Evershed outflow in dark fibrils, the inward proper motion of bright grains, the similarity between penumbral grains and umbral dots, and, of course, the average penumbral intensity, which imposes some energy constraints.

There are three classes of models which attempt to explain the filamentary structure of the penumbra, as well as the observed motions. They are still rather rudimentary and, as a matter of fact, none of them can satisfactorily explain all penumbral characteristics.

3.9.1. Convective Rolls and Lorentz-Force Models

According to Danielson (1961), penumbral fibrils are elongated convective cells, in which the gas ascends in the bright filaments and descends in the dark ones. The magnetic flux is concentrated in the dark filaments by the convective motions. The resulting excess of magnetic pressure is balanced by an outward flow in the dark filaments, in order to maintain a lateral pressure equilibrium (Galloway 1975). The Evershed motions thus created shear the magnetic field and cause it to appear flatter. But in these models the magnetic flux must be nearly horizontal ; this poses a severe problem for magnetic flux conservation (Schmidt, 1987 ; 1991). Another problem is that these models ignore the fact that filaments are made with aligned bright grains.

3.9.2. Oscillating Convective Filament Models

In these models, the filamentary structure is caused by longitudinal motions driven by instabilities in the stratified horizontal penumbra : hydrostatic equilibrium requires vertical motions resulting from the longitudinal motions. If mass is drained from a dark filament by a longitudinal flow, the filament rises ; conversely, a bright filament must fall as a result of a mass inflow (Schmidt, Spruit and Weiss 1986, Schmidt 1991, Jahn 1992). The vertical motion is oscillatory as a filament is alternatively bright and dark. When a filament rises, its cross-section with the level $\tau = 1$, moves toward the umbra, explaining the observed bright-grain motion (Spruit 1981). An Evershed flow, produced by a difference of pressure between the two ends of a bent flux tube, connecting the umbra to the photosphere (Meyer and Schmidt 1968, Thomas 1984, 1988), is superposed on the oscillatory filament motions. The resulting effect is an increased speed in the dark fibrils and a decreased speed in the bright filaments. Indeed the outward Evershed flow observed in the dark filaments is significantly larger than the inward proper motion of bright grains, and this supports the model. It must be noted, however, that a differential vertical motion in bright and dark structures has not been observed, which means that these motions are slower than a few hundred meters per second, if they exist. But this negative observationnal result does not necessarily rule out the possibility of such an oscillatory convection of filaments to exist just below the surface. This model is supported by the fact that the magnetic field is more inclined in dark falling filaments than in the bright rising ones.

3.9.3. Rising Inclined Hot Columns

Choudhuri (1986) has explained umbral dots as columns of hot gas rising in between magnetic flux tubes (see section 2.7). Similarly, he has interpreted penumbral grains as rising columns of field-free fluid, when magnetic flux tubes are inclined with respect to the vertical. If this interpretation is right, then the hot rising columns should vanish at the surface, because no significant fluctuations of the magnetic field have been observed at this level. In this picture, the inward drift of penumbral grains might be induced by external pressure effects. It also might be due to rising, bent flux tubes, as suggested by Spruit (1981), drawing the hot columns of gas which flow in between them. However, it seems difficult to explain how the hot columns can be sustained for several hours.

I believe that, owing to their similarity, a realistic model should be able to explain the properties of both umbral dots and penumbral grains ; the only difference is that dots develop in a vertical magnetic structure, while grains develop in a strongly inclined magnetic structure. It is worth remembering the similarities : same size ; same brightness (close to the average photosphere) ; gradual transition from outer umbra to inner penumbra ; intensity fluctuations not correlated with magnetic field fluctuations (the amplitudes of both fluctuations are small) ; and no significant vertical velocities.

4. Suggestions for Future Observations

Besides the evident need for observations of higher spatial and spectral resolution and of less instrumental polarisation (for magnetic field measurements), for improving the inversion techniques (see for example, Lites and Skumanich 1990), and for deriving more reliable sunspot models (average ; 2 columns ; etc.), I suggest here some observations which may be useful for understanding the fine structure of sunspots better.

Height of penetration of I, V, B. At the continuum level, strong intensity fluctuations are observed ; just above, the fluctuations of V and B are much weaker. It would be very interesting to conduct a phase and coherence analysis between fluctuations of I, V and B, observed at various levels, in order to study the penetration of these quantities into the photosphere. A downward extrapolation may be helpful for understanding the structure of the atmosphere at $\tau = 1$ and below. Of course, the level of formation of photospheric lines must be known more precisely.

In order to understand sunspots, it is of crucial importance to understand their small-scale magnetic structure. As discussed in section 2.7, the magnetic field in sunspot umbrae is supposed to be either monolithic or formed by a cluster of individual flux tubes. The fact that no significant fluctuations are observed in the low photosphere does not rule out the possibility that it is highly structured below. As these layers are not directly observable, one can learn about the magnetic structure beneath the surface from a detailed, high-resolution analysis of the birth and growth of sunspot magnetic fields, particularly at their very early stages when they are nothing but pores.

A pore is a small sunspot, a few arcseconds in size, which has not developed a penumbra. It can have umbral dots, if it is big enough. In the absence of magnetic field measurements, a dark feature is considered to be a pore if its brightness, relative to the average photosphere, is smaller than about 0.8. When a pore forms, it appears first as a darkening of an intergranular space (Schröter 1969) or an elongated dark lane in an emerging flux region (Brandts and Steenbeck 1985). Then it grows further. The growth process is not unique : according to Brants and Steenbeck (1985), a pore can grow by coalescence with other pores (the sizes of pores is 2" to 5") or micropores (≈1") ; by inflow of dark material (probably magnetic), or by a spontaneous area increase, without converging flows. On the other hand, Wang and Zirin (1991) reported a converging motion of granules toward pores ; they concluded that this provides evidence that pores are formed by the concentration of smaller magnetic flux tubes. Thus pores seem to grow either spontaneously or by accreting more magnetic flux. Unfortunally, merging of isolated flux tubes has not yet been observed to form pores, due to lack of resolution of magnetograms. That kind of observation should tell us whether a pore, and consequently a sunspot which develops by pore merging, is formed from individual flux tubes or not. High-resolution filtergrams, in which individual flux tubes can be identified as network bright points or facular points (Muller 1985), may serve as an alternate kind of observation, much easier to obtain than high-resolution magnetograms.

The magnetic structure (field strength, inclination) of light bridges. Three kinds of light bridges have been identified by Muller (1979) : granular light bridges (made of granules and facular features) ; penumbral light bridges (made of aligned penumbral grains) ; and umbral light bridges (made of dots). The differences in the magnetic structure among these classes of light bridges would be useful in understanding the formation of umbrae and penumbrae, by giving some information on the strength and inclination of a critical magnetic field, for which an umbral dot structure changes to a penumbral structure.

References

Aballe-Villero, A.M. : 1991, Paper presented at the NATO ARW on the Theory of Sunspots, Cambridge, England.

Abdussamatov, H.I. : 1976, Solar Phys. 48, 117.

Abdussamatov, H.I., Krat, V.A. : 1970, Solar Phys. 14, 132.

Adjabshirzadeh, A., Koutchmy, S. : 1978, Comptes Rendus Ac. Sc. Paris, 286, 335.

Adjabshirzadeh, A., Koutchmy, S. : 1980, Astron. Astrophys. 89, 88.

Adjabshirzadeh, A., Koutchmy, S. : 1983, Astron. Astrophys. 122, 1.

Beckers, J.M., Schröter, E.H. : 1968, Solar Phys. 4, 303.

Beckers, J.M., Schröter, E.H. : 1969, Solar Phys. 10, 384.

Bonet, J.A., Ponz, J.D., Vazquez, M. : 1982, Solar Phys. 77, 69.

Brants, J.J., Steenbeck, J.C. : 1985, Solar Phys. 96, 229.

Bray, R.J., Loughhead, R.E. : 1964, *Sunspots*, Chapman and Hall, London.

Bumba, V. : 1960, Izo. Crim. Astrophys. 23, 253.

Bumba, V. : 1991, Private communication.

Bumba, V., Hejna, L., Suda, J. : 1975, Bull. Astron. Inst. Czeschosl. 26, 345.

Bumba, V., Suda, J. : 1980, Bull. Astron. Inst. Czeschosl. 31, 101.

Bumba, V., Sobotka, M., Simberova, S. : 1990, in *The Dynamic Sun*, 6th European Solar Meeting ; Publ. Debrecen Heliophys. Obs., vol. 7, p.84.

Buurman, J. : 1973, Astron. Astrophys. 29, 329.

Collados, M., del Toro Iniesta, J.C., Vazquez, M. : 1988, Astron. Astrophys. 195, 315.

Choudhuri, A.R. : 1986, Astrophys. J. 302, 809.

Danielson, R.E. : 1961, Astrophys. J. 134, 275.

Danielson, R.E. : 1964, Astrophys. J. 139, 45.

Dialetis, D., Mein, P., Alissadrakis, C.E. : 1985, Astron. Astrophys. 147, 93.

Ewel, M.W. : 1991, Solar Phys. (in press).

Galloway, D.J. : 1975, Solar Phys. 44, 409.

Grossmann-Doerth, V., Schmidt, W. : 1981, Astron. Astrophys. 95, 366.

Grossmann-Doerth, V., Schmidt, W., Schröter, E.H. : 1986, Astron. Astrophys. 156, 347.

Hejna, L. : 1977, Bull. Astron. Inst. Czeschosl. 28, 126.

Ikhsanov, R.N. : 1972, Sol. Dannye 11, 62.

Jahn, K. : 1992, this volume.

Kawakami, H. : 1983, Pub. Astr. Soc. Japon, 35, 459.

Kitai, R. : 1986, Solar Phys. 104, 284.

Kneer, F. : 1973, Solar Phys. 28, 361.

Knobloch, E., Weiss, N.O. : 1984, Monthly Nat. Roy. Astr. Soc. 207, 203.

Koutchmy, S., Adjabshirzadeh, A. : 1981, Astron. Astrophys. 99, 111.

Krat, V.A., Karpinski, V.N., Pravdjuk, L.M. : 1972, Solar Phys. 26, 305.

Kusoffsky, U., Lindstedt, H. : 1986, Astron. Astrophys. 160, 51.

Küveler, G., Wiehr, E. : 1985, Astron. Astrophys. 142, 205.

Lites, B.W., Bida, T.A., Johannesson, A., Scharmer, G.B. : 1991, Astrophys. J. 373, 683.

Lites, B.W., Scharmer, G.B., Skumanich, A. : 1990, Astrophys. J. 335, 329.

Lites, B.W., Skumanich, A. : 1990, Astrophys. J. 348, 747.

Livingston, W. : 1991, Nature 350, 45.

Loughhead, R.E., Bray, R.J. : 1960, Austr. J. Phys. 13, 139.

Loughhead, R.E., Bray, R.J. Tappere, E.J. : 1979, Astron. Astrophys. 79, 128.

Meyer, F., Schmidt, H.U. : 1968, Zeitschrift j. Angew. Mathematik und Mechanik 48, 218.

Moore, R.L. : 1981a, Astrophys. J. 249, 390.

Moore, R.L. : 1981b, in *The Physics of Sunspots*, L.E. Cram and J.H. Thomas (ed.), Sacramento Peak Observatory summer workshop, p.259.

Muller, R. : 1973a, Solar Phys. 29, 55.

Muller, R. : 1973b, Solar Phys. 32, 409.

Muller, R. : 1976, Solar Phys. 48, 101.

Muller, R. : 1985, Solar Phys. 98, 51.

Pahlke, K.E., Wiehr, E. : 1990, Astron. Astrophys. 228, 246.

Parker, E.N. : 1979, Astrophys. J. 234, 333.

Scharmer, G. : 1991, Paper presented at the NATO ARW on the Theory of Sunspots, Cambridge, England.

Shine, R., Title, A., Tarbell, T., Topka, K., Frank, Z., Smith, R., Scharmer, G. : 1991, Paper presented at the NATO ARW on the Theory of Sunspots, Cambridge, England.

Schmidt, H.U. : 1987, in *"The role of fine-scale magnetic fields on the structure of the solar atmosphere"*, eds. E.H. Schröter, M. Vazquez and Wyller, A.A., Cambridge University Press, p.219.

Schmidt, H.U. : 1991, in *"Cowling Memorial Symposium on Solar System Plasma Physics"*, Geophys. Astrophys. Fluid Dyn.

Schmidt, H.U., Spruit, H.C., Weiss, N.O. : 1986, Astron. Astrophys. 158, 351.

Schmidt, W. : 1991, Paper presented at the NATO ARW on the Theory of Sunspots, Cambridge, England.

Schröter, E.H. : 1965, Z. Astrophys. 62, 228.

Sheeley, N.R., Bhatnagar, A. : 1971, Solar Phys. 19, 338.

Soltau, D. : 1982, Astron. Astrophys. 107, 211.

Spruit, H.C. : 1981, in *the Physics of sunspots*, ed. L.E. Cram and J.H. Thomas, Sacramento Peak Observatory summer workshop, p.359.

Stachnik, R.V., Nisenson, P., Noyes, R.W. : 1983, Astrophys. J. (Letters) 271, L37.

Stellmacher, G., Wiehr, E. : 1971, Solar Phys. 17, 21.

Stellmacher, G., Wiehr, E. : 1980, Astron. Astrophys. 82, 157.

Thomas, J.H. : 1984, in *Small-scale dynamical processes in stellar atmospheres*, ed. S.L. Keil, Sacramento Peak Observatory summer workshop, p.276.

Thomas, J.H. : 1988, Siphon Flows in isolated magnetic flux tubes. Astrophys. J., 333, 409-419.

Title, A.M., Frank, Z.A., Shine, R.B., Tarbell, T.D., Topka, K.P., Scharmer, G.B. and Schmidt, W. : 1992, this volume.

Tönjes, K., Wöhl, H. : 1982, Solar Phys. 75, 63.

Wang, H., Zirin, H. : 1991, preprint.

Weiss, N.O., Brownjohn, D.P., Hurlburt, N.E., Proctor, M.R. : 1990, Monthly Nat. Roy. Astr. Soc. 245, 434.

Wiehr, E., Balthasar, H. : 1989, Astron. Astrophys. 208, 303.

Wiehr, E., Koch, A., Knölker, M., Küveler, G., Stellmacher, G. : 1984, Astron. Astrophys. 140, 352.

Wiehr, E., Stellmacher, G., Knölker, M., Grosser, M. : 1986, Astron. Astrophys. 155, 402.

Wiehr, E., Stellmacher, G. : 1989, Astron. Astrophys. 225, 528.

Zirin, H., Wang, H. : 1989, Solar Phys. 245, 119.

Zirin, H., Wang, H. : 1990, Adv. Space Res. Vol.11, n°5, p. proceedings of 23rd COSPAR, Meeting ME9.

Zwaan, C. : 1985, in High Resolution in Solar Physics, ed. R. Muller (Berlin, Springer), p.263.

Zwaan, C., Buurman, J. : 1971, in IAU Symposium 43, *Solar Magnetic Fields*, R. Howard (ed.), p.220.

Zwaan, C., Brants, J.J., Cram, L.E. : 1985, Solar Phys. 95, 3.

HIGH RESOLUTION OBSERVATIONS OF THE MAGNETIC AND VELOCITY FIELD OF SIMPLE SUNSPOTS

Alan M. Title, Zoe A. Frank, Richard A. Shine, T. D. Tarbell,
K.P. Topka
Lockheed Palo Alto Research Laboratory
3251 Hanover Street, Palo Alto, California 94304, USA

Göran Scharmer
Swedish Royal Academy of Sciences
Stockholm Observatory, S-133000 Saltsjobaden, Sweden

Wolfgang Schmidt
Kiepenheuer Institut für Sonnenphysik
Schöneckstrasse 6, D-7800 Freiburg im Breisgau, Germany

ABSTRACT. We have observed the disk passage of relatively simple round sunspots using a narrowband filter and a large format CCD detector and have created magnetograms, Dopplergrams, and continuum images nearly simultaneously. In addition the spectral resolution of the filter (88,000) allows the construction of "spectra" for all points in the field of view. The changing patterns of the line of sight magnetic and flow fields with position around the spots and on the solar disk allowed us to infer that the mean inclination of the magnetic field increases from 45-50° to 70-75° across the penumbra and that there is a fluctuation of the inclination angle about the mean of about $\pm 18°$. The scale of the inclination fluctuations is the order of the size of the penumbral fibrils. The variation in inclination is large enough that substantial amounts of magnetic field are parallel to the solar surface from the mid to outer penumbra. We also find that the Evershed flow tends to occur in the regions where the magnetic field is horizontal. This suggests that the Evershed flow is confined to the regions of horizontal fields.

We show that a simple sunspot model with azimuthal variations in inclination but no azimuthal variations of field strength is free from azimuthal Lorentz forces. The meridional component of the currents which arise from the azimuthal variation in inclination are parallel to the field lines. This suggests that a cylindrically symmetric magnetostatic sunspot model can be perturbed into one with azimuthal variations in inclination with some adjustment in the meridional force balance.

J. H. Thomas and N. O. Weiss (eds.), Sunspots: Theory and Observations, 195–219.

1. Introduction

Even a simple round isolated sunspot is a very complicated structure (Schröter 1971, Gokhale and Zwaan 1972, Cram and Thomas 1981, Garcia de la Rosa 1984, Kalman 1991, Jahn 1992, Muller 1992). There are intensity inhomogeneities in the umbra and the penumbra, non-steady outflows in the penumbra (Evershed flows), inward motions of bright points along the bright penumbral fibrils, and various irregular wave motions in the umbra and penumbra. Given the complexity of the intensity, flow, and wave fields, it would be surprising if the magnetic field geometry were not also complex. Here we discuss a variety of high resolution observations of the line of sight and transverse magnetic and velocity fields which show that even a simple round spot has a mutually interacting spatially complex magnetic and flow structure.

High resolution magnetic measurements of sunspots are difficult to obtain. Vector magnetic field measurements made with filters and spectrographs required high precision and are sensitive to error signals introduced by telescope polarization. The interpretation of data is complex because the atmosphere of a sunspot is not well understood. In particular the local inhomogeneities of the spot may require very detailed three dimensional radiative transfer calculations to interpret spectra properly.

The above problems have long been recognized. Observers and theoreticians have generated many clever techniques to infer properties of spots. These have often used the fact that, as a spot moves across the solar disk, the projection of its vector fields on the line of sight changes progressively. In 1908, Evershed (1909) noticed that spot spectra taken with the slit oriented in an east-west direction tended to be sloped with respect to lines of constant wavelength. Center (limb) side spectral elements were blue (red) shifted. He then inferred that the changing appearance of the spot spectra was due to projection onto the line of sight of nearly radial and nearly horizontal outflow in the penumbra of sunspots. Hale *et al.* (1918) used spectra of Zeeman triplets taken in right and left circular polarization to measure where the field was perpendicular to the line of sight. By following numerous spots across the disk, they were able to generate a mean inclination of the magnetic field as a function of radius.

Kinman (1950) elaborated on the earlier projection techniques by making rasters of spectra across individual spots at several disk positions and fitting the observed line of sight flow field to models of simple spot flows. He found that, to the accuracy of his measurements, the Evershed flow was horizontal with some tendency to be inclined below the horizontal at the outer penumbral boundary. He also found that spots sometimes had a small azimuthal component in their flow field. Maltby (1964) recognized that perfectly symmetrical spots almost never occurred and calculated line of sight components of characteristic flow patterns and compared the observed flow patterns to his model spots. He established an upper limit of 0.15 km/sec for the vertical component of the Evershed flow and a maximum inclination of the flow to the surface of 6° or an inclination with respect to the spot normal of at least 84° at the outer penumbral boundary.

In 1960, Treanor developed a polarizing interferometer which allowed determination of the vector field via a fringe pattern imposed on spectra. Adam (1971, 1975, 1990) used the Treanor interferometer to measure vector magnetic fields and came to the conclusion that at the outer penumbral boundary the field strength was on the order of 300 gauss and that field inclination was about 70° with respect to the spot normal. In general, the field inclinations were consistent with a potential field geometry of a shallow buried dipole. Because of the seeing conditions at the Oxford University Observatory, the resolution of their spectra was on the order of four arcseconds. Beckers and Schröter (1969) using film spectra taken at Sacramento Peak Observatory did an analysis of Stokes V and I profiles. With one arcsecond quality data they came to the conclusion that the magnetic field strength was on the order of 1000 gauss at the outer penumbral boundary and that the inclination of the field with respect to the vertical increased linearly along a sunspot radius and reached 90° at the outer penumbral boundary. Beckers and Schröter also suggested that the magnetic field was stronger and probably more highly inclined in the dark penumbral fibrils. Wittmann (1974) also carried out a Stokes profile analysis which agreed with the results of Beckers and Schröter on the inclination of the field, but deduced lower field strengths in the outer penumbra.

Lites and Skumanich (1990) reported on a very careful analysis of photoelectric data taken with a Stokes Polarimeter. Their studies of round spots used data with six arcsecond resolution because of the long integration times required for their desired signal to noise ratios. They agreed with Adam on the inclination of the field, but had field strengths in the outer penumbra closer to those of Beckers and Schröter. Further, Lites and Skumanich saw little azimuthal variation in field strength on constant radii. Their results for the vector field of a simple round spot were consistent with the cylindrically symmetric, magnetostatic models of Pizzo (1986, 1990) and Jahn (1989), which are close to the potential field of a buried dipole.

Recently Lites, Scharmer, and Skumanich (1990) have obtained extremely high resolution spectra of spots using a CCD camera operating at video rates behind the Littrow holographic spectrograph at the Swedish Solar Observatory in La Palma. Often their spectra have a spatial resolution on the order of 0.3 arcsecond. Although these ultra high resolution spectra were made without a polarization analyzer, the intensity profiles show that there is little to no variation of the field strength on the scale of the penumbral fibrils.

The lack of agreement on the magnetic field inclination in the outer penumbra is in marked contrast to the situation for the Evershed flow, which is generally agreed to be nearly horizontal. However, if the magnetic field is not parallel to the flow field in the highly conducting atmosphere of a sunspot, there is a serious problem in our understanding of the physical processes in the sunspot. Lamb (1975) attempted to interpret the Evershed flows as wave phenomena; but, until now, there has been no satisfactory explanation of how flows and magnetic fields could exist with different geometries in a spot.

In recent years there has been a major effort at the German Vacuum Tower Telescope on Tenerife and the Swedish Solar Tower on La Palma to obtain very high resolution observations of sunspots. High quality spectra have been obtained by a number of groups. In this paper discussion we focus our attention on observations made with filtergrams with spatial resolution of about 0.5 arcsecond and spectral resolution of about 88,000 using the evaluation model of the tunable filter section of the Coordinated Instrument Package of NASA's Orbiting Solar Laboratory (OSL) (Title *et al.* 1989). These filter images allow construction of nearly simultaneous velocity, intensity, and magnetic data on sunspots. Because of the desire for detailed spatial comparisons, the flat fielded filtergrams are destretched to remove atmospheric image distortions (Topka, Tarbell, and Title 1986; November 1986) and aligned with respect to each other with an accuracy of about 1/2 pixel (0.08 arcsecond).

A great deal of the analysis of the data depends on interpretation of the projection of local vector fields in a sunspot. For this reason, section 2 is devoted to a discussion of the patterns produced by the line of sight and transverse projection of the vector field of a spot on the plane of the observer. Results obtained using several different observing techniques are contained in section 3. In section 4 we discuss a fluted magnetic field geometry for a sunspot and calculate its current distribution. Finally, in section 5 we summarize the measurements and suggest an overall structure of a simple sunspot.

2. Properties of Vector Projection Equations

Because the diameter of a spot is small compared to the diameter of the sun to a good approximation local vector fields in the spot can be described in terms of a rectangular coordinate system that is rotated with respect to the line of sight and centered on the spot with defining unit vectors $\hat{\mathbf{sx}}, \hat{\mathbf{sy}}, \hat{\mathbf{sz}}$. Another rectangular coordinate system with a common origin at the center of the spot with defining unit vectors $\hat{\mathbf{x}}, \hat{\mathbf{y}}, \hat{\mathbf{z}}$ is used to describe the observations. The spot z axis, $\hat{\mathbf{sz}}$, is in the direction of the surface normal and the angle between $\hat{\mathbf{z}}$ and $\hat{\mathbf{sz}}$ is the heliocentric angle, θ. Since the sun is far from the observer, the line of sight direction, $\hat{\mathbf{z}}$, can be either a unit vector in the direction from the center of the spot or the center of the sun to the observer. The spot x axis direction, $\hat{\mathbf{sx}}$, is in the plane containing the line of sight, $\hat{\mathbf{z}}$, and the surface normal, $\hat{\mathbf{sz}}$ – the parallel plane. The unit vector $\hat{\mathbf{x}}$ is also in the parallel plane. Both the spot y axis direction, $\hat{\mathbf{sy}}$, and the observation plane y axis direction, $\hat{\mathbf{y}}$, are perpendicular to the parallel plane and are collinear.

With the above definitions only the components of the spot field in the parallel plane contribute to the line of sight field. A vector at radius r and azimuth ψ from the spot center, with inclination to the spot z axis of γ has a line of sight component (see figure 1),

$$V_{LOS}(r_{proj}, \psi_{proj}, \gamma, \theta, \psi + \delta\psi) = V(r, \psi)(\sin\gamma \sin\theta \cos(\psi + \delta\psi) + \cos\gamma \cos\theta), \quad (1)$$

where $V(r, \psi)$ is the magnitude of the vector $\mathbf{V}(r, \psi)$, (r_{proj}, ψ_{proj}) is the projected location of the point (r, ψ) on the observation plane, and $\delta\psi$ is measured clockwise from the line connecting the center of the spot and (r, ψ). A circle centered on the spot becomes an ellipse on the observation plane with its minor axis in the direction toward the sun center. In most of the discussions here, the magnetic field of the spot will be radial and thus $\delta\psi$ will be zero.

Transverse projections to the line of sight are defined to be along the observation plane x axis, the parallel component, the observation plane y axis, the perpendicular component. The parallel component is

$$V_{\parallel}(r_{proj}, \psi_{proj}, \gamma, \theta, \psi + \delta\psi) = V(r, \psi)(\sin\gamma \cos\theta \cos(\psi + \delta\psi) - \cos\gamma \sin\theta), \quad (2)$$

while the perpendicular component is

$$V_{\perp}(r_{proj}, \psi_{proj}, \gamma, \psi + \delta\psi) = V(r, \psi)(\sin\gamma \sin(\psi + \delta\psi)). \quad (3)$$

The magnitude of the transverse component is then just

$$V_{TRAN}(r_{proj}, \psi_{proj}, \gamma, \theta, \psi + \delta\psi) = \sqrt{(V_{\parallel}^2 + V_{\perp}^2)}. \quad (4)$$

If the local vector is in the parallel plane and points in the limb direction, that is, it is at (r,180°) and $\delta\psi$ is zero; then the projection of the vector onto the line of sight will be zero when its inclination, γ, and the heliocentric angle, θ, sum to 90°. Therefore, the inclinations can in principle be mapped by observing the limb side of the spot and observing at what heliocentric position vectors at different radii have zero projection in the line of sight.

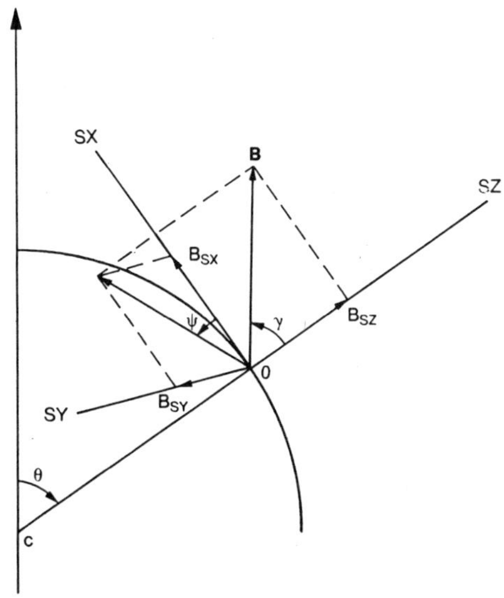

Figure 1. Coordinate system for the projection analysis. For clarity only the sunspot axes $(\hat{\mathbf{sx}}, \hat{\mathbf{sy}}, \hat{\mathbf{sx}})$ are shown. The observing coordinate system and the spot coordinate system have a common origin. The observing coordinate system has the z axis in direction of the line of sight ($\hat{\mathbf{z}}$), the x axis in the plane determined by $\hat{\mathbf{z}}$ and $\hat{\mathbf{sz}}$, and the y axis collinear with the spot y axis. In this view both the spot and observing system x axes are in the plane of the page and the y axes are both perpendicular to the page.

The observations exhibit rapid variations in the line of sight component of the magnetic field in the penumbra. Some insight into these can be gained by considering the variations due to perturbations on the magnetic field of a simple cylindrically symmetric spot. The line of sight component of a cylindrically symmetric sunspot is from equation (1)

$$B_{LOS}(r_{proj}, \psi_{proj}, \gamma, \theta, \psi) = B(r)(\sin\gamma\sin\theta\cos\psi + \cos\gamma\cos\theta). \tag{5}$$

If the line of sight field is observed on an ellipse corresponding to a circle of constant radius in the spot, then the signal as function of ψ will be a single cycle of a cosine function, due to the projection of the field in the plane of the spot onto the parallel plane, with amplitude,

$$Amp(r) = B(r)\sin\gamma\sin\theta \tag{6}$$

with an offset of,

$$Mean(r) = B(r)\cos\gamma\cos\theta, \tag{7}$$

due to the component in the spot z axis direction, $\hat{\mathbf{sz}}$.

If the signal changes in magnetograms are caused by an azimuthal variation in magnetic field strength, then $B(r)$ in equation (5) must be replaced by $(B(r) + \delta B(r,\psi))$. Because the magnetogram signal is proportional to the angular dependent part of equation (5), when the angular dependent part of the function is near zero, the entire magnetic signal including fluctuations will be near zero. Thus, a signature of a varying field strength is an absence of fluctuating signal near the neutral line.

The observed magnetogram fluctuations do not to go zero in the vicinity of the neutral line. Therefore, we must consider the effects of fluctuations in both the inclination and azimuthal angles. To generate insight into the appearance of spots we follow the path of Maltby (1964) and generate magnetogram patterns modified by rapid sine wave fluctuations assuming separately that the fluctuations are due entirely to field strength, inclination, and azimuth variations. For the field strength fluctuations, we take

$$\frac{\delta B(\psi)}{B_0} = 0.2\sin(30\psi), \tag{8a}$$

for the inclination fluctuations

$$\delta\gamma(r,\psi) = -15\sin(30\psi), \tag{8b}$$

and for the azimuthal variations

$$\delta\psi(\psi) = -20\sin(30\psi). \tag{8c}$$

The maximum signal has been set to unity in all cases. The ordinate of the graphs in figures 2 and 3 are, therefore, labeled relative signal. We have chosen thirty cycles of the fluctuation around the spot for the convenience of the display. Figure 2 shows the relative line of sight and transverse magnetic signals caused by fluctuations in field strength (top), inclination (middle), and azimuth(bottom) and without fluctuations for the case where γ and θ are 70° and 20°, while figure 3 shows similar plots the case where γ again equals 70°, but where the heliocentric angle, θ, is increased to 40°. In both figures the dashed plots are those for the cylindrically symmetric line of sight fields and the dash dotted plots are for the cylindrically symmetric transverse cases.

It should be clear from figures 2 and 3 that the line of sight and transverse signals around a spot have different characteristic patterns depending on whether they are caused by fluctuations in field strength, inclination, or azimuth. Line of sight fluctuations in the magnetic signal are large all around the spot only in the case where the inclination angle is varied. Line of sight signals modified by fluctuations in either field strength or changes in azimuth go to zero over some distance around the spot. For field strength variations these zero signal regions occur at the neutral line positions, while for azimuth variations the zeros occur in the limb and sun center directions, which are usually about ninety degrees away from the neutral line crossings. For the transverse magnetic signals only those modified by the variations in field strength are large all around the spot. The fluctuations caused by inclination and azimuth variations both disappear over ranges in angle. Even if real spots are not perfectly symmetric the overall forms caused by fluctuations in field strength, inclination, and azimuth persist.

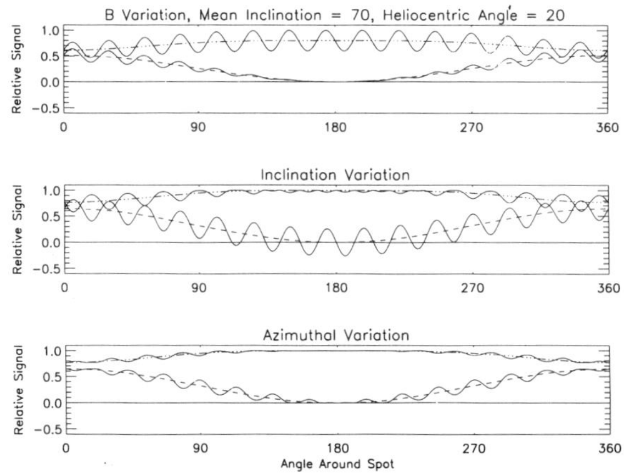

Figure 2. Plots of the relative line of sight and transverse magnetogram signals for a cylindrically symmetric spot and a spot with azimuthal perturbations for the cases where the variations are caused by fluctuations in the magnetic field strength (top), the inclination of the field (center), and the azimuth of the field (bottom) for a mean inclination angle of 70 ° and a heliocentric angle of 20 ° . The cylindrically symmetric line of sight and transverse results are shown dashed and dash-dotted.

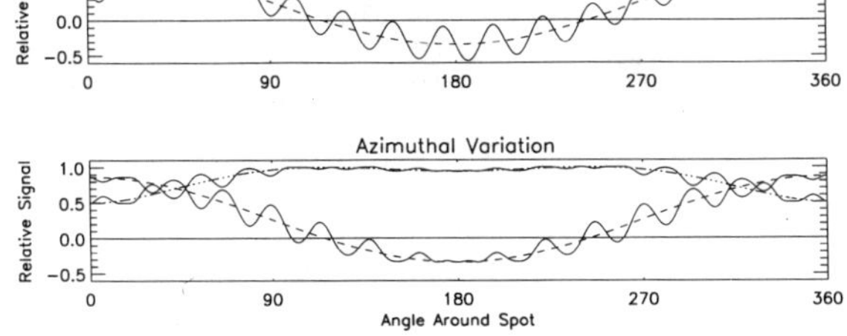

Figure 3. The same format as figure 2, but for a heliocentric angle of 40 ° .

3. Observational Results

3.1 LINE OF SIGHT MAGNETOGRAMS FROM THE TRAVERSAL OF A SIMPLE ROUND SPOT

Figure 4 shows magnetograms made from circular polarization pairs in Fe 6302.5 Å (g=2.5) of simple round spot in AR 5168 on September 14 ($\theta = 40.2^\circ$), 16 ($\theta = 20.1^\circ$), 17 ($\theta = 18.6^\circ$), and 20 ($\theta = 41.8^\circ$) 1988. The arrows on the figure indicate the direction of sun center. White (black) regions are areas where the line of sight field is positive (negative). The center of the sun side of the spots is always white. The reversals of the line of sight field due to projection effects are clearly visible as black areas on all of the days. The techniques for constructing magnetograms and Dopplergrams is discussed in Title *et al.* (1991). The existence of black areas in magnetograms from the 16th and 17th indicate that there are fields with inclinations greater than 70°. But note also that the reversal areas contain thin white lines which indicate adjacent locales where the inclination is less than 70°

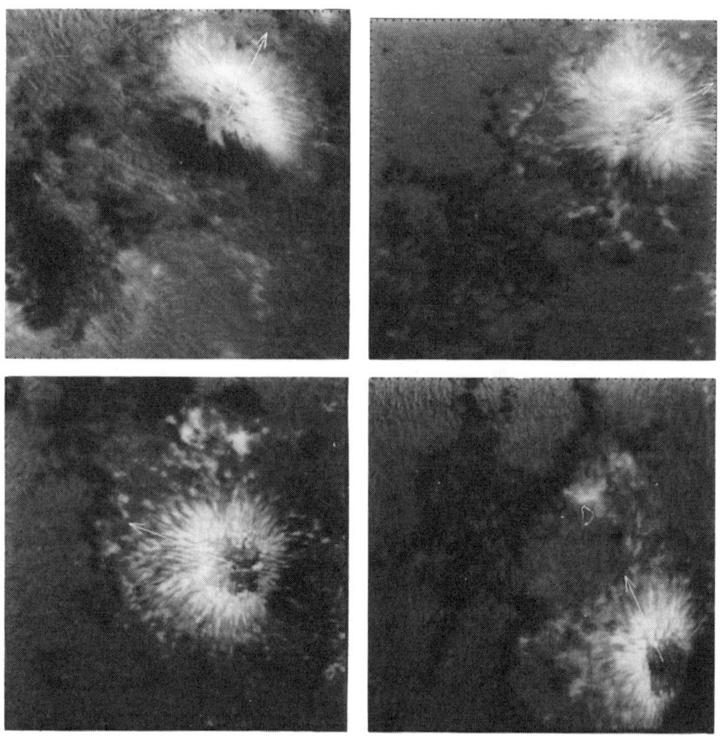

Figure 4. From upper left to lower right line of sight magnetic field maps (magnetograms) of the simple round spot in AR 5168 on 14, 16,17, and 20 September 1988. On these days the heliocentric positions of spot center were 40.2°, 20.1°, 18.6°, and 41.8°, respectively. The arrow on each map indicates the direction toward the center of the sun. The field of view of each image is 82 arcseconds and the interval between ticks is 2 arcseconds.

The spot center was determined from the mean of the best fit centers for the ellipses well contained in the penumbra. Shown in figure 5 (top) is the near 6302.5 continuum image on September 17 with the subset of ellipse paths used for the analysis indicated in white. Only a subset of the paths are shown on the figure for clarity. The complete set of paths span semi-major lengths from 40 to 180 pixels (6.48 to 29.16 arcseconds) with 10 pixel (1.62 arcsecond) spacing or from well inside the umbra to well outside the penumbra. Figure 5 (bottom) shows the ellipse paths determined from the continuum image overlayed on the corresponding magnetogram.

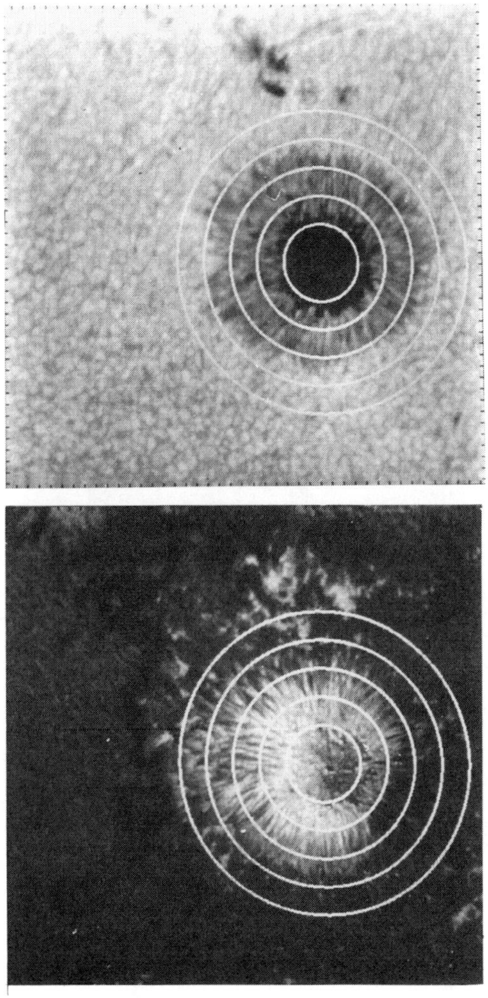

Figure 5. The constant radii paths in the observation plane shown on the continuum (top) and line of sight magnetic field (bottom) maps of one of the 17 September 15:57 UT data. Only a partial set of the paths is shown for clarity. The field of view of each image is 82 arcseconds and the interval between ticks is 2 arcseconds.

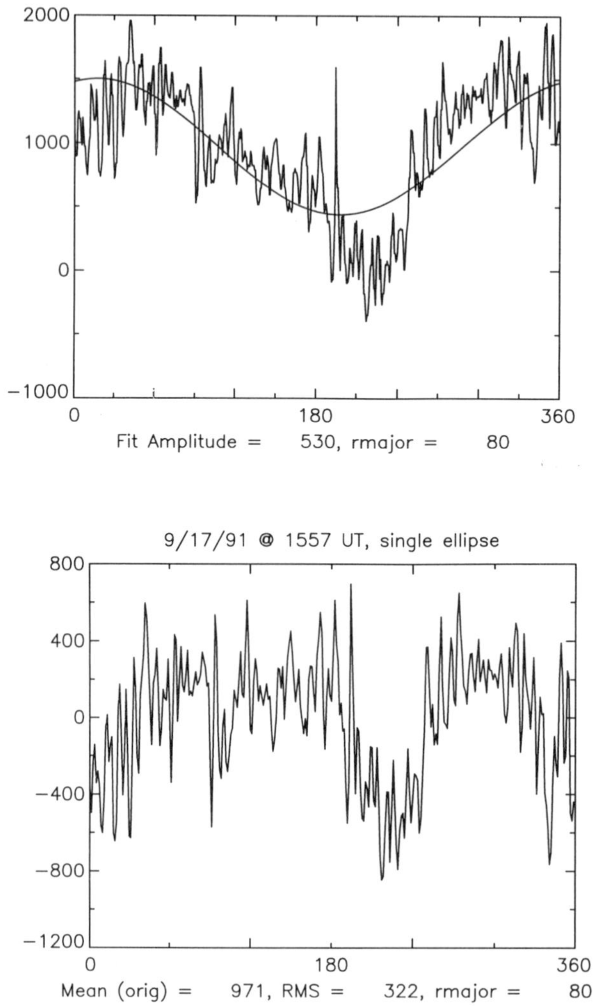

Figure 6. The magnetic field signal on a path through the midpenumbra of the 17 September 15:57 UT magnetogram (a) and the same signal with the first Fourier component, mean, and white noise removed (b) versus angle around the stop. The solid line is the best fit of an offset cosine function. Zero degrees is in the direction towards disk center. Tick marks are at 30 ° intervals which corresponds to 6.95 arcseconds along the circumference of the spot.

The best fit cosine amplitude, equation (6), and mean signal value, equation (7), were calculated from each of the elliptical magnetogram traces. In fitting the cosine to the trace, the phase was allowed to vary to accommodate slight global twists of the field. The power spectrum along each path was calculated and the white noise level of the magnetic signal estimated. The white noise and the zeroth (mean) and first Fourier components were subtracted from the Fourier expansion and the series was inverse transformed. The resulting signals are then due only to local variations in the line of sight magnetic field. As an example of the process, the magnetogram signal overlayed with the best fit offset cosine function and the magnetogram signal with the mean, cosine modulation, and noise removed versus azimuth angle are shown in figure 6a and b, respectively, for an ellipse path in the mid-penumbra on September 17 data.

By inspection of figure 6 and the arguments above, it was clear that most of the observed fluctuations occurred because of rapid inclination changes. This discrimination is largely based on the nearly constant amplitude of the fluctuations around the spot. If the fluctuations were largely due to either a variation in field strength or rapid changes in azimuth angle the fluctuations would have dropped to zero over a range in azimuth.

By measuring the amplitude and mean of a trace at a selected spot radius, r, the average inclination angle, $\gamma_{mean}(r)$, is determined (the heliocentric angle, θ, is known), so that

$$\gamma_{mean}(r) = \mathrm{arcctn}\left(\frac{Mean(r)}{Amp(r)}\tan\theta\right). \tag{9}$$

Because equation (9) is a ratio of magnetogram signals at the same radius, the direct dependence on field strength disappears. Thus, this result is relatively insensitive to calibration of the magnetogram signal. However, the calibration is still affected by line profile changes along a trace path.

The magnetic field strength follows from the above result and equation (7), so that

$$B(r) = \frac{Mean(r)}{\cos\gamma_{mean}\cos\theta}. \tag{10}$$

Since the mean value of the line of sight magnetic field is subject to the calibration of the magnetogram signal, the field strength is subject to an uncertainty of about 40%. Also because $\cos(\gamma_{mean})$ is lowest at the outer penumbral boundary the field strength is most uncertain there.

If all the fluctuations occur because of inclination variations, the value of the magnetogram signal variation caused by an inclination fluctuation of $\pm\delta\gamma(r)$ is determined by the ratio of the RMS fluctuation in the magnetogram signal relative to the mean field

$$\sin(\delta\gamma(r)) = \frac{\mathrm{RMS}_B\sqrt{2}}{B(r)}(\sin^2\gamma\cos^2\theta + \frac{1}{2}\cos^2\gamma\sin^2\theta)^{-1/2}. \tag{11}$$

Like the inclination measurement, the amplitude of the inclination variation is relatively insensitive to the calibration of the magnetogram signal.

The mean inclination angle, $\gamma_{mean}(r)$, its variation, $\delta\gamma(r)$, and $B(r)$ were calculated for line of sight magnetograms taken at 8:45 UT and 9:16 UT on September 16 and 15:45 and 15:57 UT on September 17. For each of the data sets a set of ellipse paths of widths 1 and 5 pixels (.162 and .81 arcsecond) was used. The results are shown in figure 7 for the single pixel wide data and in figure 8 for the five pixel wide data. The two data sets taken on the 16th are shown as "stars" and "pluses" while those taken on the 17th are shown as open "diamonds" and "squares". Solid lines connect the mean of the four observations at radial positions from 40 to 120 pixels. In both figures the normalized mean and the RMS continuum intensity are shown in the lower right panel. All of the functions are plotted as a function of spot radius in pixels. From the lower right panel it can be seen that umbra-penumbra and outer penumbral boundaries are at about 72 and 118 pixels (11.7 and 19.1 arcseconds), respectively.

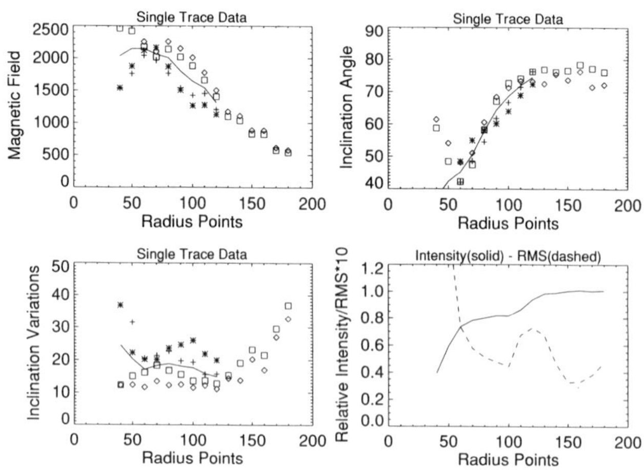

Figure 7. Plots from upper left to lower right of magnetic field strength, inclination angle, and amplitude of the inclination fluctuation versus spot radius in pixels. The stars and pluses and the open diamonds and squares indicate the results for the two data sets taken on 16 and 17 September, respectively, for the single pixel wide data. The solid lines connect the mean value at each radius. The lower right panel shows the normalized (solid) and RMS (dashed) continuum intensity fluctuations times 10 versus radius for a 17th September data set.

Two adjacent cycles of data on each day were reduced to estimate the effect of misregistration on the results. From figures 7 and 8 it is clear that the results obtained from the close data pairs are very similar. Direct estimates of the noise in the magnetograms were made by subtracting both destretched and direct filtergrams in the same circular polarization taken in the wing of Fe 6302. The magnetogram signal in these calibration tests indicate a misregistration signal of about 50 gauss in both the destretched and direct subtraction, which is considerably smaller that the 400 gauss fluctuations observed in the mid-penumbra. There exists the possibility of systematic effects which arise because the filtergrams in the different circular polarizations are affected differently by destretching.

The inclination and inclination variation results are both the most interesting and most reliable results of this analysis. We find a steady drop in mean inclination angle of the magnetic field across the penumbra, with value of about 72° at the boundary with the photosphere. These data also exhibit a 30 to 40% variation in the line of sight magnetic signal on scale of the penumbral fibrils. This can be explained with a fluctuation in inclination angle of $\pm 18^\circ$. Taken together with the mean inclination at the outer penumbral boundary, our results indicate that the magnetic field vector at the penumbra-photospheric boundary is varying between 54 and 90° on the spatial scale of the penumbral fibrils.

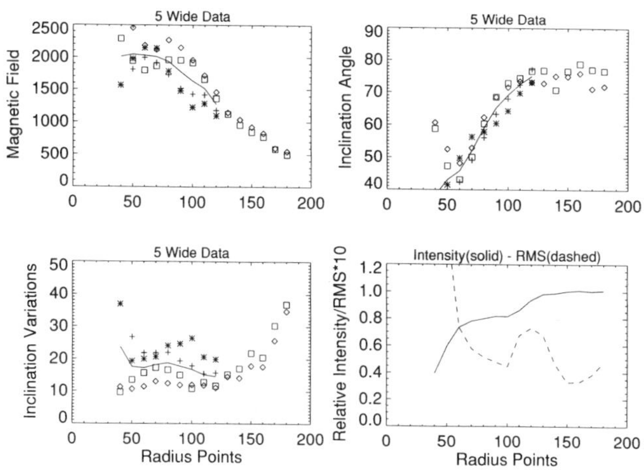

Figure 8. The format same as figure 7 for the 5 pixel wide paths.

Fluctuations in the line of sight magnetic field in spots were noted by Beckers and Schröter (1969) and associated with weakening of the field in the bright penumbral fibrils. However, the ultra high resolution data of Lites, Scharmer, and Skumanich (1990) show little to no field strength variation across fibrils. The measurements of Lites *et al.* in unpolarized light carry less information about the components of the field, especially in the outer penumbra where the magnetic splitting is small.

Because of the problems in calibration of line of sight magnetogram data, we have not attempted to measure the field strength deep into the umbra. The penumbral values may be shifted by 20% due to the uncertainly of our calibration process which does not adequately handle large changes in line profile shape. The relative calibration across the penumbra is also uncertain at the 20% level for the same reason, so that the ratio the field strengths at the umbra-penumbra to the outer penumbral boundaries are uncertain at the 20 to 40% level. Nevertheless, the field strength measurements are in agreement with those of Beckers and Schröter (1969) and Lites, Scharmer, and Skumanich (1990). Further, the magnetic field clearly extends well beyond the outer penumbral boundary, a result which is also nearly independent of calibration errors.

3.2 LINE OF SIGHT "SPECTRAL" MEASUREMENTS OF A SPOT ON THE CENTRAL MERIDIAN

The data taken of a relatively simple spot in AR 6122 on 28 June 1990 ($\theta = 28.6^\circ$) allowed the construction of line of sight velocity and line of sight magnetic maps, continuum images, and "spectra" with a resolution of about 88,000. An interactive video display program developed by one of us (RAS) allowed display of any the filtergrams, magnetograms, or Dopplergrams in a data set, where here a data set is a complete spectral sequence. Figure 9a and b shows a Fourier Dopplergram and a magnetogram made at -60 mÅ from line core. In the figure white (black) indicates the true (projection reversed) polarity and an outward (projection reversed) flow in the Dopplergram. The figure clearly shows that the flow pattern is more horizontal than the magnetic field. The Dopplergram is reversed everywhere on the limb side of the spot, whereas the magnetogram exhibits the true polarity in the inner penumbra and in fine strips that extend to the outer penumbral boundary on the limb side. Also the figure shows that the Evershed flow drops abruptly at the outer penumbral boundary, while the magnetic field extends well beyond the boundary.

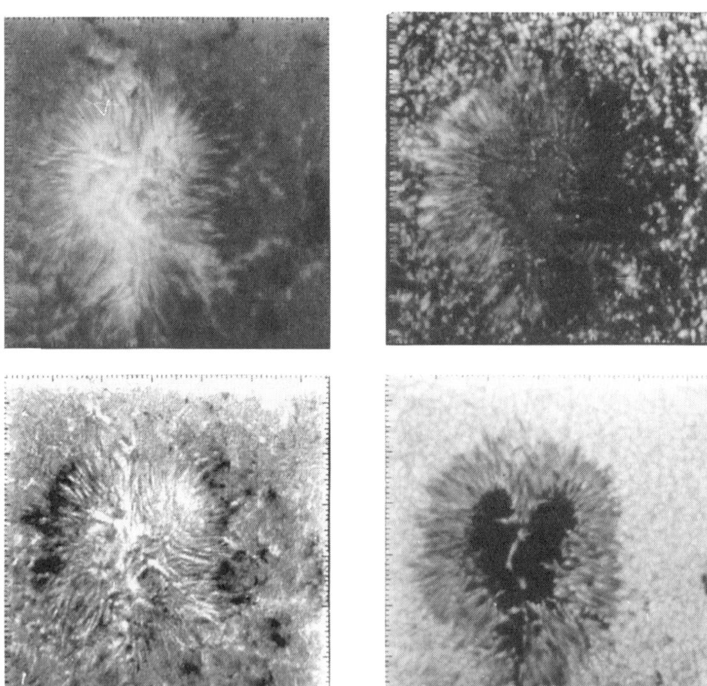

Figure 9. Fourier Dopplergram, -60 mÅ magnetogram, line center magnetogram, and continuum image for the 28 June 1990 data ($\theta = 24.6^\circ$). White (black) indicate magnetic and velocity field toward (away) from the observer. The field of view is 66 arcseconds square and the minor ticks are at 1 arcsecond intervals.

Spectra were created by choosing an image and then drawing a path on the video monitor image with a display cursor. Software had been developed which extracted the data from corresponding points on all the images which are taken at 30 mÅ intervals from −120 to +120 mÅ from the line center, so that a synthetic slit of any shape and orientation could be placed on the spot structure. An example of the slit extraction process is shown in figure 10 which contains a continuum image overlayed by 11 paths. The slits are one pixel wide (.16 arcsecond). Lines 1 through 5 are on the limb side of the spot at increasing distances

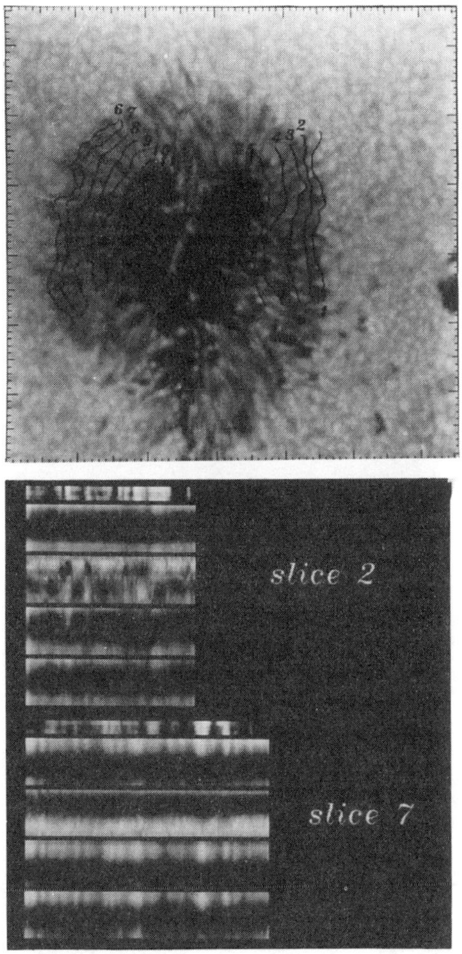

Figure 10. A continuum image of the 28 June spot overlayed with the paths used for creating spectra. Paths 1-5 and 6-11 are on the limb and center side of the spots, respectively. Also shown on the figure are arrays of spectra from paths 2 and 7. Each array consists of the Fourier Doppler velocity, spectra in Fe 5576 Å, the V-profile and profiles in right and left circular polarization in Fe 6302 Å. In all the profiles distance along the sun is in the horizontal direction and wavelength is in the vertical direction. The spacing between ticks is 1 arcsecond.

from the penumbra-photospheric boundary and 6 through 11 are a similar set of paths on the center side of the spot. Also shown on the figure are the Fourier Doppler signals, the synthesized line profiles in Fe 5576, the V-profiles (the difference between the right and left circular polarization profiles), and the right and left circular polarization profiles in Fe 6302 along slices 2 (limb) and 7 (center). In each of the profiles, space is in the horizontal and wavelength is in the vertical direction. Note that the V-profiles on the disk center side, line 7, show no reversals, while the V-profiles on the limb side, line 2, show numerous reversals.

The top panel of figures 11 and 13 show the -30 mÅ magnetogram signal, the Dopplergram signal, and the continuum intensity along paths 2 and 7, respectively. The bottom panel of the figures show the same data with a 3rd order polynomial fit removed. Figures 12 and 14 display scatter plots and correlation coefficients for the magnetogram and Dopplergram traces in the upper left, the same data with the trends removed are shown from upper right to lower right for the magnetogram and continuum, and the Dopplergram and continuum data. Also shown on the figure are the mean values of the magnetogram, Dopplergram, and continuum signals.

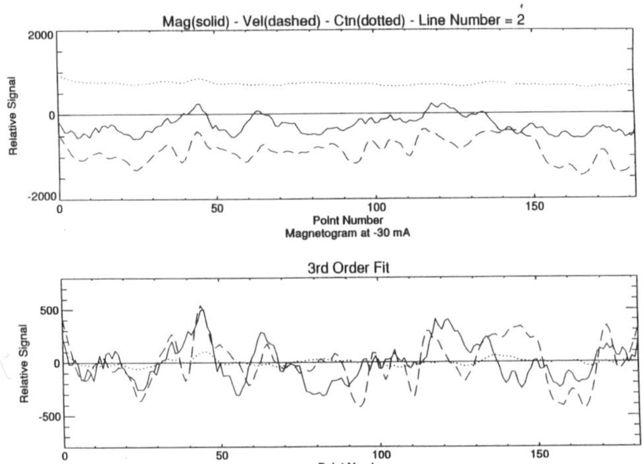

Figure 11. The upper panel contains plots of the line of sight magnetic field at -30 mÅ from line center Fe 6302 Å (solid), the Fourier Doppler velocity (dashed), and the continuum intensity (dotted) for path 2, which is on the limb side of spot. The lower panel shows the same curves with a 3rd order fit to the data removed.

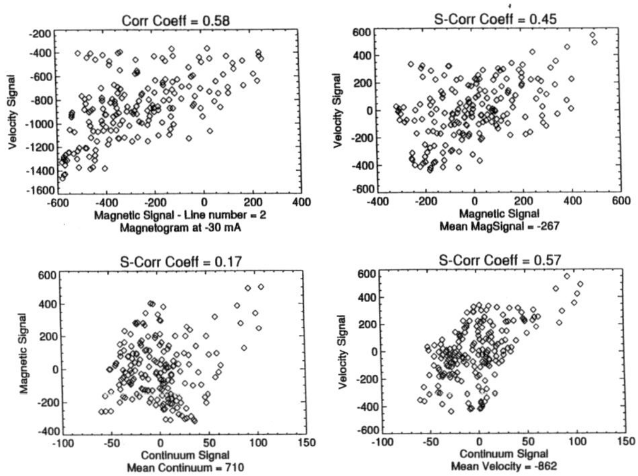

Figure 12. The panels from upper left to lower right display scatter plots of pairs of parameters for path 2. Immediately above each panel is the correlation coefficient and below the panel is the mean value of the parameter. The upper left panel shows the data for the magnetogram and Doppler signal. The rest of the panels show scatter plots for data with the 3rd order fit removed. From upper right to lower right are displayed magnetogram versus Doppler, magnetogram versus continuum, and Dopplergram versus continuum.

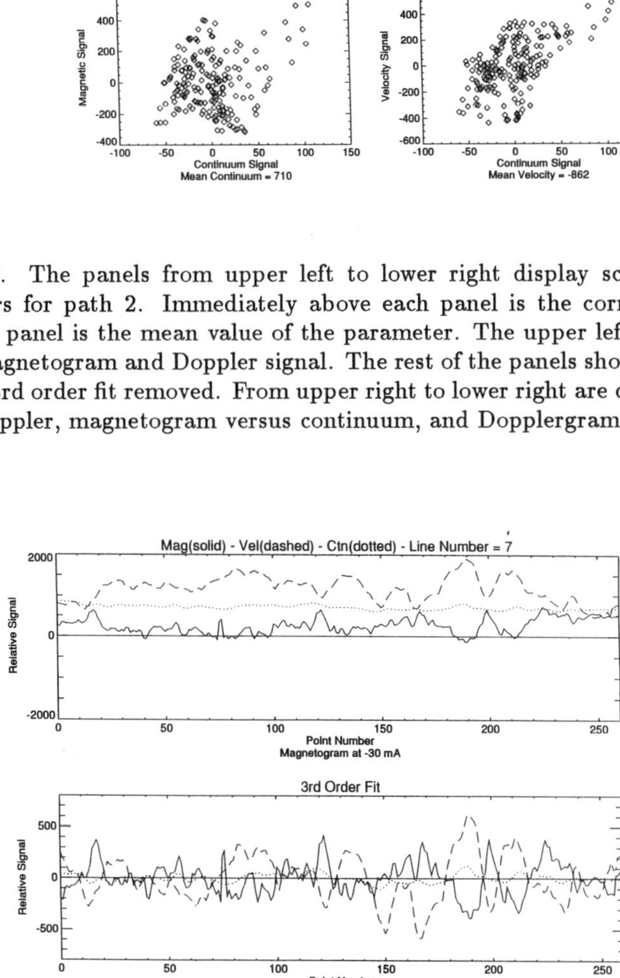

Figure 13. This figure is the same format as figure 11, but the data is from path 7.

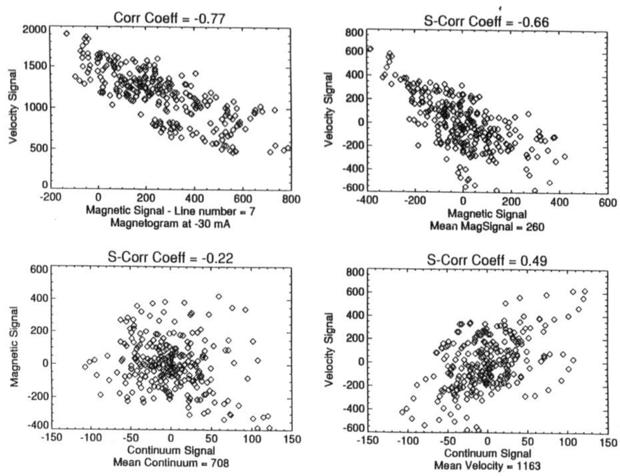

Figure 14. This figure is the same format as figure 12, but the data is from path 7.

The mean values of the Doppler velocity are negative for all the limb paths. This data is consistent with an inclination of the flow field of greater than 65° everywhere in the penumbra and greater than 85° in the outer penumbra. The data indicate that the Evershed flow has a mean speed of 1 to 3 km/s. The mean magnetic data indicates that the magnetic field is usually inclined at greater than 65° on paths 1 through 3.

On all the disk center paths the correlation coefficients between the magnetogram and Dopplergram signals are negative, which indicates that when magnetogram signal increases the Dopplergram signal tends to decrease. From the results above and those of Lites *et al.*, at a constant radius (on a single path) the magnetic field strength is nearly constant and the magnetogram fluctuations are due to inclination changes in the magnetic field vector. Increasing magnetogram signal occurs with more vertical fields. The Evershed flow is nearly horizontal, so that the negative correlation coefficients indicate the flow speeds increase (decrease) where the magnetic fields are more horizontal (vertical). On the limb paths the correlation coefficients between the magnetogram and Doppler signals are positive which indicates that the signals tend to increase and decrease together. Magnetic fields that have greater than 65.4° inclination to the local vertical on the limb paths have a negative projection along the line of sight. When the field becomes more parallel to the surface its projection along the line of sight increases in magnitude. The flow fields being nearly parallel to the surface have a negative projections along the line of sight, so that a positive correlation coefficient between the magnetogram and Dopplergram signals also indicates that horizontal magnetic fields are correlated with high outflow speeds. The correlation of velocity and magnetic field impacts our earlier estimate of the mean inclination angle; however, the effect is much less than a degree and we can neglect it.

The correlation between the detrended Dopplergram and continuum data is positive for all the paths. On the center paths this indicates that the locations of more horizontal flows occur in the darker continuum structures. While on the limb side, where the more horizontal flows have larger negative values, the positive correlation is also consistent with a correspondence between higher speeds and darker structures in the continuum. The correlation between the magnetogram and continuum data is usually small, an absolute magnitude of 0.2 or less, except near the umbra-penumbra boundary. The difference between the magnetogram and Dopplergram correlations with the continuum data may result from a flow field which is largely confined to the dark continuum structures, while the magnetic field exists everywhere in the spot.

4. The Current System of Simple Fluted Sunspots

In the sections above we have shown most of the variation in the magnetic signals result from inclination changes. In order to get an insight into the consequences of that geometry it is useful to examine how inclination variations modify the current distribution in the spot. To do this consider the effect of a sinusoidal perturbation in inclination with azimuth on a simple spot geometry. A cylindrically symmetric sunspot magnetic field configuration can be described in terms of a magnetic field strength, $B(r,z)$, and an inclination, $\gamma(r,z)$

$$\mathbf{B}(r,z) = B(r,z)(\sin\gamma(r,z), 0, \cos\gamma(r,z)). \tag{12}$$

where (r,ψ,z) indicates the components in the radial, azimuthal, and vertical directions.

Including an azimuthal variation in only the inclination, then creates the more general form

$$\mathbf{B}(r,\psi,z) = B(r,z)(\sin\gamma(r,\psi,z), 0, \cos\gamma(r,\psi,z)). \tag{13}$$

Note that the magnitude of the magnetic field strength is independent of ψ. Both equations (12) and (13) are subject to the condition $\nabla \cdot \mathbf{B}(r,\psi,z) = 0$.

For a vector field of the form described by equation (13), the current, $\frac{(\nabla\times\mathbf{B}[r,\psi,z])}{4\pi/c}$, can be written as the sum a meridional (A vector is meridional if it has no ψ component.) and azimuthal vector,

$$\mathbf{J}(r,\psi,z) = \mathbf{J}_{meridional}(r,\psi,z) + J_{azimuthal}(r,\psi,z)\hat{\psi}, \tag{14}$$

then

$$\frac{4\pi}{c}\mathbf{J}_{meridional}(r,\psi,z) = -\frac{\mathbf{B}}{r}\frac{\partial\gamma(r,\psi,z)}{\partial\psi}. \tag{15}$$

The azimuthal current is

$$\begin{aligned}\frac{4\pi}{c}\mathbf{J}_{azimuthal} = B(r,z)\hat{\psi}\Bigg(\sin\gamma(r,\psi,z)\left(\frac{1}{B(r,z)}\frac{\partial B(r,z)}{\partial z} + \frac{\partial\gamma(r,\psi,z)}{\partial r}\right)\\ - \cos\gamma(r,\psi,z)\left(\frac{\partial\gamma(r,\psi,z)}{\partial z} - \frac{1}{B(r,z)}\frac{\partial B(r,z)}{\partial r}\right)\Bigg).\end{aligned} \tag{16}$$

From equation (14) the meridional current is parallel to the field lines and therefore does not contribute to the Lorentz force $\mathbf{J} \times \mathbf{B}$. Thus, the azimuthal variation in the inclination of the field does not generate an azimuthal force field. In this situation field aligned currents exist without twisted field lines. In general, however, there will be a meridional force generated by the azimuthal current that could be balanced by gravitational forces and meridional pressure gradients.

With some definitions, equation (16) can be written in the simpler form

$$\frac{4\pi}{c}\mathbf{J}_{azimuthal} = C(r,\psi,z)B(r,z)\sin(\gamma-\phi)\hat{\psi}, \tag{17}$$

where

$$C(r,\psi,z) = \left(\left(\frac{1}{B(r,z)}\frac{\partial B(r,z)}{\partial z} + \frac{\partial\gamma(r,\psi,z)}{\partial r}\right)^2 + \left(\frac{1}{B(r,z)}\frac{\partial B(r,z)}{\partial r} - \frac{\partial\gamma(r,\psi,z)}{\partial z}\right)^2\right)^{1/2}, \tag{18}$$

and

$$\phi(r,\psi,z) = \arctan\left(\left(\frac{1}{B(r,z)}\frac{\partial B(r,z)}{\partial r} - \frac{\partial\gamma(r,\psi,z)}{\partial z}\right) \Big/ \left(\frac{1}{B(r,z)}\frac{\partial B(r,z)}{\partial z} + \frac{\partial\gamma(r,\psi,z)}{\partial r}\right)\right). \tag{19}$$

If we assume that the inclination variation is a perturbation to an original field configuration, then

$$\gamma(r,\psi,z) = \gamma_0(r,z) + \delta\gamma_n(r,z)\cos(n\psi), \tag{20}$$

where $\delta\gamma_n$ is the amplitude of a cosinusoidal azimuthal perturbation with a period n and $\gamma_0(r,z)$ is the cylindrically symmetric component of the inclination. For this situation the force free current is

$$4\pi J_{meridional} = -\frac{n\delta\gamma_n(r,z)\sin(n\psi)}{r}B(r,\psi,z). \tag{21}$$

It is easy to understand the form of this current system. When the argument of the sine function is an odd multiple of $\frac{\pi}{2}$, the current is maximal and the perturbation term in the inclination is zero, so that

$$\gamma(r,\frac{(2j+1)}{2}\pi,z) = \gamma_0(r,z), \tag{22}$$

where j is an integer. Hence, the maximal currents flow along undisplaced field lines and have a value

$$\frac{4\pi}{c}J_{max} = \frac{(-1)^{(2j+1)}n\delta\gamma_n(r,z)}{r}B(r,\psi,z), \tag{23}$$

At even multiples of π, the field inclination is at local extrema and the current is zero, and, from equation (21), the current reverses at alternate zero crossings of the perturbation.

A perturbation approach to the filamentary structure of sunspots has been examined earlier by Hu *et al.* (1983) and Hu (1984) from a mathematical point of view. Their papers focused on showing that the field was stronger in the dark fibrils, but the methods suggested are more general and revisiting the problem with the new experimental knowledge should prove very fruitful. Recently B. C. Low (1989, 1991) has examined stability of force free laminar magnetic structures in three dimensions. These laminar configurations have many of the characteristics of the fluted sunspot and are unusual in they can often be proved stable. There are no laminar solutions in a simple cylindrical coordinate system.

5. Summary

In summary, the measurements described above indicate:

1) There is a RMS variation in inclination of the magnetic field in the penumbra of about ± 15 to 20° with a spatial scale of the penumbral fibrils.
2) The magnetic field strength at a constant radius varies slowly compared to the spatial scale of the penumbral fibrils.
3) The average field inclination at the outer penumbral boundary is 70 to 75° to the sunspot normal.
4) The Evershed flow is on average more horizontal than the mean magnetic field at all radii in the penumbra.
5) The more horizontal magnetic fields occur in the regions of higher Evershed speed.
6) The Evershed flow is structured on the scale of the penumbral fibrils.
7) There is a tendency of the Evershed flow to occur in darker penumbral structures.
8) The magnetic field of the spot extends beyond the outer penumbral boundary.
9) The Evershed flow speed increases toward the edge of the penumbra and ends abruptly at the outer boundary of the penumbra.

In part to verify these results, Topka *et al.* (1991) have taken sequences of filtergrams in linear and circular polarizations and constructed line of sight, transverse field, and vector magnetograms of spots. They see fluctuations in both the line of sight and transverse fields that are consistent with results described above. In particular the patterns of fluctuations in the line of sight and transverse fields on the disk center and limb sides of a spot are consistent with the majority of the variation in magnetic signal being caused by inclination fluctuations. Keller *et al.* (1991) using filtergrams taken simultaneously in right and left circular polarization in a Zeeman triplet have also seen fluctuations in the line of sight magnetic field on the scale of the penumbral fibrils. At the Vacuum Tower Telescope on Tenerife in 1991 we also have taken filtergrams simultaneously in right and left circular polarization. Magnetograms made from these images look essentially identical to those made from sequential pairs. Scharmer (1991, 1992) has observed Stokes profiles with a spatial resolution approaching 0.3 arcseconds. He reports that the line profiles indicate rapid changes in field inclination on the scale of penumbral fibrils.

From the body of measurements on sunspot fields, we conclude that the inclination variation of the magnetic field is great enough throughout the mid to outer penumbra to allow the more horizontal magnetic fields to be essentially parallel to the solar surface. This leads us to suggest that the Evershed flow occurs only in those magnetic field lines that are nearly parallel to the surface. The electric current calculations for a simple fluted sunspot suggest that the observed field inclinations do not require significant modifications to the concepts of force balance in spots in magnetostatic equilibrium.

It is very satisfying that our observations are consistent with the two apparently different views of the inclination of the magnetic field across a penumbra found in the literature. The original measurements of Hale *et al.* (1918) suggested that the magnetic field is nearly horizontal at the outer penumbral boundary. However, the measurements described in the papers of Hale *et al.* (1918) and Hale and Nicholson (1938) were not made sufficiently close to disk center to establish field inclinations at the outer penumbral boundary. Measurements using the location of the reversal line by Beckers and Schröter (1969) and Wittmann (1974) have more strongly supported the concept that the field was nearly horizontal at the outer penumbral boundary. In contrast the vector magnetic field observations of Adam (1971, 1975, 1990) and Lites and Skumanich (1990) have indicated that the field has a mean inclination of 70-75° at the boundary of the spot with the photosphere. Detailed hydrostatic models of sunspots by Pizzo (1986, 1990) and Jahn (1989, 1992) yield stable configurations with field geometry not far from the potential field of a buried magnetic dipole, which implies field inclinations of 70-75° at the outer penumbral boundary.

Here we have shown that the mean inclination at the boundary is about 70-75° in agreement with the Adam-Lites interpretation of the data, while the inclination fluctuations of $\pm$15-20° are sufficient to put a fraction of the field nearly horizontal, so that reversals are detectable on substantial segments of the penumbral boundary even near disk center, which can explain the results of Hale *et al.*, Beckers and Schröter, and Wittmann. In addition, the observations seem to resolve the serious difficulty caused by the simultaneous existence of inclined magnetic fields and horizontal Evershed flows, while still allowing inclined fields in regions without Evershed flow.

During the last few years normal incidence X-ray telescopes have obtained very high resolution X-ray images of the sun. On the most recent flight of NIXT (Normal Incidence X-ray Telescope) one of the filters that reduce the visible light failed (Golub 1991). Because of the visible leak the location of X-ray loops with respect to a sunspot can be easily seen. Separate thin X-ray loops are observed to arise in the penumbra with inclination on the order 45°. The X-ray image suggests that penumbral loops are coincident with the tops of the magnetic flutes.

Although our measurements seem to be consistent with a relatively simple model in which the field lines vary in inclination around the spot being parallel to the surface in the darker fibrils and more inclined than a buried dipole field in the brighter ones, the picture completely neglects the possibility that the spot is an intrinsically dynamic phenomenon. Wentzel (1991) has suggested that the Evershed flow is the result of density fluctuations in magnetic flux bundles which causes them to fall toward the surface and flow outward along the bundle. Shine *et al.* (1991) have studied the Evershed flow in time sequences of Dopplergrams and have found that the flow tends to occur in moving clouds which are large compared to the penumbral fibrils. That is, a cloud represents a coordinated Evershed flow in adjacent fibrils. By tracking the motion of individual clouds they measured velocities of 1.5 to 3 km/s which is consistent with their Doppler shifts. Dopplergrams also show that the flow ends abruptly (in less than an arcsecond) at the outer penumbral boundary, and time sequences of Dopplergrams show the clouds are non-steady with repetition rate of about ten minutes. The appearance of clouds in the Doppler movies is consistent with the bright penumbral fibrils overlaying clouds. The path of the clouds tends to be coincident with dark fibrils and the fibrils tend to brighten during the passage of the clouds. The bright penumbral structures do not have a detectable Doppler signature, although brightness fronts in them move toward the umbra with a velocity of about 300 meters/s.

The development of the OSL evaluation tunable filter system represented a substantial effort of a team of hardware and software engineers and scientists at Lockheed Palo Alto Research Laboratory; we would especially like to acknowledge Mike Levay, Stuart Ferguson, Darrel Torgerson, Roger Rehse, Kathy Hwu, and Bill Rosenberg. The CCD camera, computer interface, and data acquisition computer were a joint effort of LPARL and the Kiepenheuer Institut für Sonnenphysik under the direction of Dexter Duncan, Chris Edwards, Mike Levay, and Thomas Rimmele. The logistics of the transportation and set up at the Swedish Solar Observatory was greatly aided by the efforts of Mike Morrill, Kermit Smith, Paco Armas and Rolf Kever. We would like to especially thank Lennard Fisk, Bill Wagner, and David Bohlin of NASA Headquarters and Jerri Hartman and Daniel Spicer of Goddard Space Flight Center for their strong and continued support of scientific aspects of the engineering tests of the OSL evaluation system. We greatly appreciate helpful comments on data analysis made by Boon Chye Low, Henk Spruit, Jack Thomas, Nigel Weiss, and Kees Zwaan. Financial support for this project was provided by NASA contracts NAS 8-32805 (SOUP) and NAS 5-26813 (OSL), by the National Science Foundation by grant ATM 8912841, the BMFT Förder Nr. 010E86059, the Swedish Royal Academy of Sciences, and Lockheed Independent Research.

References

Adam, M.G. 1971, *Mon. Not. Roy. Astr. Soc.,* **145**, 1.

Adam, M.G. 1975, *Mon. Not. Roy. Astr. Soc.,* **155**, 169.

Adam, M.G. 1990. *Solar Physics*, **125**, 37.

Beckers, J.M. and Schröter, E.H. 1969, *Solar Physics*, **10**, 384.

Cram, L., and Thomas, J. eds. 1981, Observatory Conference, The Physics of Sunspots, Sunspot, NM: Sacramento Peak Observatory.

Evershed, J. 1909, *Mon. Not. Roy. Astr. Soc.,* **69**, 454.

Garcia de la Rosa, J.I. 1987, The Role of Fine-Scale Magnetic Fields on the Structure of the Solar Atmosphere, eds. H. Schröter, M. Vazquez, and A. A. Wyller, Cambridge University Press, Cambridge, England, p. 136.

Gokhale, M. H., and Zwaan, C. 1972, *Solar Physics*, **26**, 52.

Golub, L. 1991, *Nature,* **344**, 6269, 842.

Hale, G. E., Ellerman, F., Nicholson, S. B., and Joy, A. H., 1918, Contributions from the Mt. Wilson Observatory, No. **165**, 153.

Hale, G. E. and Nicholson, S. B. 1938, Papers of the Mt. Wilson Obs., **5**, Part 1, 1.

Hu, W.R., Hu, Y.Q., and Low, B. C. 1983 *Solar Physics*, **83**, 195.

Hu, Y.Q. 1984, *Chin. Astron. Astrophys.,* **8**, 41.

Jahn, K. 1989, *Astr. Astrophy.*, **222**, 264.

Kalman, B., 1991, *Solar Physics*, **135**, 299.

Keller, C.U., Stenflo, J.O., von der Lühe, O. 1991 (submitted), *Astr. Astrophy.*, **000**, 000.

Kinman, T.D. 1950, *Mon. Not. Roy. Astro. Soc.,* **112**, 425.

Lamb, S. A. 1975, *Mon. Not. Roy. Astr. Soc.,* **172**, 205.

Lites, B., and Skumanich, A. 1990, *Ap. J.*, **355**, 329.

Lites, B., Scharmer, G., and Skumanich, A. 1990, *Ap. J.*, **348**, 747.

Low, B.C. 1989, *Ap. J.*, **330**, 992.

Low, B.C. 1991, *Ap. J.*, **370**, 427.

Maltby, P. 1964, *Astrophysica Norvegica,* **8**, 205.

November, L. J. 1986, *App. Opt.*, **25**, 392.

Pizzo, V.J. 1986, *Ap. J.*, **302**, 785.

Pizzo, V.J. 1990, *Ap. J.*, **365**, 764.

Scharmer, G.B. 1989, in Solar and Stellar Granulation, eds. R.J. Rutten and G. Severino, Kluwer Academic Publishers, Dordrecht, Holland, p. 161.

Scharmer, G. B. 1991, Private Communication.

Shine, R., Smith, K.,Tarbell, T., Title, A., 1990, *Bull. Amer. Astro. Soc.,* **22**, 2, 878.

Schröter, E.H. 1971, in Solar Magnetic Fields, I. A. U. Symp. 43, ed. R. Howard, Springer-Verlag New York Inc., New York, p.167.

Title, A. M., and Rosenberg, W. 1981, in Proceedings of the Sacramento Peak Observatory Conference, Solar Instrumentation: What's Next, ed. R. B. Dunn (Sunspot, NM: Sacramento Peak Observatory), p. 326.

Title, A.M., Tarbell,T.D., Topka, K.P., Ferguson, S.H., Shine, R.A. 1989, *Ap. J.*, **336**, 475.

Title, A.M., Topka, K.P., Tarbell,T.D., Schmidt, W., Balke, C., Scharmer, G. B. 1991 (in press), *Ap. J.*, **000**, 000.

Topka, K. P., Tarbell, T.D., Title, A.M. 1986, *Ap. J.*, **306**, 304.

Topka, K. P.,Frank, Z. A.,Tarbell, T. D., Title, A. M. 1990, *Bull. Amer. Astro. Soc.,* **23**, 2, 1052.

Treanor, P.J. 1960, *Mon. Not. Roy. Astr. Soc.*, **120**, 412.

Wentzel, D. 1991 (submitted), *Ap. J.*, **000**, 000.

Wittmann, A. 1974, *Solar Physics*, **36**, 29.

MAGNETOCONVECTION

M.R.E. PROCTOR
Department of Applied Mathematics and Theoretical Physics
University of Cambridge
Silver Street
Cambridge CB3 9EW, U.K.

ABSTRACT. Numerical and analytical work on magnetoconvection in a compressible atmosphere is reviewed. Attention is focused on the identification of features of the solutions that depend upon compressibility, as opposed to those that can be understood by solving the simpler Boussinesq equations. Recent results on time dependent (oscillatory) convection are described in some detail.

1. Introduction

Review papers have a peculiarly important place in solar physics meetings. In consequence many reviews on the same topic appear in the literature over a period of two or three years. While it is not necessary to be apologetic about producing yet another for the present volume of invited reviews it is, I think, prudent to be selective in presenting material. I shall in fact confine myself to a discussion of the simulation of convection in the presence of an imposed magnetic field, and particularly work that involves a compressible atmosphere; though, as I hope to show, many important phenomena can be understood in the Boussinesq approximation appropriate to a very shallow layer. Although these studies involve simplified models of the solar convection zone and sunspot atmospheres their proper understanding is an essential prerequisite for getting to grips with the detailed magnetohydrodynamics of sunspots. Indeed the simulations with an imposed vertical field that are discussed below would seem to have direct relevance to umbral convection, and have provided new insights into the governing mechanisms. The Boussinesq theory has been extensively reviewed by Proctor & Weiss (1982); more recent reviews, focusing on the effects of compressibility, have been presented by Hughes & Proctor (1988) and by Weiss (1992), while Weiss (1990) focuses on the relation between simulations and the actual solutions of the governing P.D.E.'s. I shall not include the effects of rotation, although its presence can lead to interesting new phenomena (see for example Jones & Roberts 1990).

It is helpful, before becoming immersed in detail, to stand back a little and ask what one can hope to learn from a detailed (almost invariably numerical) calculation of an idealized problem in a simplified geometry, with relatively poor resolution. Here there are two rather different philosophies: on the one hand, one can try to include all the relevant physics appropriate to the top of the convection zone (e.g. detailed discussion of radiative transfer) and hope to produce results that can be compared directly with the observations. This is the approach adopted by Nordlund and his collaborators (Nordlund 1984, 1985; Nordlund & Stein 1989). There are, I believe, difficulties with this point of view. Not only can it sometimes be difficult to disentangle the effects of the various competing mechanisms,

J. H. Thomas and N. O. Weiss (eds.), Sunspots: Theory and Observations, 221–241.

but also there remains doubt as to whether in fact the most accurate simulation of this kind is presently capable of producing correct quantitative information; in these circumstances it can be misleading to try to relate the results too closely to the observational data. On the other hand, one can try to isolate the effects that are due primarily to the presence of a magnetic field in the convective region. This can be done with idealized geometries and a simplified treatment of the non-magnetic physics. The price to be paid is that direct comparison with the observations is not possible. Both approaches have much to offer, but their unification must await significant developments in computing power.

The review is organised as follows: In §2 I discuss the basic kinematic and dynamic effects associated with the magnetic field. §3 gives, for the record, a typical set of governing equations that form the prototype for many simulations. In §4 I describe results on concentration and amplification of fields by convection, while in §5 I treat oscillatory convection and its possible relevance to the physics of sunspot umbrae and penumbrae. Finally in a conclusion I discuss the prospects for future high resolution three-dimensional simulations.

2. Kinematic and Dynamical Fundamentals

2.1. KINEMATICS

The influence of a velocity field $\mathbf{u}$ on a solenoidal magnetic field $\mathbf{B}$ is described (ignoring relativistic effects) by the induction equation

$$\frac{\partial \mathbf{B}}{\partial t} = \nabla \times (\mathbf{u} \times \mathbf{B}) - \nabla \times (\eta \nabla \times \mathbf{B}) \tag{2.1}$$

where the magnetic diffusivity η may depend on position (as it is a function of temperature). Here we have implicitly assumed an isotropic Ohm's Law, and neglected such features of plasma charge transport as the Hall effect, though this may be significant near the very top of the solar atmosphere. The importance of magnetic diffusivity is conveniently measured by the magnetic Reynolds number $R_m \equiv UL/\eta$, where U and L are typical velocity and length scales for the flow.

There is no doubt that if we adopt molecular values for η of the order $10^2 - 10^3$ $\mathrm{m^2s^{-1}}$ in the upper regions of the convection zone, though perhaps rather higher, of the order of 10^6 $\mathrm{m^2s^{-1}}$ in the umbra itself (Meyer *et al.* 1974), then the diffusivity is only important (for the timescales appropriate to sunspot evolution) on very short length scales of the order of $1 - 10$ km. When diffusivity can be neglected, (2.1) leads to *Alfvén's theorem* (adumbrated by Alfvén (1942), though the full analogy with Kelvin's theorem for vortex lines was not made explicit until later), that magnetic field lines move with the fluid. Thus one might expect magnetic fields to be concentrated in regions of converging flow, as observations suggest. The final distribution of magnetic field in a prescribed flow will depend on diffusion, as numerical computations in simple geometries show. (Arter, Proctor & Galloway 1982, Arter 1983, Galloway & Proctor 1983). Figure 1, from the last paper, shows that for a steady flow with hexagonal symmetry (crudely modelling conditions in a granule) the initial tendency is for all vertical flux to be concentrated in converging flow regions, while in the final, diffusively limited steady state only about half the flux remains near points of flow convergence.

Numerical simulations, whether kinematic or dynamic, cannot hope to include diffusion at its actual very low level. Either artificially high values of η are used, or large diffusion

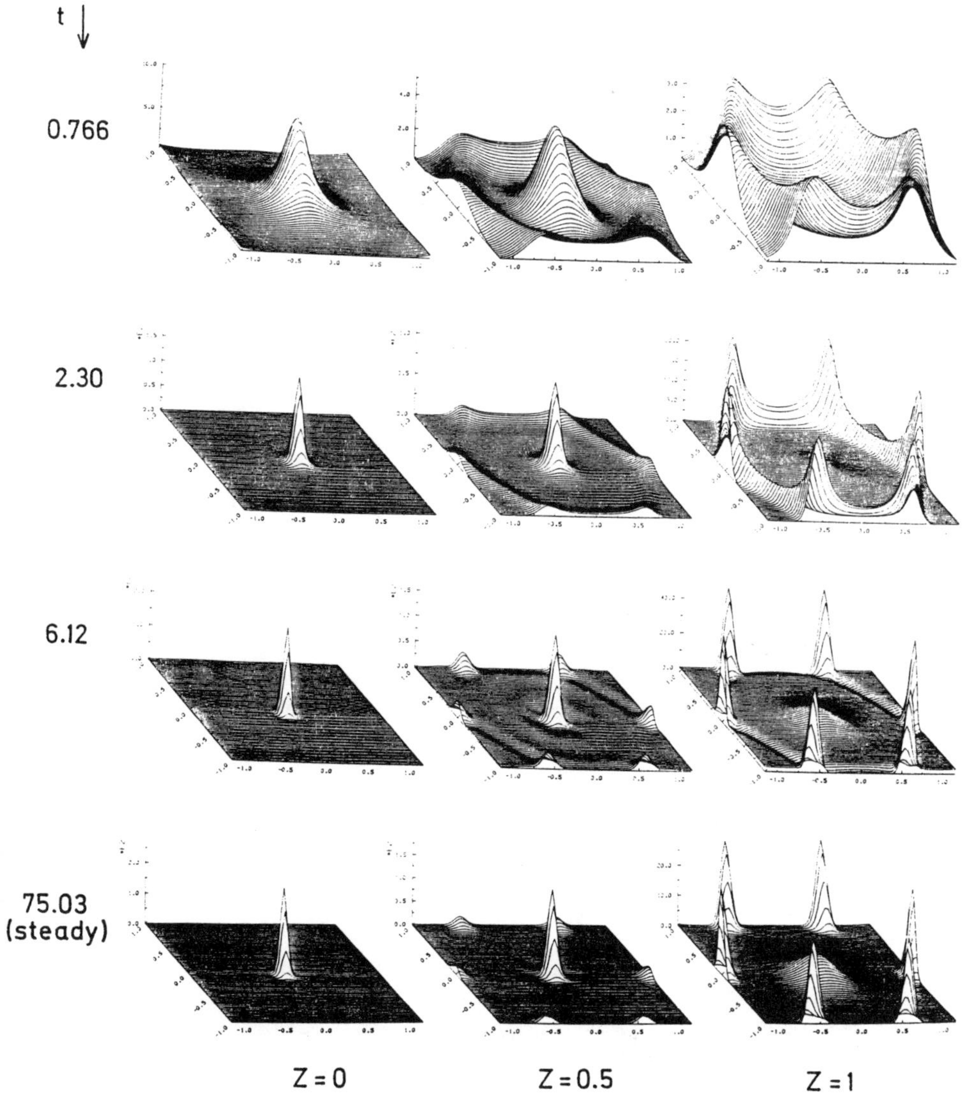

Fig. 1. Evolution of the vertical field in a flow with hexagonal symmetry (from Galloway & Proctor 1983). Here the top of the layer is at $z = 1$. The magnetic Reynolds number is 400.

is only put in 'where it is needed' using subgrid scale modelling (see e.g. Fox, Theobald & Sofia 1991), or else numerical diffusion is relied upon. Although in simple recirculating flows diffusion at even the smallest levels becomes effective at times of order the turnover time (see e.g. Hughes & Proctor 1988) in practice for dynamically significant fields concentration of flux will be halted by Lorentz forces. The concomitant presence of small scale turbulence in the solar atmosphere invites modelling with a 'turbulent diffusivity' but, as Parker has pointed out, the meaning of such a quantity for a magnetic field is less obvious than for the diffusion of a passive scalar. More work is needed in this direction.

2.2. THE LORENTZ FORCE

The magnetic field reacts with the flow in two ways: through an ohmic heating term in the energy balance, which does not seem to be of much importance in most applications, and through a body force acting directly on the fluid. This (Lorentz) force has the form

$$\mathbf{J} \times \mathbf{B} = \frac{1}{\mu_0}(\nabla \times \mathbf{B}) \times \mathbf{B} = \frac{1}{\mu_0}(\mathbf{B} \cdot \nabla \mathbf{B} - \nabla(\frac{1}{2}|\mathbf{B}|^2)) \tag{2.2}$$

The term $\mathbf{B} \cdot \nabla \mathbf{B}$ on the r.h.s. of (2.2) is only effective when the field lines are curved, and takes the form of a restoring force that tends to straighten the lines. The other, gradient term is the *magnetic pressure* and demonstrates the tendency of a concentration of field to expand. Both these parts of the Lorentz force inhibit flux concentration, the former by reducing local fluid velocities and the latter by reducing the local gas pressure and inducing a buoyancy force that acts against the converging flow. In addition, the restoring (curvature) forces allow the system to support waves if diffusivities are sufficiently low; thus in a superadiabatically stratified layer new forms of convection, involving oscillatory motion, can occur. All these effects are discussed below by reference to detailed numerical simulations.

3. Governing equations

It is, I think, helpful in focusing ideas to set down at this point a set of equations typical of those used for numerical simulation. Different authors will choose models that differ in detail from that exhibited here, but, provided that the diffusion approximation for heat transfer is adopted, the only important differences will concern the role of boundary conditions at the top and bottom of the convective layer, and the dependence of the diffusive coefficients on temperature and pressure. The particular choices made here are those utilised by Hurlburt and his collaborators (Hurlburt & Toomre 1988, Hurlburt *et al.* 1989, Weiss *et al.* 1990).

We shall describe the equations governing fully compressible convection in a perfect monatomic gas occupying a layer of depth d. In the absence of motion the equilibrium solution is taken to be a polytrope of index $m = (gd/\Re T) - 1$, where g is the gravitational acceleration, acting in the positive z-direction and $\Re$ is the gas constant. We take the temperature difference across the layer to be ΔT, the top of the layer to be at $z_0 \equiv T_0 d/\Delta T$ and $\rho_0 \equiv \rho(z_0)$ to be the density at $z = z_0$.

We cast the equations into dimensionless form by measuring lengths in terms of the layer depth d, times in terms of the reduced acoustic travel time $d/(\Re \Delta T)^{\frac{1}{2}}$, density in terms of ρ_0, temperature in terms of ΔT and magnetic field in terms of B_0, where the imposed field is $\mathbf{B}_0 = B_0 \hat{\mathbf{z}}$. The electrical conductivity σ_c, the thermal conductivity K, the shear viscosity μ, the magnetic permeability μ_0 and the heat capacities c_p and c_v are all constant for computations in the above references, though of course more realistic dependences on the thermodynamic variables could be included. $\gamma \equiv c_p/c_v$ is the usual ratio of specific heats. Then the magnetic, thermal and viscous diffusivities are given by

$$\eta \equiv (\mu_0 \sigma_c)^{-1} \quad , \quad \kappa \equiv K/c_p \rho \quad \text{and} \quad \nu \equiv \mu/\rho \tag{3.1}$$

respectively, so that both κ and ν vary inversely with ρ while η remains constant. Hence the Prandtl number $\sigma \equiv \nu/\kappa = c_p \mu/K$ is constant but the diffusivity ratio

$$\zeta \equiv \frac{\eta}{\kappa} = \frac{c_p \rho}{\mu_0 \sigma_c K} = \zeta_0 \left(\frac{\rho}{\rho_0} \right) \quad , \tag{3.2}$$

where $\zeta_0 = c_p \rho_0 / \mu_0 \sigma_c K$, is proportional to ρ and varies with depth across the layer. Then the equations governing the thermodynamic variables, the velocity $\mathbf{u}$ and the magnetic field $\mathbf{B}$ can be written as

$$P = T\rho \quad , \quad \frac{\partial \rho}{\partial t} = -\nabla.(\rho \mathbf{u}) \quad , \tag{3.2}$$

$$\nabla.\mathbf{B} = 0 \quad , \quad \frac{\partial \mathbf{B}}{\partial t} = \nabla \times (\mathbf{u} \times \mathbf{B}) + \zeta_0 \bar{K} \nabla^2 \mathbf{B} \quad , \tag{3.3}$$

$$\frac{\partial}{\partial t}(\rho \mathbf{u}) = -\nabla.(\rho \mathbf{u}\mathbf{u} + F\mathbf{B}\mathbf{B}) - \nabla(P + \tfrac{1}{2}F|\mathbf{B}|^2) + (m+1)\rho \hat{\mathbf{z}} + \nabla.\boldsymbol{\tau} \tag{3.4}$$

and

$$\frac{\partial}{\partial t}[\rho\{T/(\gamma-1) + \tfrac{1}{2}|\mathbf{u}|^2 - (m+1)z\} + \tfrac{1}{2}F|\mathbf{B}|^2] =$$

$$-\nabla.[\rho\{\gamma T/(\gamma-1) + \tfrac{1}{2}|\mathbf{u}|^2 - (m+1)z\}\mathbf{u} + F\mathbf{B} \times (\mathbf{u} \times \mathbf{B} - \zeta_0 \bar{K} \nabla \times \mathbf{B}) - \bar{K}\nabla T + \mathbf{u}.\boldsymbol{\tau}] \quad , \tag{3.5}$$

where the viscous stress tensor

$$\tau_{ij} = \sigma \bar{K} \left(\frac{\partial u_i}{\partial x_j} + \frac{\partial u_j}{\partial x_i} - \tfrac{2}{3} \delta ij \frac{\partial u_\kappa}{\partial x_\kappa} \right) \quad . \tag{3.6}$$

The dimensionless thermal conductivity $\bar{K} = K/[c_p \rho_0 d (\Re \Delta T)^{\frac{1}{2}}]$ while the field strength is measured by the dimensionless quantity

$$F = B_0^2 / (\mu_0 \Re \rho_0 \Delta T) = \sigma \zeta_0 \bar{K}^2 Q \quad , \tag{3.7}$$

where Q is the Chandrasekhar number. Once the properties of the reference atmosphere are fixed, the two parameters $\bar{K}$ and F define the state of the system. The static polytrope is given by

$$T = z \quad , \quad \rho = (z/z_0)^m \quad , \quad p = z^{m+1}/z_0^m \tag{3.8}$$

and the corresponding ratio of the gas pressure to the magnetic pressure is

$$\beta = 2z^{m+1}/(Fz_0^m) \quad , \tag{3.9}$$

while the superadiabatic gradient is measured by the Rayleigh number

$$R = (m+1)^2 (\nabla - \nabla_{ad}) \frac{z^{2m+1}}{\sigma \bar{K}^2 z_0^{2m}} \quad , \quad (\nabla - \nabla_{ad}) \equiv \frac{1}{m+1} - \frac{\gamma - 1}{\gamma} \quad . \tag{3.10}$$

Since R, β and ζ depend on z, it is convenient to define the constants $\bar{R} = R(z_0 + \frac{1}{2})$, $\bar{\beta} = \beta(z_0 + \frac{1}{2})$ and $\bar{\zeta} = \zeta(z_0 + \frac{1}{2})$. The model is thus defined by the above equations together with boundary conditions. It will be seen that the diffusive approximation has been assumed for the heat flux term in the energy equation, though of course the effects of radiation could be included here at the cost of having to solve some additional equations (see e.g. Nordlund 1984, 1985). For most purposes it is convenient to use idealized boundary conditions such that the vertical velocity, the horizontal field and the tangential component of the viscous stress vanish at the top and bottom of the layer. (These boundary conditions are suited to astrophysical applications and correspond to those adopted for mathematical convenience in the Boussinesq problem). Nordlund and his collaborators have used a compliant boundary condition, permitting normal fluxes of mass, at the bottom of the layer. This inhibits the return flows that are a feature of the problem with impermeable boundaries, though in a deep enough layer one might hope for similar results in each case (but see discussion after Nordlund 1985). Most computations have kept the temperature fixed at both horizontal boundaries, though as discussed in §5 below the upper temperature condition may have a significant effect on the flow and one should perhaps substitute constant temperature with a flux-temperature relation of Stefan type. When appropriate periodicity conditions are applied to the lateral boundaries, the state of the system is described by the aspect ratio of the computational box and by six physical parameters: m and z_0 define the reference atmosphere, σ and ζ_0 (or $\bar{\zeta}$) fix the properties of the gas and $\bar{R}$ and Q are measures of the superadiabatic gradient and the imposed magnetic field.

4. Steady Flows: inhibition of motion and flux concentration by strong fields

4.1. BOUSSINESQ THEORY

Many of the fundamental ideas concerning the inhibition of flows by strong magnetic fields can be comprehended within the Boussinesq approximation. If we consider a shallow layer of fluid with almost uniform material properties, and a temperature difference across the layer that is small compared with the absolute temperature, then the fluid may be treated as effectively incompressible except as regards the buoyancy force. The pressure gradient then plays the passive role of ensuring the incompressibility condition $\nabla \cdot \mathbf{u} = 0$. In the Boussinesq limit the pressure can be eliminated from the momentum equation (3.4) by taking the curl; this means that magnetic pressure forces are irrelevant. Only the curvature forces then contribute to the dynamical effect of the field on the flow. Full details of the results summarised below can be found in the review of Proctor & Weiss (1982).

Many numerical experiments have been undertaken in a two-dimensional geometry by Weiss (1981*a*,*b*). The initial field $\mathbf{B}_0$ is vertical. By varying the Rayleigh number and diffusivity ratio ζ (both unambiguously defined as constant in the Boussinesq theory) one can observe a sequence of transitions. In appropriate parameter ranges (see §5 below) the initial transition is to steady convection; even if the initial bifurcation takes the form of oscillations, then overturning convection supervenes at higher values of the Rayleigh number. Although at high values of R the solutions are time dependent, there is usually an interval in R for which there is steady convection. Thus it makes sense to investigate the latter as an example of the effects of flows whose velocities have a non-zero average.

For large enough velocities the field is weak in the cell interiors and becomes confined to

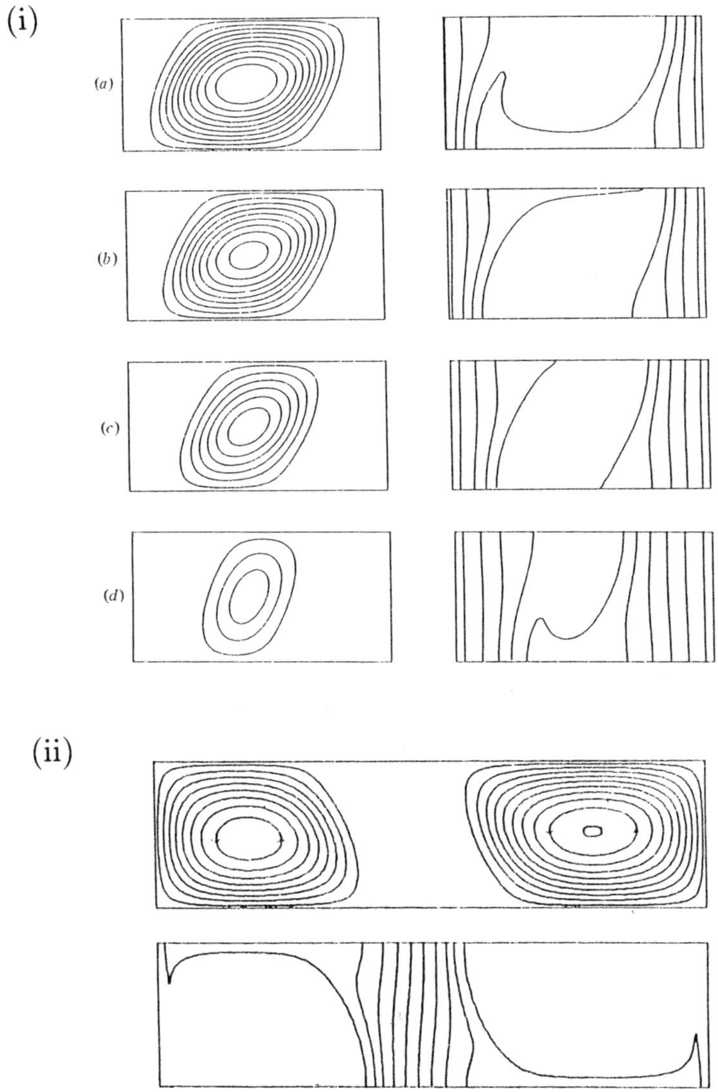

Fig. 2. Field lines and streamlines for nonlinear steady two-dimensional Boussinesq magnetoconvection (from Weiss 1981*b*, Proctor & Weiss 1982). (i) Development of static flux sheets: (a) $Q = 1000$, (b) $Q = 1400$, (c) $Q = 1900$, (d) $Q = 2300$. Here $R = 10^4$, $\zeta = 0.2$. (ii) Example of corotating cells for a strong field. Here $R = 10^4$, $Q = 1000$, $\zeta = 0.1$.

thin 'flux sheets' of thickness $O(R_m^{-\frac{1}{2}})$ compared with the cell size. Although the geometry is rather specialised, the Galloway & Proctor (1983) results referred to earlier do suggest that similar concentration occurs for three-dimensional motions, too. For larger field values (larger values of Q) however, the curvature forces interfere with the convection in the neighbourhood of the flux sheets, and thus the latter thicken. Figure 2 shows typical configurations for various Q; so stagnant is the sheet when the field is strong that it is not difficult to have *corotating* motions on either side, in defiance of intuition. A simple asymptotic theory (Galloway, Proctor & Weiss, 1977, 1978) shows that the peak field B^* in a sheet has a maximum value such that the Alfvén speed $V_A^* \equiv B^*/\sqrt{\mu_0 \rho}$ satisfies $(V_A^*/U)^2 \approx (\nu/\eta)R_m^{\frac{1}{2}}$; while this can be larger than the "equipartition limit" ($V_A^* \approx U$) that might have been anticipated on the basis of diffusionless theory, it is of the same order if we adopt the idea of "turbulent diffusivities" for ν and η though the existence of a turbulent magnetic diffusivity is problematical (Parker, these Proceedings). Computations have also been conducted with an axisymmetric geometry (Galloway & Moore 1979). Here concentration into axisymmetric 'ropes' of small volume leads to much more effective field enhancement and the back-reaction due to the Lorentz forces is considerably more localised. Calculations analogous to those for the two-dimensional case show that the peak field in an axisymmetric geometry is such that $(V_A^*/U)^2 \approx (\nu/\eta)(\ln R_m)^{\frac{1}{2}}$ (Galloway *et al.* 1977, 1978).

Another important nonlinear effect that can be understood within the framework of the Boussinesq theory is that of *subcritical convection*; when the field is concentrated into sheets or ropes vigorous convection may occur at Rayleigh numbers rather less than required for the onset of convection in an initially uniform field. Figure 3 shows some typical dependences of Nusselt number (measuring the vigour of convection) on Rayleigh number; subcritical behaviour can appear when $\zeta << 1$ (as appropriate in the upper photosphere where the effective thermal diffusivity is greatly enhanced by radiative effects). Thus the predictions of linearized theory (see e.g. Chandrasekhar 1961, Proctor & Weiss 1982) on the conditions for convection to occur may be misleading. In addition, the linear result that the preferred mode of convection in strong fields takes the form of tall thin cells, with the horizontal wavenumber increasing as $Q^{\frac{1}{6}}$, should be confronted with the existence of finite amplitude convection at aspect ratios of order unity at much lower Rayleigh numbers.

4.2. EFFECTS OF COMPRESSIBILITY

In a fully compressible atmosphere, in contrast to the Boussinesq situation, the magnetic pressure term comes into play, causing lower densities where the fluid is strong. In addition, hot rising fluid occupies more space than colder falling material, so downdrafts tend to be concentrated, while upwellings are diffuse. Since the pressure must be maximal over both rising and falling fluid regions the associated increase in density interferes with thermal buoyancy over the upwellings (buoyancy braking) and cooperates in the downwellings, enhancing the asymmetry (Massaguer & Zahn 1980, Hurlburt, Toomre & Massaguer 1984). This tendency leads, when fields are weak, to greater field amplification in downdrafts than upwellings. For larger imposed fields, however, the action of the curvature forces reduces the vigour of the flow, while the pressure forces lead to partial evacuation of the tube. The region of strong magnetic field then acquires buoyancy due to its reduced density, and this further inhibits concentration. The phenomenon is strikingly demonstrated in Figure 4, from Hurlburt and Toomre (1988) , who were the first to conduct a systematic study of magnetic convection in a compressible atmosphere. The example for a weak

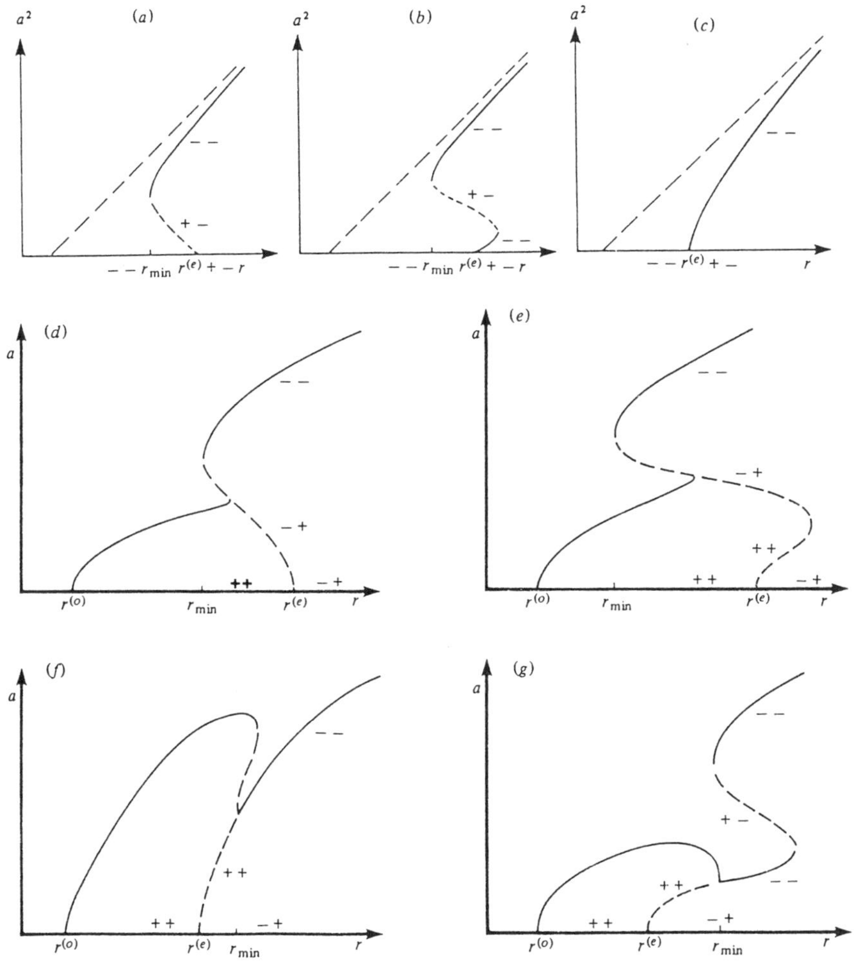

Fig. 3. Amplitude-Rayleigh number dependences for Boussinesq magnetoconvection (from Knobloch *et al.* 1981). r is a reduced Rayleigh number. The top three sketches illustrate situations where there is no Hopf bifurcation and there is a steady state bifurcation at $r = r^{(e)}$ (stable solutions are full lines, unstable ones broken lines). In the bottom four pictures, an oscillatory branch bifurcates at $r = r^{(o)}$. Sketches (d) and (f) show the canonical behaviour that occurs when $r^{(e)}$ and $r^{(o)}$ are very close (Knobloch & Proctor 1981).

imposed field shows the above-mentioned asymmetry, while for a larger value of Q the Lorentz forces and induced buoyancy actually lead to weak counter cells within the flux sheet, causing a bimodal distribution of vertical field. For persistently driven flows, it is possible to achieve greatly reduced densities, so that the theoretical upper limit $B_p \approx \sqrt{2p_e\mu_0}$ (where p_e is the local external pressure) for a static field concentration (Parker 1976) can be approached. This has been demonstrated by Cattaneo (1984a) using a simple model in which compressible flows are driven by a prescribed temperature gradient. Fine details of the structure of the 'steady state' flux sheet or rope are hard to tease out; a preliminary attempt by Proctor & Weiss (1984), though giving some insight, contained many simplifications. Very recently, Kerswell & Childress (1992) have produced a more sophisticated theory, which nonetheless relies on some parametrized effects. However, their work has perhaps begun to build a bridge between the detailed simulations, where the concentration must be seen primarily as an effect of the external flow, and the ideas behind the 'flux tube instability', developed by Webb & Roberts (1978), Unno & Ando (1979), Spruit (1979), Spruit & Zweibel (1979), and Hasan (1985) where collapse to a strong field is essentially a local effect, not critically dependent on the external flow.

5. Oscillatory Convection

5.1. TRAVELLING AND STANDING WAVES

The principal new ingredient provided by the interaction of magnetic fields and convection is the possibility of oscillatory behaviour. As previously mentioned, the curvature forces endow magnetic field lines with some of the character of strings under tension, and so oscillations might be expected provided that damping effects are not too strong. In a compressible fluid these waves fall into three classes (for a discussion see Priest 1982). We shall ignore the torsional (true) Alfvén waves, since these cannot be destabilized by superadiabatic temperature gradients when the initial field is vertical. The other so-called 'fast' and 'slow' modes have a character determined by the value of the 'plasma beta' (Equation (3.9)); the midlayer value $\bar{\beta}$ can be expressed up to constants of order unity in the alternative form

$$\bar{\beta} = \frac{2\mu_0 p_0}{B_0^2} = \frac{2V_S^2}{\gamma V_A^2} \tag{5.1}$$

where p_0 is the ambient pressure, and V_A and V_S are the Alfvén and sound speeds. For small $\bar{\beta}$ the slow mode is essentially longitudinal and acoustic, using the field lines as waveguides, while for $\bar{\beta} >> 1$ the slow mode is essentially transverse, relying for its restoring force on the curvature of field lines. In almost all cases of interest it is the slow modes that seem to be easiest to destabilize. Syrovatskii & Zhugzhda (1967) and Zhugzhda (1970) studied the small $\bar{\beta}$ case, where the magnitude of B_0, if sufficiently large, has little effect on the form of the instability. The large $\bar{\beta}$ case can be well understood in terms of the Boussinesq approximation.

Antia & Chitre (1979) and Cattaneo (1984*a,b*) have carried out full surveys for a wide range of $\bar{\beta}$'s, assuming an initial polytropic atmosphere. An interesting cooperative mechanism manifests itself when $\bar{\beta}$ is of order unity, when fast and slow modes are very similar and can combine to destabilize even a slightly subadiabatic atmosphere (Cattaneo 1984*a,b*).

The results of Boussinesq theory (Weiss 1981*a,b*) which are backed up by extensive integrations of systems of low order o.d.e.'s derived from the full Boussinesq system by

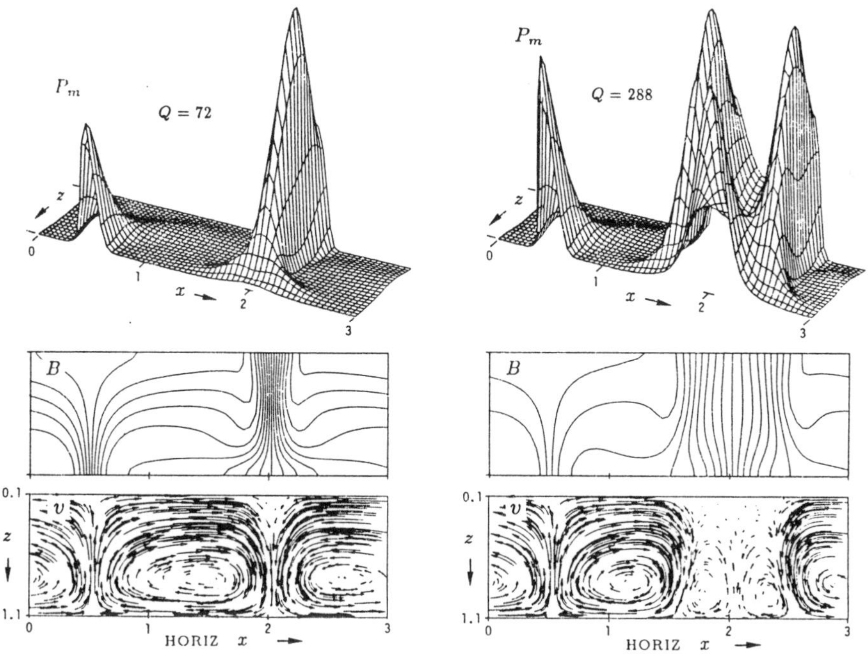

Fig. 4. The effect of Lorentz forces on compressible magnetoconvection (from Hurlburt & Toomre 1988). The density contrast across the layer is 11 and $\bar{R} = 10^5$. P_m is the magnetic pressure ($|\mathbf{B}|^2/2\mu_0$).

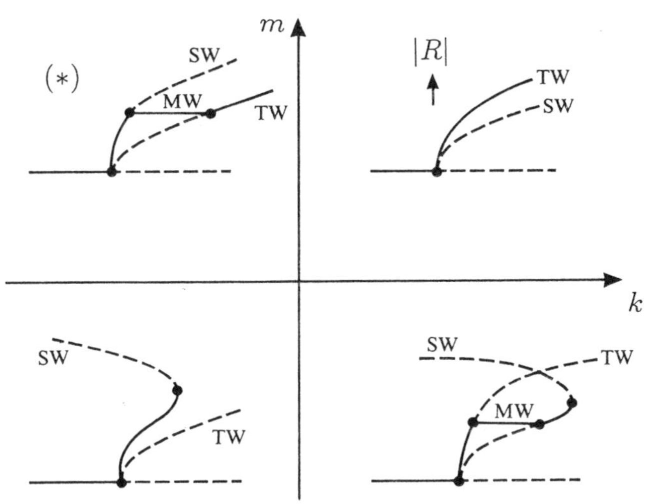

Fig. 5. Relation of the different steady solutions to equation (5.3) in the four quadrants of the (k, m) plane. Stable solutions are full lines, unstable ones dashed. *SW*= standing waves, *TW*= travelling waves, *MW*= modulated waves (from Hurlburt *et al.* 1989).

truncation (Knobloch *et al.* 1981) are described in detail by Proctor & Weiss (1982). The most important conclusion is that if the diffusivity ratio ζ is less than unity and if the field imposed is sufficiently strong, then branches of oscillating finite amplitude solutions can occur for lower values of $\bar{R}$ than necessary for steady motion. [It can actually be shown (Rucklidge *et al.* 1992) that even when conditions for linear oscillatory instability are not realised oscillatory convection can still occur in the nonlinear regime.] These solution branches can, however, exist only in moderately unstable situations, and will eventually give way to overturning (steady) convection as the temperature gradient is increased. Certain aspects of the interaction can be captured analytically when the Rayleigh numbers for steady and oscillatory onset are very close (or, alternatively, when Q is just greater than its minimum value for oscillation to be possible) (Knobloch & Proctor 1981; Dangelmayr & Knobloch 1987; Rucklidge 1992) and examples are shown in Figure 3, but in the general case the fate of the oscillatory branch is much more complicated, with secondary and tertiary bifurcations leading to aperiodic solutions with more than one dominant spatial scale. Extensive computations have been carried out by Hurlburt *et al.* (1989) on fully compressible 2–D convection with an imposed vertical field, according to the prescription of §3. These computations are, like most others, performed with periodic boundary conditions in the horizontal direction. In consequence, the system possesses translational as well as reflexional invariance and there is some degeneracy in the nature of the oscillations that bifurcate from the steady state. We can see this by noting that the linear eigensolution for the temperature, say, is always expressible in the form

$$A_1 e^{i\omega t - ikx} f(z) + A_2 e^{i\omega t + ikx} f(z) + c.c. \tag{5.2}$$

where A_1 and A_2 are complex amplitudes undetermined by linearized theory. If (say) $A_2 = 0$ then the solution is in the form of a *travelling wave*, moving to the right at speed ω/k without change of form, while if $|A_1| = |A_2|$ we have a *standing wave* where the perturbations change sign without propagating. Finally if $|A_1| \neq |A_2| \neq 0$ we have a *modulated wave*, which is quasiperiodic in a stationary frame but periodic in an appropriately translating one. The competition between these three solutions involves nonlinear interactions. There is a large body of formal theory that shows how such competitions are governed by relatively simple systems of low order ordinary differential equations, with coefficients that can in principle be determined in a straightforward way from the governing p.d.e.'s (for details see e.g. Hurlburt *et al.* 1989). In the present case, the moduli $R_{1,2}$ of the amplitudes $A_{1,2}$ can be shown to be governed in the nonlinear regime by the system

$$\left.\begin{aligned} \dot{R}_1 &= R_1[\mu - (R_1^2 + R_2^2) - kR_2^2 - mR_2^4] \\ \dot{R}_2 &= R_2[\mu - (R_1^2 + R_2^2) - kR_1^2 - mR_1^4] \end{aligned}\right\} \tag{5.3}$$

where here μ represents the difference between the actual value of the Rayleigh number and its value at the oscillatory bifurcation, and k, m depend on the parameters of the problem. Figure 5 shows the possible ways that the solution branches mentioned above can interact in different portions of the (k, m) plane. Note that while modulated waves cannot occur at onset, they may be present as transitional structures connecting the standing and travelling wave branches. It should be emphasised once more that all this diversity is due to the degeneracy in the underlying bifurcation problem, and that the transitions shown can take place at very low amplitudes.

Now it has been established that in the Boussinesq approximation, the oscillations appear (except perhaps for extreme geometries) to take the form of standing waves (Dangelmayr

& Knobloch 1986). This is perhaps not too surprising on physical grounds since for very large $\bar{\beta}$ the slow waves have a vertical phase velocity in the absence of boundaries. The horizontal boundaries act as waveguides, making horizontal propagation possible, but there is no mechanism for horizontal energy transfer. In a compressible atmosphere the waves are no longer completely transverse and the phase velocity can have a horizontal component. This fact would seem to make travelling waves easier to excite, and indeed they have been observed for strong fields. Figure 6 (from Hurlburt *et al.* 1989) shows standing, modulated and travelling waves that can occur in the same numerical experiment simply by changing one parameter ($\bar{R}$ in this case). The transition from stable standing waves, near onset, to stable travelling waves, is more complex than the transitions pictured in Figure 5 since there is an additional change of spatial scale. However, analogous numerical experiments with an imposed *horizontal* field, for which much of the preceding discussion applies, reveal that standing rather than travelling waves are stable at onset. This differs from the result for Boussinesq convection (Knobloch 1986), for which, as one might expect physically from the fact that there is now a horizontal component to the Alfvén velocity, travelling waves are stable. However at larger temperature gradients these standing waves lose stability to travelling waves (via a transitional branch of modulated waves), in the same way as shown on the starred section of Figure 5 (Brownjohn *et al.* 1992). Figure 7 shows three runs for different Rayleigh numbers ($\bar{R}$ increases to the right), showing the behaviour of the lines of force for standing, modulated and travelling waves, respectively.

5.2. MODELS FOR UMBRAL CONVECTION IN SUNSPOTS

The occurrence of umbral dots presents a challenge for the theorist; while 'dots' near the umbral edge appear to emanate from the penumbra, there is reason to think that the central dots are manifestations of umbral oscillatory convection. In contrast to the results presented above, it is now necessary to carry out calculations in a deep layer simulating the sunspot atmosphere. Furthermore, calculations of the effective value of ζ in a typical umbra (Meyer *et al.* 1974; Cowling 1976; Weiss *et al.* 1990) show that while $\zeta < 1$ near the surface, so permitting oscillations according to linearized theory, it drops below unity deeper down. Indeed, a linear stability analysis for a model with just this property (Weiss *et al.* 1990) which is achieved by taking constant values of η and the thermal conductivity so that ζ increases with the density of the polytropic basic state, shows that the initial bifurcation is to steady convection (see also Moreno Insertis & Spruit 1989). Numerical experiments show, however, that in the nonlinear regime the system can reconcile the two different types of motion with a mixed mode oscillation (Figure 8). It can be seen that the top of the layer oscillates while at the bottom the velocity does not reverse. Yet there is no evidence of 'layering' and vertical coherence is maintained. The results are relevant to the phenomenon of 'umbral dots'; bright points that appear in sunspot umbrae, with a typical lifetime of a few minutes. They contribute $\approx 15\%$ of the umbral energy flux and are believed to be manifestations of convection, pushing field-free plasma through the radiative layer at the top of the photosphere (see, e.g. R. Muller, these Proceedings). While similar phenomena near the umbra/penumbra boundary would appear to have a different origin, the calculations described above give some confidence that a picture of umbral dots as a convective phenomenon in a reasonably coherent deep seated spot can be sustained.

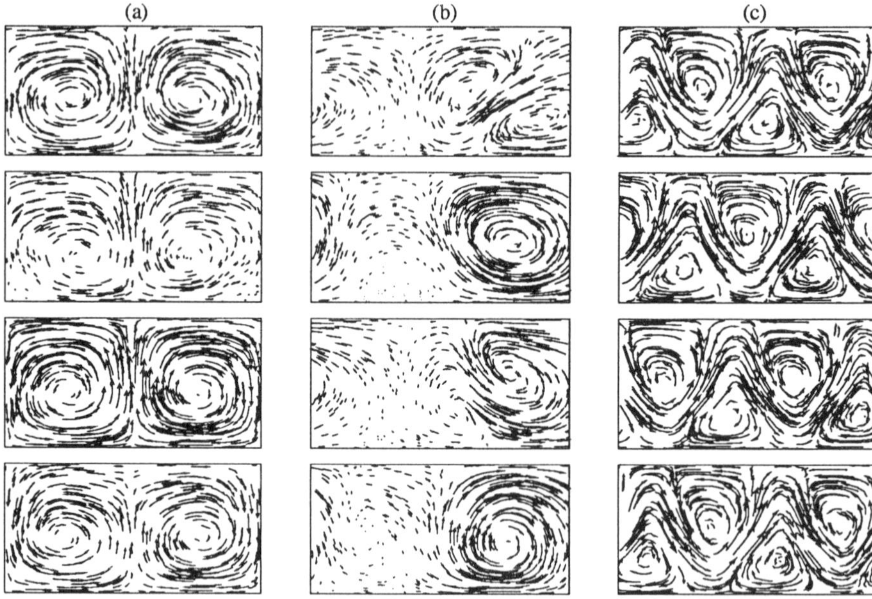

Fig. 6. Different forms of convection in a shallow layer. Here the density contrast across the layer is 1.63; $\bar{\zeta} = \sigma = 0.1$, $\bar{\beta} = 8$ (from Hurlburt *et al.* 1989). (a) Standing waves; , $\bar{R} = 5400$. (b) mixed mode travelling waves with two dominant scales; $\bar{R} = 10800$. (c) travelling waves with half the original scale; $\bar{R} = 43200$.

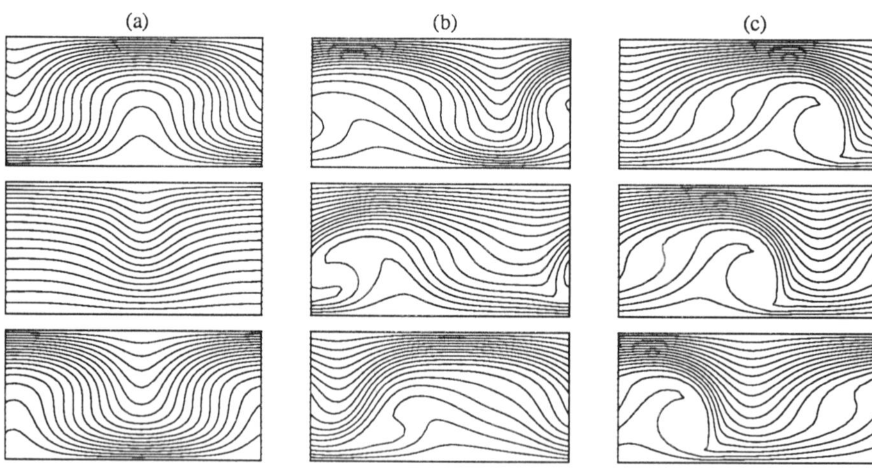

Fig. 7. Different forms of convection in an imposed *horizontal* field (from Brownjohn *et al.* 1992). Here $\bar{\beta} = 32$. About half a period is shown in each case. (a) standing waves for $\bar{R} = 4000$. (b) modulated waves for $\bar{R} = 32000$. (c) travelling waves for $\bar{R} = 128000$.

5.3. VARIATIONS ON THE THEME

In this subsection are collected together some disparate results that are tangential to the main thoughts of the paper, but nonetheless of significance. Fox, Theobald and Sofia (1991) have followed up previous non-magnetic convection calculations with a 2-D magnetoconvection code that uses subgrid scale modelling of the diffusive coefficients. This has the advantage of introducing significant diffusion only 'when needed' by the numerical scheme, but the disadvantage that quantities such as ζ, whose values are crucial for deciding the form of any diffusive instabilities, cannot be properly determined. This unfortunately makes it impossible to give sensible criteria for deciding when convection will be direct or oscillatory. They report, among other things, on the rise and ejection of an initially horizontal field by convection and self-buoyancy, which has interesting features, though a longer calculation might have revealed additional effects. Further, three-dimensional computations are in prospect.

In quite another direction, Matthews *et al.* (1992) and Hurlburt *et al.* (1992) have shown by a linear stability analysis and by direct numerical simulation, that when convection takes place in a layer permeated by an *oblique* field, all bifurcations take the form of travelling waves, since there is not enough symmetry to permit steady convection or standing waves. The direction of travel for the preferred mode seems to depend crucially on the precise boundary conditions applied to the field at the top and bottom of the layer, and it is not clear which conditions (fixed angle, anchored field lines, external insulator) best represent reality. Though the computations are two-dimensional, it can be shown that convection rolls with axes perpendicular to the field tilt are actually preferred over those with aligned axes, at least if the obliquity is not too large. This may have implications for convection in sunspots, with a transition taking place between the central region, where the field is nearly vertical, and the outer region (which may be identified with the penumbra) where the field is nearly horizontal and the theory suggests that rolls with nearly radial axes are the stable form of convection.

Finally in this section we should discuss the consequences of employing the diffusion approximation for heat transfer, even very high in the convective zone. Many computations, including those of Hurlburt, Weiss and their collaborators, assume constant temperature boundary conditions,together with the diffusion approximation. There is no doubt that these conditions may give a quantitatively incorrect description of the thermal effects of flux concentration, since they do not allow the surface of a concentrated region of flux to cool. Nordlund (1984, 1985) has called for (and implemented) a proper treatment of radiative transfer in convection simulations. This detailed approach is clearly necessary if one is attempting to obtain information on emission spectra etc. for comparison with observations, but may not be needed for understanding global processes. Indeed, Edwards (1992) has carried out simulations using the intermediate Eddington approximation and finds little significant change from previous diffusive calculations, except for stronger countercells in field concentrations caused by the relative ineffectiveness of radiative heat transfer at small scales.

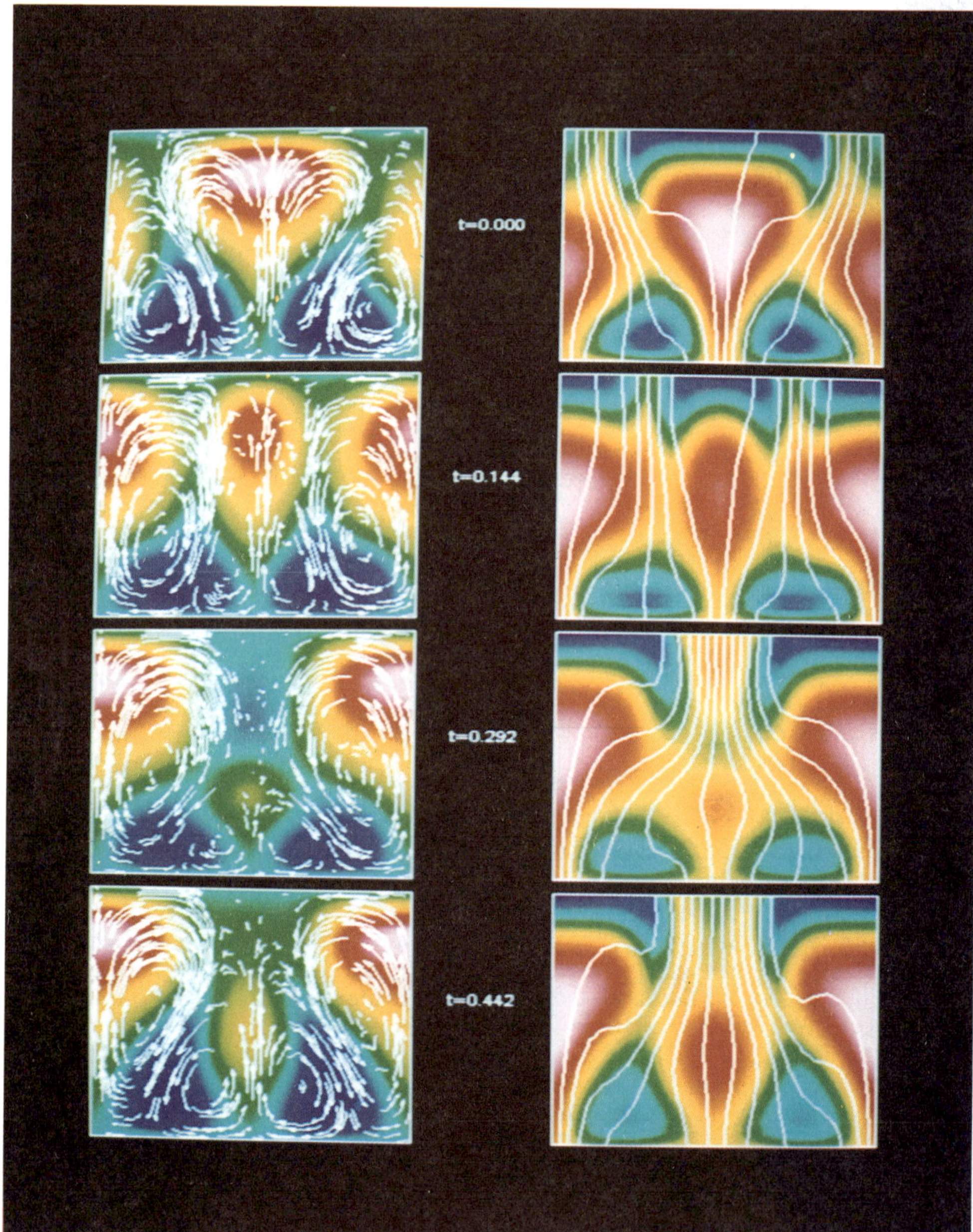

Fig. 8. Mixed mode convection in a deep layer (data from Weiss *et al.* 1990). Here the density contrast across the layer is 11, and $\bar{\zeta} = 1.2$ so that ζ passes through unity in the middle of the layer. $\bar{R} = 100000$. Note that the velocity only reverses near the top of the layer, and merely fluctuates slightly at the bottom. The left hand series shows velocity streak lines and colour contours of temperature while the right hand series shows magnetic field lines and colour contours of pressure. Times shown are expressed as fractions of the period of the oscillation.

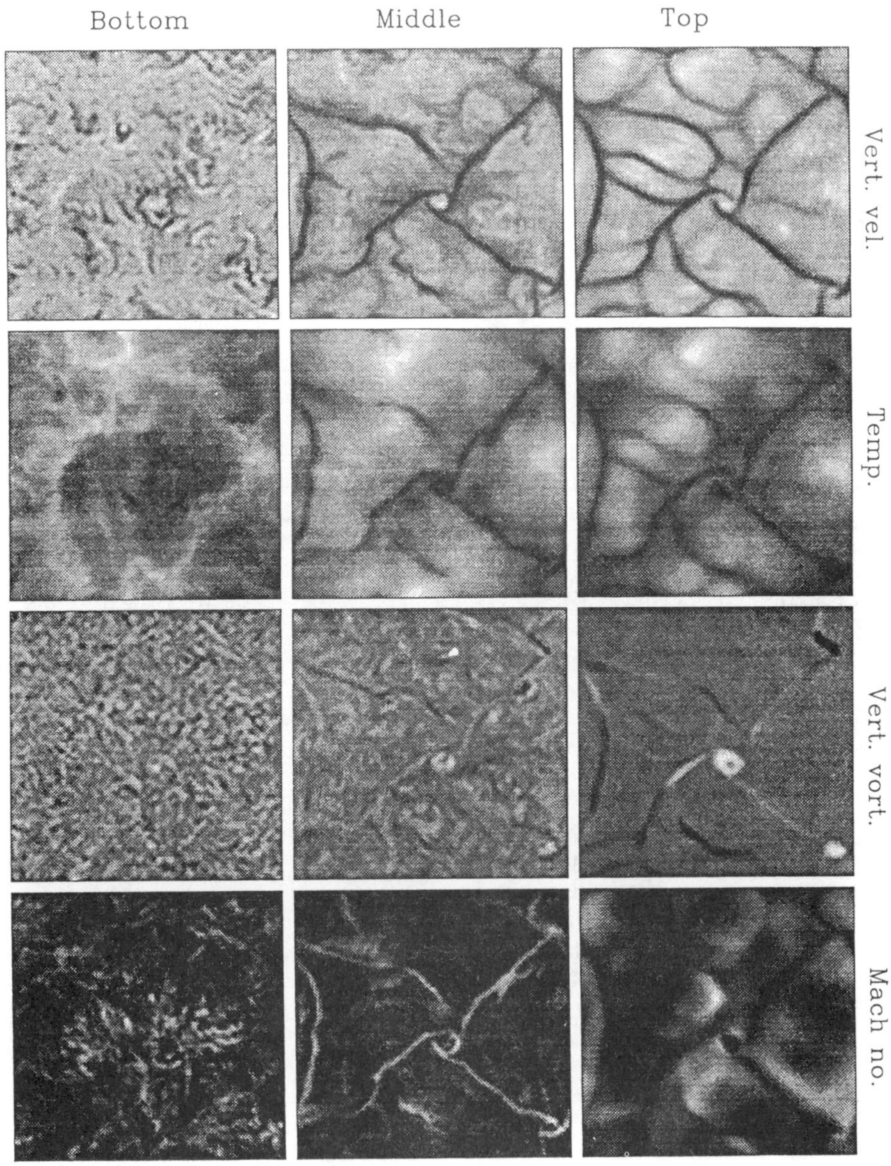

Fig. 9. Three-dimensional non-magnetic convection (from Cattaneo *et al.* 1991). Various quantities are shown at three different levels in the layer (near top, middle, near base). Note that the asymmetry between upwellings and downdrafts is much more pronounced near the top of the layer than at the bottom.

6. Conclusion: Towards Three-dimensional Simulations

Constraints in computing power have up until now meant that three-dimensional studies of compressible magnetoconvection have not been possible at high resolution. There has however been considerable work done on the non-magnetic problem (Stein & Nordlund 1989, Cattaneo *et al.* 1991). These computations, conducted in relatively deep atmospheres, show that while near the surface downflows are typically more concentrated than upwellings there is less asymmetry at greater depths (though this may be a result of the forms of viscosity and thermal conductivity chosen). An example is given in Figure 9. There is also evidence that the typical scale of convection increases with depth (and with the local scale height), as mixing length theories suppose, but as two-dimensional computations have failed to demonstrate. Thus the simpler two-dimensional models provide some, but not all of the correct physics.

Nordlund (1984) has carried out computations in a three-dimensional geometry with an imposed vertical field, using the anelastic approximation (thus filtering out sound waves). Further preliminary results on the fully compressible problem have been reported by Nordlund & Stein (1989). These computations differ in spirit from those of Hurlburt and his collaborators in that fluid is permitted to enter and leave the computational domain. Magnetic diffusivity is also neglected, although there is numerical diffusion due to the finite spatial resolution. More recently, Brandenburg *et al.* (1991) and Nordlund *et al.* (1992) have begun a study of rotating compressible magnetoconvection with the ultimate aim of simulating a dynamo at the base of the solar convection zone. Fascinating results have emerged showing the stretching and twisting of flux ropes, but a full description is beyond the scope of this review. These computations place the convectively unstable region adjacent to a stably stratified layer; perhaps this is the best way of solving the problem of the correct boundary conditions to implement. The only study of three-dimensional convection with an imposed horizontal field (Arter 1985) utilizes the Boussinesq approximation. Three-dimensional codes are in the final stages of development in Cambridge and Palo Alto, but no results have emerged yet.

What, then, is the state of play regarding the useful lessons that can be learned about umbral convection from formal numerical studies? The chief message, I think, is that time dependence is to be almost universally expected, even for relatively feeble convection, and that there is a great variety of possible time dependent motion encompassing both standing waves (umbral dots?) and travelling waves (apparent inward transport of wave-like objects in the penumbra?) The simpler two-dimensional theories postulate (and produce) strong magnetic field concentrations that are coherent in depth. Only more detailed, high resolution three-dimensional computations will resolve the dichotomy between monolith and tight cluster sunspot models. In a certain sense there may be no conflict since penetration of fluid into the sunspot umbra can occur either through 'turbulent diffusion' or (which is really the same thing) detailed folding of the boundary between field free and magnetized plasma. It will also be of great importance to pay attention to the appropriate thermal boundary conditions at the surface of the convection zone, with the aim of better simulating the effects of radiation without having to solve the transfer equations in detail.

Acknowledgement

I am grateful for support from the U.K. Science and Engineering Research Council, and am indebted to Nigel Weiss, Neal Hurlburt, Paul Matthews, Alastair Rucklidge and Axel Brandenburg for stimulating conversations. Derek Brownjohn has given invaluable assistance in the preparation of the Figures.

References

Alfvén, H. (1942) "On the existence of electromagnetic-hydrodynamic waves," *Arkiv mat. astron. fys.* **29B,** No.2 .

Antia, H.M. and Chitre, S.M. (1979) "Waves in the sunspot umbra," *Solar Phys.* **63,** 67–78.

Arter, W. (1983) "Nonlinear convection in an imposed horizontal magnetic field", *Geophys. Astrophys. Fluid Dyn.* **25,** 259–292.

Arter, W. (1985) "Magnetic-flux transport by a convecting layer including dynamical effects," *Geophys. Astrophys. Fluid Dyn.* **31,** 311–344.

Arter, W., Proctor, M.R.E. and Galloway, D.J. (1982) "New results on the mechanism of magnetic flux pumping by three-dimensional convection," *Mon. Not. R. Astron. Soc.* **201,** 57P–61P.

Brandenburg, A., Jennings, R.L., Nordlund, Å. and Stein, R.F. (1991) "Magnetic flux tubes as coherent structures," in: *Spontaneous Formation of Space-time Structures and Criticality* (eds. T. Riste and D. Sherrington) Kluwer, Dordrecht.

Brownjohn, D.P., Hurlburt, N.E., Proctor, M.R.E. and Weiss, N.O. (1992) "Nonlinear compressible magnetoconvection Part 3. Travelling waves in a horizontal field," in preparation.

Cattaneo, F. (1984*a*) "Compressible magnetoconvection," *Ph.D. Dissertation, University of Cambridge.*

Cattaneo, F. (1974*b*) "Oscillatory convection in sunspots," in: *The Hydrodynamics of the Sun,* pp. 47–50, ESA SP-220, Noordwijk.

Cattaneo, F., Brummell, N.H., Toomre, J., Malagoli, A. and Hurlburt, N.E. (1991) "Turbulent compressible convection," *Astrophys. J.* **370,** 282–294.

Chandrasekhar, S. (1961) *Hydrodynamic and Hydromagnetic Stability,* Clarendon Press, Oxford.

Cowling, T.G. (1976) "On the thermal structure of sunspots," *Mon. Not. R. Astron. Soc.* **177,** 409–414.

Dangelmayr, G. and Knobloch, E. (1986) "Interaction between standing and travelling waves and steady states in magnetoconvection," *Phys. Lett.* **117A,** 394–398.

Dangelmayr, G. and Knobloch, E. (1987) "The Takens-Bogdanov bifurcation with O(2) symmetry," *Phil. Trans. Roy. Soc.* **A322,** 243–279.

Edwards, J.M. (1992) "A brief survey of radiative magnetoconvection," *Mon. Not. R. Astr. Soc.* (submitted).

Fox, P.A., Theobald, M.L. and Sofia, S. (1991) "Compressible magnetic convection: formulation and 2-D models," *Astrophys. J.* **383,** 860-881.

Galloway, D.J. and Moore, D.R. (1979) "Axisymmetric convection in the presence of a magnetic field," *Geophys. Astrophys. Fluid Dyn.* **12,** 73–105.

Galloway, D.J. and Proctor, M.R.E. (1983) "The kinematics of hexagonal magnetoconvection," *Geophys. Astrophys. Fluid Dyn.* **24,** 109–136.

Galloway, D.J., Proctor, M.R.E. and Weiss, N.O. (1977) "Formation of intense magnetic fields near the surface of the Sun," *Nature* **266,** 686–689.

Galloway, D.J., Proctor, M.R.E. and Weiss, N.O. (1978) "Magnetic flux ropes and convection," *J. Fluid Mech.* **87,** 243–261.

Hasan, S.S. (1985) "Convective instability in a solar flux tube II. Nonlinear calculations with horizontal radiative heat transport and finite viscosity," *Astron. Astrophys.* **143,** 39–45.

Hughes, D.W. and Proctor, M.R.E. (1988) "Magnetic fields in the solar convection zone: magnetoconvection and magnetic buoyancy," *Ann. Rev. Fluid Mech.* **20,** 187–223.

Hurlburt, N.E. and Toomre, J. (1988) "Magnetic fields interacting with nonlinear compressible convection," *Astrophys. J.* **327,** 920–932.

Hurlburt, N.E., Toomre, J. and Massaguer, J.M. (1984) "Two-dimensional compressible convection extending over multiple scale-heights," *Astrophys. J.* **282,** 557–573.

Hurlburt, N.E., Matthews, P.C. and Proctor, M.R.E., "Nonlinear compressible magnetoconvection in an oblique field," in preparation (1992).

Hurlburt, N.E., Proctor, M.R.E., Weiss, N.O. and Brownjohn, D.P. (1989) "Nonlinear compressible magnetoconvection. Part 1. Travelling waves and oscillations," *J. Fluid Mech.* **209,** 587–628.

Jones, C.A. and Roberts, P.H. (1990) "Magnetoconvection in rapidly rotating Boussinesq and compressible fluids," *Geophys. Astrophys. Fluid Dyn.* **55,** 263–308.

Kerswell, R. and Childress, S. (1992) "Equilibrium of a magnetic flux tube in a compressible flow," *Astrophys. J.,* in press.

Knobloch, E. (1986) "On convection in a horizontal magnetic field with periodic boundary conditions," *Geophys. Astrophys. Fluid Dyn.* **36,** 161–177.

Knobloch, E. and Proctor, M.R.E. (1981) "Nonlinear periodic convection in double-diffusive systems," *J. Fluid Mech.* **108,** 291–316.

Knobloch, E., Weiss, N.O. and DaCosta, L.N. (1986) "Oscillatory and steady convection in a magnetic field," *J. Fluid Mech.* **113,** 153–186.

Massaguer, J.M. and Zahn, J.-P. (1980) "Cellular convection in a stratified atmosphere," *Astron. Astrophys.* **87,** 315–327.

Matthews, P.C., Hurlburt, N.E., Proctor, M.R.E. and Brownjohn, D.P. (1992) "Compressible magnetoconvection in oblique fields: linearised theory and simple nonlinear models," *J. Fluid Mech.,* in press.

Meyer, F., Schmidt, H.U., Weiss, N.O. and Wilson, P.R. (1974) "The growth and decay of sunspots," *Mon. Not. R. astr. Soc.* **169,** 35–57.

Moreno Insertis, F. and Spruit, H.C. (1989) "Stability of sunspots to convective motions I. Adiabatic instability," *Astrophys. J.* **342,** 1158–1171.

Nordlund, Å. (1984) "Magnetoconvection: the interaction of convection and small scale magnetic fields," in: *The Hydromagnetics of the Sun,* pp. 37–46, ESA SP-220, Noordwijk.

Nordlund, Å. (1985) "The 3-D structure of the magnetic field, and its interaction with granulation," *Proc. Workshop Theor. Probl. High Resolution Sol. Phys.*(ed. H.U.Schmidt), pp. 101–123, Max-Planck Inst. Phys. u. Astrophys., Munich.

Nordlund, Å. and Stein, R.F. (1989) "Simulating magnetoconvection," in: *Solar and stellar granulation* (eds. R.J. Rutten and G. Severino) pp. 453-470, Kluwer, Dordrecht.

Nordlund, Å., Brandenburg,A., Jennings, R.L., Rieutord, M., Ruokokainen, J., Stein, R.F. and Tuominen, I. (1992) "Dynamo action in stratified convection with overshoot," *Astrophys. J.,* in press.

Parker, E.N. (1976) "Hydraulic concentration of magnetic fields in the solar photosphere III. Fields of one or two kilogauss," *Astrophys. J.* **204,** 259–267.

Priest, E.R. (1982) "Solar Magnetohydrodynamics," Reidel, Dordrecht.

Proctor, M.R.E. and Weiss, N.O. (1982) "Magnetoconvection," *Rep. Prog. Phys.* **45,** 1317–1379.

Proctor, M.R.E. and Weiss, N.O. (1984) "Amplification and maintenance of thin magnetic flux tubes by compressible convection," in: *The Hydrodynamics of the Sun,* pp. 77–80, ESA SP-220, Noordwijk.

Rucklidge, A.M. (1992) "Chaos in low-order models of double convection," *J. Fluid Mech.,* in press.

Rucklidge, A.M., Weiss, N.O., Proctor, M.R.E. and Brownjohn, D.P., (1992) "Subcritical oscillatory magnetoconvection," *Geophys. Astrophys. Fluid Dyn.*, submitted.

Spruit, H.C. (1979) "Convective collapse of flux tubes," *Sol. Phys.* **61,** 363–378.

Spruit, H.C. and Zweibel, E.G. (1979) "Convective instability of thin flux tubes," *Sol. Phys.* **62,** 15–22.

Stein, R.F. and Nordlund, Å. (1989) "Topology of convection beneath the solar surface," *Astrophys. J.* **342,** L95–L98.

Syrovatskii, S.I. and Zhughzda, Yu.D. (1967) "Oscillatory convection of a conducting gas in a strong magnetic field," *Astron. Zh.* **44,** 1180–1190 (*Sov. Astron.* **11,** 945–952).

Unno, W. and Ando, H. (1979) "Instability of a thin magnetic tube in the solar atmosphere," *Geophys. Astrophys. Fluid Dyn.* **12,** 105–115.

Webb, A.R. and Roberts, B. (1978) "Vertical motions in an intense magnetic flux tube. II. Convective instability," *Sol. Phys.* **59,** 249–274.

Weiss, N.O. (1981*a*) "Convection in an imposed magnetic field. Part 1. The development of nonlinear convection," *J. Fluid Mech.* **108,** 247–272.

Weiss, N.O. (1981*b*) "Convection in an imposed magnetic field. Part 2. The dynamical regime," *J. Fluid Mech.* **108,** 273–289.

Weiss, N.O. (1990) "Solar and stellar convection zones," *Comp. Phys. Rep.* **12,** 233–245.

Weiss, N.O. (1992) "Magnetoconvection," *Geophys. Astrophys. Fluid Dyn.* **54,** in press.

Weiss, N.O., Brownjohn, D.P., Hurlburt, N.E. and Proctor, M.R.E. (1990) "Oscillatory convection in sunspot umbrae," *Mon. Not. R. Astron. Soc.* **245,** 434–452.

Zhughzda, Yu.D. (1970) "The properties of low-frequency oscillatory convection in a strong magnetic field," *Astron. Zh.* **47,** 340–350 (*Sov. Astron.* **14,** 274–282.

THE CLUSTER MODEL OF SUNSPOTS

ARNAB RAI CHOUDHURI
Department of Physics
Indian Institute of Science
Bangalore - 560012. India

1. Introduction

More than one decade ago, Parker (1979a) began his famous paper on the cluster model of sunspots with the statement: "The sunspot phenomenon, with its reduced temperature and intense magnetic field, is at present without a scientific explanation." The aim of the present review is to assess whether the cluster model of sunspots (perhaps more popularly known by the less respectable name "spaghetti model") has been able to contribute something significant to change the situation since that time. The central idea of the cluster model is that a sunspot is not a single monolithic flux tube as in Figure 1a, but a collection of many smaller flux tubes herded together as in Figure 1b. The first question before us is whether the cluster model is in some sense a "correct" description of a sunspot. If it is so, then the second question is at what depth the individual flux tubes start separating. A third question is how the flux tubes making up a sunspot connect to the global solar magnetic field deep down. Hale's polarity law clearly tells us that sunspots are just outward manifestations of the global magnetic fields. When the cluster model was first proposed, perhaps the time was still not ripe to raise this third question. Now we may start preliminary enquiries on this vital and difficult question. Apart from these three morphological questions, there remains the challenging dynamical question why a cluster of flux tubes come together to form a sunspot. This set of basic questions is taken to define the scope of this review. In other words, we are going to look at the efforts by both theorists and observers to come up with answers to these four (three morphological and one dynamical) questions.

At the very outset, it should be pointed out that the cluster model is not a "model" in the sense of being a coherent, well-organized and self-consistent body of scientific ideas. Rather, under the name of the cluster model, ideas have been proposed which are radically divergent and sometimes almost orthogonal. The cluster model as a scientific model somewhat resembles Hinduism as a religion! While reviewing such a subject, following the historical approach is perhaps as good as any approach. So we shall begin by recapitulating the original arguments of Parker in support of the cluster model when it was first proposed.

J. H. Thomas and N. O. Weiss (eds.), Sunspots: Theory and Observations, 243–257.

Then we shall discuss in what way the theoretical and observational developments of the intervening decade have strengthened or weakened those original arguments. We also have to examine whether there are new arguments for or against the cluster model. As we go along, our positions with respect to the four basic questions mentioned above should become clear. We shall try to steer a middle course and to give a balanced account of different viewpoints. The personal prejudices of the present author, however, are bound to colour his presentation to some extent.

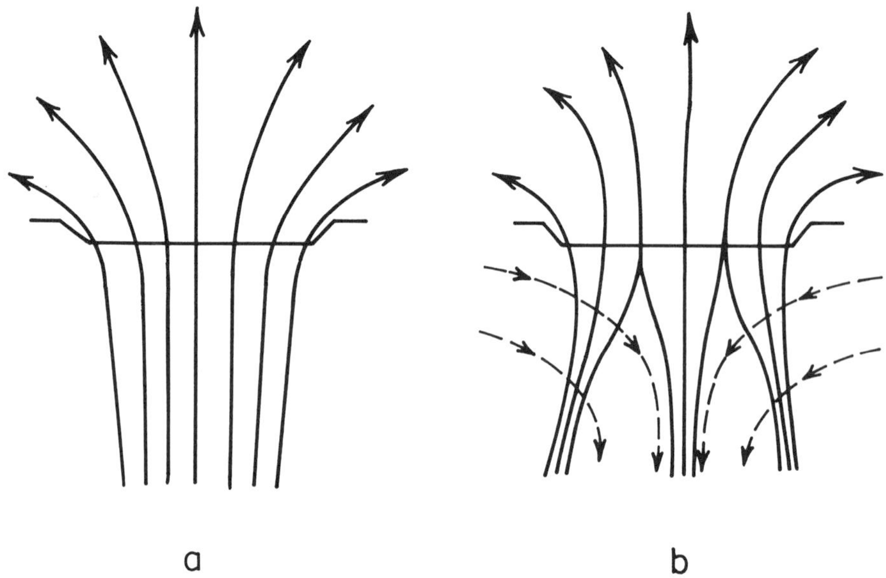

Fig. 1. Subsurface structure of sunspots. (a) monolithic flux tube. (b) Cluster model with the downdraft shown by dashed lines. (After Parker 1979a)

2. Original Arguments for the Cluster Model

Though the idea that a sunspot is a collection of magnetic knots was discussed briefly in a paper by Obridko (1975), the credit for clearly stating the basic assumptios of the cluster model and making it an attractive paradigm for solar research goes to Parker (1979a). Let us briefly summarize his main arguments.

I. Parker (1979a) pointed out that the surface temperature and the magnetic field strength are independent of the spot size, being roughly equal for larger and smaller spots. Figure 2 showing a large and a small spot according to a single flux tube viewpoint makes it clear that it will be difficult to explain this simple fact if sunspots were single flux tubes. One would expect a larger spot to be cooler. On the other hand, if a sunspot is a loose collection of many flux tubes and convection can take place in the field-free regions between them, then the puzzle disappears - at least qualitatively. Parker's argument was purely qualitative. The obvious question is: does this qualititative argument stand a quantitative scrutiny? Since it is extremely difficult to do a proper magnetoconvection calculation with a realistic sunspot geometry, the merits of the above argument still remain somewhat uncertain, and we shall not discuss this point any further.

II. High-resolution observations show a considerable amount of fine structure in sunspot umbrae, the bright umbral dots being the most notable. Parker (1979a, 1979d) suggested that these dots are intrusions of field-free convecting fluids in the gaps between the flux tubes making up the sunspot. The last decade has seen considerable new observational and theoretical efforts to understand fine structures of sunspots. We shall summarize them in the next Section. Though Parker's idea may be approximately correct, one probably needs much more sophisticated theoretical modelling than what Parker envisaged.

III. Since magnetic fields in the quiet photospheric region also exist in the form of thin flux tubes, Parker (1979a) suggested that sunspots may form by these flux tubes coming together and decay due to their falling apart. Section 4 will be devoted to the question whether the growth and decay of sunspots provide any evidence for the cluster structure. The possible relation between the quiet photospheric flux tubes and the flux tubes making up sunspots will also be discussed in that Section.

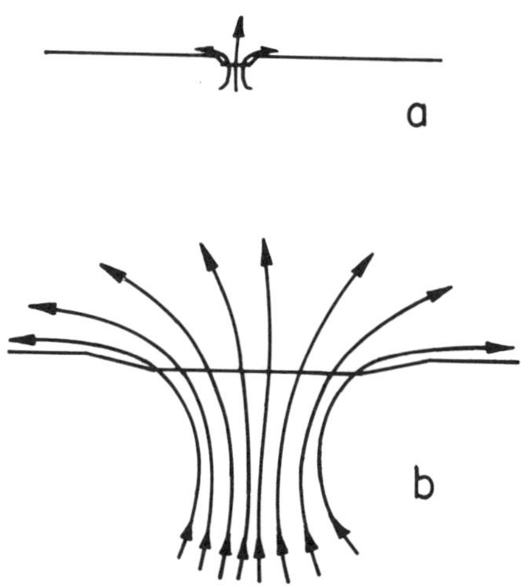

Fig. 2. A sketch of the field configurations of two sunspots differing in diameter by a factor of 10, according to the monolithic flux tube model. (From Parker 1979a)

IV. Even if a sunspot is produced by several thin flux tubes collecting together, it does not necessarily follow that those flux tubes will retain their identities even in a fully grown sunspot. The arguments I and II certainly point to the existence of composite structures in mature sunspots. Parker (1979a) uses the stability condition obtained by Meyer, Schmidt and Weiss (1977) to reinforce his argument that individual flux tubes will separate out below the photosphere. It is well-known that a magnetic field, which is concave to a region of unmagnetized plasma, is subject to fluting instability. According to Meyer, Schmidt and Weiss (1977), when the magnetic field diverges at a large angle with respect to the vertical (as in the penumbra), the fluting instability is suppressed by magnetic buoyancy and bundles of flux cannot separate out from the sunspot. However, at the wasp waist of the sunspot below the photosphere, the magnetic field no longer makes a sufficiently large angle with the vertical and there is nothing to stop fluting instability. Parker (1979a) argued that this will lead to the separation of the flux tubes below the wasp waist of the sunspot and estimated that this would happen at a depth of about 1000 km below the photosphere.

Parker did not consider the question how sunspot fields may be connected to the global magnetic field of the sun. As for the dynamical question of what brings the flux tubes together to form a sunspot, Parker (1979b) pointed out that there can be aerodynamic attractions between flux tubes when they rise. However, this attraction stops when the flux tubes reach the photosphere, and Parker (1979a) postulated a hypothetical converging downdraft as shown in Fig. 1b in order to bring the flux tubes together to form the sunspot. We shall see in Section 4 that there are other scenarios to explain the herding together of flux tubes without invoking the downdraft.

Apart from bringing the flux tubes together, Parker (1979a) pointed out that this downdraft could solve another long-standing problem. A sunspot blocks about 80% of the energy flux which must be coming from the solar interior. What happens to this blocked energy flux is a question that has puzzled theorists for several decades. One expects the blocked flux to produce a bright ring around the sunspot or even to disrupt the sunspot itself if sufficient heat collects underneath it. Parker (1979a, 1979c) showed that a downdraft would very efficiently carry away the blocked heat downwards. The subject of heat flow around sunspots has been reviewed in this Workshop by Spruit (1991), who argues that one would need a rather small temperature difference inside and outside at the bottom of a sufficiently deep sunspot in order to explain the depletion of the energy flux, and hence there is no problem with blocked flux for a sufficiently deep sunspot. Thus, postulating a downdraft may not be the only way of solving the blocked heat problem.

Though there was no observational evidence for the downdraft then or now, Parker (1979a) postulated the downdraft as an integral part of the original cluster model for the theoretical reason of solving the twin problems of bringing the flux tubes together and taking away the blocked heat. As we have become aware of other ways of solving these problems, the reasons for postulating a downdraft appear much less compelling and there has not been any other work on it after Parker's original work (Parker 1979a, 1979c). So we shall not discuss the downdraft any further.

In the next two Sections, we look at the current status of the arguments II and III mentioned above.

3. Fine Structures of Sunspots

Since the observational properties of sunspot fine structures ha been reviewed thoroughly in this Workshop by Muller (1992), we merely summarize the main results. The bright dots seen in the sunspot umbrae typically have sizes smaller than the resolution possible under the best seeing conditions. They were inferred to be regions of near photospheric intensity with diameters of the order 300 km and having an average lifetime of about an hour (Beckers and Schröter 1968; Adjabshirzadeh and Koutchmy 1980). Some recent observations, however, suggest that the temperatures of umbral dots can sometimes be less than the photospheric temperature (Grossmann-Doerth, Schmidt and Schröter 1986). Since umbral dots are smaller in size and longer-lived than photospheric granules, there has been a debate whether the umbral dots are just granules modified by the presence of the sunspot. Claims have been made both for (Bumba and Suda 1980) and against (Loughhead, Bray and Tappere 1979) this hypothesis by different observers. Some of the early observations suggested that umbral dots are regions of significantly weaker magnetic field and high upward velocity (Kneer 1973). Though it now appears that these results were probably based on uncertain analysis of data obtained at the limit of resolution, these results influenced theoretical thinking for about a decade. Let us look at the main trends of theoretical research before discussing the more current observations which are still not assimilated in the body of theory.

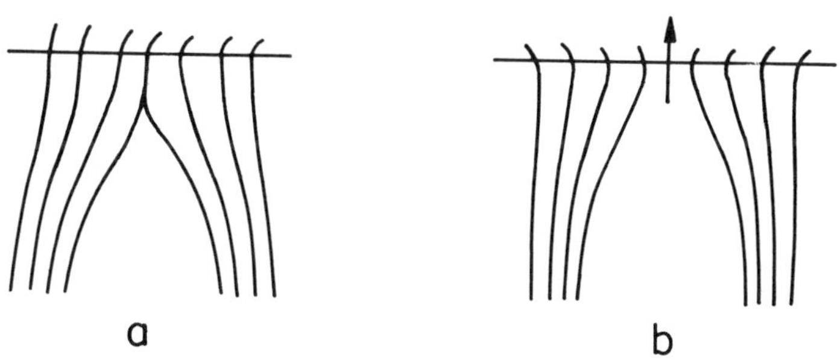

Fig. 3. Choudhuri's model of field-free fluid column. (a) Before reaching the photosphere. (b) After reaching the photosphere. (From Choudhuri 1986)

Within the framework of the cluster model, the umbral dots lead to the natural explanation as intrusions of field-free hotter fluid between the flux tubes darkened due to the suppression of convection by the magnetic field (Obridko 1975; Parker 1979a,1979d; Choudhuri 1986). Parker (1979d) studied the convective instability in a thin vertical column of field-free fluid sandwiched between magnetic fields on the two sides and obtained growth times of the order 1-3 hours, comparable to the lifetime of umbral dots. Elaborating on these ideas, Choudhuri (1986) investigated how the surrounding magnetic field may try to close on the field-free gas from the top and control the dynamic behavior of the system. A pocket of field-free gas surrounded by a vertical magnetic field in the presence of gravity takes up the shape of a tapering column ending at a vertex at the top, as shown in Figure 3a. If the magnetic field around such a tapering column inhibits heat flow, then the heat coming from below will tend to pile up in the field-free tapering column, causing a local increase in temperature and pressure. Choudhuri (1986) found that the field-free column would rise vertically as a whole due to this heat pile-up. Eventually, when the tip of the tapering column reaches the photospheric surface where the pressure suddenly drops, Choudhuri (1986) showed that the surrounding magnetic field may no longer be able to close on the hot field-free gas which comes out through a "magnetic valve" as shown in Figure 3b. The earlier observations of umbral dots as regions of weak magnetic fields and upward velocities could then easily be understood if umbral dots were identified as such open magnetic valves. After sufficient gas had flown out of the open magnetic valve, the pressure of the field-free column would decrease sufficiently so that the surrounding magnetic field would be able to close on it again, causing the umbral dot to decay away.

The other school of theory regarded the umbral dots as tops of convective cells present in a fairly homogeneous vertical magnetic field which modifies the nature of convection substantially to make the umbral dots look sufficiently different from granules in the quiet photosphere (Knobloch and Weiss 1984). The ratio of magnetic to thermal diffusivity, which determines the nature of convection, is <1 in the upper layers of a sunspot and >1 in the deeper layers. Hence one expects oscillatory convection in the upper layers and steady convection in the deeper layers. Weiss *et al.* (1990) carried out a 2D simulation for magnetoconvection in such a system. They found that the oscillatory convection in the upper layers gets coupled to the steady convection in the deeper layers in such a way that one gets hot regions of upwelling gas with somewhat dispersed magnetic field lasting for about an Alfven time. If umbral dots are regarded as such regions, then there are some obvious resemblances between the two theoretical models (cluster model and magnetoconvective model) which start from very different presuppositions. In either of these models, umbral dots are caused by intrusions of upwelling hot gases which tend to weaken the magnetic fields. A possible connection between the two theoretical models was suggested by Moreno-Insertis and Spruit (1989), who studied the linear stability of a convectively unstable stratified atmosphere with a vertical magnetic field. They found that small horizontal wavelengths are preferred, indicating that magnetoconvection may tend to fragment a uniform vertical field into a cluster-like structure. Thus there remains the theoretical possibility that the cluster structure is the ultimate end-product of magnetoconvection. Observationally, however, one does not see a uniform vertical field fragmenting into cluster structure, but rather flux tubes collect together to form sunspots as we discuss in the next Section. Hence, it seems unlikely that the instability discussed by Moreno-Insertis and Spruit (1989) is the origin of the cluster structure, but it probably plays a role in making sure that the cluster structure is retained after the formation of the sunspot.

One would have felt very happy at such a remarkable convergence of theoretical models starting from very different ideas, if it were not for the recent observations of Lites *et al.* (1991). These observations, done at La Palma with CCD and video processing techniques, achieved resolutions about 2-3 times better than the previous ones. They indicate that magnetic fields are not substantially weaker in umbral dots and there is no evidence of upward motion. As we have seen, both types of theoretical models (cluster model and magnetoconvective model) suggest upward motions in umbral dots. Theoretical calculations for both types of models, however, so far neglected radiative transfer. It was pointed out by Lites *et al.* (1991) that radiative tranfer would play an important role in the upper layers of a sunspot where the opacity becomes small, and the upward motion is probably damped out in the upper layers due to considerable amount of energy being lost by radiative transfer. The lack of upward flow rules out the possibility that the umbral dot is a magnetic valve as shown in Figure 3b. However, the umbral dot can be the upper part of the tapering column (Figure 3a) from which energy is leaking due to efficient radiative transfer. Since the magnetic field closes on the field-free fluid from the above, one may not see the reduction of the magnetic field unless the magnetic measurements were done by using a line formed below the vertex of the field-free column. If the brightness of the umbral dot is due to energy diffusing radiatively from the vertex, then the temperature of the dot is also expected to be somewhat less than the photospheric temperature as observed (Grossmann-Doerth, Schmidt and Schröter 1986). One feels optimistic that more detailed theoretical calculations to extend the model of Parker (1979d) and Choudhuri (1986) by incorporating radiative tranfer and by analyzing depths of line formation carefully would close the gap between theory and observations. We hope that such calculations will be done in the near future.

The penumbra of a sunspot also has a rich fine structure. Recent observations show a fairly complicated pattern of velocity and magnetic fields in the penumbra (Title *et al.* 1991). To begin with, we never had a completely satisfactory theory for the penumbra. More detailed observations of penumbra seem to make the job of a theorist more challenging rather than providing any easy clue on how to proceed. Choudhuri (1986) extended the model of umbral dots to suggest that the penumbral grains are also pockets of field-free gas trapped inside an inclined magnetic field. It is not clear whether such a simple model would be able to explain penumbral physics adequately without serious modifications. At present we have to accept the penumbra as something completely mystifying which we are yet to understand.

4. Growth and Decay of Sunspots

We now consider the question whether the growth and the decay of sunspots provide any evidence for cluster structure. It has been known for about two decades that magnetic field outside sunspots exists in the form of fibril flux tubes in the photosphere. As we look at the growth and decay of sunspots, we shall also discuss the possible relations between the photospheric fibril flux tubes and the flux tubes making up a sunspot. Since the evolution of sunspots including growth and decay is reviewed by Zwaan (1992) in this Workshop, we merely summarize the relevant features without getting into details. As the growth and the decay of sunspots are not symmetric processes, we have to consider them separately. Let us first look at the growth of sunspots.

Sunspots seem to form in regions where fresh flux emerges. First some pores appear, and then they coalesce to form sunspots. Since sunspots form out of freshly emerged flux, we can rule out the possibility that the ubiquitous photospheric flux tubes, which were already present, somehow mysteriously came together to form a sunspot. Perhaps the most satisfactory explanation of why the pores coalesce comes from the "tree structure" scenario shown in Figure 4. This scenario was developed by Zwaan (1978, 1985) following a suggestion of Vrabec (1974). Pores in this scenario are thought to be the photospheric sections of flux tubes of which the roots are joined together deep down. Just as several gas balloons of which the strings are held together tend to gather together, the pores also seem to coalesce. Though the tree structure may explain the coalescence of pores, we are then faced with the question how such a remarkable tree structure may come into existence in the first place. Piddington (1976) suggested a model of sunspots as strongly twisted flux ropes which are stabilized against fluting instability by the presence of twist. According to Piddington (1976), the upper part of a sunspot can get untwisted and a tree structure may result from fluting instability in that part. Since there is no observational evidence for such strong twists in sunspots, this theory can more or less be ruled out (Zwaan 1978), and we have to look for alternative schemes for producing a tree structure. Though we do not have a complete theory how the tree structure may be created, we discuss below some theoretical ideas which are highly suggestive that tree structures may indeed form. Many of us find this tree structure scenario rather attractive, because, in addition to providing an explanation for the coalescence of pores, it rather nicely ties with our current ideas of the solar dynamo and the rise of magnetic flux in the convection zone, as we shall see below. It should, however, be kept in mind that there are still many gaps in the theoretical arguments we are going to give, and one should entertain alternative models also seriously. Recently Parker (1991) has attempted to explain the coalescence of pores by suggesting that the rising flux tubes form vortex rings around them which produce mutual attractions.

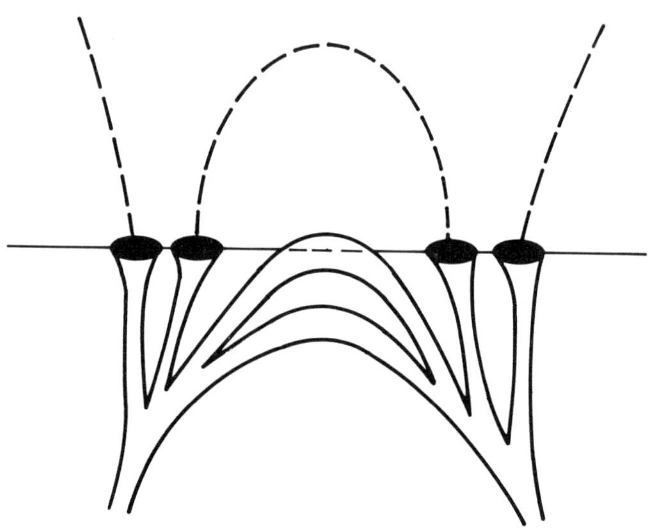

Fig. 4. Tree structure scenario for the emergence of flux. (After Zwaan 1985)

It is currently believed that the solar dynamo operates at the base of the convection zone in a thin layer where magnetic buoyancy is somehow suppressed (see Choudhuri 1990 and references therein). If that is true, then the magnetic flux is first created in that thin layer. Afterwards, strands of magnetic flux presumably come out of this stable layer and rise through the convection zone to reach the photospheric surface eventually (see Hughes 1992 and Moreno-Insertis 1992). It was found by Choudhuri and Gilman (1987) and Choudhuri (1989) that the Coriolis force tends to make the rising flux move parallel to the rotation axis and emerge at high latitudes significantly poleward of where sunspots are seen. Hence the Coriolis force somehow has to be suppressed if we want the flux to appear at the typical sunspot latitudes. Various mechanisms for suppressing the Coriolis force through interactions of the rising flux with the surrounding convection zone are being considered now (Choudhuri and D'Silva 1990; D'Silva and Choudhuri 1991). It is always found that the suppression of the Coriolis force leading to the emergence of flux at lower latitudes is possible only if the flux rises in the form of thin tubes (cross-sectional radius not more than a few hundred kilometers in the interior of the convection zone). The thinness of the tube essentially increases the efficiency of its interactions with the surrounding fluid. If all the flux making up a sunspot rose in the form of a monolithic flux tube, it would have been diverted to polar latitudes due to the Coriolis force. Hence we suggest that the flux rises through the convection zone in the form of thin tubes carrying just enough flux to make pores.

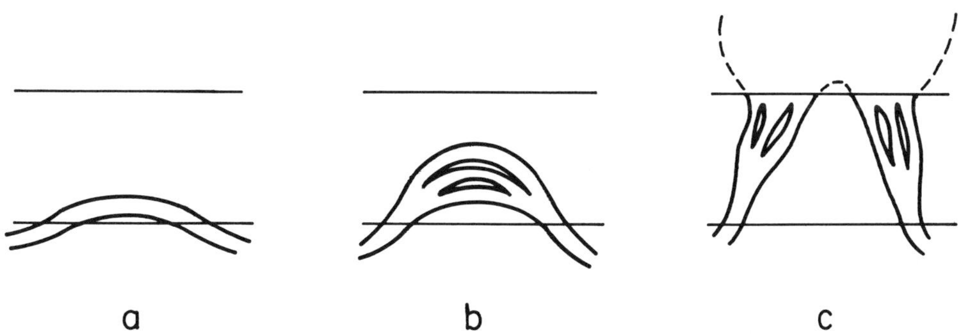

Fig. 5. Various stages of a flux strand rising through the convection zone.

We now come to the question of what causes the tree structure. A sunspot has two typical scales for flux - the scale of the whole sunspot (about 10^{21} Mx) and the scale of individual flux tubes making up the spot (about 10^{19} Mx). In the language of the tree structure, they may be called the scale of the stem and the scale of the branches. The process of the breakup of flux from the stable dynamo layer is still not well understood, and it is difficult to predict theoretically the sizes of flux tubes which would result from such a breakup (see Hughes 1992). It would, however, be logical to identify the larger scale (the stem) as resulting from this breakup process. Figure 5a shows such a breakaway flux tube of which a part has come out of the stable dynamo layer. This part will rise due to magnetic buoyancy, and we can understand the production of the tree structure if this rising part were to fragment into smaller flux tubes with the smaller scale (the branches) as shown in Figure 5b. There are indeed theoretical reasons to expect that such a fragmentation takes place, though estimating the scale of fragmentation theoretically is rather non-trivial. An early simulation by Schüssler (1979) showed the fragmentation of a rising flux tube. Tsinganos (1980) found that the flow around a rising flux tube would give rise to Kelvin-Helmholtz instability. This instability will try to fragment the flux tube, while the twist in the flux tube will have a stabilizing effect opposing the tendency to fragment. The fragmentation of rising flux tubes is a topic which has not been studied at all for over a decade and urgently needs revisiting - perhaps through numerical simulations. Since the fragmented flux tubes would be thin enough to resist the Coriolis force, they would rise through the convection zone radially to reach the photosphere and eventually produce the configuration of Figure 5c.

Thus one can understand the production of the tree structure at least qualitatively. As we have pointed out, there are still many gaps in the theoretical arguments. For example, we do not yet have any satisfactory quantitative theory for the production of the two scales. One of the attractive aspects of the tree structure scenario is that it provides a rather natural explanation how the cluster structure arises when a sunspot is produced. The next question is whether the structure sketched in Figure 5c survives after the formation of the sunspot, or whether the flux tubes under the surface also collapse together to form a single flux tube. One can use either the fluting instability discussed by Meyer, Schmidt and Weiss (1977) or the convective instability discussed by Moreno-Insertis and Spruit (1989) to argue against the formation of a single flux tube underneath the surface. But is the configuration shown in Figure 5c stable enough to last over the lifetime of a sunspot? Meyer, Schmidt and Weiss (1977) concluded that smaller flux tubes, especially if they are inclined to the vertical, tend to be unstable. Hence the individual flux tubes in the cluster under the surface may become unstable, unless they possess some amount of twist (Piddington 1976). The difficulty of invoking twist to stabilize a flux tube with non-uniform cross-section is that the twist would propagate to the expanded parts of the flux tubes as torsional Alfven waves until an eventual equilibrium configuration is reached in which the twist is mainly concentrated in the expanded parts (Parker 1979e). For the configuration of Figure 5c, however, the expanded parts of different flux tubes merge with each other, and the resultant topological connectivities may not allow the twist to propagate as torsional Alfven waves to the expanded parts above. Hence it is conceivable that some amount of twist is trapped in the thin tubes below the surface which helps to stabilize the configuration of Figure 5c.

Finally we come to a discussion of the decay of sunspots. There are some cases of flux mysteriously disappearing from sunspots, which probably implies the submergence of the flux underneath the photospheric surface (Parker 1984). More typically, a moat forms around a sunspot with an outward flow, and magnetic flux is chipped from the edges of the sunspot to be carried away in this flow. One observes magnetic features of both polarities in the outward flow of the moat. The most popular explanation of this remarkable observation is due to Harvey and Harvey (1973), who suggest that a flux tube is carried in an undulated form as shown in Figure 6. Recently Spruit, Title and van Ballegooijen (1989) have explored other possible scenarios to explain these observations. One wonders whether the original flux tubes, which formed the sunspot, retained their identities throughout the lifetime of the sunspot and whether it is these same flux tubes which are coming apart to be carried by the outward flow. Perhaps the observations are more consistent with a random chipping of flux from the edges of the sunspot, although

Meyer *et al.* (1974) have argued that it is necessary to include diffusion in addition to random chipping so that the decay rate is consistent with observations. It is not very clear how diffusion and random chipping from the edges may affect a cluster structure if it still persists. However, the essential point is that the flux seems to be chipped away from the edges of the sunspot to be carried outward. What happens to this flux afterwards? Certainly the flux spreads around in the surrounding photosphere where supergranular flows are expected to carry the flux to supergranular boundaries. Processes such as convective collapse may operate there to produce the photospheric fibril flux tubes (see Schüssler 1990). Thus the possible relation between sunspots and photospheric fibril flux tubes may be that these photospheric tubes can be end-products of the deaths of sunspots.

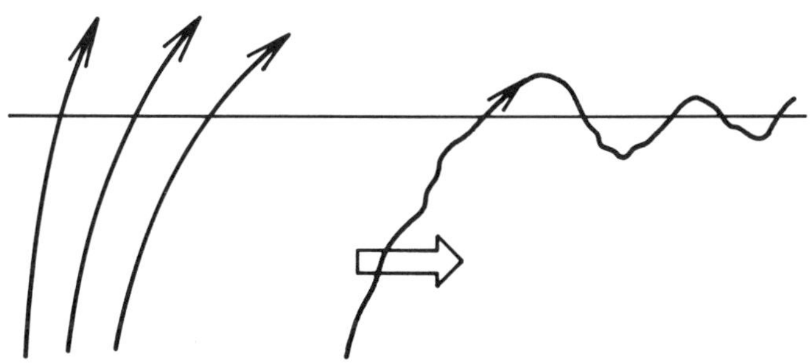

Fig. 6. A flux tube being chipped away from the edge of a decaying sunspot. (After Harvey and Harvey 1973)

One would like to know whether all the fibril flux tubes in the quiet photosphere are formed of the flux which has gone through a sunspot at one time. Sheeley and his coworkers have been studying this problem for several years (see Wang, Nash and Sheeley 1989 and references therein). They find that the magnetogram data of the quiet region flux can be modelled by assuming that the active regions are the only sources of flux and by treating the subsequent evolution of the magnetic flux on the solar surface through a 2-dimensional diffusion equation in which the magnetic field is essentially taken as a scalar residing on the solar surface. They even model the polar fields by using the tilts of active regions to diffuse one polarity preferentially at the higher latitudes and by incorporating a meridional flow to convect and concentrate this flux at the poles. Certainly the magnetic field is not a scalar residing on a spherical solar surface, but a 3-dimensional vector having considerable amount of subsurface physics associated with it. Hence one may legitimately raise worries about treating the magnetic field as a scalar while treating the polar fields. But a scalar model is probably adequate for studying the diffusion of the flux from a sunspot in the surrounding photosphere and throws much light on how the decay of a sunspot spreads the flux due to supergranular diffusion.

To summarize, current observations of sunspot growth seem consistent with the idea of a flux tree rising from the deeper layers of the convection zone and automatically producing the cluster structure. The decay of a sunspot can give rise to photospheric fibril flux tubes in the surrounding region.

5. Waves and Oscillations in Sunspots

We have seen in the last two Sections how some of the original arguments for the cluster model developed and transformed over the years. We now come to the discussion of any new evidence for or against the cluster model. Much attention has been focussed on waves and oscillations in sunspots during the last few years as possible probes for subsurface structures of sunspots.

It has been known for some time that a sunspot can support a variety of waves and oscillations. We see both oscillations which appear to be normal modes of a sunspot and oscillations forced by p-modes impinging on the sunspots from the surrounding region. Theoretical research on waves and oscillations in sunspots also has a long history (see Thomas 1981 for a review of the early work). It is, however, relatively recently that serious efforts are being made to use the observations of oscillations in order to infer theoretically the subsurface structure of sunspots. The existence of waves and oscillations suggest that a sunspot should have enough coherence to support waves and oscillations. It is, however, a purely qualitative statement to say that a sunspot should have enough coherence. The difficulty lies in making this loose and qualitative statement quantitive. How much coherence should a sunspot have and up to what depth? Are the observations consistent with the existence of a cluster structure underneath the visible surface of a sunspot? If so, can we say something about the volume filling factor for the magnetic regions in the cluster structure? Considerable amount of theoretical research is now being carried on in order to address these questions. This is, however, a fairly technical subject and the present author does not possess the necessary expertise to review it. Happily waves and oscillations in sunspots are being reviewed in this Workshop by some of the leading researches in this difficult field (Chitre 1992; Roberts 1992; Bogdan 1992). We have said just a few words about this important topic for the sake of completeness and refer the reader to these other reviews. The remarkable discovery of the absorption of p-modes by sunspots (Braun, Duvall and LaBonte 1988) raised a short-lived hope amongst many that we are now ready for a seismology of sunspots which would reveal the subsurface structure. Careful and detailed works by many (see Bogdan 1992) seem to have dispelled the hope that we can get a quick and unambigious solution for the subsurface structure of sunspots by inverting waves and oscillations data. Though we are unable to draw any firm conclusions at present, one still expects that future research on waves and oscillations will provide an important probe for subsurface structure of sunspots.

6. Concluding Remarks

It is now clear that the cluster model has come to stay and has acquired a life of its own. If we look at the original arguments advanced at the time of the proposal of this model, it is clear that many of those arguments have to be modified and reworked substantially. The idea of a downdraft, suggested by Parker (1979a) as an integral part of the original model, has quietly slipped out of people's minds. But the central idea that a sunspot is a cluster of many flux tubes proves invaluable as we try to understand a sunspot more and more.

The different theoretical models for sunspots can be broadly classified under three categories: (i) magnetohydrostatic model, (ii) magnetoconvective model and (iii) the cluster model. The magnetohydrostatic and magnetoconvective models have been reviewed in this Workshop by Jahn (1992) and Proctor (1992) respectively. In a way, the cluster model is less developed mathematically compared to the other two types of theoretical models. Both in magnetohydrostatic and magnetoconvective models, one begins with well-defined sets of mathematical equations and the theory is developed on the basis of them. By contrast, the cluster model is not the kind of "model" in which all calculations begin with a particular set of equations. It is rather a kind of qualitative and pictorial model which helps us make sense out of diverse varieties of observational data. Even though all calculations in the model do not begin with a given set of equations, one still has to carry out detailed mathematical analyses to elaborate different aspects of the model. We have summarized many of these calculations and also tried to identify gaps which should be bridged by detailed calculations in future.

In the past, one often heard the rather naive discussion about which of the different sunspot models was correct. We are now gradually learning to accept the fact that a sunspot is a very complicated object of which all aspects cannot probably be studied with the help of any one simple mathematical model. All the three models mentioned above are different kinds of appoximations of the "reality", and each of them helps us understand certain aspects of sunspots. So, instead of bothering about the question of the "correctness" of a particular model, the relevant issue should be to understand the nature of assumptions and approximations in the model and to figure out its regime of validity. In other words, we should try to form a clear understanding when a particular model should be applied and when it should not be applied. Perhaps such an understanding often comes more from intuition and experience rather than from logic and strict rules. Hence we do not attempt to write down any rules of that kind. One should regard these various approaches for sunspot modelling as complementary to each other rather than being mutually exclusive. Arguments amongst the sunspot model builders may often remind one of the blind men who visited an elephant. Each blind man touched a part of the elephant's body and thought that the whole elephant was like that. We certainly have to compare notes with each other if we want to form even a dim notion of what the whole elephant may look like.

Acknowledgements

It is a great pleasure to thank Nigel Weiss and Jack Thomas for inviting me to this Workshop and for finding the financial assistance to make my trip possible. I am also grateful to several participants of the Workshop who made important points during and after my talk.

References

Adjabshirzadeh A., Koutchmy S., 1980, A&A 89, 88
Beckers J.M., Schröter E.H., 1968, Solar Phys. 4, 303
Bogdan T.J., 1992, This Volume
Braun D.C., Duvall T.L., LaBonte B.J., 1988, ApJ 335, 1015
Bumba V., Suda J., 1980, Bull. Astr. Inst. Czechoslovakia 31, 101
Chitre S.M., 1992, This Volume
Choudhuri A.R., 1986, ApJ 302, 809
Choudhuri A.R., 1989, Solar Phys. 123, 217
Choudhuri A.R., 1990, ApJ 355, 733
Choudhuri A.R., D'Silva S., 1990, A&A 239, 326
Choudhuri A.R., Gilman P.A., 1987, ApJ 316, 788
D'Silva S., Choudhuri A.R., 1991, Solar Phys. (in press)
Grossmann-Doerth U., Schmidt W., Schröter E.H., 1986, A&A 156, 353
Harvey K., Harvey J., 1973, Solar Phys. 28, 61
Hughes D.W., 1992, This Volume
Jahn K., 1992, This Volume
Kneer F., 1973, Solar Phys. 28, 361
Knobloch E., Weiss N.O., 1984, MNRAS 207, 203
Lites B.W., Bida T.A., Johannesson A., Scharmer G.B., 1991, ApJ 373, 683
Loughhead R.E., Bray R.J., Tappere E.J., 1979, A&A 79, 128
Meyer F., Schmidt H.U., Weiss N.O., 1977, MNRAS 179, 741
Meyer F., Schmidt H.U., Weiss N.O., Wilson P.R., 1974, MNRAS 169, 35
Moreno-Insertis F., 1992, This Volume
Moreno-Insertis F., Spruit H.C., 1989, ApJ 342, 1158
Muller R., 1992, This Volume
Obridko V.N., 1975, SvA 18, 758
Parker E.N., 1979a, ApJ 230, 905
Parker E.N., 1979b, ApJ 231, 270
Parker E.N., 1979c, ApJ 232, 291
Parker E.N., 1979d, ApJ 234, 333
Parker E.N., 1979e, Cosmical Magnetic Fields, Clarendon Press, Oxford
Parker E.N., 1984, ApJ 281, 839
Parker E.N., 1991, in preparation
Piddington J.H., 1976, Ap. Space Sci. 41, 79
Proctor M.R.E., 1992, This Volume
Roberts B., 1992, This Volume
Schüssler M., 1979, A&A 71, 79
Schüssler M., 1990, in Solar Photosphere: Structure, Convection and Magnetic Field, IAU-Symp. No. 138. Stenflo J.O. (ed.) Kluwer, Dordrecht, p. 161
Spruit H.C., 1992, This Volume
Spruit H.C., Title A.M., van Ballegooijen A.A., 1987, Solar Phys. 110, 115
Thomas J.H., 1981, in The Physics of Sunspots. Cram L.E., Thomas J.H. (eds.) National Solar Observatory, Sunspot, p. 345

Title A.M. *et al.*, 1991, preprint
Tsinganos K.C., 1980, ApJ 239, 746
Vrabec D., 1974, in Chromospheric Fine Structure, IAU-Symp. No. 56. Athay R.G. (ed.) Reidel, Dordrecht, p. 201
Wang Y.-M., Nash A.G., Sheeley N.R., 1989, ApJ 347, 529
Weiss N.O., Brownjohn D.P., Hurlburt N.E., Proctor M.R.E., 1990, MNRAS 245, 434
Zwaan C., 1978, Solar Phys. 60, 213
Zwaan C., 1985, Solar Phys. 100, 397
Zwaan C., 1992, This Volume

V. WAVES AND OSCILLATIONS IN SUNSPOTS

As Bernard Roberts says in his chapter in this section, "Sunspots must surely be the location *par excellence* in which to study magnetohydrodynamic phenomena and magnetic oscillations in particular". A sunspot provides a medium for the propagation of waves that comes close to satisfying all of the conditions of ideal magnetohydrodynamics, conditions that are very difficult to achieve in the laboratory. The processes of turbulent and convective excitation ensure that there is a rich spectrum of propagating and resonant modes of oscillation in each sunspot. But the somewhat complicated geometry of the magnetic field and the strong vertical stratification of the gas under solar gravity provide enough complexity to make the theoretical modelling of sunspot oscillations challenging, even within the realm of linear wave theory.

Recent interest in sunspot oscillations has been fuelled by the prospect of using them as a probe of sunspot structure. The three–minute umbral oscillations, which apparently are resonant modes of oscillation trapped within coupled resonant cavities in the sunspots, provide a diagnostic of the structure of the stratified umbral atmosphere. The five–minute oscillations in a sunspot, excited by incident *p*–mode oscillations from the surrounding convection zone, offer the possibility of probing the structure of a sunspot beneath its visible surface. The results of this method of "sunspot seismology" may, for example, eventually settle the controversy over the extent to which the magnetic flux of a sunspot is fragmented into separate, smaller flux tubes beneath the surface of the Sun.

This section contains four authoritative reviews of different aspects of sunspot oscillations. B. W. Lites presents a comprehensive review of the observations and their implications for the theory. B. Roberts reviews the mathematical theory of magnetohydrodynamic waves in structured magnetic fields, which underlies all of the theoretical modelling of sunspot oscillations. S. M. Chitre discusses specific theoretical models of umbral oscillations and penumbral waves. T. J. Bogdan gives a detailed summary of developments in the new field of sunspot seismology.

In addition to these four invited reviews there were five contributed papers on sunspot oscillations at the Workshop. M. A. Aballe Villero presented observational results concerning the different behaviour of umbral oscillations in the dark core of the umbra, in the outer, brighter part of the umbra, and in an umbral light bridge. He found a negative correlation between magnetic field strength and power in the three–minute oscillations in the umbra. Yu. D. Zhugzhda spoke on the coupled photospheric and chromospheric resonant cavities that are responsible for the character of the three–minute umbral oscillations. J. Staude described his use of observations of umbral oscillations as a diagnostic test of the temperature profile in semi–empirical models of the umbral atmosphere. Two different mechanisms were proposed for the absorption of *p*–modes by sunspots. C. S. Rosenthal argued in favour of resonance absorption of waves at the Alfvén resonance surface in a model based on a twisted axisymmetric flux tube in an unstratified atmosphere. M. P. Ryutova proposed instead that the absorption is due to scattering of *p*–mode waves by strong, small–scale inhomogeneities.

SUNSPOT OSCILLATIONS: OBSERVATIONS AND IMPLICATIONS

B. W. LITES
High Altitude Observatory
National Center for Atmospheric Research[1]
P.O. Box 3000
Boulder, Colorado 80307, U.S.A.

ABSTRACT. This review summarizes the current state of observational knowledge of oscillations within sunspots. Three classes of oscillations are discussed: (1) oscillations measured in the photosphere of sunspot umbrae which are dominated by the apparent response of the umbral photosphere to the 5-minute p-mode oscillations, (2) the energetic oscillations in the umbral chromosphere, and (3) oscillations observed in the penumbral chromosphere. Observational results are critically examined. The theory of sunspot oscillations is discussed in light of the observations, and new observational tests are proposed.

1. Introduction

The chromosphere above sunspots participates in several apparent modes of oscillatory behavior that, from an observational standpoint, appear more vigorous and spatially ordered than chromospheric oscillations in the surrounding quiet Sun. In contrast to the oscillatory behavior of the sunspot chromosphere, the 5-minute p-mode oscillations in the sunspot photosphere, and indeed also in photospheric facular regions where the average photospheric magnetic field is strong, are very much weaker than those of the surrounding quiet Sun. These substantial differences between the oscillatory behavior of the solar atmosphere inside and outside of sunspots motivate the study of sunspot oscillations, not only to learn about the nature of the oscillations themselves, but also to discover what they can tell us about the structure of sunspots.

The presence of the strong magnetic field of a sunspot and the spatial order that it imposes locally upon the solar atmosphere conspire to alter the oscillatory character of the atmosphere, permitting some modes of oscillation not present in the quiet Sun and substantially altering the 5-minute modes. What do we stand to learn from these differences in oscillatory behavior? A few of the more obvious applications are given below.

- *Identification of Magneto-Atmospheric Wave Modes*

  The comparatively large spatial extent of sunspot magnetic fields provides us with an opportunity to observe the response of the atmosphere to wave modes which actually exist in

[1]The National Center for Atmospheric Research is sponsored by the National Science Foundation.

J. H. Thomas and N. O. Weiss (eds.), Sunspots: Theory and Observations, 261–302.

magnetized stellar atmospheres, provided their spatial coherence length is large enough to be observable. Throughout the lower part of the atmosphere above a sunspot the "turbulent" line broadening is small compared to that of the quiet Sun (Figure 1), so that wave modes characterized by coherence lengths smaller than those of observed oscillations have very small amplitudes, if they are present at all. The reduced turbulence within the sunspot atmosphere also allows one to measure oscillations with a higher signal-to-noise ratio than is possible elsewhere on the solar surface. Thus, sunspots provide us with a handy "laboratory" for investigation of magneto-atmospheric waves in stellar atmospheres.

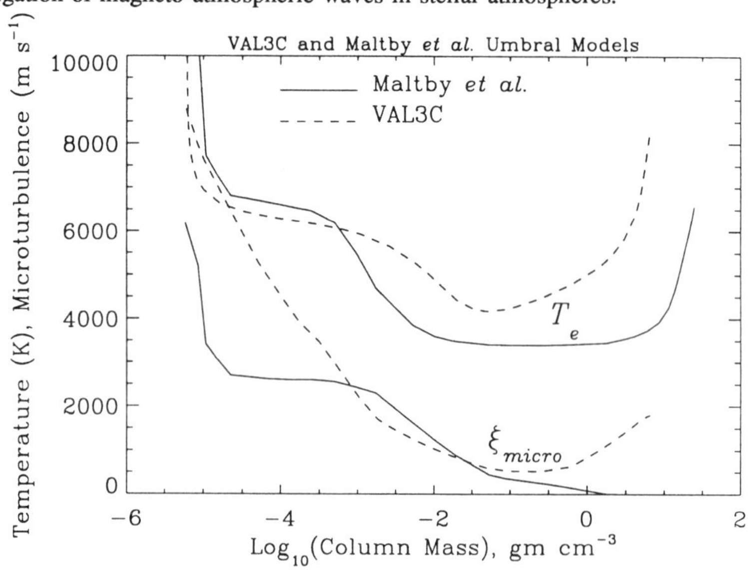

FIGURE 1. The temperature (T_e) and microturbulence ($\xi_{\rm micro}$) are plotted as a function of column mass for the average quiet Sun model "C" of Vernazza, Avrett, and Loeser (1981) and the umbral model "M" of Maltby et al. (1986). Note that the microturbulence is insignificant in the umbral photosphere, implying a very quiescent state. Even though the oscillatory amplitude is small in the umbral photosphere, small oscillatory amplitudes are easily detected because they are not masked by large convective motions.

- *Assessment of the Energetic Consequences of Wave Modes for the Heating and Dynamics of Stellar Atmospheres*

The sunspot "laboratory" may eventually permit us to gauge in a rather accurate way the amount of energy that is propagated by waves upward into the chromosphere and corona, and the amount of energy that is deposited at each level. Detailed theoretical modeling of the energy balance in sunspot atmospheres (for example, including the backwarming by conduction from the corona) is a formidable task, but it is nonetheless greatly simplified from the case of the quiet Sun because of its relatively simple physical state: an atmospheric structure which is highly ordered by the magnetic field and a quiescent umbral photosphere. In hindsight it is thus not surprising that the first observations indicating shock wave dissipation of regular three-minute oscillations of the chromosphere were found over sunspot umbrae.

- *Oscillations as Probes of Atmospheric Structure*

  When oscillatory behavior is present, the propagation characteristics of the waves carry unique information about the structure of the atmosphere which supports them. Traditional methods of exploring atmospheric structure which use spectral features formed at differing heights usually rely (in general through complex non-LTE line formation) upon the variation with height of density and temperature in order to translate observed properties into a model as a function of height. In contrast, an upward propagating slow MHD wave in the umbral chromosphere behaves much like a sound wave. The phase of such a propagating wave has a height variation $\delta\phi = \frac{\omega}{v_s}\delta z$. To first order the difference in phase measured in spectral lines formed at different heights represents the height difference δz, because v_s has a weak temperature dependence. Similarly, the propagation characteristics of waves and oscillations may be modified strongly by large gradients in physical parameters. The temperature gradient at the transition region between the chromosphere and corona or the gradient in strength of the magnetic field (the "magnetic canopy") above active regions are examples of gradients which may reflect some wave modes. If measurable, the degree of reflectivity provides an indication of the steepness of the gradient. In some cases the reflectivity depends upon frequency and horizontal wavelength of the oscillation, so that resonant cavities for discrete frequencies and wavenumbers may be present in the atmosphere.

- *Oscillations as Indicators of Conditions Below the Surface of Sunspots*

  The inference of subsurface structure from reflection, scattering, or absorption of p-mode oscillations *as measured outside of sunspots* is discussed by Bogdan elsewhere in this volume. The observed behavior of oscillations within sunspots also provides clues to their subsurface structure. The character of the modes deep in the sunspot photosphere (e.g., their temporal and spatial spectra) are particularly useful in this regard.

At present we have observational knowledge of three oscillatory phenomena within sunspots. The first is the internal linear response of the sunspot to the 5-minute p-mode oscillations; that is, the observed oscillatory behavior of the lower sunspot atmosphere at periods around 5 minutes. The second is the chromospheric umbral flash, or umbral oscillation phenomenon which may be driven by weak oscillations observed in the photosphere within the same frequency band. The third observed mode of sunspot oscillations is the penumbral "running wave" phenomenon which has an apparent observed link to the umbral oscillations. In the following sections I will review in turn the current state of observational knowledge of these phenomena, and relate these observations to theoretical scenarios that have been proposed to explain them.

Several previous reviews considered observations of sunspot oscillations: Moore 1981, Moore and Rabin 1985, Thomas 1985. Since the time of these reviews, new observational studies that have added significantly to our understanding. It is not uncommon that observational results on sunspot oscillations are misunderstood or misinterpreted. The objectives of this review are then (1) to clarify the observational picture of the oscillations within sunspots through critical examination of the literature, (2) to indicate the areas where our knowledge of the observations is incomplete, (3) to suggest directions for future observations, particularly those which can work together with theoretical studies to advance our understanding of the phenomena, and (4) to indicate how proposed theoretical models relate to the observations.

2. Oscillatory Behavior of the Umbral Photosphere

This review begins not with the dramatic oscillatory behavior of sunspot chromospheres, but rather with the subdued oscillations that have been measured deep in the photospheres of sunspot umbrae. There are several important reasons for starting here:

- In spite of their relatively low amplitudes, the oscillations in umbral photospheres represent a likely driving source for the oscillations observed higher in the atmosphere.
- Observations deep in the sunspot photosphere expose the oscillatory character of the atmosphere at a height where $\beta = P_{\text{gas}}/P_{\text{magnetic}} \approx 1$, that is, where wave modes having the dominant character of fast modes below the surface are "transformed" (Zhugzhda 1984) into waves having the dominant character of slow modes in the upper layers.
- The wave amplitudes are small enough in the photosphere that the linearized MHD equations are applicable. Higher in the atmosphere this is certainly not the case. Coupling of wave modes through nonlinear interactions (shocks) may be significant in the chromosphere, but not in the deep photosphere.
- The deepest observable layers afford the closest access to MHD modes which resonate below the surface.

2.1. OBSERVED CHARACTER OF THE UMBRAL PHOTOSPHERIC OSCILLATIONS

It now appears to be an established fact that the oscillatory motions measured in spectral lines formed in the photosphere of umbrae show a dominant power peak centered around 5-minute periods. However, for several reasons this picture did not become clear until recently. Beckers and Schultz (1972) reported measurements of umbral "photospheric" oscillations in the Zeeman-insensitive line Fe I 543.4 nm with periods of about 180 seconds and peak-to-peak amplitudes of 1 km s^{-1}. This paper is often cited as the first evidence for oscillations in the umbral photosphere. Given what we now know about formation of this line in sunspots (Bruls et al. 1991) and its oscillatory nature relative to chromospheric oscillations as described below, the Doppler signature at the center of this line most certainly has a contribution from the chromosphere above sunspot umbrae. The 3-minute oscillations are usually larger in amplitude than the 5-minute oscillations measured in this line. The 5-minute oscillations dominate the spectrum of lines formed at lower heights within the umbra.

A number of the earlier papers reporting analysis of time series of Doppler velocities measured in the umbral photospheric lines suggest that the amplitude of the 5-minute oscillation is greatly reduced there (Howard, Tannenbaum, and Wilcox 1968; Bhatnagar 1971; Rice and Gaizauskas 1973). The reports of identification of oscillatory periods of umbrae from such data were frequently challenged based upon uncertainty of the contribution of scattered light from outside of the umbra: the Doppler signal from these lines was not unique to umbra.

The contamination by stray or scattered light from the surrounding photosphere may be minimized either by measuring Doppler shifts of molecular lines formed only within umbrae, by measuring Doppler shifts in low-excitation lines of neutral species having low ionization potentials (such as Ti I) that are greatly strengthened in the cooler regions of sunspot umbrae and penumbrae, or by measuring the Doppler shift of the polarized component of a Zeeman-sensitive line. Earlier results using these techniques (Bhatnagar, Livingston, and Harvey 1972; Schröter and Soltau 1976, Soltau, Schröter, and Wöhl 1976) did not yield a consistent picture of the oscillatory periods that are present, and periods both longer and shorter than 5 minutes, as well as within the 5-minute band, were reported.

Improved measurement techniques in recent years permitted larger amounts of high-quality data to be taken, so that now similar power spectra are emerging from these data (Livingston and Mahaffey 1981; Thomas, Cram, and Nye 1982, 1984; Balthasar and Wiehr 1984; Soltau and Wiehr 1984; Abdelatif, Lites, and Thomas 1984, 1986; Abdelatif 1985; Lites and Thomas 1985; Thomas et al. 1987; Balthasar, Küveler, and Wiehr 1987; Aballe Villero et al. 1990; Kobanov 1990). I compile the following picture from this group of papers:

- *Most data sets show a peak in the distribution of power within the 5-minute band (2.5-4.5 mHz),* with an average rms amplitude of 75 m s^{-1} (Table 1); that is, oscillatory amplitudes 2-3 times weaker than those of quiet regions. Typical power spectra averaged over an umbra and an equivalent area outside of the sunspot are shown in Figure 2. The sampled areas are shown in Figure 3.

TABLE I: OBSERVED AMPLITUDES OF PHOTOSPHERIC UMBRAL OSCILLATIONS

Authors	Spectral Lines	RMS Power (m/s) 2.5 – 4.5 mHz	5 – 8 mHz
Thomas, Cram, Nye 1982	FeI 630.2 nm	50	25*
Soltau and Wiehr 1984	FeI 709.0 nm	80*	33*
Balthesar and Wiehr 1984	TiI 616.0 nm	96*	47*
Lites and Thomas 1985	TiI 630.4 nm	80+	39+
Abdelatif, Lites, & Thomas 1986	FeI 630.2 nm	65+	24+
Thomas et al. 1987	FeI 396.9 nm	62	39
Aballe Villero et al. 1990	TiI 839.7 nm	100*+	54*+

* Inferred from published power spectra.
+ Average value for several sunspots.

- *The umbral photosphere oscillates with considerably less power in the 2-3 minute band (5-8 mHz).* The rms oscillatory amplitude (above the noise and integrated over this band) falls in the range 25-50 m s^{-1} (Table 1). Individual isolated peaks above 5 mHz are sometimes observed, c.f., the peak identified 'U1' in Figure 2. Authors who call attention to such peaks include: Thomas et al. 1982, Lites and Thomas 1985, Abdelatif et al. 1986, and Aballe Villero et al. 1990. Equally, some data sets, even in the papers mentioned above, show no evidence

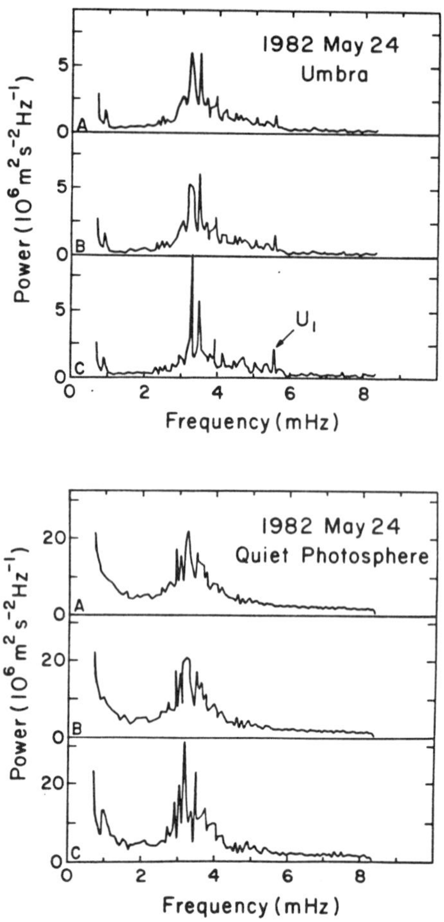

FIGURE 2. From Abdelatif et al. 1986. Mean temporal power spectra of velocity oscillations measured in Ti I 630.4 nm are shown. In the upper panel the power spectra A, B, and C refer to successively smaller nested areas of the sunspot umbra, as shown in Figure 3. The lower panel displays similar power spectra of the nested areas outside of the sunspot in Figure 3. The 5.5 mHz oscillation (peak U_1) is more pronounced in the dark central part of the umbra.

of enhanced power at these frequencies. When considering the range of observed amplitudes given in Table 1, one must remember that the oscillations are observed in a number of lines formed over a range of heights in the umbral photosphere. Lines formed at larger heights will probably experience larger amplitudes, especially at the higher frequencies.

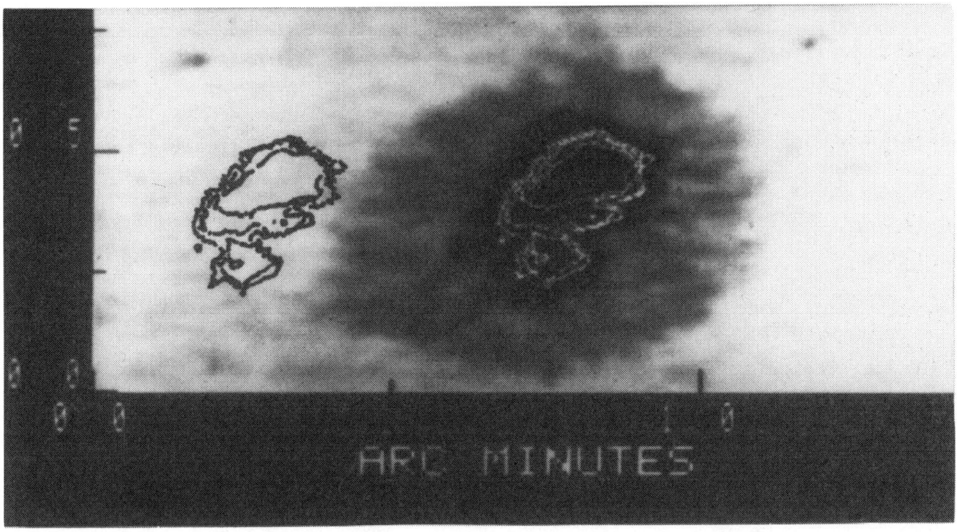

FIGURE 3. From Abdelatif et al. 1986. Images of a sunspot in the continuum at 630.2 nm are constructed from scanned spectral measurements corresponding to the power spectra of Figure 2 and the relative $k_\perp - \omega$ power spectrum given in Figure 4. The spatial masks used to construct Figure 4 are given by the outer contours.

- *Coherently oscillating elements with 5-minute periods appear to cover a large fraction of the umbral area.* Abdelatif, Lites, and Thomas (1985) produced a movie of the oscillatory Doppler shifts in and around the sunspots observed by Abdelatif et al. 1986 (see also Abdelatif 1985). The time series was filtered so that only oscillations having power within the 5-minute band were visible. This analysis demonstrated that the power within the 5-minute band within umbral photospheres is characterized by small values of horizontal wavenumber $k_\perp$. In other words, the coherently oscillating elements appeared to cover large fractions of the umbra. This conclusion has been confirmed by Kobanov (1990) using an independent observational technique.

  Abdelatif et al. (1986) attempted to quantify the spatial behavior of the oscillations within umbrae using the powerful diagnostics afforded by $k_\perp - \omega$ power spectra. The spatial extent of a sunspot umbra is so small that very little resolution in $k_\perp$ is possible from the raw power spectra. These authors highlight changes of the spatio-temporal power spectra within sunspots by displaying the ratio of the $k_\perp - \omega$ power spectra inside the sunspot umbra to that outside (Figure 4). An example of a spatial mask used to define the 'umbral' and 'outside' regions is given by the outer contours of Figure 3. This analysis suggests that the oscillatory power, relative to that of the regions outside the sunspot, is restricted to low $k_\perp$, roughly delineated by the dashed line corresponding to a horizontal phase speed of 25 km s^{-1}. Abdelatif (1985) and Abdelatif and Thomas (1987) proposed the following explanation of this shift to smaller $k_\perp$: Acoustic oscillations outside the sunspot are transformed into fast magnetoacoustic waves within the umbra – waves which propagate at nearly the Alfvén speed which is approximately 25 km s^{-1} in the umbral photosphere.

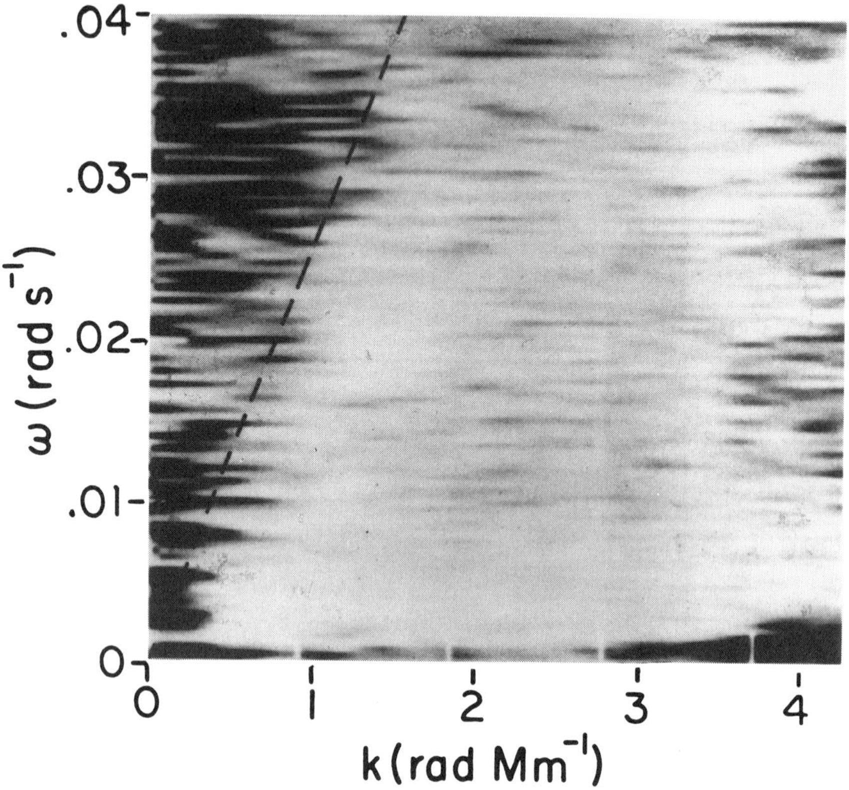

FIGURE 4. From Abdelatif et al. 1986. The ratio of the $k_{\perp} - \omega$ power spectra inside the umbra (measured in Ti I 630.4 nm) to that outside (measured in Fe I 630.3 nm) is shown as a function of frequency and horizontal wavenumber. The outer contours of Figure 3 define the 'umbral' and 'outside' regions. Black regions in the image correspond to a ratio of unity (complete transmission into the umbra), and white represents complete attenuation. The dashed line separating distinct regions of transmission and attenuation has a slope of 25 km s^{-1}, roughly equal to the propagation speed of fast mode waves in the umbral photosphere.

- *Often the 5-minute oscillations within the umbral photosphere exhibit enhanced power at several frequencies in the power spectrum.* An example of these peaks may be found in Figure 2. Note also that the power spectrum measured outside of the sunspot, but otherwise equivalently, shows peaks, albeit at different frequencies. It has long been recognized that power spectra of the p-modes measured within a restricted region of the solar atmosphere show such fine structure. The detailed structure of the power spectrum arises from the interference of the many modes each with amplitude and phase that is unique to that particular time interval and location on the disk. Therefore, any interpretation of discrete frequencies of oscillation observed within sunspots, either for photospheric or chromospheric oscillations, should be met with a great deal of skepticism, as such behavior is a natural consequence of mode interference when several modes are present.

2.2. NATURE OF THE UMBRAL 5-MINUTE OSCILLATIONS

What can we conclude about the oscillations from these observations? First, the interpretation of Thomas et al. (1982) or of Zhugzhda, Locans, and Staude (1987) regarding the selection mode structure based on observed peaks in the power spectrum does not take into consideration the mode interference described above. The analysis of Thomas et al. (1982) is not compatible with the interpretation of Abdelatif and Thomas (1987) regarding the transformation to oscillations to larger $k_\perp$ within the umbra, in the sense that Thomas et al. (1982) associated observed frequency peaks with horizontal wavelengths corresponding to the (nonmagnetic) p-mode spectrum outside of the sunspot.

On the other hand, the general notion that the umbra responds in a passive way to the 5-minute oscillations outside, acting as a selective filter and transforming the oscillations to the appropriate magneto-atmospheric modes, is a natural and attractive explanation of the largest part of the oscillatory power observed in umbral photospheres. This explanation was first posed by Thomas (1981). The subsequent observational and theoretical papers (Abdelatif 1985, Abdelatif et al. 1986, Abdelatif and Thomas 1987) demonstrate that this scenario is in reasonable agreement with existing observations.

More recently Zhugzhda, Locans, and Staude (1987) have proposed a "sub-photospheric fast mode-resonator" as an alternate explanation for the 5-minute oscillations in the photosphere of umbrae. They argue that the trapping mechanism is essentially the same as that of the non-magnetic p-modes because fast-mode oscillations behave like sound waves in the subphotospheric layers where $\beta > 1$. They suggest that a cavity appropriate to the p-mode oscillations in the quiet Sun is not highly modified by the magnetic field because the upper reflecting layer in the photosphere for acoustic waves (the rapid outward drop in temperature produces a layer that is evanescent to sound waves of five-minute periods) is at approximately the same location where fast-mode waves become evanescent due to the rapid decrease in β with height. However, the upward reflection at the base of their cavity is more problematic. Acoustic waves in an unmagnetized region suffer upward reflection at great depth for horizontal wavelengths appropriate to sunspot umbrae (see Table 1 of Zhugzhda et al. 1987). Even if sunspot flux tubes did not constrict below the surface, so that they extend with constant strength and diameter to the base of the convection zone, the geometry would not allow refractive turnaround of the wave within the flux tube. Almost certainly the subphotospheric magnetic field does not behave this way – it is likely to strengthen, constrict, and form kinks within the convection zone well above the bottom reflection heights indicated by Zhugzhda et al. (1987). For such a mechanism to be operative the magnetic field must aid in establishing an upward reflective layer not far below the umbral photosphere, as it apparently does for trapped fast-modes in the calculations of Scheuer and Thomas (1981 - see below).

2.3. THE REDUCED POWER OF THE 5-MINUTE OSCILLATIONS IN UMBRAE

In sunspots and active regions alike, the observed power of the 5-minute oscillations is considerably reduced from that present over the rest of the solar surface. Three explanations have been offered. Abdelatif and Thomas (1987) point out that the sunspot acts as a filter to incident acoustic waves, permitting the transmission of only selected modes into the interior and altering their character in the process. However, their analysis does not consider the highly stratified nature of the sunspot atmosphere. Alternatively (or additionally) the Wilson depression of sunspots allows one to measure oscillations several scale heights below the visible surface of the surrounding photosphere – a level where the external velocity amplitudes of the p-mode oscillations are much reduced.

Finally, Bogdan (private communication) points out that reduction in amplitude of individual eigenfunctions is likely to occur when exterior p-modes couple to the magneto-acoustic modes inside the sunspot, particularly in the case of transformation to fast mode waves inside the sunspot.

2.4. UMBRAL PHOTOSPHERIC OSCILLATIONS AT HIGHER FREQUENCIES

The oscillatory power in the photosphere at periods shorter than 5 minutes, and especially the isolated peaks that are occasionally measured, are of considerable significance to the vigorous oscillations observed in the umbral chromosphere. When photospheric umbral oscillations are measured with enough sensitivity (Thomas et al. 1982; Balthasar and Wiehr 1984; Soltau and Wiehr 1984; Lites and Thomas 1985; Abdelatif et al. 1986; Aballe Villero et al. 1990) one detects power at frequencies up to at least 7 mHz. The upper envelope of these power spectra usually decrease with increasing frequency above 4 mHz (viz. Figure 2). An obvious and reasonable explanation for this envelope of the high frequency oscillatory spectrum is that it arises from the high frequency tail of the p-mode spectrum outside of the sunspot.

The isolated peaks in umbral power spectra most often appear around a frequency of 5.5 mHz. Note in Figure 5 that this frequency corresponds to the acoustic cutoff frequency of quiet Sun models, but *it is significantly lower than the acoustic cutoff frequency of a sunspot umbra.* Present evidence suggests that these peaks are genuine features of the oscillatory spectrum in the umbral photosphere. Analysis of observations of the photosphere cited above and simultaneous observations of umbral photospheres and chromospheres discussed in the next section support the reality of these peaks.

Thus, the observations beg a theoretical explanation of oscillatory power confined to a narrow range of frequencies. The model for a resonance of fast-mode waves trapped in a thin layer just below the umbral photosphere (Uchida and Sakurai 1975; Scheuer 1980; Scheuer and Thomas 1981;Thomas and Scheuer 1982) neatly predicts an isolated resonant peak at about this frequency. As pointed out e.g., by Zhugzhda et al. (1985,1987), one difficulty with the computations of Thomas and Scheuer (1982) is the somewhat arbitrary specification of a lower reflecting boundary defined by a layer in the sub-photosphere where overstable oscillations (i.e., oscillatory convection) become efficient. The overstable oscillations would then act as a driving "piston" at this lower boundary. Aside from the uncertainty as to precisely which depth overstable oscillations take effect (and hence the specification of the resonant period for the fast-mode oscillations), the physical reason for fast mode waves suffering a reflection when overstable oscillations become efficient has not been adequately justified, even in more recent work (e.g., Hasan 1991). However, this assumption is not present in the computations of Scheuer and Thomas (1981) where the upward reflection of fast mode waves arose in a natural way from the inward increase of temperature in their numerical model.

Most theoretical treatments of umbral oscillations assume that the the side walls of the flux tube below the photosphere of sunspots is perfectly reflecting to the waves within. This assumption

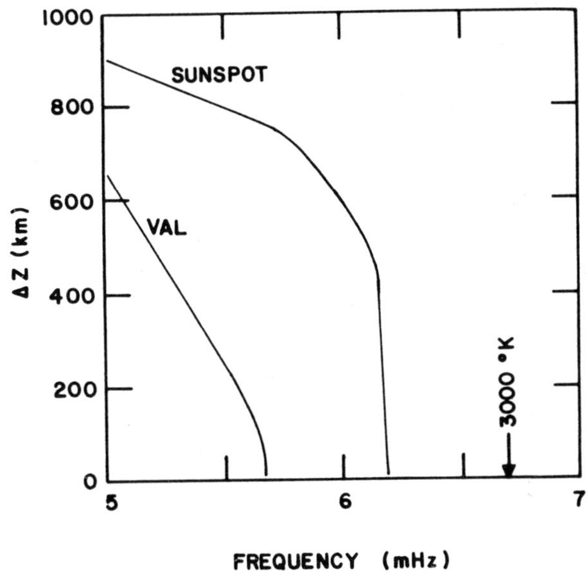

FIGURE 5. From Lites 1984. The thicknesss of the layer Δz over which acoustic waves are evanescent is plotted as a function of frequency of oscillation for the quiet Sun model "C" (Vernazza et al. 1981) and the sunspot model (Avrett 1981). The acoustic cutoff frequency for a plasma at 3000 K is also indicated.

needs to be justified on the basis of theoretical computations similar to those of Abdelatif (1985; Abdelatif and Thomas 1987), but including the effects of atmospheric stratification.

In view of these uncertainties accompanying fast-mode resonance theory, it is appropriate to examine other possibilities for production of the isolated power peaks near 5.5 mHz. New observations and theoretical analyses open one possible avenue of investigation. Recently it has been shown that the oscillations above 5.5 mHz are not trapped oscillations in the normal sense (Kumar et al. 1990) as they exist above the acoustic cutoff frequency of the temperature minimum region of the solar atmosphere. However, they can be refracted at the lower boundary. Hence an "acoustic source" (Brown 1991) for the high frequency wave (such as a vigorous granule or a local shock created at rapid downflows near a fluxtube) can cause a disturbance both locally near its source and at some distance, because the downward-directed part of the acoustic wave front suffers a refraction at depth and reemerges at some distance from the source itself. New observations of Brown et al. (1992) indicate that high frequency power sources outline and are intermingled with regions of concentrated magnetic field. These observations indicate that the increase in power, relative to the average power spectrum, above the magic frequency 5.5 mHz is accompanied by a local decrease in power in the 5-minute band, showing that the local enhancement in emission is not just a local enhancement in oscillatory power for all frequencies. If it turns out that acoustic sources are enhanced around active regions then it is possible that the high frequency waves may be able to penetrate into sunspots and produce the power peaks that are observed. Again, theoretical investigation of this possibility appears to be quite tractable in view of the computations by Abdelatif cited above.

One final point regarding the presence of isolated power peaks near 5.5 mHz should be stressed. I believe it quite unlikely that these peaks result from some combination of broad-band excitation coupled with the resonant transmission properties of the chromosphere. The acoustic cutoff frequency of the umbral photosphere is significantly higher than most of the measured frequencies, so it is unreasonable to assume that the lower umbral atmosphere will pick this frequency to resonate. The isolated peaks in the photosphere then should result either from an internal resonance of the photosphere and below, or from external sources.

3. Chromospheric Umbral Oscillations

Since the discovery of "umbral flashes" in 1969 by Beckers and Tallant a great deal has been learned about the phenomenon, both from observational and theoretical viewpoints. Henceforth I will refer to the phenomenon as "chromospheric umbral oscillations," and I will argue that these oscillations provide us with an excellent observational paradigm for purely hydrodynamic processes that are capable of heating the chromosphere away from magnetic activity. Indeed, these oscillations have provided us with the first solid example of a periodic nonlinear wave phenomenon in the solar chromosphere. In addition to being one of the most dramatic dynamic phenomena associated with sunspots, the chromospheric umbral oscillations are a playground for testing theories of wave generation, propagation, and dissipation in magnetized stellar atmospheres.

The historical literature on this subject has been covered in a number of previous reviews (Thomas 1981; Moore 1981; Moore and Rabin 1984; Thomas 1985). This review will focus on observed properties of the chromospheric oscillations as gleaned from the existing literature, and the relationship of these observed properties to the theories that have been put forward to explain them.

3.1. THE OBSERVED CHARACTERISTICS OF CHROMOSPHERIC UMBRAL OSCILLATIONS

- *The Chromospheric Oscillations are Nonlinear.* Perhaps the single most important attribute of chromospheric umbral oscillations is the observed behavior indicating the presence of shocks. Figure 6 illustrates the typical time series behavior of the core of the Ca II K-line. Often the cores of the Ca II resonance lines show a single emission reversal in sunspot umbrae, as is the case for much of the umbral sequence shown in Figure 6. The excursions in wavelength of this emission core are often used as a diagnostic for Doppler shifts within the umbral chromosphere. The striking sawtooth waveform of the oscillation – violent shift from red to blue followed by a more gradual drift back to the red – and the large amplitude of the excursion (peak-to-peak often as large as 6 km s^{-1}) are highly suggestive of nonlinear, or shock, behavior of the oscillation.

  Radiative transfer computations of the Ca II resonance lines in static (therefore highly suspect) umbral model chromospheres suggest the formation of these lines is effectively thick, so that phases and amplitudes of the oscillation measured from the spectra are not easily interpreted. One might then question whether the Z-shaped waveform visible in Figure 6 results from line formation rather than from nonlinear behavior of the oscillation itself. However, the simultaneous measurement of these chromospheric umbral oscillations in both the Ca II H-line and

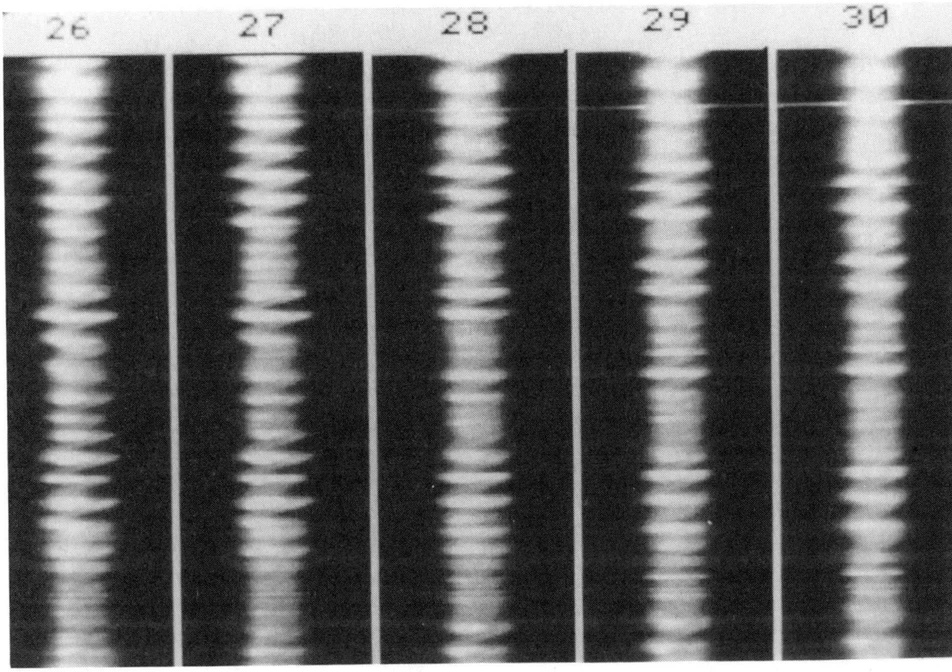

FIGURE 6. Spectra of the core of the Ca II K-line is shown as a function of time at five locations within an umbra. Time increases in the vertical direction; wavelength increases to the right. Note that the K-line core usually shows an isolated emission peak which undergoes a brightening and rapid shift to the blue followed by a gradual return to the red. The sawtooth form of this behavior is suggestive of shocks. Observations courtesy of Thomas, Cram, and Nye.

the He I line at 1083.0 nm by Lites (1986b) strongly reinforce the notion of nonlinear oscillations. At the same time they emphasize the difficulties of interpreting phase and amplitude information derived from the Ca II resonance lines. Figure 7, from that paper, demonstrates that excursions in the optically-thin He I absorption were of larger amplitude (up to 11 km s^{-1} peak-to-peak) than had been observed before, and that the measured Doppler shifts in He I take on the sawtooth 'Z'-shaped waveform when the amplitude of the excursion becomes large.

- *The Dominant Period of Oscillation in the Chromosphere is Usually Around 5.5 mHz.* Figure 8 from Lites (1986b) presents power spectra for both Doppler shift and intensity fluctuations in He I 1083.0 nm and the Ca II H-line. Power spectra of umbral oscillations measured in chromospheric lines (the Na I D-lines: von Uexküll, Kneer, and Mattig 1983; the Ca II infrared triplet (IRT) lines: Lites, White, and Packman 1982, von Uexküll et al. 1983, Lites 1984, 1986a, Thomas et al. 1987; the Ca II H and K resonance lines: von Uexküll et al. 1983; Lites 1986b, Thomas et al. 1987; the Mg II k-line: Gurman 1987; the He I 1083.0 nm line:

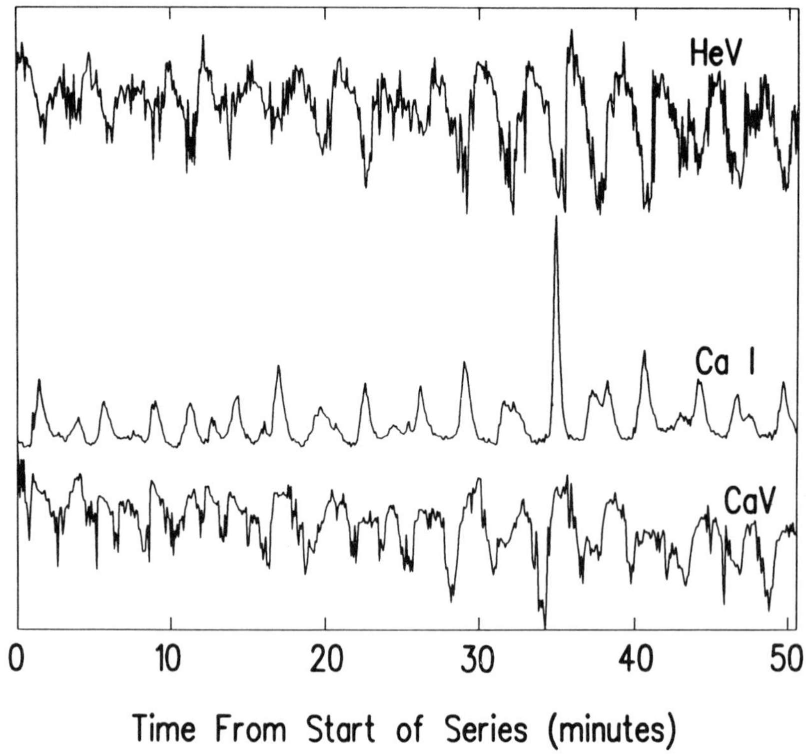

FIGURE 7. From Lites 1986b. A time series of simultaneous measurements of the Ca II H-line core intensity (Ca I) and measured Doppler shift (Ca V) as well as the Doppler shift in He I 1083.0 nm (He V) is shown. Blueshift is positive. The peak-to-peak Doppler shifts measured in the He I line are as large as 11 km s^{-1}. The He V waveform exhibits its most distinct sawtooth behavior when the amplitude of oscillation is largest. This time corresponds to especially large brightening in the H-line.

Lites 1986b, Thomas et al. 1987) do not show significant power peaks in the 5-minute band. Instead, the power rapidly rises to a dominant peak at about 5.5 mHz. Above this frequency one often finds several peaks of lesser amplitude, and significant power is present out to and above 10 mHz.

The low chromosphere/temperature minimum region is spanned by Fe I 543.4 nm. This is a great line for study of sunspot oscillations because it faithfully echos velocity oscillations in both the umbral photosphere and low chromosphere. It allows us to bridge between the photospheric and chromospheric umbral oscillations as measured in lines forming at lower and higher altitudes. Oscillations in this line have been measured simultaneously with a photospheric line (Lites and Thomas 1985, Figure 9), and with chromospheric lines (Lites 1984, 1986a, Figure 10). Note in Figures 9 and 10 that Fe I line core reacts to chromospheric

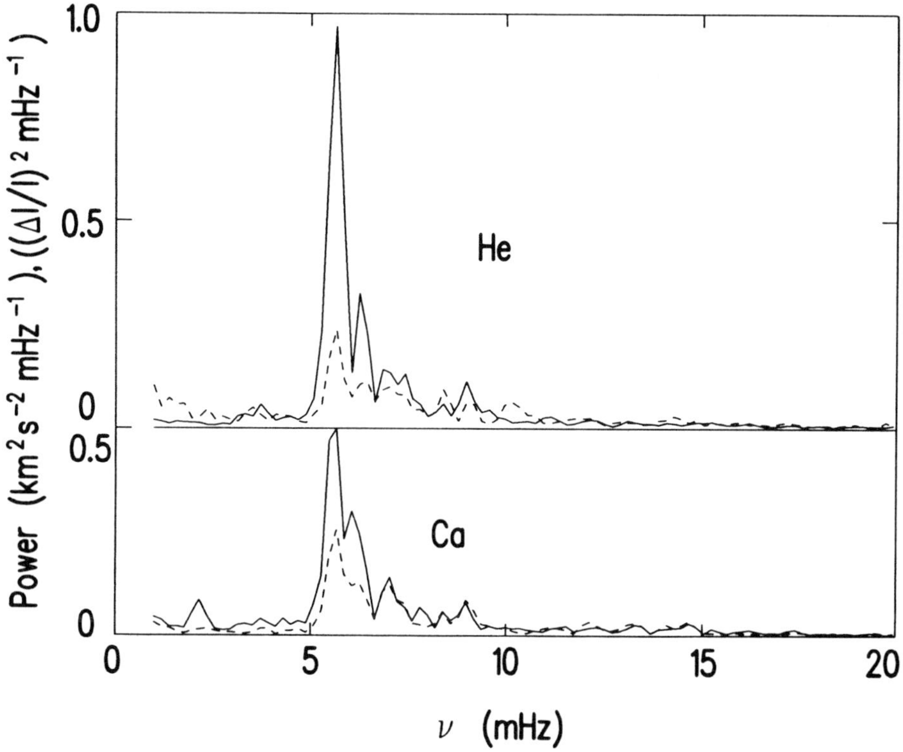

FIGURE 8. From Lites 1986b. Average umbral power spectra for He I 1083.0 nm (top) and the Ca II H-line (bottom) are shown. The simultaneous observations show correspondence of frequency of oscillation, but not amplitude of the individual peaks. Power spectra for Doppler shifts are given by solid curves, for intensities by dashed curves.

oscillations with highest amplitude, but the photospheric 3 mHz oscillations are also clearly present. I suspect that this line is sensitive to Doppler shifts over a wide range of heights, so that it picks up some Doppler signal from the 3 mHz oscillations in the photosphere along with its dominant sensitivity to the chromospheric oscillations at line center. From Figure 10 one may see that there is not a one-to-one correspondence between power peaks observed in Fe I 543.4 and Ca II 849.8 nm, but such a correspondence is present when comparing power spectra of Ca II 396.9 and He I 1083.0 nm in Figure 8.

Power spectra of umbral oscillations observed in transition region lines sometimes show isolated peaks, and these peaks often occur at frequencies higher than 5.5 mHz (Gurman et al. 1982).

- *The 5-Minute Oscillations are not Observed in the Umbral Chromosphere.* The low coherence of fluctuations in the 5-minute band among lines formed in the middle chromosphere (von

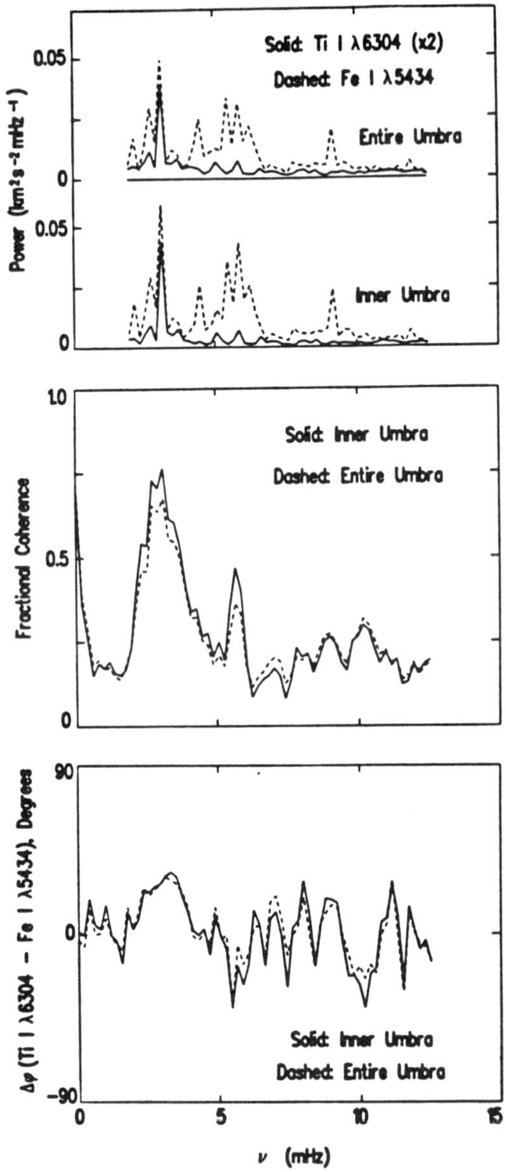

FIGURE 9. From Lites and Thomas 1985. From top to bottom, power, coherence, and phase delay spectra averaged over the umbra of a sunspot are shown for simultaneous observations of Ti I 630.4 nm and Fe I 543.4 nm. The Ti I power spectra have been multiplied by a factor of 2 for clarity.

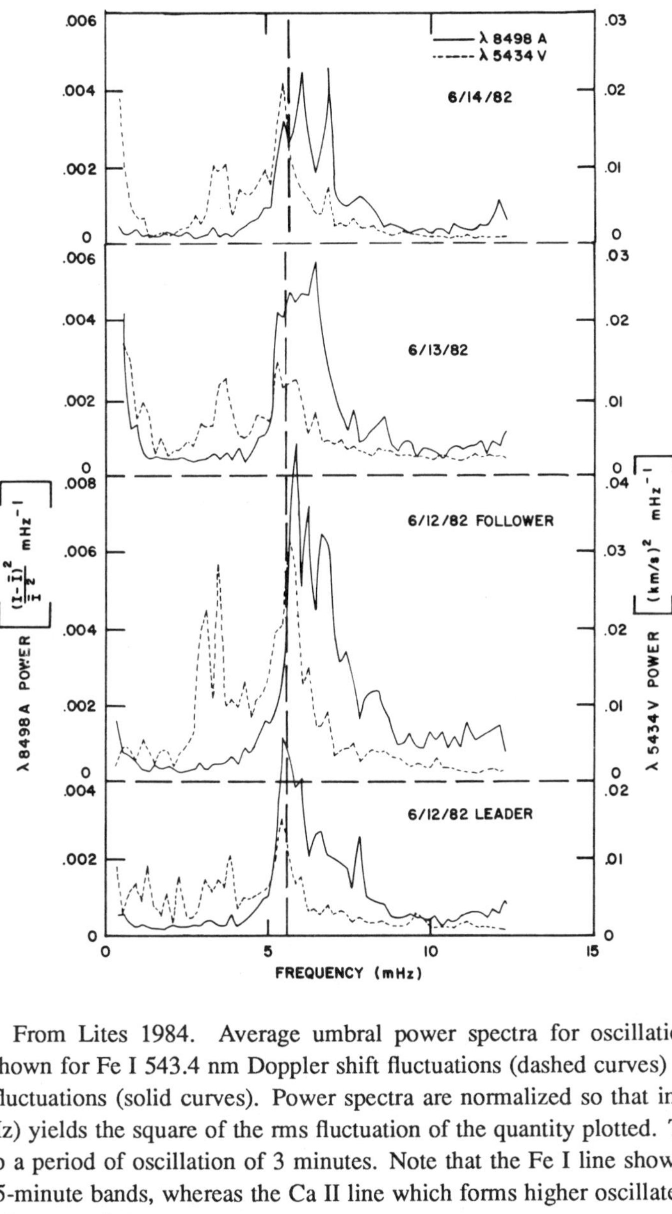

FIGURE 10. From Lites 1984. Average umbral power spectra for oscillations within four sunspots are shown for Fe I 543.4 nm Doppler shift fluctuations (dashed curves) and Ca II 849.8 nm intensity fluctuations (solid curves). Power spectra are normalized so that integration of the curves (in mHz) yields the square of the rms fluctuation of the quantity plotted. The vertical line corresponds to a period of oscillation of 3 minutes. Note that the Fe I line shows similar power in the 3- and 5-minute bands, whereas the Ca II line which forms higher oscillates exclusively at periods of 3 minutes and shorter.

Uexküll et al. 1983) demonstrates that the 5-minute oscillations have very low amplitudes there, if they are present at all. The low amplitude of the 5-minute oscillations in the umbral photosphere couples with the high acoustic cutoff frequency of the umbral temperature minimum to heavily suppress the 5-minute oscillations in the umbral chromosphere.

- *Several peaks are often present in the power spectra of chromospheric lines, indicating the presence of several cospatial modes of oscillation.* Lites (1986a) demonstrated that the multiple frequencies of chromospheric oscillations averaged over an entire umbra do not result from spatially separated elements oscillating independently at different frequencies.
- *In spite of recent claims to the contrary (Campos 1989) there appears to be no preferred set of frequencies common to most umbrae.* Figure 11 illustrates the highly variable oscillatory spectrum of the chromospheric umbral oscillations (see also Zhugzhda and Makarov 1982). Distinct peaks in umbral power spectra most likely result from this variability. In Lites (1986a) I listed several possibilities for the temporal variation of the power spectrum: spatio-temporal interference of closely spaced modes in frequency much as with the p-modes in the quiet photosphere, temporal switching among modes, drifts in frequency, and disruption of the phase of the oscillation. Any of these possibilities could arise from either changes in chromospheric structure or changes in a driving source in the photosphere. However, spatial correlations of the amplitudes of the chromospheric oscillations on a frequency-by-frequency basis strongly point to *variable mode structure* of the driving source rather than variable chromospheric structure as a cause for the multiple peaks.
- *The horizontal extent of a typical oscillating element within the chromosphere is 3-4".* This contasts with the much larger size of the photospheric umbral elements oscillating at 5-minute periods, which typically cover a large fraction of umbrae in average-sized sunspots (see §2.1 above). Horizontal phase speeds are large (Lites et al. 1982 estimate 60-70 km s^{-1}), and movies of the oscillations sometimes give the impression of swirling motion.

  The umbral magnetic field must guide fluid motions in the chromosphere, thereby isolating laterally adjacent elements from acoustic or slow MHD waves. Any spatial ordering of the oscillations in the chromosphere ultimately must be traced to spatial ordering of the excitation source at a level in the atmosphere low enough that lateral communication is possible. However, the disparity of spatial scales of oscillating elements from photosphere to chromosphere does not eliminate the the photospheric 5-minute oscillations as a driving source for the chromospheric oscillations. It is conceivable that the nonlinear response of the umbral chromosphere could transform the low order 5-minute oscillations into the more highly structured order that is observed.
- *The amplitudes of the chromospheric oscillations in the 2-3 minute band are spatially uncorrelated with the amplitude of oscillations observed in Fe I 543.4 nm in the 5-minute band (Lites 1986a).* This observation poses a serious problem for theories which rely upon the photospheric oscillations with 5-minute periods to drive the chromospheric oscillations. It is difficult to imagine how oscillations restrained laterally in the atmosphere by the magnetic field could divorce their amplitudes from the amplitude of the driving source below.
- *When high frequency peaks are observed in the photospheric power spectrum, simultaneous measurements of chromospheric oscillations indicate a strong correlation (Lites and Thomas 1985).* The coherence spectra shown in Figure 9 give one example of an isolated peak in the coherence of the oscillations at 5.5 mHz

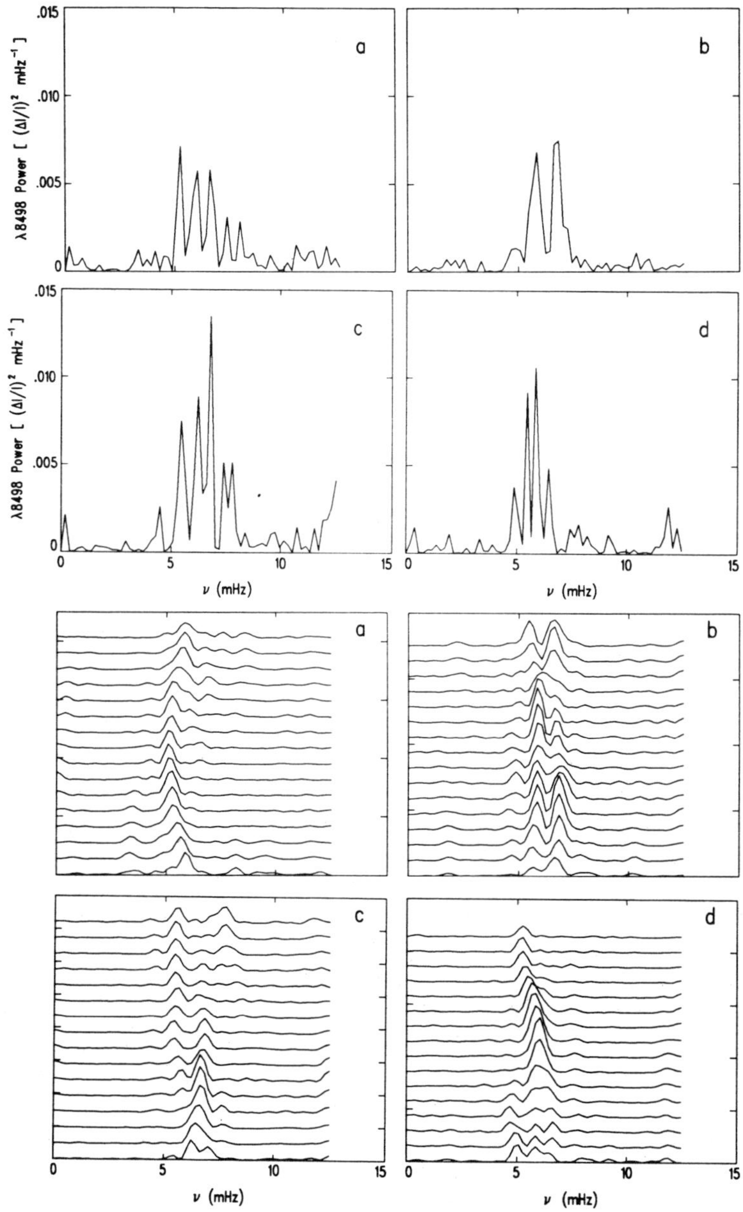

FIGURE 11. From Lites 1986a. This figure demonstrates the temporal variability of umbral chromospheric oscillations at four locations within an umbra. The top four panels are power spectra of the full 83-min. time series of intensity fluctuations in Ca II 849.8 nm at each of the four locations *a* - *d*. The bottom four panels illustrate a sequence of power spectra at each of the four spatial positions. Each of these curves represents the power spectrum for a restricted region of the entire time sequence, resulting from application of a temporal mask 30 minutes in duration. The mask was displaced along the time series by 200-second increments to give the curves from bottom (earliest time) to top (latest time) within each panel.

- *The kinetic energy density of the oscillatory power in the 2-3 minute band measured in the photosphere is typically greater by at least a factor of 5 than the power in the chromosphere within the same frequency band (Lites and Thomas 1985).* This measurement demonstrates that the power peaks observed in the photosphere do not result from self-excited chromospheric oscillations. That is, purely chromospheric mechanisms such as the CO instability could not be responsible for the power observed in the photosphere because they would be unable to drive such energetic oscillations in the photosphere through the evanescent layers of the umbral temperature minimum.
- *The amplitude of umbral oscillations measured in the low chromosphere relative to that measured in the middle-upper chromosphere decreases with increasing frequency of oscillation.* This result given by Lites (1986a) may be used as a specific test of theoretical models of the oscillation (Zhugzhda et al. 1987).
- *There is some suggestion for advancing phase (hence upward propagation) of the waves with increasing height in the chromosphere.*

3.2. PHASE BEHAVIOR OF CHROMOSPHERIC UMBRAL OSCILLATIONS

Perhaps the most powerful diagnostic of solar oscillations are the phase delays as a function of frequency of oscillation derived from simultaneous measurements of two or more spectral lines. This diagnostic technique has the potential of clearly indicating the phase speed of a wave traveling along or away from the line of sight (if phase differences progressively increase or decrease with frequency, respectively, for compressive waves). Thus, the phase diagnostic allows us to distinguish between traveling and standing waves, a character of the oscillation which is often a crucial distinguishing feature of proposed theoretical models. For example, if the oscillations are principally trapped waves in the chromospheric cavity, the phase of the velocity oscillations will be rather constant with height except in the vicinity of nodes (where the phase changes abruptly by π: see phase vs. height curves of Zhugzhda , Locans, and Satude 1985). Such a phase behavior should be a readily identifiable trait of the models which invoke strong trapping of waves by a highly reflective temperature gradient of the transition region.

Provided we know the wave mode (i.e., a slow mode in the chromosphere which propagates essentially along field lines at the sound speed), the observed phase speeds allow us to determine the group velocity of the wave. This information, along with the observed amplitude of the fluctuation, leads to specific estimates of the energy flux transmitted by the waves. Thus, it is possible to infer a great deal of extremely useful information from observed phases of oscillations if we are only able to properly interpret the observations. Even though the phase results to date are inconclusive, the potential of this diagnostic is great. It is important then to clarify the present observational situation and highlight the many pitfalls that plague the analysis of phase measurements.

Unfortunately most of the optically thick spectral lines which form in the middle chromosphere, i.e., those which clearly show umbral oscillations, also exhibit an oscillatory line core emission behavior resulting from the oscillations. The most notable lines showing this behavior are the Ca II resonance and IRT lines (Figure 12). The line core emission lends a great deal of confusion to the interpretation of most measures of either Doppler shift or intensity in these lines. In the following I will summarize the observational evidence for the phase of the umbral chromospheric oscillations with height, but the reader must remember both new observations and detailed numerical simulations will be needed to properly understand what observed phases are telling us.

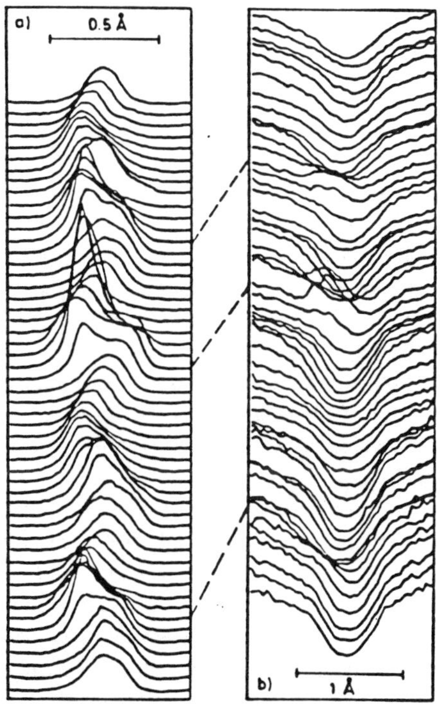

FIGURE 12. From Kneer, Mattig, and von Uexküll 1981. A simultaneous time series of umbral profiles of the Ca II K-line (left) and 854.2 nm infrared triplet line (right) is shown. The dashed lines indicate the correspondence of profiles for times of stronger line emission.

3.2.1. Phase Delays among Chromospheric Lines. In an early effort to interpret the phase of umbral oscillations (Lites, White, and Packman 1982) we indicated a number of the pitfalls which may have compromised our tentative interpretation. In retrospect, most of these pitfalls probably do apply to those observations. Our naive interpretation of downward propagation of waves based upon the observed phase relationships between Doppler shift and intensity measured in Ca II 854.2 nm almost surely is incorrect as a result of several of the pitfalls listed.

Von Uexküll et al. (1983) analyzed time series of umbral spectra from the famous HIRKHAD observing mode (Beckers et al. 1972) to extract phase and coherency spectra for various combinations of H_α, the Na I D-lines, and the Ca II resonance and IRT lines. They found moderate upward phase speeds within the frequency range 5.5 - 8.7 mHz from Doppler shifts of the intensity minima of Na I D_1, D_2 and H_α lines. However, they found that the wavelength shift of maximum emission in the Ca II K-line progressively leads the Doppler shift measured in the Na I lines and H_α with increasing frequency. They proposed a complicated (as yet unsubstantiated) scenario to explain the behavior of the Ca II K-line based upon assumptions regarding the optically thick transfer in the Ca II K-line with self-reversed emission and a rapidly varying response of this emission with oscillatory frequency.

In Lites (1984) I reported further observations involving the Ca II IRT lines along with Fe I 543.4 nm. The phase delay between intensity disturbances in two of the IRT lines, as shown in Figure 13, indicate a slight tendency for upward propagation above 5 mHz. Similar phase diagrams for quiet Sun disturbances by Lites, Chipman, and White (1982) indicate a negative phase delay in the 5-minute band, becoming more negative with increasing frequency; i.e., the opposite of what is shown by Figure 13. Deubner and Fleck (1990) found smaller phase delays of the same sign in the quiet Sun, probably as a result of the measurement of intensity oscillations in the line core rather than in the wings. The magnitude of the phase shift in Figure 13 is much smaller than would be expected of a slow mode disturbance propagating vertically along field lines provided the line intensities respond according to simple estimates of heights of formation for the IRT lines (Lites and Chipman 1979, von Uexküll et al. 1983). The credence of such formation heights for highly dynamical phenomena has been questioned (e.g., Rutten and Uitenbroek 1991) on the basis of the interlocking of these lines with the Ca II resonance lines. Estimates of the formation height of the IRT lines could also be improved by the calculation of velocity response functions in dynamical models.

At lower heights in the chromosphere the delay in phase of oscillation measured between Fe I 543.4 nm Doppler shift and the core intensity of Ca II 849.8 nm is shown in Figure 14. Here we see a much more dramatic progression of phase with increasing frequency than that noted above for the IRT lines. One must remember that this is the phase difference between a reliable Doppler shift observed in the Fe I line and an unreliable measure of velocity (the core intensity) in the IRT line. In order to clarify the situation I relate the phase of oscillation in Fe I 543.4 nm to that in He I 1083.0 nm. The He I 1083.0 nm line forms somewhere in the upper chromosphere. It is a weak absorption feature over sunspot umbrae so that the line formation is indeed optically thin. In this line we now have a reliable diagnostic of the true Doppler shift in chromospheric lines. It clearly demonstrates the nonlinear character of the oscillations in the upper chromosphere (Lites 1986 - see above). Figure 15 suggests that one may "calibrate" the oscillatory phase of the Doppler shift in optically thick chromospheric lines through a comparison of phases measured in the He I line with phases of the core intensity of the Ca II H-line. This graph shows that the phase of the Ca II core intensity leads the measured He I blue shift by about 130-150° within the frequency range 5-10 mHz. Von Uexküll et al. (1983) claim that intensity fluctuations in the K-line lag those in the IRT line at 854.2 nm by about 15°, but their Figure 6 shows no significant phase shift between these lines for their sunspot no. 2. Thus the K-line intensity fluctuations are roughly in phase with the intensity fluctuations in 854.2. I use this observation along with the phase differences shown in Figures 13 and 14 to relate the phase of the Doppler shift measured in Fe I 543.4 to those measured in the He I line. The result is the dashed line shown in Figure 14. This inference suggests that the waves still may be characterized as propagating rather than standing waves between the levels of formation of Fe I 543.4 and He 1083.0 nm, and the slope of the advancing phase difference with frequency is even somewhat greater than that given by

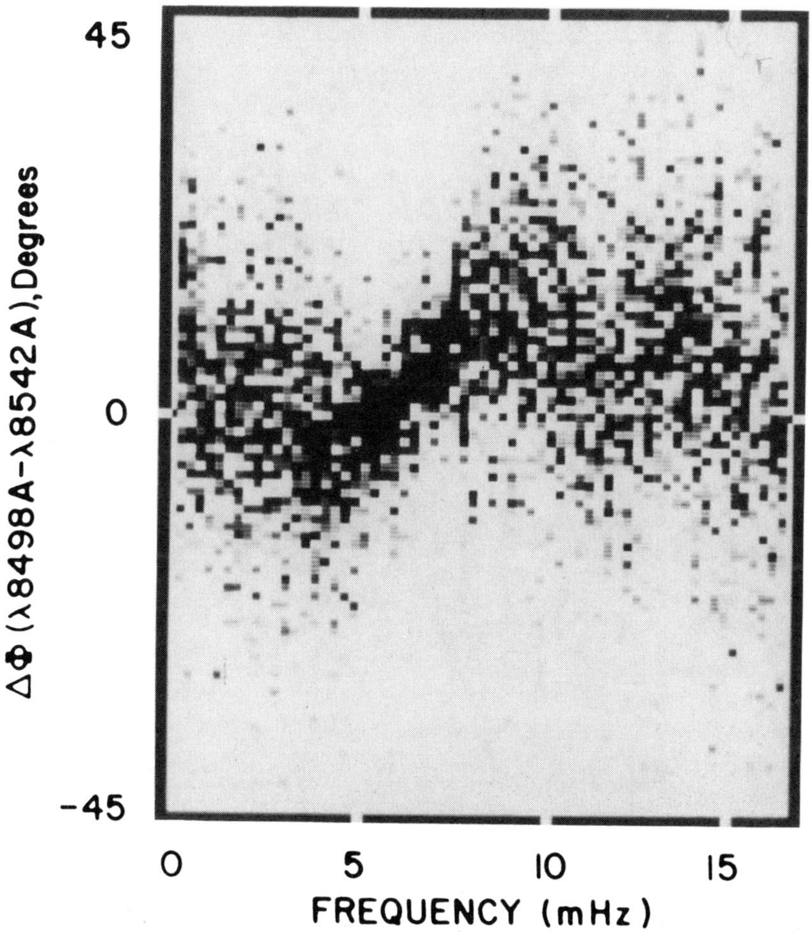

FIGURE 13. From Lites 1984. The difference in phase as a function of frequency is shown for umbral intensity oscillations observed in the cores of two of the Ca II infrared triplet lines at 849.8 and 854.2 nm. Here the square root of the product of oscillatory power in each line is shown as a gray scale, with dark corresponding to high power at a given frequency. The total power in each vertical frequency column is normalized in order to clearly display the behavior of phase difference at each frequency.

the original solid line. However, the neat coincidence of the projection of the phase delay back to zero phase at zero frequency is now gone.

3.2.2. Phase Delays Between the Photosphere and Chromosphere Lites and Thomas (1985) measured simultaneous time series of Doppler shifts in an umbral photospheric Ti I line (630.4 nm) and Fe I 543.4 nm. They found coherent power peaks at 5.5 mHz at these two heights with the

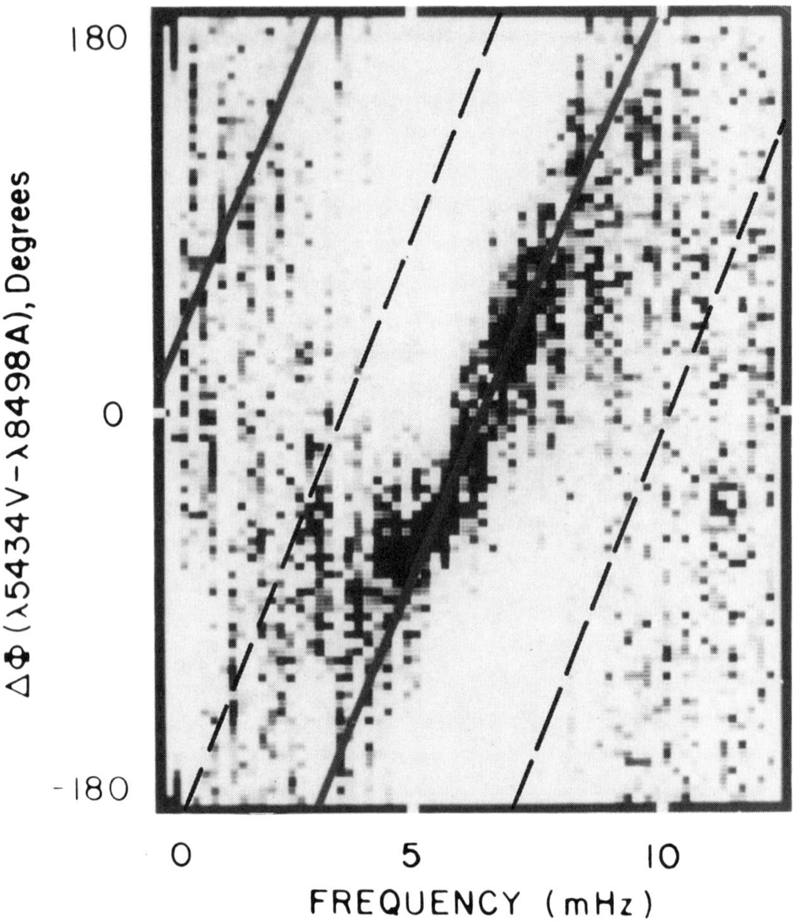

FIGURE 14. From Lites 1984. A phase difference diagram similar to that shown in Figure 13 is given for the line pair Fe I 543.4 nm and Ca II 849.8 nm. The solid line extrapolates the trend in phase difference with frequency back to zero frequency. The dashed line indicates the inferred phase difference between this Fe I line and He I 1083.0 nm (see text).

oscillation nearly in phase. In contrast, the phase delays were slightly positive in the 5-minute band, in general agreement with measurements of lines formed at similar heights in the quiet Sun (see e.g., Lites and Chipman 1979). It is expected that the oscillations at frequencies below about 6.2 mHz will not propagate through the temperature minimum of the umbral atmosphere (Figure 5). Consequently one would expect small phase delays at these frequencies in lines which form below the chromosphere. However one also expects that oscillations at 5.5 mHz would show phase behavior similar to that within the 5-minute band. At present we do not know whether this behavior is simply a deficiency of the observations or if it is indicative of some yet unknown

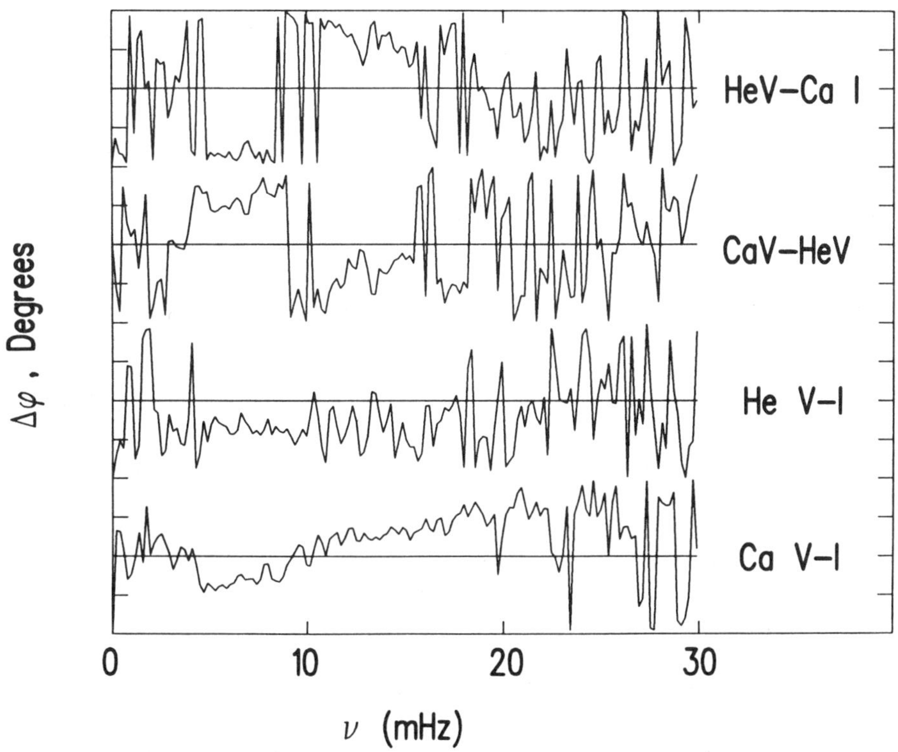

FIGURE 15. From Lites 1986b. Phase difference as a function of frequency for various combinations of Doppler shift and intensity measurements in the Ca II H-line and He I 1083.0 nm are shown. These results are averaged over the one-dimensional slice through an umbra of a spectrograph slit. These relationships allow one to "calibrate" the phase of oscillation observed in the optically thick Ca II line using the optically thin He I line Doppler shifts.

property of the oscillations.

What may one conclude from the observed phases? At this time the observational picture remains unclear. It is not surprising that many theoreticians are able to find phase measurements in the literature in agreement with their model predictions! Nodes of the oscillation may be present at some heights in the chromosphere. Limitations imposed by the observational technique and inhomogeneity of the umbral chromosphere may smear the observed phase relationship near a node, thereby producing phase relationships like those shown in Figure 14. However the extant data do not support the notion of purely standing oscillations: it appears that the waves do propagate with a modest group velocity, at least in the low chromosphere.

3.3. COMPETING MODELS OF UMBRAL CHROMOSPHERIC OSCILLATIONS

The merits of two theoretical pictures of the chromospheric umbral oscillations have been debated in the literature. In the first picture one assumes that the umbral oscillations are the resonant, i.e., linear, response of the solar atmosphere to forcing by a broad-band power source below. This picture has been put forward by Zhugzhda, Locans, and Staude (1983). The second picture, due to Scheuer and Thomas (1981) (see also Thomas and Scheuer 1982), has the oscillations driven by a sub-photospheric resonance to fast-mode waves. The literature contains a dialogue comparing these two models (Thomas 1984, 1985; Zhugzhda et al. 1985, 1987; Lites 1986a; Abdelatif 1990). For a brief comparison, see Table 3 in Lites (1986a).

The picture that emerges from this work, and more recent numerical modeling efforts (Hasan 1991), is that features exclusive to both models are present in the observations. The fast-mode resonance is needed to explain the isolated peaks in the photospheric power spectra which are correlated with oscillations in the chromosphere, yet some other mechanism must be responsible for the rather broad spectrum of power above 5 mHz observed in the chromosphere.

I am convinced that none of the eigenmode analyses based upon linear magnetohydrodynamic treatment of model umbral atmospheres are viable in the face of the observational evidence. The nonlinear *hydrodynamic* nature of the oscillations in the chromosphere contains the essence of the phenomenon, and theoretical models that incorporate nonlinear effects are probably the only valid way to make progress in understanding the observations.

Gurman and Leibacher (1984) illustrated the importance of nonlinear effects when they produced 3-minute oscillations in a model chromosphere from a photosphere driven by 5-minute oscillations. The recent one-dimensional radiative/hydrodynamic models of disturbances in the quiet Sun developed by Stein and Carlsson (1991) are illustrated by theoretical time series of profiles in Figure 16. Compare these simulations with actual observations presented in Figure 17. The simulations show an astounding degree of realism. Now they may be tested in detail with observations. An important result from these simulations is that the chromosphere responds to sinusoidal driving deep in the atmosphere with periods of 60 sec or below, or 180 sec or above, by repeated shocks in the chromosphere with periods around 3 minutes. This response cannot be the "chromospheric resonance," because the undisturbed model has neither a transition region nor a corona! Some property of the hydrodynamics itself tends to produce the resonance near 3 minutes. It could, for example, be the dispersion of the atmosphere leading to reflection of upcoming shock waves by the disturbance which already exists at higher levels from previous shock waves. It is plausible that this mechanism is responsible for umbral oscillations as well. One-dimensional hydrodynamic simulations are especially appropriate for sunspot umbrae where the magnetic field plays a dominant role in confining fluid motions to the predominantly vertical direction.

3.4. CHROMOSPHERIC STRUCTURE FROM UMBRAL OSCILLATIONS?

Noting that there are often multiple peaks in the power spectra of chromospheric lines, some authors have inferred the extent of the umbral chromosphere from the observed frequencies of the peaks assuming that the peaks result from a chromospheric resonance (Zhugzhda and Makarov 1982; Campos 1989). This inference is highly questionable in view of the variability of the power spectrum with time and position within the umbra (Figure 11).

The relative amplitude of the umbral oscillation observed at two heights in the atmosphere reveals that the oscillations are sensitive to chromospheric structure (Lites 1986a). Within very

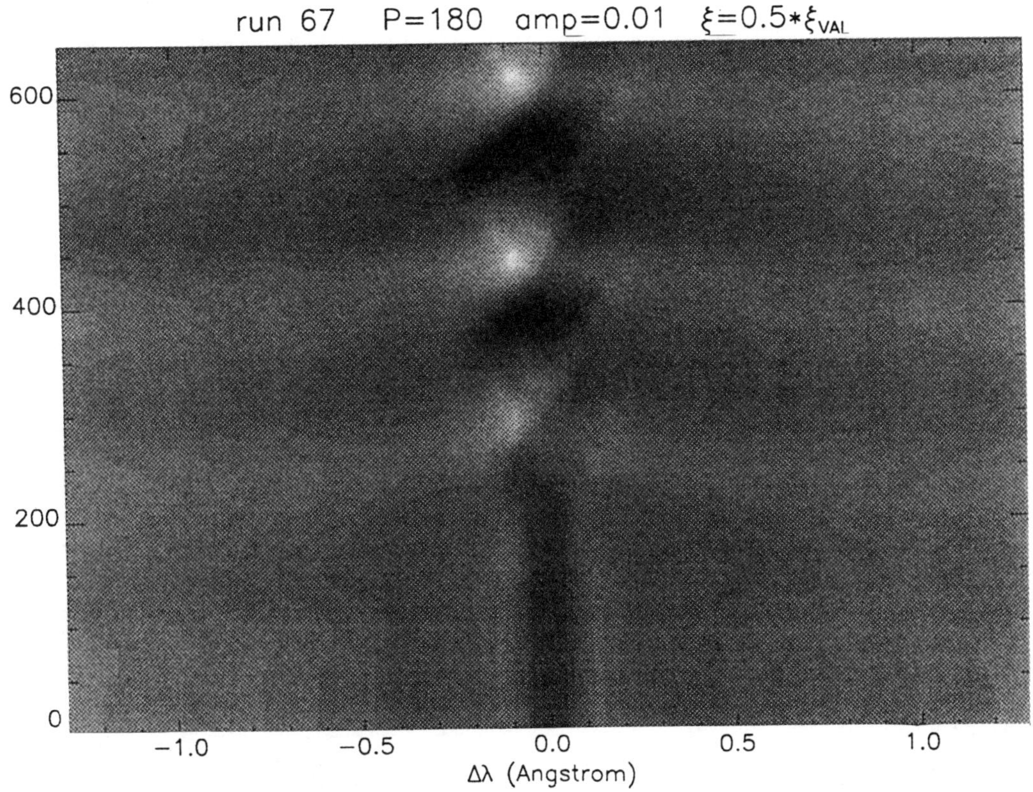

FIGURE 16. Simulations of quiet Sun chromospheric dynamics coupling fully compressible hydrodynamics with one-dimensional non-LTE radiative transfer (Stein and Carlsson 1991) predicts the time series of spectra of the Ca II K-line shown here. In this case the atmosphere is driven at the bottom by a piston with a 180-second sinusoidal vertical motion. Time increases from bottom to top, wavelength increases from left to right. These simulations bear a striking resemblance to observations, such as those shown in Figure 17. This figure is courtesy of R. F. Stein and M. Carlsson (paper in preparation).

narrow frequency bands one finds that the amplitude of oscillation measured in the low chromosphere (Fe I 543.4 nm) is proportional to the amplitude measured at greater heights (Ca II 849.8 nm) from a point-by-point correlation of amplitudes within distinct umbrae. This proportionality varies rapidly with frequency however, and one finds differing proportionality within separate umbrae. This finding suggests the following: (1) individual umbral elements have some homogeneity of chromospheric structure, otherwise no strong correlation of amplitudes would be expected, (2) if the chromospheric structure is rather homogeneous, then the rapid variability of the correlation with frequency probably arises from the interference pattern of horizontal mode structure of the driving source in the umbral photosphere, and (3) separate umbrae have differing average chromospheric structure, otherwise one would not find highly correlated amplitudes of

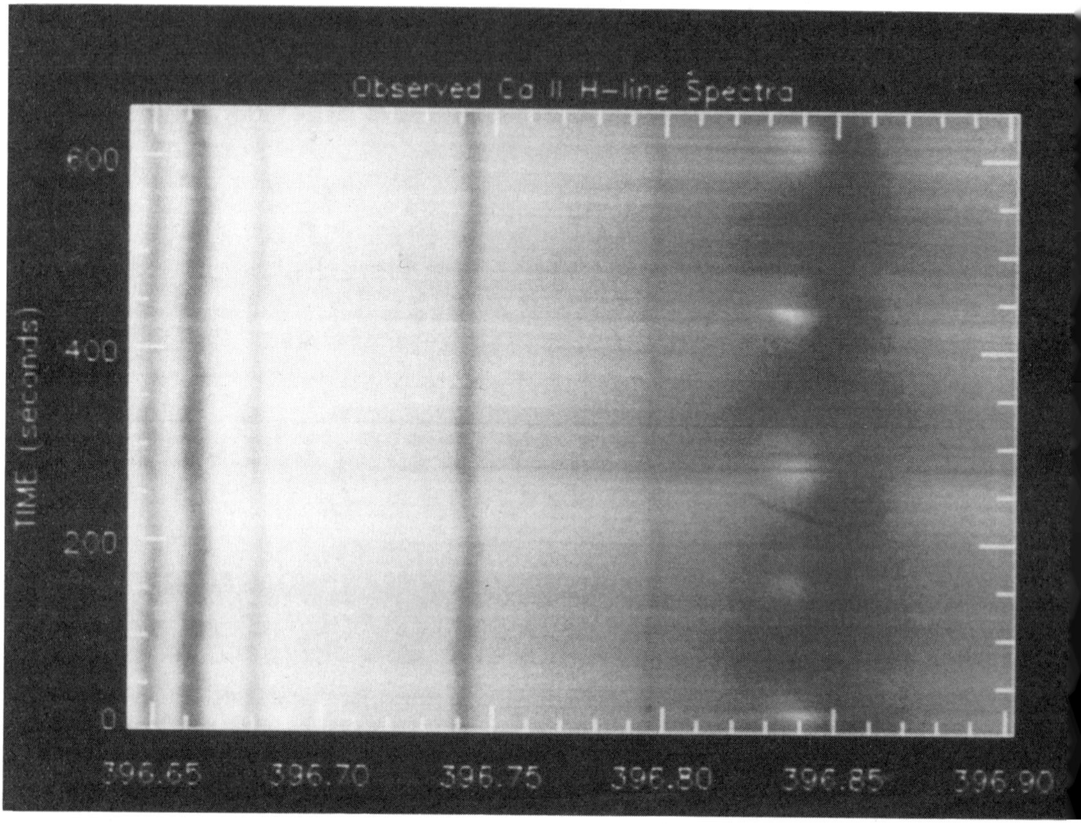

FIGURE 17. The observed variation with time of the Ca II H-line profile is shown for a location on the Sun showing particularly strong H_{2V} grains. Note the similarity with the simulations presented in Figure 16. The observations are from work by Lites, Kalkofen, and Rutten (in preparation).

oscillation whose proportionality differs from one umbra to the next. Very little theoretical or interpretive work has been carried out to exploit the information about chromospheric structure contained in these amplitude correlations.

3.5. THE EXCITATION OF THE UMBRAL OSCILLATIONS

Regardless of the mechanism that eventually accounts for the umbral oscillations, the umbral 5-minute oscillations do not appear to be the driving source, as was suggested by Gurman and Leibacher (1984). First, the photospheric amplitudes of these oscillations are much lower than in the quiet Sun, and they must tunnel through a much cooler temperature minimum region to reach the chromosphere. Second, the amplitude of the 5-minute oscillations is so low in the middle chromosphere that they have not been detected there. Third, the size scale of the chromospheric oscillations is much smaller than that of the five-minute oscillations in the photosphere. Fourth, there is no spatial correlation of oscillatory amplitudes between chromospheric oscillations and the five-minute oscillations. The latter two reasons depend upon the character of the waves in the chromosphere: they must be compressive slow-mode waves guided along the direction of the magnetic field.

Note that none of these arguments hold if the broad-band, high frequency tail of the p-mode spectrum drives the chromospheric oscillations. The observed fact that, except for isolated power peaks, there is little coherence of power between the photosphere and chromosphere at these frequencies suggests that the broadband tail of the p-modes contributes little to the excitation mechanism. Lites and Thomas (1985) measured oscillations in one sunspot which had unusually strong five-minute oscillations in the photosphere, yet no detectable oscillations in the 3-minute band in the low chromosphere. This observation reinforces arguments against the 5-minute excitation of the chromospheric oscillations. It also suggests that they are not driven by the broadband tail of the 5-minute oscillations, provided that the power at higher frequencies is proportional to the power in the 5-minute band. Alternatively, for some unknown reason, this particular sunspot may be highly transmissive to oscillations in the 5-minute band. From this single observation one should not eliminate the high frequency tail of the p-mode spectrum from contention as a driver for umbral oscillations. New measurements of high sensitivity in the photosphere with a significantly larger statistical sample of sunspots are needed to resolve questions regarding the driving source of the umbral oscillations.

4. Penumbral Oscillations

The penumbral oscillations, often called "running penumbral waves," were discovered by Giovanelli (1972) and Zirin and Stein (1972) as disturbances moving radially outward from the umbra in H_α movies. Often the disturbance appears to be coherent around a substantial fraction of the umbra-penumbra border. These oscillations are not as energetic as the chromospheric umbral oscillations, but their importance as a wave mode which is unique to a highly magnetized atmosphere cannot be minimized. Even more importantly, they may eventually provide us with information about the structure of the magnetic field below the surface of penumbrae – information that may not be available by other means.

4.1. OBSERVED PROPERTIES OF PENUMBRAL OSCILLATIONS

Much of the information presented below comes from an observational paper by Lites (1988). The reader is referred to that paper and to the reviews cited above for more detailed summary of the observations. Unfortunately, there has not been much observational work regarding this interesting phenomenon since the time of that paper.

- *The disturbances move outward from umbra at sonic velocities or higher.* There is quite a spread in measured horizontal phase speeds for the waves: 20-35 km s^{-1} by Lites et al. 1982, 12-25 km s^{-1} by Giovanelli (1974), 8-12 km s^{-1} by Zirin and Stein (1972), and 10-18 km s^{-1} by Musman, Nye, and Thomas (1976). The measured velocities of Lites et al. are considerably higher the others, and are therefore suspect. Phase speeds in the range 10-20 km s^{-1} appear to be commonplace in the observations. If the observed phase speed indicates a group velocity for disturbances, and if they propagate at speeds greater than 15 km s^{-1} (i.e., considerably greater than the local sound speed) then the wave mode must be Alfvénic in character. On the other hand, phase speeds larger than the sound speed are possible for an essentially acoustic disturbance propagating obliquely to the line of sight: i.e., for a wavefront rising along the obliquely inclined penumbral field lines.

- *No positive identification of penumbral oscillations has been made in the low photosphere.* Musman, Nye, and Thomas (1976) searched for the penumbral oscillations using simultaneous measurements in H_α and the Zeeman insensitive line Fe I 557.6 nm. In the Fe I line, they found intermittent running disturbances with periods comparable to those seen in the chromosphere, but with much larger horizontal phase speeds. Association of these disturbances with the penumbral oscillations is dubious. Beckers (1975) identified penumbral oscillations in the intensity fluctuations at the core of the Mg I b-lines, which form in the temperature minimum or low chromosphere. They are also clearly visible in the inner penumbra as viewed in Fe I 543.4 nm (Lites 1988), but this line also forms in the temperature minimum region or the low chromosphere in penumbrae (Bruls et al. 1991). Clearly new observations of high sensitivity are needed to clarify the observed nature of penumbral oscillations in the photosphere.
- *Compared to the chromospheric umbral oscillations, the velocity excursions associated with the penumbral oscillations are moderate in amplitude.* The rms amplitudes are typically 200 m s^{-1} in high photosphere and 1 km s^{-1} in the chromosphere. However, observations in the high chromosphere, in He I 1083.0 nm for example, could confirm if the oscillations are linear or if they develop into shocks.
- *When sunspots are observed away from disk center, the oscillatory Doppler amplitude measured in the upper photosphere of the inner penumbra is largest on the side of the sunspot closest disk center.* Figure 18 shows this effect clearly in the map of oscillatory amplitude measured in the Fe I 543.4 nm line. The ring of low amplitude encircling the penumbra about half-way between the inner and outer edge of the penumbra (the "attenuation ring") always crowds the umbra on the side away from disk center. This result suggests that the oscillatory motions in the upper photosphere have a preferential alignment with the vector magnetic field in the inner penumbra.
- *The oscillatory amplitude measured in chromospheric lines appears largest in the outer penumbra on the side of sunspots closest to disk center.* Figure 18 demonstrates this effect in the amplitude of oscillation measured in the Ca II 854.2 nm line. Maltby (1975) found a similar behavior in H_α observations of sunspots observed away from disk center. This result suggests that the motions are also field-aligned in the outer penumbra at chromospheric heights. However, this interpretation must be tempered by the fact that the oscillatory frequency decreases with distance from the umbra.

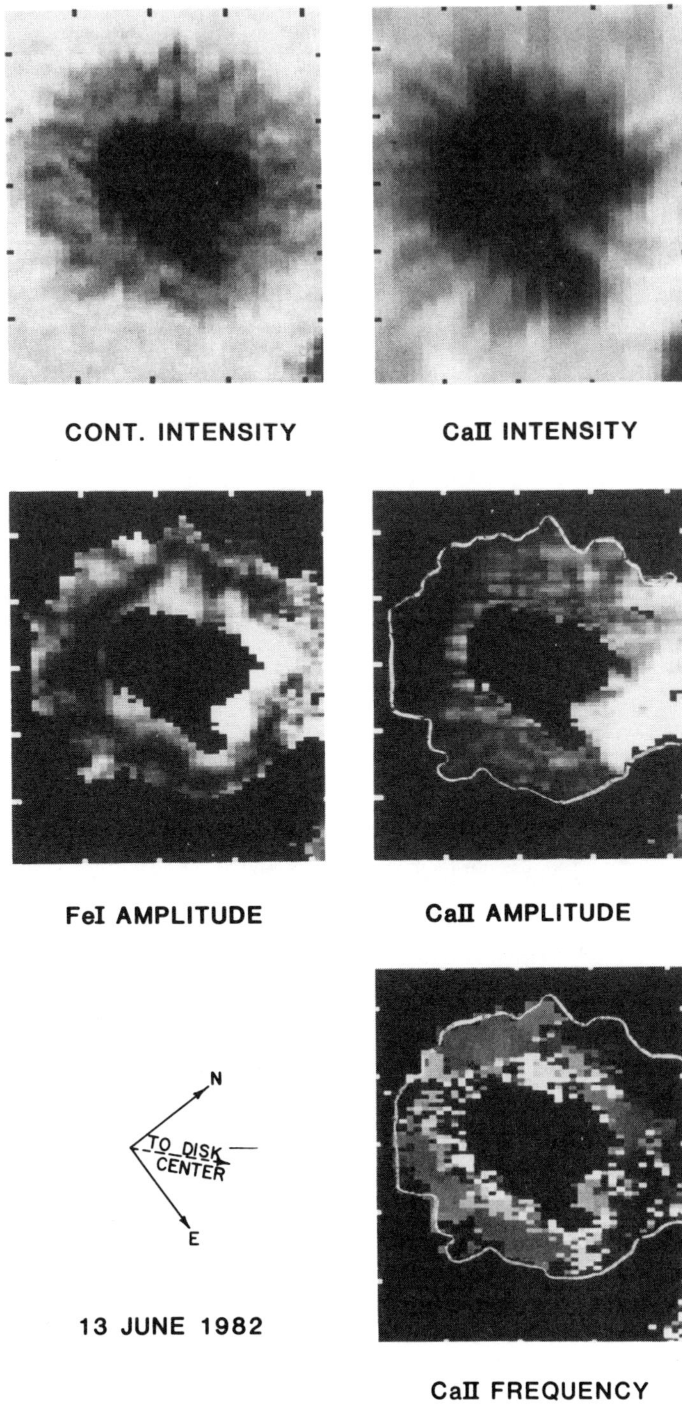

FIGURE 18. From Lites 1988. The spatial distribution of oscillatory amplitudes and frequencies of oscillation in a sunspot penumbra are shown. The upper left image shows the continuum near 543.4 nm, and the upper right image shows the intensity in the core of Ca II 849.8 nm. The left middle image shows, as a gray scale, the logarithm of the rms amplitude of oscillation measured in Fe I 543.4 nm, and the right middle image gives the corresponding rms amplitude for the Ca II line. The dominant frequency of oscillation in Ca II is coded as a gray scale in the bottom image, with darkest and lightest shades corresponding to frequencies of ≤ 1.67 and ≥ 5.0 mHz, respectively. Areas outside of the penumbra and within the umbra are assigned black in the amplitude and frequency images. The penumbral boundary is sketched onto the Ca II amplitude and frequency images. Note the "attenuation ring" surrounding the sunspot umbra and crowding it on the side away from disk center in the Fe I amplitude image. Note also the decrease in frequency of oscillation toward the outer penumbra as measured in the Ca II line.

- *The oscillatory frequency in the penumbral chromosphere decreases from 4 mHz near the umbral boundary to* $\leq$ 1.5 *mHz at the outer penumbral boundary.* Beckers and Schultz (1972) and Lites (1986a) noted a decrease of frequency by more than a factor of two from the inner to the outer edges of the penumbra. Figures 18 and 19 illustrate this significant property of penbumbral oscillations. If the oscillations throughout the penumbra arise from a common physical mechanism, the term "running penumbral waves" is a misnomer. *More likely we observe the effects of distinct physical origins of the oscillations in the inner and outer penumbra, and perhaps associated with fluctuations in the Evershed flow.*

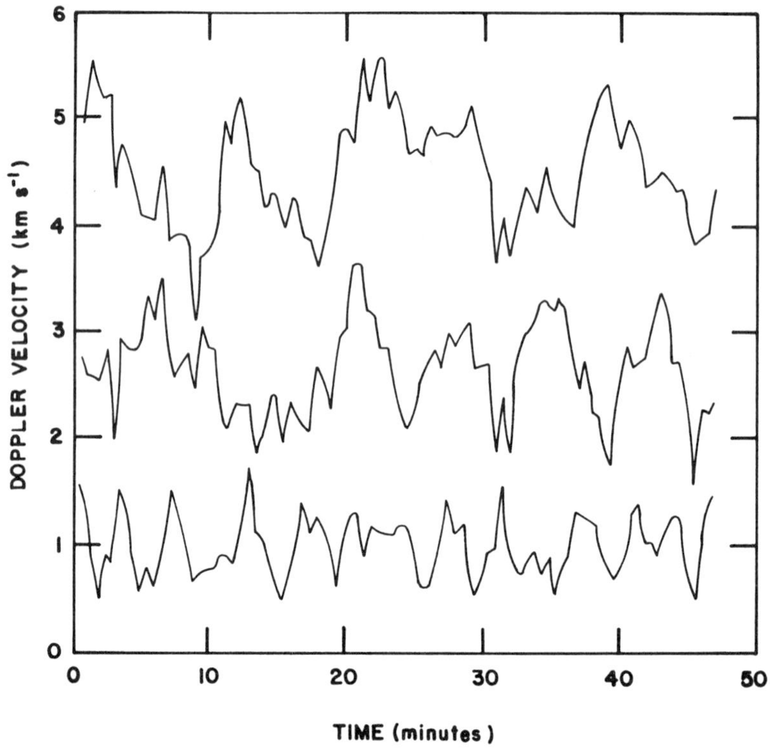

FIGURE 19. From Lites 1984. Selected time series of line center displacement of the Ca II 854.2 nm line are shown for three locations in a sunspot penumbra. The top curve is near the outer edge of the penumbra and shows a period of about 10 minutes. The second and third curves from the top represent time series halfway between the umbra and outer edge of the penumbra, and close to the umbra, respectively.

- *The 5-minute p-mode oscillations dominate the oscillatory spectrum in the outer penumbra in the temperature minimum region of the atmosphere.* The "attenuation ring" shown in Figure 18 illustrates how the oscillatory amplitude measured in Fe I 543.4 nm *increases* near the outer edge of the penumbra. Doppler movies show that the spatial pattern of the p-mode oscillations outside of the sunspot encroaches upon the outer penumbra. In contrast, Dopplershifts measured in the outer penumbra in lines which form higher in the chromosphere show no evidence of the 5-minute p-modes.

- *The vertical phase velocity in the 5-minute band is much smaller than that of the quiet Sun.* Figure 20 shows power and phase difference spectra for oscillations measured in Fe I 543.4 nm and Ca II 849.8 nm. The 30-40° delay in phase between these two heights is several times the 7-10° phase difference for the quiet Sun. Even though there are some uncertainties in the measurements of the phases, the differences from the quiet Sun results are substantive and they likely herald differences in propagation characteristics between the evanescent p-modes outside of sunspots and the magneto-atmospheric modes in the penumbra.
- *The oscillations in the penumbra frequently are coherent around a substantial arc of the umbra-penumbra boundary, and they appear to be "spawned" by the umbral oscillations.* This observational result is difficult to present in a quantitative way in print. It is however, quite obvious in cinematographic presentations of observations which show simultaneously the umbral and penumbral oscillations. If the penumbral oscillations were a global mode of oscillation of much of the penumbral arc around the umbra, then the coherence of oscillatory phase along this arc would demand a very fast communication of points along the arc. The transit speed for such communication should then be much greater than the 15 km s^{-1} outward phase speed of the oscillations. Communication could therefore proceed only via Alfvén (or fast mode) waves in the chromosphere. Chromospheric excitation of wave modes present other problems when faced with the large energy of the oscillations at lower levels in the atmosphere. Given the connection with umbral oscillations, it is possible that the penumbral oscillations are excited by the umbral oscillations, which themselves would then be constrained to have uniform phase at the outer edge of the subsurface umbral flux tube.

4.2. IMPLICATIONS FOR THEORETICAL MODELS OF PENUMBRAL OSCILLATIONS

Most theoretical treatments of penumbral oscillations assume horizontal magnetic field structure (Nye and Thomas 1974, 1976b; Antia, Chitre, and Gokhale 1978; Small and Roberts 1984), and often this field is assumed to float upon an unmagnetized convective layer immediately below (Nye and Thomas 1976b; Antia et al. 1978; Small and Roberts 1984). The penumbral oscillations are most visible in the inner penumbra where, at this workshop, we have seen that the field is inclined significantly with respect to the horizontal. As a result, the concept of a "thin" horizontal penumbra floating on a field-free region below should be discarded. The corrugation of the field also suggested by observations at this conference applies mainly to the outer penumbra. The concept of horizontal surface waves at the interface of magnetized and unmagnetized regions (Small and Roberts 1984), or the trapping of fast mode waves (Nye and Thomas 1974, 1976b; Antia et al. 1978) by the increasing sound speed in the subsurface layers and the increasing Alfvén speed in the photosphere (similar to the fast mode trapping in sunspot umbrae) must also be rethought in view of the current picture of the penumbral magnetic field structure.

The theory should also be capable of predicting the observed range of horizontal phase speeds, and the fact that the observed motions appear to be aligned along the field rather than perpendicular

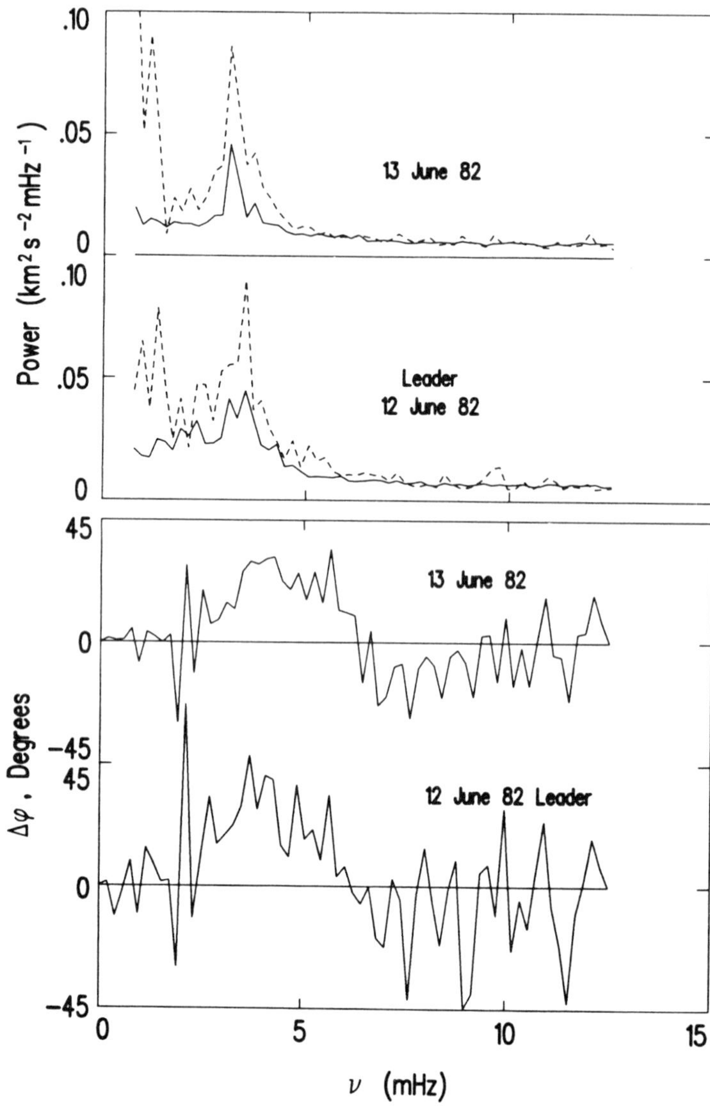

FIGURE 20. From Lites 1988. Average power and phase spectra are given for Doppler shifts of Fe I 543.4 nm and Ca II 849.8 nm as measured in two sunspot penumbrae. The Fe I power spectra have been multiplied by a factor of 10. The phase differences are positive when oscillations in the Fe I line lead those in the Ca II line.

to it. Although the mode of oscillation may be Alfvénic at depth, it appears to be compressive high in the solar atmosphere where it is observed. It should also predict the observed phase delays in the chromosphere, which differ considerably from those of the quiet Sun.

The change in oscillatory frequency near the outer edge of penumbrae may be related in some way to the corrugation of the field there, but we have no evidence that the magnetic field in the chromosphere (where the penumbral oscillations are observed) experiences this same corrugation. In any case one should not insist that the oscillatory modes of the inner and outer penumbra arise from the same physical mechanism. The change in the character of the oscillation may be related to the presence of 5-minute oscillations at lower levels of the atmosphere in the outer penumbra. It may not be coincidental that oscillations of similar long periods are observed in the quiet chromospheric network (Orrall 1966; Damé, Gouttebroze, and Malherbe 1984; Lites, Kalkofen, and Rutten 1992). Perhaps the low frequency oscillations in both the quiet network and the outer penumbrae are driven by buffeting either by the 5-minute oscillations or by convective motions. Another possibility is that the Evershed flow, which now appears to be confined to channels coincident with the dark penumbral fibrils, could drive some instability in the outer penumbra. New high-resolution observations of the Evershed flow presented by Shine at this workshop (see also Shine et al. 1990) show a quasi-periodic transit of the outer penumbra by cloud-like structures having an apparent velocity of a few km s^{-1}.

It is clear that significant modification of existing theories for penumbral oscillations will be needed before one may identify with confidence the nature and driving source for the oscillations. In particular, realistic field geometry should be an integral part of this theory – including the variable inclination and strength of the field with distance from the umbra and the possible corrugation of the field inclination in the outer penumbra. Full treatment of this problem may require time-dependent MHD calculations in 3-dimensional geometry, although Nye and Thomas (1974) argue that their trapping mechanism will work in an inclined field also. Zhugzhda and Dzhalilov (1984; also a short contribution by Zhugzhda at this workshop) describe in a schematic way some effects of oscillations in an inclined magnetic field. Their ideas regarding the initiation of penumbral oscillations by p-mode oscillations in the convection zone warrant further scrutiny. Such external, large-scale driving of the penumbral oscillations may result in the large-scale coherence of the penumbral oscillations along the umbral-penumbral boundary. One problem with such a scenario is the apparent lack of correlation of the 5-minute and 3-minute umbral oscillations, whereas the penumbral and 3-minute umbral oscillations appear to be related.

5. What New Observations are Needed to Clarify the Nature of Sunspot Oscillations?

5.1. THE UMBRAL PHOTOSPHRE AND BELOW

A firm identification of the driving source of the umbral chromospheric oscillations is still lacking. Probably the most urgent need is for observations of umbral oscillations in the photosphere with *high sensitivity* with simultaneous measurements of the chromospheric oscillations. A filter instrument or spectrograph having exceptional wavelength stability is needed for this task. Such observations could verify the existence of the narrow-band oscillations that have been observed at photospheric heights, and they could quantify the relationship between the broadband power at high frequencies in the photosphere and the chromospheric oscillations. An adequate sample of sunspots should be measured in this way.

The observational analysis of Abdelatif (1985) regarding the spatio- temporal behavior of the oscillations should also be repeated with higher sensitivity. Stable filter measurements of umbral Doppler shifts would be most appropriate for this task.

Time series measurements of sunspots and their surroundings at high angular resolution may be helpful to establish the relationship (if it exists) between the "acoustic sources" responsible for the high frequency p-modes and photospheric umbral oscillations. Such observations should be carried out with a stable imaging device rather than a spectrograph in order to maintain the integrity of the image. The time series could be filtered to show only the high frequency component of the oscillation. One could then search for continuity of observed high frequency disturbances from outside into the sunspot.

5.2. THE UMBRAL CHROMOSPHERE

A fundamental property of umbral chromospheric oscillations is the behavior of the phase of oscillation with height from the photosphere up to the corona. Believable phase measurements could elucidate the nature of the chromospheric oscillations: Are they propagating or standing waves? Are there nodes of oscillation? What are their energetic consequences? Once such phase relationships have been established it might be possible to use umbral oscillations for studying the formation of optically thick lines in the dynamical chromospheres.

The optically thin He I 1083.0 nm line is ideally suited for measurement of sunspot oscillations. Its absorption over sunspots is strong enough to allow unambiguous determination of Doppler velocity, and hence the phase of oscillation. The observations of Lites (1986b) suffered from the severe fringing properties of the CCD detector used at that time: the amplitude of the fringes dwarfed the depth of the He I line. For that reason, it is important to verify the large Doppler shifts, sawtooth waveforms, and phase of oscillation using newer detectors with significantly weaker fringes. Several new data sets were recently obtained (Lites, Rutten, and Thomas 1991) at the National Solar Observatory/ Sunspot with this aim in mind. It is hoped that these new data will be satisfactory to determine phases of oscillation in a number of lines including He 1083.0, Si I 1082.7, Ca II H, Al I 396.1, Fe I 396.9, and a number of other lines in the wings of the Ca II H line which form at various heights in the photosphere.

There exists also a clear need for simultaneous observations of lines formed in the chromosphere and in the transition region. Such observations would indicate the reflectivity of the transition region for the oscillations, which depends sensitively upon its temperature gradient. Most of the chromospheric and transition region lines in the ultraviolet (120-250 nm) are either effectively thin or optically thin over sunspots. Interpretation of Doppler shifts measured in these lines is then not clouded by complex radiative transfer effects. Ultraviolet observations of sunspot oscillations promise to yield precise information regarding the phases and relative amplitudes of oscillations in the upper umbral chromosphere. In this regard it is extremely unfortunate that the Orbiting Solar Laboratory has been severely delayed.

5.3. PENUMBRAL OSCILLATIONS

The behavior of penumbral oscillations at levels below the chromosphere remains poorly determined. Simultaneous measurements in chromospheric and photospheric lines, with instrumentation which is highly sensitive to very small Doppler shifts in the photospheric lines, may point both to the mode of oscillation and to the excitation mechanism of the "running waves" in the inner penumbra.

As with umbral oscillations, measurements of the He I 1083.0 nm line could clarify the phase behavior of penumbral oscillations in the chromosphere.

Oscillations of very high angular resolution over a period of an hour or more are needed to investigate the association (if any) of penumbral oscillations with penumbral filamentation. The magnetic field in the penumbral photosphere appears to be more nearly horizontal in the dark filaments, and the outward Evershed flow seems to be concentrated there as well (see the review by Title et al. in this volume). The flow reverses direction in the chromosphere. Little is know about the small-scale behavior of either the flow or the magnetic field structure in the penumbral chromosphere. A sound observational study of chromospheric fine structure in penumbrae, comprising also the relationship with structures in the underlying photosphere, is an essential ingredient of theoretical understanding of this phenomenon. Such an observational study now appears within grasp of modern solar instrumentation.

6. The Intersection of Theoretical and Observational Understanding

I conclude this review with a group of questions for the theoretician. These questions are founded upon those observational results which appear to be sound at the time of writing. The interested theoretician should referto the body of the review for other observations of a less certain nature which also pose interesting challenges for theoretical study.

6.1. THE UMBRAL PHOTOSPHERE AND BELOW

The 5-minute p-mode oscillations penetrate into the umbral photosphere, but in so doing they are reduced in amplitude and perhaps they are restricted in horizontal wavenumber. Occasionally, the umbral photosphere also resonates at isolated frequencies above 5 mHz.

- Why quantitatively is the oscillatory amplitude, integrated over the 5-minute band, 2-3 times lower than that observed in the surrounding quiet photosphere?
- Can one explain the isolated power peaks above 5 mHz as a resonance of the sub-photospheric cavity of a sunspot umbra, when physically realistic boundary conditions for the sides of the flux tube and a flaring geometry of the tube are imposed?
- How would an inhomogeneous sub-photospheric magnetic field structure affect the above resonance?
- Could acoustic radiation from nearby "acoustic sources" penetrate laterally into sunspots to excite the observed high frequency oscillations?

6.2. THE UMBRAL CHROMOSPHERE

The nonlinear nature of the chromospheric umbral oscillations presents the greatest current challenge to theoretical understanding, but it also affords a tremendous opportunity for fruitful theoretical investigation.

- Can the oscillations contribute substantially to the heating of the mean, i.e., time-averaged, umbral chromosphere?

- Can oscillations observed at different frequencies couple effectively through the nonlinearity?
- Can fully nonlinear numerical modeling help us to understand the formation of optically thick chromospheric lines in a highly dynamic atmosphere, thereby allowing a determination of the phase of oscillation in both umbral and quiet Sun contexts?
- Why are umbral oscillations strong relative to the quiet Sun, when the umbral photosphere is so tranquil?
- Why is the dominant frequency of oscillation in the chromosphere about 5.5 mHz; that is, well below the acoustic cutoff frequency of the umbral temperature minimum, if the oscillations are driven by the umbral photosphere or below?
- Is there a quantitative explanation for the behavior of the rms amplitude of oscillation with frequency: the sharper decline of power with increasing frequency in the low chromosphere relative to that at higher levels?

6.3. THE PENUMBRAL OSCILLATIONS

The mode of oscillation peculiar to the penumbra does not appear to have a convincing theoretical explanation given the "thick" magnetic field structure of the penumbra which has been discussed in this workshop. Most theoretical treatments to date have assumed an essentially thin penumbral field. With this in mind, the following questions may be posed:

- What modes of oscillation produce field-aligned motions at the heights where oscillations are observed?
- What mode(s) explains the low frequency observed in the outer penumbra?
- Do low frequency oscillations observed both in the outer penumbra and in the quiet chromospheric network share a common physical mechanism?
- What is the influence of 5-minute p-mode oscillations on the oscillatory character of the outer penumbra?
- Can simulations of penumbral oscillations be carried out with realistic field geometry and atmospheric stratification?
- What influence does the fine structure of flows and magnetic fields in the photosphere have upon the oscillations observed in the chromosphere?
- Can theoretical models lead us to an understanding of the connection between umbral and penumbral oscillations?

7. Acknowledgments

I wish to thank all of the colleagues with whom I have collaborated over the years in studying oscillations of sunspots. In particular I thank Jack Thomas for his interest and involvement in several programs, and his advice in preparation of this manuscript. I also wish to thank the staff of the National Solar Observatory. They provided observing time at the Vacuum Tower Telescope at Sunspot, observing support, and data analysis support without which much of my own work on sunspot oscillations would not have been possible. I am deeply grateful to Tom Bogdan and Richard Shine for careful reading of this manuscript and important suggestions regarding its content. Finally, I thank the organizers of this workshop for providing the timely opportunity to meet and discuss the important issues in the light of exciting new observations.

References

Aballe Villero, M. A., Garciá de la Rosa, J. I., Marco, E., and Vázquez, M. 1990, "Oscillatory behaviour in sunspot umbrae," *Astrophys. Space Sci.*, **170**, 121.

Abdelatif, T. E. 1985, "Umbral oscillations as a probe of sunspot structure," Ph.D. Thesis, University of Rochester.

__________ 1990, "Magneto-atmospheric waves," *Solar Phys.*, **129**, 201.

Abdelatif, T. E., Lites, B. W., and Thomas, J. H. 1984, "Oscillations in a sunspot and the surrounding photosphere," in *Small-Scale Dynamical Processes in Quiet Stellar Atmospheres*, ed. S. L. Keil (Sunspot, NM: National Solar Observatory), p. 141.

__________ 1985 (unpublished movie of sunspot oscillations.)

__________1986, "The interaction of solar p-modes with a sunspot. I. Observations," *Ap. J.*, **311**, 1015.

Abdelatif, T. A. and Thomas, J. H. 1987, "The interaction of solar p-modes with a sunspot. II. Simple Theoretical Models," *Ap. J.*, **320**, 884.

Antia, H. M., Chitre, S. M., and Gokhale, M. H. 1978, "Waves in the sunspot penumbra," *Solar Phys.*, **60**, 31.

Avrett, E. H. 1981, "Reference model atmosphere calculation: the Sunspot sunspot model," in *The Physics of Sunspots*, ed. L. E. Cram and J. H. Thomas (Sunspot, N. M.: Sacramento Peak Observatory). p. 235.

Balthasar, H., Küveler, G., and Wiehr, E. 1987, "A comparison of the oscillations in sunspot umbrae, penumbrae, and the surrounding photosphere," *Solar Phys.*, **112**, 37.

Balthasar, H. and Wiehr, E. 1984, "Umbral oscillations measured in the Stokes-V inversion point," *Solar Phys.*, **94**, 99.

Beckers, J. M. 1975, "New views of sunspots," *Sacramento Peak Observatory Contribution*, 249, Sunspot, NM.

Beckers, J. M., Mauter, H. M., Mann, G. R., and Brown, D. R. 1972, "High resolution spectroscopy of the disk chromosphere. I. Observing procedures," *Solar Phys.*, **25**, 81.

Beckers, J. M. and Schultz, R. B. 1972, "Oscillatory motions in sunspots," *Solar Phys.*, **27**, 61.

Beckers, J. M. and Tallant, P. E. 1969, "Chromospheric inhomogeneities in sunspot umbrae," *Solar Phys.*, **7**, 351.

Bhatnagar, A. 1971, "On the oscillatory velocity field in sunspot atmosphere," *Solar Phys.*, **18**, 40.

Bhatnagar, A., Livingston, W. C., and Harvey, J. W. 1972, "Observations of sunspot umbral velocity oscillations," *Solar Phys.*, **27**, 80.

Brown, T. M. 1991, "The source of solar high-frequency acoustic modes: theoretical expectations," *Ap. J.*, **371**, 396.

Brown, T. M., Bogdan, T. J., Lites, B. W., and Thomas, J. H. 1992, (in preparation).

Bruls, J. H. M. J., Lites, B. W., and Murphy, G. A. 1991, "Non-LTE formation heights of Stokes profiles of Fe I lines," in *Solar Polarimetry*, ed. L. J. November (Sunspot, NM: National Solar

Observatory), p. 444.

Campos, L. M. B. C. 1989, "On oscillations in sunspot umbrae and wave radiation in stars," *Mon. Not. R. Astr. Soc.*, **241**, 215.

Damé, L., Gouttebroze, P., and Malherbe, J.-M. 1984, "Observation and analysis of intensity oscillations in the solar K-line," *Astron. Astrophys.*, **130**, 331.

Deubner, F.-L. and Fleck, B. 1990, "Dynamics of the solar atmosphere. III. Cell-network distinctions of chromospheric oscillations," *Astron. Astrophys.*, **228**, 506.

Giovanelli, R. G. 1972, "Oscillations and waves in a sunspot," *Solar Phys.*, **27**, 71.

__________ 1974, "Waves and oscillations in the chromosphere in active and quiet regions," in I.A.U. Symposium No. 56, *Chromospheric Fine Structure*, ed. R. G. Athay (Dordrecht: Reidel), p. 137.

Gurman, J. B. 1987, "Sunspot umbral oscillations in MgII k," *Solar Phys.*, **108**, 61.

Gurman, J. B. and Leibacher, J. W. 1984, "Linear models of acoustic waves in sunspot umbrae," *Ap. J.*, **283**, 859.

Gurman, J. B., Leibacher, J. W., Shine, R. A., Woodgate, B. E., and Henze, W. 1982, "Transition region oscillations in sunspots," *Ap. J.*, **253**, 939.

Hasan, S. S. 1991, "Magnetoatmospheric oscillations in sunspot umbrae," *Ap. J.*, **366**, 328.

Howard, R., Tannenbaum, A. S., and Wilcox, J. M. 1968, "A new method of magnetograph observation of the photospheric brightness, velocity, and magnetic fields," *Solar Phys.*, **4**, 286.

Kneer, F., Mattig, W., and von Uexküll, M. 1981, "The chromosphere above sunspot umbrae. III. Spatial and temporal variations of chromospheric lines," *Astron. Astrophys.*, **102**, 147.

Kobanov, N. I. 1990, "On spatial characteristics of five-minute oscillations in the sunspot umbra," *Solar Phys.*, **125**, 25.

Kumar, P., Duvall, T. L., Jr., Harvey, J. W., Jefferies, S. J., Pomerantz, M. A., and Thompson, M. J. 1990, "What are the obseved high frequency solar acoustic modes?," in *Progress of Seismology of the Sun and Stars*, ed. Y. Osaki and H. Shibahashi (Berlin: Springer), 87.

Lites, B. W. 1984, "Photoelectric observations of chromospheric sunspot oscillations. II. Propagation characteristics," *Ap. J.*, **277**, 874.

__________1986a, "Photoelectric observations of chromospheric sunspot oscillations. III. Spatial distribution of power and frequency in umbrae," *Ap. J.*, **301**, 992.

__________1986b, "Photoelectric observations of chromospheric sunspot oscillations. IV. The Ca II H line and He I λ10830," *Ap. J.*, **301**, 1005.

__________ 1988, "Photoelectric observations of chromospheric sunspot oscillations. V. Penumbral oscillations," *Ap. J.*, **334**, 1054.

Lites, B. W. and Chipman, E. G. 1979, "The vertical propagation of waves in the solar atmosphere. I. Observations of phase delays," *Ap. J.*, **231**, 570.

Lites, B. W., Chipman, E. G., and White, O. R. 1982, "The vertical propagation of waves in the solar atmosphere. II. Phase delays in the quiet chromosphere and cell-network distinctions," *Ap. J.*, **253**, 367.

Lites, B. W. Kalkofen, W., and Rutten, R. J. 1992, (in prepartion).

Lites, B. W. and Thomas, J. H. 1985, "Sunspot umbral oscillations in the photosphere and low chromosphere," *Ap. J.*, **294**, 682.

Lites B. W., Rutten, R. J., and Thomas, J. H. 1992 (in preparation).

Lites, B. W., White, O. R., and Packman, D. 1982, "Photoelectric observations of propagating sunspot oscillations," *Ap. J.*, **253**, 386.

Livingston, W. C. and Mahaffey, C. 1981, "Umbral oscillations at the photospheric level," in *The Physics of Sunspots*, eds. L. E. Cram and J. H. Thomas (Sunspot, NM: National Solar Observatory), p. 312.

Maltby, P. 1975, "Mode propagation of penumbral waves," *Nature*, **257**, 468.

Maltby, P., Avrett, E. H., Carlsson, M., Kjeldseth-Moe, O., Kurucz, R. L., and Loeser, R. 1986, "A new sunspot umbral model and its variation with the solar cycle," *Ap. J*, **306**, 284.

Moore, R. L. 1981, "Dynamic phenomena in sunspots," in *The Physics of Sunspots*, eds. L. E. Cram and J. H. Thomas (Sunspot, NM: National Solar Observatory), p. 259.

Moore, R. and Rabin, D. 1985, "Sunspots," *Ann. Rev. of Astronomy and Astrophysics*, **23**, 239.

Musman, S., Nye, A. H., and Thomas, J. H. 1976, "Observations of penumbral waves in the photosphere," *Ap. J. Lett.*, **206**, L175.

Orrall, F. Q. 1966, "Observational study of macroscopic inhomogeneities in the solar atmosphere. VIII. Vertical chromospheric oscillations measured in K_3," *Ap. J.*, **143**, 917.

Nye, A. H. and Thomas, J. H. 1974, "The nature of running penumbral waves," *Solar Phys.*, **38**, 399.

__________ 1976a, "Solar magneto-atmospheric waves. I. An exact solution for a horizontal magnetic field," *Ap. J.*, **204**, 573.

__________ 1976b, "Solar magneto-atmospheric waves. II. A model for running penumbral waves," *Ap. J.*, **204**, 582.

Rice, J. B. and Gaizauskas, V. 1973, "The oscillatory velocity field observed in a unipolar sunspot region," *Solar Phys.*, **32**, 421.

Rutten, R. J. and Uitenbroek, H. 1991, "Ca II H_2V Cell Grains," *Solar Phys.*,(in press).

Scheuer, M. A. 1980, Ph.D. Thesis, University of Rochester.

Scheuer, M. A. and Thomas, J. H. 1981, "Umbral oscillations as resonant modes of magneto-atmospheric waves," *Solar Phys.*,**71**, 21.

Schröter, E. H. and Soltau, D. 1976, "On the time behaviour of oscillations in sunspot umbrae," *Astron. Astrophys.*, **49**, 463.

Shine, R., Smith, K., Tarbell, T., Title, A., and Scharmer, G. 1990, "Penumbral flows and magnetic fields," *Bull. Am. Astron. Soc.*, **22**, 878.

Small, L. M. and Roberts, B. 1984, "On running penumbral waves," *The Hydrodynamics of the Sun* , in Proc. Fourth European Meeting on Solar Physics, (Paris: European Space Agency SP-220), p. 257.

Soltau, D., Schröter, E. H., and Wöhl, H. 1976, "On velocity oscillations in sunspot umbrae," *Astron. Astrophys.*, **50**, 367.

Soltau, D. and Wiehr, E. 1984, "Time-dependence of umbral oscillations," *Astron. Astrophys.*, **141**, 159.

Stein, R. F. and Carlsson, M. 1992 (in preparation).

Thomas, J. H. 1981, "Theories of dynamical phenomena in sunspots," in *The Physics of Sunspots*, eds. L. E. Cram and J. H. Thomas (Sunspot, NM: National Solar Observatory), p. 345.

Thomas, J. H. 1984, "Umbral oscillations in sunspots," *Astron. Astrophys.*, **135**, 188.

Thomas, J. H. 1985, "Oscillations in sunspots," *Aust. J. Phys.*, **38**, 811.

Thomas, J. H., Lites, B. W., Gurman, J. B., and Ladd, E. F. 1987, "Simultaneous measurements of sunspot umbral oscillations in the photosphere, chromosphere, and transition region," *Ap. J.*, **312**, 457.

Thomas, J. H., Cram, L. E., and Nye, A. H. 1982, "Five-minute oscillations as a subsurface probe of sunspot structure," *Nature*, **297**, 485.

__________ 1984, "Dynamical phenomena in sunspots. I. Observing procedures and oscillatory phenomena," *Ap. J.*, **285**, 368.

Thomas, J. H. and Scheuer, M. A. 1982, "Umbral oscillations in a detailed model umbra," *Solar Phys.*, **79**, 19.

Uchida, Y. and Sakurai, T. 1975, "Oscillations in sunspot umbrae due to trapped Alfven waves excited by overstability," *Publ. Astron. Soc. Japan*, **27**, 259.

Vernazza, J. E., Avrett, E. H., and Loeser, R. 1981, "Structure of the chromosphere. III. Models of the EUV brightness components of the quiet sun," *Ap. J. Suppl.*, **45**, 635.

von Uexküll, M., Kneer, F., and Mattig, W. 1983, "The chromosphere above sunspot umbrae. IV. Frequency analysis of umbral oscillations," *Astron. Astrophys.*, **123**, 263.

Zhugzhda, Y. D. 1984, "Resonance oscillations in sunspots and transformation of magneto-acoustic-gravity waves," *Mon. Not. Roy. Astron. Soc.*, **207**, 731.

Zhugzhda, Y. D. and Dzhalalov, N. S. 1984, "Magneto-acoustic-gravity waves on the sun III. The theory of running penumbral waves," *Astron. Astrophys.*, **133**, 333.

Zhugzhda, Y. D., Locans, V., and Staude, J. 1983, "Seismology of sunspot atmospheres," *Solar Phys.*, **82**, 369.

__________ 1985, "Oscillations in the chromosphere and transition region above sunspot umbrae. A photospheric or a chromospheric resonator?," *Astron. Astrophys.*, **143**, 201.

__________ 1987, "The interpretation of oscillations in sunspot umbrae," *Astron. Nachr.*, **308**, 257.

Zhugzhda, Y. D. and Makarov, V. I. 1982, "On the dynamics of the chromosphere above sunspots," *Solar Phy.*, **81**, 245.

Zirin, H. and Stein, A. 1972, "Observations of running penumbral waves," *Ap. J.* (Letters), **178**, L85.

MAGNETOHYDRODYNAMIC WAVES IN STRUCTURED MAGNETIC FIELDS

B. ROBERTS
Department of Mathematical and Computational Sciences
University of St Andrews
St Andrews
Fife KY16 9SS
Scotland

ABSTRACT. The nature of magnetohydrodynamic waves in structured magnetic fields is discussed, with particular reference to sunspots. The modes of oscillation of an isolated flux tube are reviewed. We consider the nature of surface waves on an interface between two isothermal gases, one of which is embedded within a horizontal magnetic field.

1. Introduction

Sunspots must surely be the location *par excellence* in which to study magnetohydrodynamic phenomena and magnetic oscillations in particular. By virtue of their size they are easy to observe; they are relatively quiescent, yet dynamic enough to exhibit a wide range of phenomena; and they are strongly magnetic (with a low plasma β) in their upper layers, though only weakly magnetic (with a high β) in the layers below the visible surface. There are spots that are broadly circular in appearance, so as to conform with the theorists' elementary notions, yet –and more frequently–spots are also found in a variety of shapes, offering the opportunity for the theorist to test the geometrical aspects of his or her theories. They are observed to have oscillations that are quite distinct from those in the spot's environment, and yet sunspots are also observed to strongly interact with the sound waves in their surroundings. Indeed, spots are observed to absorbed well over half the flux of p-modes in their vicinity (Braun, Duvall and Labonte (1987,1988), Braun, Labonte and Duvall (1990)).

Despite these clear attractions, the modes of oscillation of a sunspot are not fully understood. The most notable reason for this is our lack of understanding of the precise equilibrium make-up of a spot's magnetic field, both in the visible layers and in the regions below. Is a spot essentially a single magnetic structure, a monolithic tube, albeit with some inhomogeneity? Or is a spot more properly considered to be an *ensemble* of individual tubes that are pressed together in the visible layers, though still retaining some of their identity, but are distinct and separated from one another in the layers below the solar surface (Parker (1979a))?

The sunspot, then, provides us with a marvellous example of a magnetically structured plasma through the study of which we may hope to understand a variety of magnetohydrodynamic phenomena. What are its modes of oscillation? The complex-

J. H. Thomas and N. O. Weiss (eds.), Sunspots: Theory and Observations, 303–331.

ity of the geometry of a spot, with some evidence of inhomogeneity within an umbra and plain evidence of inhomogeneity in the penumbra, has served as a block to the full development of a theory of sunspot oscillations. Ultimately, a detailed numerical simulation of the oscillations of a spot will be needed but at our present stage of understanding, with uncertainty in the very structure of a spot, such a development is for the future. Instead, we here examine the modes of oscillation of two simple structures: an isolated flux tube, representing the umbra of a spot, and a magnetic interface to represent its penumbra and field-free environment below. Complementary to this view are the studies of waves in *rigid* but stratified flux tubes (see, for example, Antia and Chitre (1979), Scheuer and Thomas (1981)) and *vertical* but unbounded magnetic fields (see, for example, Ferraro and Plumpton (1958), Zhugzhda, Staude and Locans (1984), Zhugzhda, Locans and Staude (1987), Moreno-Insertis and Spruit (1989)). The review by Chitre (1992) gives a detailed discussion of these aspects. These studies provide much insight into the complex array of modes that a spot may possess and, taken altogether, suggest how in the future we may progress by combining the various elements of the several complementary views so as to form a whole. For the present, though, such a construction has not been made and must await future developments.

Progress in our understanding of sunspots would, of course, be considerably accelerated if an observational study of its modes of oscillation and their interaction with the environment and its sea of p-modes could be used for diagnostic information. Such sunspot seismology (Thomas, Cram and Nye (1982), Zhugzhda, Locans and Staude (1983); see the reviews by Davila (1990) and Bogdan (1992)), if fully developed, could resolve our uncertainties about the sub-surface structure of sunspots, just as with helioseismology we have be able to learn about the sub-photospheric layers of the sun down to the base of the convection zone and perhaps below. Progress in this direction depends upon our understanding of the role of stratification on the modes of a sunspot residing in an environment which supports p-modes.

To begin our study of magnetically structured plasmas, recall that in a uniform unbounded medium there are principally three modes of oscillation open to a magnetized fluid: an incompressible, inaudible disturbance in which field-lines are vibrated transversely–this is the *Alfvén wave*–and two other disturbances, the *fast* and *slow magnetoacoustic waves*. The magnetoacoustic waves are compressional, arising from combinations of gas pressure and magnetic pressure forces. (For a recent review of the properties of these modes, see Roberts (1985).) Magnetic structuring seriously complicates our description of these modes, introducing new features not present in a uniform medium. Gravitational stratification, important in sunspots, adds further to the complexity of modes available to the plasma.

The simplest new feature that a structured magnetic field introduces is the possible occurrence of *surface waves*. Wherever there occurs a sharp change in the magnetic field (or, more importantly, the Alfvén speed), thereby defining a magnetic interface, there arises the possibility that a magnetic surface wave may run along that interface. In fact, depending upon circumstances, either one or two surface waves may occur; these are the so-called *fast* and *slow magnetoacoustic surface waves* (Roberts (1980; 1981a-c)). They are compressible, springing from the magnetoacoustic branch of

MHD waves. They are not related to the Alfvén wave. The term 'Alfvén surface wave', commonly used in the literature, is accordingly a misnomer.

As well as supporting surface waves, a structured field may reflect and refract magnetoacoustic waves. Refraction of sound waves, whereby a sound wave is bent away from the region of higher sound speed, is responsible for the cavity that gives rise to the distinctive parabolic ridges of p-modes in frequency-wavenumber space. Whether a similar effect is important in sunspots is not clear. In any case, observations show the existence of oscillations in sunspots with periods close to five-minutes, the characteristic period of p-modes. These modes are found only in the photospheric and sub-photospheric layers of a sunspot. By contrast, oscillations with periods around three-minutes (more precisely, in the range 100-200s) are found both in the photospheric umbra and in the chromosphere above. Observational evidence of sunspot oscillations is reviewed in Moore and Rabin (1985), Lites, White and Packman (1982) and Lites (1986a,b, 1988, 1992).

In the general case of a structured magnetic field, the magnetohydrodynamic equations possess *continuous spectra*. In fact, generally there are two continuous spectra, one asssociated with the Alfvén wave and the other with the slow magnetoacoustic wave; these modes share the property in a uniform medium of being unable to propagate across the magnetic field, and both have their speed of energy propagation–their group velocity–directed along (in the case of the Alfvén wave) or very close to (in the case of the slow wave) the direction of the applied magnetic field, whatever the direction of wave propagation. Continuous spectra signal an anisotropic behaviour in disturbances, with the component of motion perpendicular to the applied field behaving differently from the component in the direction of the field. Small spatial scales are rapidly built up in disturbances and efficient damping (with subsequent heating) likely to take place. In sunspots, such processes may occur on and close to the edges of spots (Hollweg (1988)), just as it may on the boundaries of small-scale intense flux tubes (Roberts (1986)). Similar effects may also take place throughout the body of a spot, if spots are strongly inhomogeneous (Ryutova and Persson (1984)). Certainly there is strong evidence for inhomogeneity, most notably in the occurrence of umbral dots and in the fibril structure of the penumbra. Recently, there have been several attempts to investigate the theoretical consequences of such inhomogeneities, leading to possible explanations of the observed p-mode absorption by sunspots (Hollweg (1988), Lou (1988,1990), Rosenthal (1990), Sakurai, Goossens and Hollweg (1991a,b), Chitre and Davila (1991), and Braun *et al.* (1991); see the reviews by Davila (1990) and Bogdan (1992)).

Next to the magnetic interface, the simplest example of a structured magnetic field, of central importance in solar physics, is the isolated magnetic flux tube. Such a structure, consisting of a concentration of magnetic field–typically assumed to be a cylinder or slab of magnetic flux–confined by a field-free external gas, acts as an elastic wave guide for magnetoacoustic waves (see Roberts (1990a,b)). The precise nature of the modes that the tube supports depends upon the particular ordering of sound and Alfvén speeds that pertain. In the stratified atmosphere of a sunspot, this ordering changes with height, which has implications for the modes that are ducted by the tube (Evans and Roberts (1990a)). It turns out that the slow mode is robust, in that it is a guided mode of the tube both where the field is relatively weak (as in the layers below the surface of a spot) and where it is strong (as in the uppers reaches of a spot). This consideration, taken together with other aspects, led Evans and Roberts (1990a) to view the three-minute umbral oscillation as essentially a slow magnetoacoustic wave.

2. Magnetohydrodynamic Waves

Consider a basic state in which there is an applied unidirectional magnetic field $B_0\hat{\mathbf{z}}$ aligned with the z-axis in an appropriate coordinate system. The equilibrium gas density ρ_0, temperature T_0, pressure p_0, and field strength B_0 may all vary in a direction perpendicular to the applied magnetic field. This treatment includes, of course, the more elementary case of a *uniform* medium (for which B_0, p_0, T_0 and ρ_0 are simply constants).

The governing system of differential equations are the usual ones of ideal magnetohydrodynamics (e.g., Cowling (1976)). We consider isentropic disturbances. Denote by p, ρ and T the pressure, density and temperature perturbations of an ideal gas with ratio of specific heats γ and magnetic permeability μ. The perturbed magnetic (induction) field is $\mathbf{B}$ and the fluid motion is $\mathbf{v}$.

To begin with, we will ignore the effects of stratification. This simplification allows us to describe rather fully the basic modes of oscillation of a flux tube (see Section 3) without the attendant complexities of a gravitationally stratified atmosphere. Of course, in the real atmosphere of a sunspot the complexities of stratification are everywhere present. Nonetheless, it is of pedagogical value to explore the simpler (though still complicated) problem of an unstratified but magnetically structured object.

In the absence of gravity, the equilibrium state of a structured unidirectional magnetic field is one of constant total hydromagnetic pressure:

$$\mathrm{grad}\left(p_0 + \frac{B_0^2}{2\mu}\right) = 0. \tag{1}$$

To examine small amplitude (linear) disturbances about this equilibrium, it proves convenient to introduce the perturbation in total pressure, p_T:

$$p_T \equiv p + \frac{1}{\mu}\mathbf{B}_0 \cdot \mathbf{B}. \tag{2}$$

Then, splitting the velocity field $\mathbf{v}$ into components perpendicular and parallel to the applied magnetic field, the linearized equations of ideal magnetohydrodynamics yield

$$\rho_0\left(\frac{\partial^2}{\partial t^2} - c_A^2\frac{\partial^2}{\partial z^2}\right)\mathbf{v}_\perp + \nabla_\perp\left(\frac{\partial p_T}{\partial t}\right) = 0, \tag{3}$$

$$\rho_0\left(\frac{\partial^2}{\partial t^2} - c_t^2\frac{\partial^2}{\partial z^2}\right)v_z + \left(\frac{c_s^2}{c_f^2}\right)\frac{\partial}{\partial z}\left(\frac{\partial p_T}{\partial t}\right) = 0, \tag{4}$$

where pressure variations are related to the flow through

$$\frac{\partial p_T}{\partial t} = \rho_0 c_A^2\frac{\partial v_z}{\partial z} - \rho_0 c_f^2 \mathrm{div}\ \mathbf{v}. \tag{5}$$

Here we have written $\mathbf{v} = \mathbf{v}_\perp + v_z\hat{\mathbf{z}}$, and $\nabla_\perp$ denotes the component of the gradient operator that is perpendicular to the direction of the applied magnetic field (i.e., perpendicular to the z-axis). The four characteristic speeds, c_s, c_A, c_t and c_f, appearing in the above are defined as follows:

$$c_s = \left(\frac{\gamma p_0}{\rho_0}\right)^{1/2}, \qquad c_A = \left(\frac{B_0^2}{\mu\rho_0}\right)^{1/2},$$

$$c_f^2 = c_s^2 + c_A^2, \qquad c_t^{-2} = c_s^{-2} + c_A^{-2}. \tag{6}$$

These are the sound speed c_s, the Alfvén speed c_A, the slow speed c_t, and the fast speed c_f. Each speed is a function of the coordinate perpendicular to the applied magnetic field. The slow speed c_t, which is both sub-sonic and sub-Alfvénic, is also commonly referred to as the *tube* speed (because of its intimate association with certain tube waves) or the *cusp* speed (because of its association with the cusp point in the group velocity diagram of the slow magnetoacoustic wave).

It may be noted that equations (4) and (5) combine to give

$$\frac{\partial^2 v_z}{\partial t^2} = c_s^2\frac{\partial}{\partial z}(\text{div }\mathbf{v}). \tag{7}$$

Before going further it is perhaps useful to pause at this point and examine the reduction of the above equations in the case of a uniform medium. For a *uniform* medium, equations (3)-(5) reduce to the familiar result (Lighthill (1960))

$$\frac{\partial^4\triangle}{\partial t^4} - (c_s^2 + c_A^2)\frac{\partial^2}{\partial t^2}\nabla^2\triangle + c_s^2c_A^2\frac{\partial^2}{\partial z^2}\nabla^2\triangle = 0, \tag{8}$$

where $\triangle \equiv \text{div }\mathbf{v}$. Applied to a Cartesian coordinate system (x, y, z), Fourier analysis then yields the dispersion relation

$$\omega^4 - (c_s^2 + c_A^2)k^2\omega^2 + c_s^2c_A^2k_z^2k^2 = 0, \tag{9}$$

for wavevector $\mathbf{k} = (k_x, k_y, k_z)$ with magnitude $k = (k_x^2 + k_y^2 + k_z^2)^{1/2}$.

The dispersion relation (9) applies to modes that are compressive ($\triangle \neq 0$); these are the *magnetoacoustic waves*, designated as *slow* or *fast* in accordance with the relative magnitude of the two roots (in ω^2) of the dispersion relation (9). If, however, we consider the solution $\triangle = 0$ of equation (8), then from equation (7) it follows that $v_z = 0$ and so motions are transverse to the applied field. Furthermore, by equation (5), these incompressible disturbances have $p_T = 0$, and obey the dispersion relation (following from equation (3))

$$\omega^2 = k_\parallel^2c_A^2, \tag{10}$$

where $k_\parallel = \mathbf{k}\cdot\hat{\mathbf{B}}_0 (= k_z)$ is the component of the wavevector $\mathbf{k}$ in the direction of the applied magnetic field. These are the *Alfvén waves.*

Returning to the general equations (3)-(5), we consider now their form in a cylindrical geometry, appropriate for a tube. In a cylindrical coordinate system (r, θ, z), we may write equation (3) in the component form

$$\rho_0(r)\left(\frac{\partial^2}{\partial t^2} - c_A^2(r)\frac{\partial^2}{\partial z^2}\right)v_r + \frac{\partial^2 p_T}{\partial r\partial t} = 0, \tag{11}$$

$$\rho_0(r)\left(\frac{\partial^2}{\partial t^2} - c_A^2(r)\frac{\partial^2}{\partial z^2}\right)v_\theta + \frac{1}{r}\frac{\partial^2 p_T}{\partial\theta\partial t} = 0. \tag{12}$$

There is a simple solution of equations (4), (11) and (12) that we may note immediately. If $\partial/\partial\theta = 0$, corresponding to symmetric disturbances of the tube, then equation (12) reduces to

$$\frac{\partial^2 v_\theta}{\partial t^2} = c_A^2(r)\frac{\partial^2 v_\theta}{\partial z^2}, \tag{13}$$

showing that a *torsional* oscillation v_θ propagates as an Alfvén wave (see Spruit (1982), Parker (1983) and Hollweg (1986, 1990a,b, 1991)), the disturbance moving with the local value of the radially variable Alfvén speed $c_A(r)$. Equation (13) illustrates the phenomenon of *phase mixing*, whereby a wave rapidly becomes of large amplitude and highly sheared as it adjusts to the variation in the basic speed of propagation (see Heyvaerts and Priest (1983), Nocera, Leroy and Priest (1984), Hollweg (1986, 1990a), Roberts (1988) and Parker (1991a,b) for recent discussions). The D'Alembert solution of equation (13) is

$$v_\theta(r,\theta,z,t) = f_+(r, z + c_A(r)t) + f_-(r, z - c_A(r)t), \tag{14}$$

where $f_\pm(r,\xi)$ are arbitrary functions of two variables (r,ξ) and are determined by initial conditions.

It is clear from equation (14) that radial gradients in v_θ grow secularly in time t. To be specific, suppose we have a propagating wave of the form

$$v_\theta = \sin[z - c_A(r)t]. \tag{15}$$

Then

$$\frac{\partial v_\theta}{\partial r} = -(\frac{dc_A}{dr})t\cos[z - c_A(r)t], \tag{16}$$

which grows secularly on a time-scale of $(dc_A/dr)^{-1}$. This can be very short. For example, on the edge of a spot where the Alfvén speed changes from perhaps 10 km s^{-1} in the spot to essentially zero outside it, over a distance of (say) 100 km, this time-scale is 10 seconds. We may expect, then, that such modes, if generated, would rapidly be dissipated by non-adiabatic processes. This suggests that the edges of sunspots ought to be excessively hot, producing a bright ring (as suggested long ago by Parker (1974) for quite different reasons); however, no such bright ring is observed. Presumably, in reality sunspot conditions are not sufficiently symmetric as to permit a mode that is independent of θ, as envisaged here, or there is too little energy flux in such waves to have any appreciable effect on the observable layers of a spot.

Returning to the general case of disturbances in a cylindrical geometry, with $B_0 = B_0(r)$ and $\rho_0 = \rho_0(r)$, it is convenient to Fourier analyze thus:

$$v_r(r,\theta,z,t) = v_r(r)\exp i(\omega t + n\theta - k_\parallel z) \tag{17}$$

for mode number $n = 0, 1, 2, \ldots$. Then equations (4), (11) and (12) yield

$$\frac{d}{dr}\left\{\frac{\rho_0(r)(k_\parallel^2 c_A^2(r) - \omega^2)}{(m^2(r) + n^2/r^2)}\,\frac{1}{r}\,\frac{d}{dr}(rv_r)\right\} = \rho_0(r)(k_\parallel^2 c_A^2(r) - \omega^2)v_r, \tag{18}$$

with m^2 given by

$$m^2(r) = \frac{(k_\parallel^2 c_s^2 - \omega^2)(k_\parallel^2 c_A^2 - \omega^2)}{(c_s^2 + c_A^2)(k_\parallel^2 c_t^2 - \omega^2)}. \tag{19}$$

Note that m^2 may be positive or negative. Equation (18) is sometimes referred to as the *Hain-Lüst equation*, since an equation of related form first appeared in a paper by Hain and Lüst (1958) dealing with considerations of stability of a cylindrical tube. In fact, Hain and Lüst included a twisted field ($B_\theta \neq 0$) in the equilibrium state, and this complicates the appearance of the equation; more-over, they did not exhibit their equation in an attractive and useful form. More recent discussions of this equation include Goedbloed and Hagenbeuk (1972), Goedbloed (1983), Edwin and Roberts (1983), Roberts (1986,1990a-c, 1991b) and Goossens (1991).

The differential equation satisfied by p_T is also of interest:

$$\rho_0(r)(k_\parallel^2 c_A^2(r) - \omega^2)\frac{1}{r}\frac{d}{dr}\left\{\frac{1}{\rho_0(r)(k_\parallel^2 c_A^2(r) - \omega^2)} r\frac{dp_T}{dr}\right\} = \left(m^2(r) + \frac{n^2}{r^2}\right)p_T. \tag{20}$$

Observe that equations (18) and (20) possesses singularities at $\omega^2 = k_\parallel^2 c_A^2$ and $\omega^2 = k_\parallel^2 c_t^2$; these singularities generate the *Alfvén continuum* and the *slow continuum*, respectively. (The presence of singularities is in fact clearer from the pair of first order equations that under-lie (18) and (20); see Appert, Gruber and Vaclavik (1974), Goedbloed (1983) and Adam (1982, 1986)). The possibility of heating the active corona and coronal holes (Parker (1987), Browning (1991)) are commonly explored in terms of these equations (see Davila (1991), Hollweg (1990a,b, 1991), Poedts, Goossens and Kerner (1990), Goossens (1991), and Parker (1991a,b)).

3. Waves In A Magnetic Flux Tube

We are here interested in the waves in an isolated flux tube. Accordingly, then, we consider an equilibrium with

$$B_0(r) = \begin{cases} B_0, & r < a, \\ 0, & r > a, \end{cases} \qquad \rho_0(r) = \begin{cases} \rho_0, & r < a, \\ \rho_e, & r > a, \end{cases} \tag{21}$$

corresponding to a uniform tube of radius a embedded in a field-free gas. See Figure 1. The density and also the temperature inside the tube are in general different from those in the tube's environment. The sound speed in the tube's environment is denoted by c_e, that in the tube by c_s.

Since the media both inside and outside the tube are uniform, we may simplify the basic equations (18) and (20). In particular, for a medium in which ρ_0, c_A and m are constants equation (20) is simply Bessel's equation,

$$\frac{d^2 p_T}{dr^2} + \frac{1}{r}\frac{dp_T}{dr} - \left(\frac{n^2}{r^2} + m^2\right)p_T = 0. \tag{22}$$

Solving equation (22) for the region inside the tube gives

$$p_T = A_0 I_n(m_0 r), \qquad r < a. \tag{23}$$

Here I_n denotes a modified Bessel function of order n, and m_0^2 is $m^2(r)$ (see equation (19)) evaluated within the tube:

$$m_0^2 = \frac{(k_\parallel^2 c_A^2 - \omega^2)(k_\parallel^2 c_s^2 - \omega^2)}{(c_s^2 + c_A^2)(k_\parallel^2 c_t^2 - \omega^2)}. \tag{24}$$

There is no restriction on the sign of m_0^2; modes with $m_0^2 > 0$ have an *evanescent* (non-propagating) character, whereas modes with $m_0^2 < 0$ (for which I_n may be

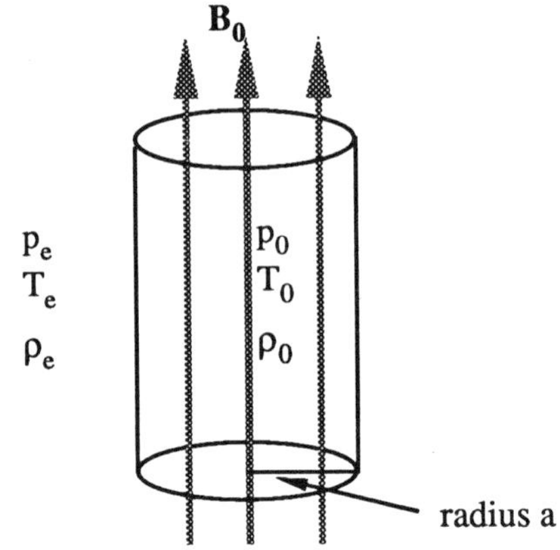

Figure 1. The equilibrium configuration of an isolated magnetic flux tube.

replaced by the Bessel function J_n, applied to argument $|m_0|r$) have an *oscillatory* character. We refer to modes with $m_0^2 > 0$ as *surface* waves, and those with $m_0^2 < 0$ as *body* waves (Roberts (1980, 1981a-c)).

The environment of an isolated flux tube is field-free and so governed by equation (22) with $c_A = 0$ and p_T now simply the perturbed gas pressure. Requiring that p_T declines to zero far from the tube, we select the modified Bessel function K_n as our solution of equation (22):

$$p_T = A_e K_n(m_e r), \quad r > a, \tag{25}$$

where

$$m_e^2 = k_\parallel^2 - \frac{\omega^2}{c_e^2} \tag{26}$$

is the value of m^2 in the environment of the tube. The requirement of a declining pressure field outside the tube means that we are imposing the constraint $m_e^2 > 0$, and so $\omega^2 < k_\parallel^2 c_e^2$.

The solutions (23) and (25) are matched by requiring that the radial velocity component v_r and the pressure perturbation p_T be continuous across $r = a$. The result is the dispersion relation (McKenzie (1970), Roberts and Webb (1978, 1979), Wilson (1980), Spruit (1982), Edwin and Roberts (1983), Cally (1985,1986), Abdelatif (1988), Evans and Roberts (1990a))

$$\rho_e \omega^2 m_0 \frac{I_n'(m_0 a)}{I_n(m_0 a)} + \rho_0 (k_\parallel^2 c_A^2 - \omega^2) m_e \frac{K_n'(m_e a)}{K_n(m_e a)} = 0, \tag{27}$$

where the dash denotes the derivative of the Bessel function ($I_n'(x) \equiv dI_n(x)/dx$, etc.).

The form of the dispersion relation (27) is convenient for surface waves ($m_0^2 > 0$).

For body modes, it is more convenient to use

$$\rho_e \omega^2 n_0 \frac{J_n'(n_0 a)}{J_n(n_0 a)} + \rho_0 (k_\parallel^2 c_A^2 - \omega^2) m_e \frac{K_n'(m_e a)}{K_n(m_e a)} = 0, \tag{28}$$

where $n_0^2 = -m_0^2$.

We should note that dispersion relations of structurally similar form arise in the discussion of the stability of cylindrical magnetic fields surrounded by a vacuum within which is a twisted magnetic field (Tayler (1957)). Also, waves of much the same kind arise in the study of astrophysical jets, with of course the added complication of a flow within the tube (jet) (see, for example, Roberts (1987), Bodo *et al.*(1989)).

Notice that the various body modes of oscillation of a tube, described by the integer $n(= 0, 1, 2, ...)$, give rise to patterns of oscillation across the circular cross-section of the tube. These patterns of oscillation–known as Chladni figures (see Rayleigh (1877) and Coulson (1955)) –are similar to the modal patterns on a sphere that helioseismologists construct. The patterns simply delineate regions of motion that are towards the observer (i.e., parallel to the tube axis) from those that are directed away (i.e., anti-parallel to the tube). It is tempting to associate such patterns with the observed cells of oscillation in the umbra of a spot. If such an association is correct, then it suggests that such cells of oscillation are not indicative of any particular underlying inhomogeneous structure within a spot but rather are simply a reflection of an essentially random excitation process which, for a short time, just happens to pick out and accentuate a particular array of modes.

The solutions of dispersion relations (27) and (28) are displayed in Figure 2, based upon the investigations of Edwin and Roberts (1983), Evans (1990) and Evans and Roberts (1990a). The figure has been computed for two possible orderings in the basic speeds, namely for (a) $c_e > c_s > c_A$ and for (b) $c_A > c_e > c_s$. Case (a) is appropriate for the sub-photospheric layers of a spot and case (b) for the upper reaches of a spot.

There are a number of conclusions to be drawn from Figure 2. Firstly, we note the ubiquitous nature of the slow body waves, which arise for both parameter cases discussed here and have phase-speeds lying between c_t and min (c_s, c_A). These results suggest that the slow mode, being strongly field-aligned, is able to adjust to changes in tube geometry (such as tube flaring) and thus likely to be present in more complicated configurations arising when the effects of stratification are allowed for. This simple observation lead Evans and Roberts (1990a) to identify the observed 3-minute oscillation of the sunspot umbra with the slow mode. The 5-minute oscillation, which is observed to exist in the photospheric layers of a spot but not in the higher chromospheric layers, is then identified in Figure 2 with the fast body modes. The fast body modes arise in case (a), appropriate for the lower regions of a spot, and have phase-speeds $\omega/k_\parallel$ between c_s and c_e. The fast body modes, which in case (a) are channelled by the tube, are unable to exist in the upper parts of a spot (case (b)) but instead either leak from the tube into the environment or convert to fast surface waves. Thus, in this picture the p-modes of the non-magnetic sun penetrate into the lower reaches of a spot where they become fast body modes; these modes are then trapped in the photosphere and below.

Evidently, for the same longitudinal wavenumber $k_\parallel$, the phase-speed of a fast wave is greater than that of a slow wave. Thus, the identification of slow modes with the 3-minute oscillation and the fast modes with the 5-minute one requires that slow

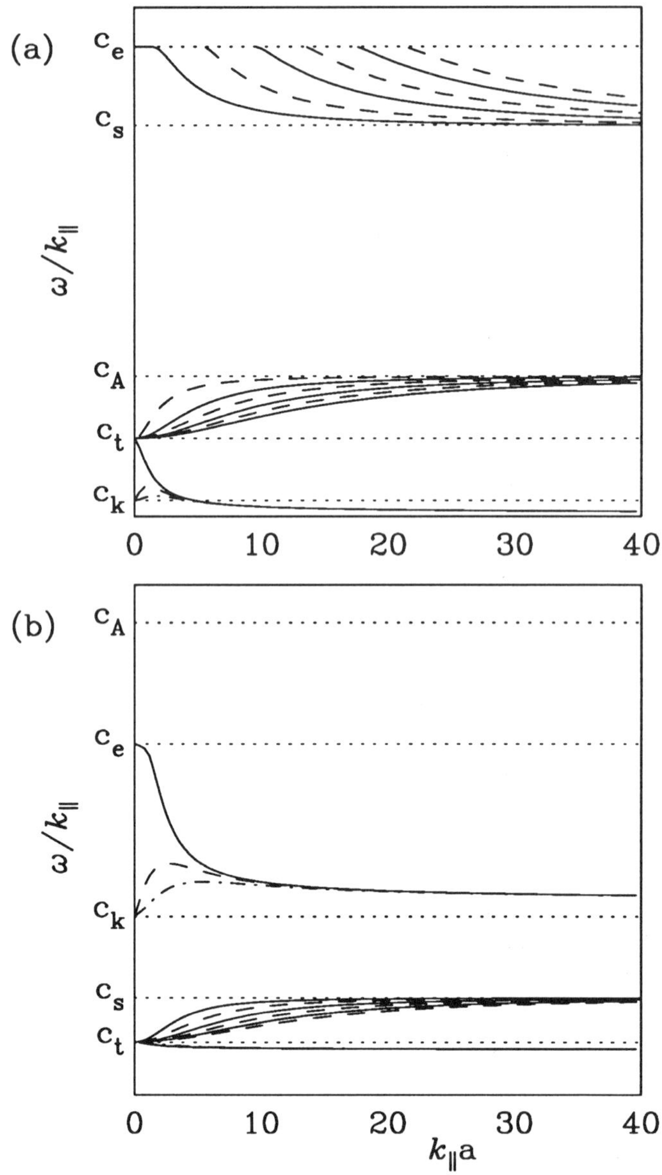

Figure 2. The phase-speed $\omega/k_\parallel$ as a function of $ak_\parallel$ for waves in an isolated magnetic flux tube, illustrating the various modes that arise for two orderings of the basic speeds, namely (a) $c_e > c_s > c_A$ and (b) $c_A > c_e > c_s$. Full curves are sausage ($n = 0$) modes and dashed curves are kink ($n = 1$) modes. Some fluting modes ($n = 2$) are also shown (as dot-dashed curves). (After Evans (1990); see also Edwin and Roberts (1983) and Evans and Roberts (1990a).)

modes have higher wavenumbers $k_{\|}$ (and so lower longitudinal wavelengths) than those generated in the fast mode. This suggests that the 3-minute mode is generated at shorter wavelengths than the 5-minute mode. Consistent with this ordering in wavelength is the suggestion that the 5-minute mode of a sunspot is driven by the p-modes of the quiet sun and that the 3-minute mode is driven by overstability within the spot (see the discussion in Evans and Roberts(1990a)).

Besides body waves, Figure 2 shows the occurrence of surface waves (which run along the sides of a flux tube). In case (a) only a slow surface wave (with phase-speed below c_t) can propagate without leakage, but where the field is stronger (as in case (b)) both a fast and a slow surface wave occur. The fast surface waves have phase-speeds between c_s and c_e; the slow surface waves have phase-speeds below c_t. What happens to these surface waves when a flux tube flairs out, as in a sunspot? Presumably, being strongly field-guided, the surface wave or waves would simply run round the flairing edge of the tube, though perhaps suffering some leakage in the process. In which case, surface waves generated on the sub-photospheric edges of a spot would run upwards and downwards along the field; upwardly propagating surface waves would, guided by the field-lines, eventually run outwards along the outermost field-lines and propagate almost horizontally along the lower penumbral boundary of the spot. We return to this topic in the next section.

Note in Figure 2 the occurrence of a speed c_k, describing the kink mode of oscillation of a tube; the speed c_k is defined by

$$c_k = \left(\frac{\rho_0}{\rho_0 + \rho_e}\right)^{\frac{1}{2}} c_A, \tag{29}$$

and so is sub-Alfvénic. In the case of a thin tube, the vibration of the tube disturbs the surroundings by about as much as the tube itself, giving rise to the enhanced inertia represented by $(\rho_0 + \rho_e)$; see Ryutov and Ryutova (1976), Parker (1979b) and Spruit (1981). In the case of a tube of arbitrary thickness, represented by cases (a) and (b) above, the kink mode gains some dispersion. The speed c_k arises also in the description of surface waves (see Section 4).

We turn now to a brief consideration of what happens if a magnetic flux tube is embedded in a strong magnetic field, rather than in a field-free atmosphere as we have so far discussed. In fact, this is the circumstances arising in coronal tubes (loops) and has been extensively investigated in Edwin and Roberts (1983) and Roberts, Edwin and Benz (1984). It is of relevance to sunspots if we regard umbral dots as tubes of magnetic field within which the Alfvén speed is locally depressed below its value in the umbra surroundings. Similarly, fibril waves in the penumbra of a spot may be described from the view-point of a magnetic flux tube in a strongly magnetized environment if we represent penumbral fibrils as local tube-like depressions in Alfvén speed. The point is that a region of reduced Alfvén speed, be it a coronal loop, an umbral dot or a penumbral fibril, acts as a wave-guide for fast magnetoacoustic body waves, and these body waves are closely analogous to Pekeris (sound) waves in ocean layers, Love (elastic) waves in seismology, and the electromagnetic modes of a fibre optic (Edwin and Roberts (1983), Roberts, Edwin and Benz (1984)). Additional to the fast body modes, the ubiquitous slow body waves are also present. In the circumstances where the Alfvén speed c_{Ae} in the magnetic environment of a tube (umbral dot, penumbral fibril) is in excess of the Alfvén speed c_A within the tube, with both Alfvén speeds exceeding the sound speeds, then fast body waves exist with phase-speeds $\omega/k_{\|}$ between c_A and c_{Ae}. These modes are simply fast magnetoacoustic waves which bounce along the inside of the flux tube, which thus forms a wave-guide; the propagation is strongly dispersive. The fast modes would leak from the wave guide if $c_{Ae} < c_A$. The slow body waves have phase-speeds between c_t and the sound speed c_s (for the case $c_e < c_s$). The fast and slow body waves are attractive candidates for explaining H_α fibril waves (Edwin and Roberts (1983)), but little has been done to pursue the case.

So far in our discussion of the modes of oscillation of a tube we have ignored the effects of gravitational stratification. In fact, stratification is a principal ingredient in the recipe of sunspot oscillations, for in the visible layers of a spot the density scale height reduces to the order of 100 km and so is comparable with the scale of the smallest observed features of a spot and far below the diameter of a spot. It would be logical, then, for us now to describe the effects of gravity on the tube modes we have catalogued. Unfortunately, it has not yet proved possible to carry out such a study on a stratified but wide tube. Nontheless, we can approach our problem from a different side, seeing various aspects of it but not as yet viewing the whole. To begin with, note that gravity adds considerably to the *equilibrium* structure of a spot. A magnetic flux tube has a natural tendency to fan out with height as it responds to the fall-off in the now stratified external gas pressure that confines the tube. Only by imposing an artificial temperature difference between the tube and its surroundings, or by containing the tube within rigid walls, can we circumvent this effect. The splaying of a stratified flux tube with height is most conveniently demonstrated in the case of a *thin* flux tube. This case is of direct concern to a description of the intense magnetic flux tubes that make up over 90% of the photospheric magnetic flux outside of sunspots. But it is possibly also a guide to how the modes of a stratified flux tube of arbitrary size may be described. Waves in thin flux tubes have been reviewed elsewhere in some detail (see Spruit and Roberts (1983), Thomas (1985), Roberts (1986, 1990a,b), Hollweg (1990b) and Ryutova (1990a,b)) so we will here keep our remarks to a minimum.

A thin magnetic flux tube supports sausage and kink modes which are described by the Klein-Gordon equation (Roberts (1981c), Rae and Roberts (1982)). The dispersion relation associated with the Klein-Gordon equation is

$$\omega^2 = k_{\|}^2 c^2 + \Omega_{tube}^2, \tag{30}$$

where Ω_{tube} is a cutoff frequency and c a propagation speed of the mode in question. For the sausage mode, $c = c_t$ and Ω_{tube} is close to the familiar acoustic cutoff frequency of a sound wave; for the kink mode, $c = c_k$ and Ω_{tube} is about a third of the acoustic cutoff (Spruit and Roberts (1983)). In the absence of stratification, $\Omega_{tube} = 0$.

We see from equation (30) that high frequencies propagate much as in the absence of gravity but low frequencies lead to evanescence (non-propagation). If *impulsively excited*, then a tube mode propagates with the speed c but trails a *wake* which oscillates with the frequency Ω_{tube} (Rae and Roberts (1982)). The amplitude of the mode grows with height as the wave propagates higher into the atmosphere. Consequently, at some stage *shocks* may be expected to form; this would occur as a *series* of shocks as first the wave front forms a shock and then the trailing wake adds its series of shocks of diminishing strength. The picture is much as envisaged by Hollweg (1982) for the spicule phenomenon. Kalkofen (1990) has argued that much the same type of event, a shock followed by a sequence (typically three) of weaker trailing shocks, is present in the chromospheric network.

What evidence is there of such shock phenomenon in sunspots? One immediately thinks of *umbral flashes*, the first detected oscillatory phenomenon in sunspots (Beckers and Tallant (1969), Wittmann (1969), Thomas, Cram and Nye (1984), Lites (1986a,b)). It is perhaps natural, then, to associate umbra flashes with slow waves which have steepened to form shocks and trains of shocks.

4. Surface Waves In A Gravitationally Stratified Atmosphere

We have seen that surface waves may propagate on the edges of a magnetic flux tube. The magnetic field of a sunspot flairs out with height as the field expands in response to the fall-off with height of the confining external gas pressure. This suggests that surface waves may run along the lower penumbral magnetic boundary of a spot. Could such surface waves be the observed running penumbral wave (Small and Roberts (1984))? Running penumbral waves were first seen by Giovanelli (1972) and Zirin and Stein (1972), and have the appearance of propagating wave-fronts which travel outwards in the penumbra (see Moore and Rabin (1985) and Lites (1988, 1992) for recent reviews).

To discuss surface waves on a penumbral magnetic field we consider the MHD equations as applied to a horizontal magnetic field $B_0(z)\hat{\mathbf{x}}$ aligned with the x-axis of a Cartesian coordinate system. We include the effect of gravitational stratification.

Gravity, like magnetic structuring, imposes a preferred direction and a spatial scale (the atmosphere's scale-height) on all wave modes of the atmosphere. Consequently, in the presence of both magnetic structuring and gravitational stratification, a complicated interplay arises between these two elements (Roberts 1980, 1981a-c, 1985, 1991a,b). Again, dispersion arises and stratification in the sound speed introduces the effect of wave refraction, so important in the physics of p-modes. Added to this, is the importance of wave cutoff. Thus, taken altogether, a description of the MHD modes in a magnetically structured, gravitationally stratified atmosphere is a complex business with a rich variety of phenomena on display.

Consider, then, a horizontal magnetic field, the strength of which may, in general, vary with depth z. In the presence of gravity $\mathbf{g} = g\hat{\mathbf{z}}$, the equilibrium in the gas is achieved by satisfying the equation

$$\frac{d}{dz}\left(p_0(z) + \frac{B_0^2(z)}{2\mu}\right) = g\rho_0(z), \tag{31}$$

where $g(= 274 \text{ m s}^{-2}$ in the solar atmosphere) is the local gravitational acceleration. We have adopted a coordinate system (x, y, z) with the z-axis pointing downwards. See Figure 3.

Consider two-dimensional disturbances of the form

$$\mathbf{v} = (v_x(z), 0, v_z(z))\exp[i(\omega t - k_\parallel x)] \tag{32}$$

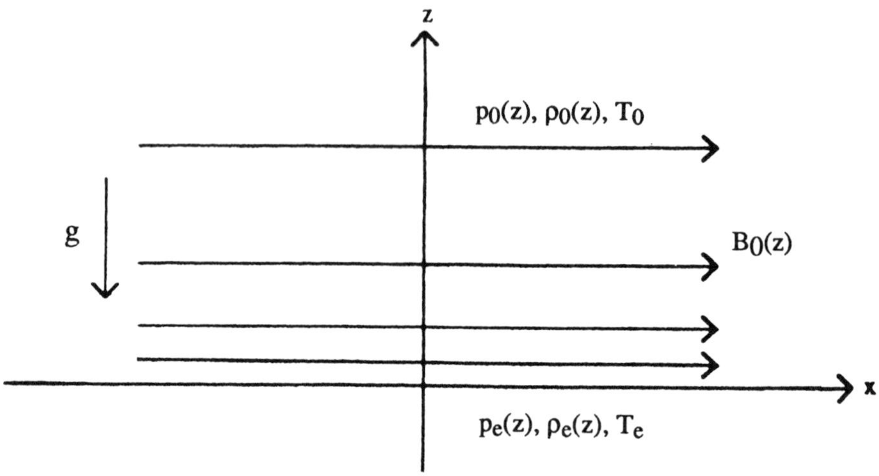

Figure 3. The equilibrium configuration of a magnetic interface between two isothermal atmospheres.

for horizontal wavenumber $k_{\|}$. The restriction to the case of propagation parallel to the applied magnetic field decouples the Alfvén wave from the system; for convenience, we have taken $v_y = 0$ and so the Alfvén wave is removed from consideration. Then, the governing second order ordinary differential equation (Adam (1977), Roberts (1985); cf. Goedbloed (1971)) is

$$\begin{aligned} &\frac{d}{dz}\left[\rho_0(z)(c_s^2+c_A^2)\left(\frac{\omega^2-k_{\|}^2c_t^2}{\omega^2-k_{\|}^2c_s^2}\right)\frac{dv_z}{dz}\right] \\ &= \left[\frac{\rho_0 g^2 k_{\|}^2}{\omega^2-k_{\|}^2c_s^2}-\rho_0(\omega^2-k_{\|}^2c_A^2)-gk_{\|}^2\frac{d}{dz}\left(\frac{\rho_0 c_s^2}{\omega^2-k_{\|}^2c_s^2}\right)\right]v_z. \end{aligned} \tag{33}$$

Notice that although we have now lost the Alfvén continuum (because of our assumptions of parallel propagation and $v_y = 0$) we still have the slow mode one.

Suppose now that at some location within the stratified atmosphere there is a discontinuity in a physical variable, so that the sound speed, the Alfvén speed, or the density changes discontinuously at that location, thereby defining an interface. Across such an interface the velocity component v_z must be continuous. Additionally, it follows from equation (33) that

$$\rho_0(z)(c_s^2+c_A^2)\left(\frac{\omega^2-k_{\|}^2c_t^2}{\omega^2-k_{\|}^2c_s^2}\right)\frac{dv_z}{dz}+g\rho_0(z)\left(\frac{k_{\|}^2c_s^2}{\omega^2-k_{\|}^2c_s^2}\right)v_z$$

must be continuous also. This is equivalent to the requirement that $p_T + g\rho_0(z)\zeta$ be continuous, where $z = \zeta(x,t)$ is the vertical displacement of the interface from its equilibrium place.

We are interested in the application of the above to the propagation of surface waves on the lower boundary of a sunspot's penumbral magnetic field. To set the scene, though, it is instructive to examine the incompressible limit of our equations, even though this is not directly applicable to the compressible medium of a penumbra. We can obtain the incompressible form of our equations directly from the above by taking the limit $c_s \to \infty$. Thus, in the *incompressible* case the governing differential equation is

$$\frac{d}{dz}\left[\rho_0(z)(\omega^2 - k_\parallel^2 c_A^2(z))\frac{dv_z}{dz}\right] - k_\parallel^2\left[\rho_0(z)(\omega^2 - k_\parallel^2 c_A^2(z)) - g\frac{d\rho_0}{dz}\right]v_z = 0. \tag{34}$$

The structure of equation (34), complicated by the occurrence of a regular singularity at locations where $\omega^2 = k_\parallel^2 c_A^2(z)$, has not been fully explored. The equation has a *slow* mode continuum (albeit in the limit of an incompressible medium, for which $c_s \to \infty$). However, the special case of zero gravity, which is analogous to similar equations arising in plasma physics (Uberoi (1972)), has attracted considerable attention (e.g., Goedbloed (1971, 1984), Sedlacek (1971), Tataronis and Grossmann (1973), Grossmann and Tataronis (1973), Southwood (1974), Appert, Gruber and Vaclavik (1974), Chen and Hasegawa (1974), Ionson (1978), Rae and Roberts (1981), Lee and Roberts (1986), Davila (1987), Einaudi and Mok (1987), Hollweg (1987a,b), Grossmann and Smith (1988)). See Adam (1982, 1986), Goedbloed (1983), Hollweg (1990a), Goossens (1991) and Browning (1991) for reviews, the recent contributions giving special attention to solar applications. The interest in this equation springs from the phenomenon of *resonant absorption*, by which wave motions become highly anisotropic in the vicinity of a region of rapidly changing Alfvén speed. In such a region, local oscillations become highly sheared and a collective (surface wave) disturbance decays as energy is transferred into body wave motions. The process is an ideal one so there is no loss of energy from the system, only a transfer from one kind of motion to another. The conversion proceeds at a rate $\hat{\gamma}$; for a magnetic interface with one side field-free and with equal gas densities either side of the interface, this rate is given by (see Sedlacek (1971), Ionson (1978), Abels-van Maanen and Weenik (1979), Rae and Roberts (1981), Goedbloed (1983), Lee and Roberts (1986), Cally (1991))

$$\hat{\gamma} = \frac{\pi k_\parallel a}{8\sqrt{2}} k_\parallel c_A, \tag{35}$$

for a layer of thickness a (within which the squared Alfvén speed changes linearly from c_A^2 to zero). Of course, with the inclusion of dissipative processes this rapid build-up of highly sheared motions within layers of strongly inhomogeneous magnetic field makes dissipation particularly effective (see, for example, Ionson (1978), Einaudi and Mok (1987) and Poedts *et al.* (1990)). It may be noted that the compressible equivalent of equation (35) has not, to our knowledge, been obtained; such a problem raises the interesting question of the role of the square roots that enter through the definitions of m_0 and m_e.

Applying the idea of resonant absorption to sunspots, it seems likely that the edges of a spot and the lower penumbra boundary are likely to experience excessive heating from the dissipation of slow waves and Alfvén waves. Moreover, inhomogeneities within sunspots may equally be subject to such processes (Ryutova and Persson (1984)). The time-scale $\hat{\gamma}^{-1}$ for mode conversion under sunspot conditions is short. For the same sunspot values as we used in our illustration of phase mixing, namely a layer of thickness 100 km and an Alfvén speed of 10 km s^{-1}, we obtain a time-scale of order $(\lambda^2/10^4)$ seconds for a wavelength $\lambda(= 2\pi/k_\parallel)$ expressed in km. Thus, for a disturbance of wavelength 10^3 km (as observed in running penumbral waves) we obtain a time-scale of 10^2 seconds. The effect, then, is rapid, taking place in about one period of the wave!

Despite the theoretical attractions of resonant absorption, there is at present no direct observational evidence that this process occurs in sunspots to any appreciable degree. It may well be that such evidence can only be found with very high resolution observations of (say) the edges of spots, and such observations have not yet been made. There again, other physical effects–such as radiative processes–may completely mask any enhancement in dissipation of MHD waves in inhomogeneous regions of a sunspot, unless those waves are generated to a particularly large amplitude. However, one observation, the ability of spots to absorb p-modes in their vicinity (Braun *et al.*(1987, 1988, 1990)), raises the question of whether resonant absorption, due to internal inhomogeneity within a spot or to the strong inhomogeneity on the boundary of a spot, is responsible for sunspots appearing as sinks for sound waves in their neighbourhoods (see Hollweg (1988), Lou (1988, 1990), Chitre and Davila (1990), Sakurai *et al.*(1991a,b), and Braun *et al.*(1991)). Developments in this area are likely to shed light on this important question in the near future (see the discussion in Bogdan (1992)).

We return now to equation (34) with gravity non-zero and consider a fluid with constant density ρ_0 and uniform magnetic field B_0 (in $z < 0$) resting upon a field-free fluid of constant density ρ_e (in $z > 0$), so that

$$\rho_0(z) = \begin{cases} \rho_e, & z > 0, \\ \rho_0, & z < 0, \end{cases} \qquad c_A(z) = \begin{cases} 0, & z > 0, \\ c_A, & z < 0. \end{cases} \tag{36}$$

Then $z = 0$ is a current sheet (interface) across which we require v_z and

$$\rho_0(z)[(\omega^2 - k_\parallel^2 c_A^2(z))\frac{dv_z}{dz} + k_\parallel^2 g v_z]$$

to be continuous.

Now, in both $z > 0$ and $z < 0$, $d\rho_0(z)/dz$ is zero and c_A^2 is a constant (possibly zero) so equation (34) has solution of the form

$$v_z(z) = \begin{cases} A \ \exp(-|k_\parallel|z), & z > 0, \\ A \ \exp(|k_\parallel|z), & z < 0. \end{cases} \tag{37}$$

We have selected the solution that decays as $|z| \to \infty$, rejecting the exponentially growing solution.

The choice of the same arbitrary constant A in $z > 0$ and $z < 0$ ensures continuity of v_z across $z = 0$. Application of the second boundary condition then yields the familiar dispersion relation (Kruskal and Schwarzschild (1954), Chandrasekhar (1961))

$$\omega^2 = k_\parallel^2 c_k^2 + \left(\frac{\rho_e - \rho_0}{\rho_e + \rho_0}\right) g|k_\parallel|. \tag{38}$$

We thus see that surface waves on a uniform incompressible fluid in the presence of a uniform magnetic field confined to the upper region propagate *dispersively*: the phase-speed $\omega/k_\parallel$ of the wave is a function of $k_\parallel$. If a lighter fluid rests on top of a heavier fluid (so that $\rho_0 < \rho_e$), then dispersion results in a hydromagnetic surface wave that propagates faster at long wavelengths (small $k_{\parallel}$) than at short wavelengths (large $k_{\parallel}$). In any case, provided a lighter fluid rests upon a heavier fluid, the hydromagnetic surface wave propagates with a speed that is in excess of the mean Alfvén speed c_k. In the limit of arbitrarily large wavelength the phase-speed of the wave becomes arbitrarily large.

We may illustrate dispersion relation (38) for penumbral parameters. For example, taking $p_e = 2\times10^5$ dynes cm^{-2} as the pressure in the field-free region immediately below the penumbral field and a magnetic field strength of $B_0 = 1000$G within the penumbra, we obtain from pressure balance a penumbral gas pressure of 1.6×10^5 dynes cm^{-2}. Then, from the ideal gas law, we obtain a density ratio of $\rho_0/\rho_e = 24/25$ for temperatures of 6000K just below the penumbra and 5000K within the field. Thus, for a photospheric density of $\rho_e = 2 \times 10^{-7}$ gm cm^{-3} in the field-free region, we obtain a density of $\rho_0 = 1.92 \times 10^{-7}$ gm cm^{-3}, which yields an Alfvén speed c_A of 6.4 km s^{-1} and a mean Alfvén speed c_k of 4.6 km s^{-1}. Equation (38) produces a phase-speed of 6 km s^{-1} at a wavelength ($\equiv 2\pi/k_{\parallel}$) of 10^3 km, a speed of 13 km s^{-1} at a wavelength of 10^4 km, and a speed of 27 km s^{-1} at a wavelength of 5×10^4 km. These phase-speeds are reasonably in accordance with reported speeds of running penumbral waves.

However, the assumptions of incompressibility and uniformity that underlie the above results are too severe for us to apply them uncritically in the solar context. So what effect does compressibility have? The first observation to make is that for a gas *two* surface modes may propagate, a fast surface wave and a slow surface wave (Roberts (1980, 1981a,c), Miles and Roberts (1989), Jain and Roberts (1991a)). The assumption of incompressibility ($c_s \to \infty$) sends the fast surface off to infinity, leaving the slow surface mode.

To see this, consider for a moment the case of zero gravity. In a uniform compressible medium, with gravity ignored, equation (33) becomes

$$\frac{d^2 v_z}{dz^2} - m^2 v_z = 0, \tag{39}$$

with m^2 defined as in equation (19). Thus, for the equilibrium (36), we have

$$v_z(z) = \begin{cases} A \exp(-m_e z), & z > 0, \\ A \exp(m_0 z), & z < 0, \end{cases} \tag{40}$$

with m_0 and m_e given by equations (24) and (26). The densities either side of the interface are related by

$$\frac{\rho_e}{\rho_0} = \frac{c_s^2 + \frac{1}{2}\gamma c_A^2}{c_e^2}. \tag{41}$$

In writing solution (40) we have required m_0^2 and m_e^2 to be positive and selected the decaying solutions (with $m_0 > 0$ and $m_e > 0$). This requires that

$$\frac{\omega^2}{k_{\parallel}^2} < c_t^2, \quad \text{or} \quad \min(c_s^2, c_A^2) < \omega^2/k_{\parallel}^2 < \max(c_s^2, c_A^2),$$

with additionally $\omega^2 < k_{\parallel}^2 c_e^2$. The dispersion relation then follows on application of the continuity conditions on v_z and p_T (Wentzel (1979), Roberts (1980, 1981a-c),

Miles and Roberts (1989), Cadez and Okretic (1989), Jain and Roberts (1991a,b)):

$$\frac{\omega^2}{k_\|^2} = \frac{\rho_0 m_e}{\rho_0 m_e + \rho_e m_0} c_A^2. \tag{42}$$

Equation (42) describes magnetoacoustic surface waves propagating parallel to a surface of discontinuity one side of which is field-free. It applies subject to the conditions that $m_0 > 0$ and $m_e > 0$. Evidentally, the phase-speeds of the surface waves are less than the Alfvén speed c_A. Notice that in the incompressible limit (for which $m_0 \to |k_\||, m_e \to |k_\||$) equation (42) reduces to equation (38), with $g = 0$.

The transcendental form of dispersion relation (42) disguises the fact that it possesses possibly *two* solutions, rather than the one solution of the more familiar incompressible case. There is always a *slow* surface wave, with $\omega/k_\| < \min(c_t, c_e)$; and there is also a *fast* surface wave, with $\omega/k_\|$ lying between c_s and min (c_e, c_A), provided the field region is *cooler* than the environment (i.e., $c_s < c_e$) and the magnetic field is sufficiently strong ($c_A > c_s$).

Finally, we turn to the inclusion of gravity in the compressible case. Two cases have been investigated, namely that of a uniform magnetic field and that of a constant Alfvén speed c_A, assuming isothermal atmospheres both within the field and below it (Miles, Allen and Roberts (1991), Miles and Roberts (1991a,b), Gonzalez and Gratton (1991)). These investigations are a natural continuation of earlier studies (notably Nye and Thomas (1974, 1976a,b), Summers (1976), Cally and Adam (1983), Small and Roberts (1984), Zhugzhda and Dzalilov (1984), Campbell and Roberts (1989), and Evans and Roberts (1990b, 1991)) which allow for a non-isothermal atmosphere in the medium below the field. This is closer to the circumstances of a sunspot but only at the expense of a gain in complexity, which tends to obscure the nature of surface waves in a stratified atmosphere. The studies by Campbell and Roberts (1989) and Evans and Roberts (1990b, 1991) concentrated on p-modes, ignoring the surface waves. For ease of presentation we outline just one case here, the case of constant Alfvén speed.

In an isothermal atmosphere with a constant Alfvén speed (and therefore a declining magnetic field strength) the equilibrium density profile is exponential in form:

$$\rho_0(z) = \begin{cases} \rho_e \exp(z/H_e), & z > 0, \\ \rho_0 \exp(z/H_B), & z < 0, \end{cases} \tag{43}$$

where $H_e(= c_e^2/\gamma g)$ is the density scale-height in the field-free region and $H_B(= c_s^2/\Gamma g)$ is the scale-height in the field. The magnetically-modified exponent Γ is defined by

$$\Gamma = (\frac{c_s^2}{c_s^2 + \frac{1}{2}\gamma c_A^2})\gamma. \tag{44}$$

(In the absence of a magnetic field, $\Gamma = \gamma$.) The assumption of a constant Alfvén speed implies that the field is partially supporting the gas in $z < 0$, as is evident from the fact that H_B increases with increasing field strength.

The differential equation (33) reduces to (Yu (1965), Thomas (1983), Campbell and Roberts (1989))

$$\frac{d^2 v_z}{dz^2} + \frac{1}{H_B}\frac{dv_z}{dz} + A_B v_z = 0, \quad z < 0, \tag{45}$$

where

$$A_B = -m_0^2 + \frac{(\Gamma - 1)g^2 k_\parallel^2}{c_f^2(\omega^2 - k_\parallel^2 c_t^2)}. \tag{46}$$

A similar equation holds in the non-magnetic region ($z > 0$), with H_e replacing H_B, c_e replacing c_s, and Γ becoming γ.

The differential equations are then solved subject to the requirement that the kinetic energy density $\frac{1}{2}\rho_0(z)v_z^2$ tends to zero at infinity and that v_z is continuous across $z = 0$. Thus,

$$v_z = \begin{cases} A\exp[-\frac{1}{2H_e} - M_e]z, & z > 0, \\ A\exp[-\frac{1}{2H_B} + M_0]z, & z < 0, \end{cases} \tag{47}$$

where

$$M_0 = \frac{(1 - 4A_B H_B^2)^{\frac{1}{2}}}{2H_B}, \quad M_e = \frac{(1 - 4A_e H_e^2)^{\frac{1}{2}}}{2H_e}; \tag{48}$$

M_0 and M_e are taken to be positive.

Application of the matching condition on the pressure perturbation then yields the dispersion relation (Miles and Roberts (1991a,b))

$$\begin{aligned} &\frac{\omega^2}{k_\parallel^2}\left[\rho_0\left(M_0 - \frac{1}{2H_B}\right)m_e^2 + \rho_e\left(M_e + \frac{1}{2H_e}\right)m_0^2\right] \\ &\quad = \rho_0\left(M_0 - \frac{1}{2H_B}\right)m_e^2 c_A^2 + g\left[\frac{\rho_e c_e^2}{(k_\parallel^2 c_e^2 - \omega^2)} - \frac{\rho_0 c_s^2}{(k_\parallel^2 c_s^2 - \omega^2)}\right]m_0^2 m_e^2. \end{aligned} \tag{49}$$

Evidently, the structure of equation (49) is too complicated to be immediately unravelled. Nonetheless, we can see something of this structure by examining its special cases as complementary to a full numerical investigation (displayed in Figure 4). For example, in the absence of gravity ($g = 0$) we have $H_B^{-1} = H_e^{-1} = 0$ and so $M_0^2 = -A_B = m_0^2$ and $M_e^2 = -A_e = m_e^2$. Thus equation (49) reduces to equation (42), and so we can immediately conclude that the dispersion relation (49) possesses at least two modes when the Alfvén speed exceeds the sound speed in the magnetic atmosphere, taken to be cooler than the field-free gas below (i.e., when $c_A > c_s$ and $c_e > c_s$). (The special case when $c_s = c_A$ also admits a fast surface wave when $g \neq 0$.) The unstratified case of $g = 0$ in fact governs the behaviour of equation (49) in the limit $k_\parallel H_e \to \infty$, corresponding to waves of extremely short wavelength. For in this limit, a surface wave is tightly confined to the region close to the interface and thus 'sees' an unstratified gas with density and magnetic field given by their values local to the interface. This is the configuration described by the case of zero gravity.

Another special case of equation (49) that is instructive to consider is that arising in the absence of a magnetic field. With $c_A = 0$, the scale-height H_B becomes $H_0(= c_s^2/\gamma g)$ and equation (49) reduces to

$$\frac{\omega^2}{k_\parallel^2}\left[\rho_0\left(M_0 - \frac{1}{2H_0}\right)m_e^2 + \rho_e\left(M_e + \frac{1}{2H_e}\right)m_0^2\right] = gk_\parallel^2(\rho_e - \rho_0). \tag{50}$$

Interestingly enough, it turns out that equation (50) may be recast (after considerable algebra) in the form of a polynomial (Bernstein and Book (1983), Miles (1991), Miles

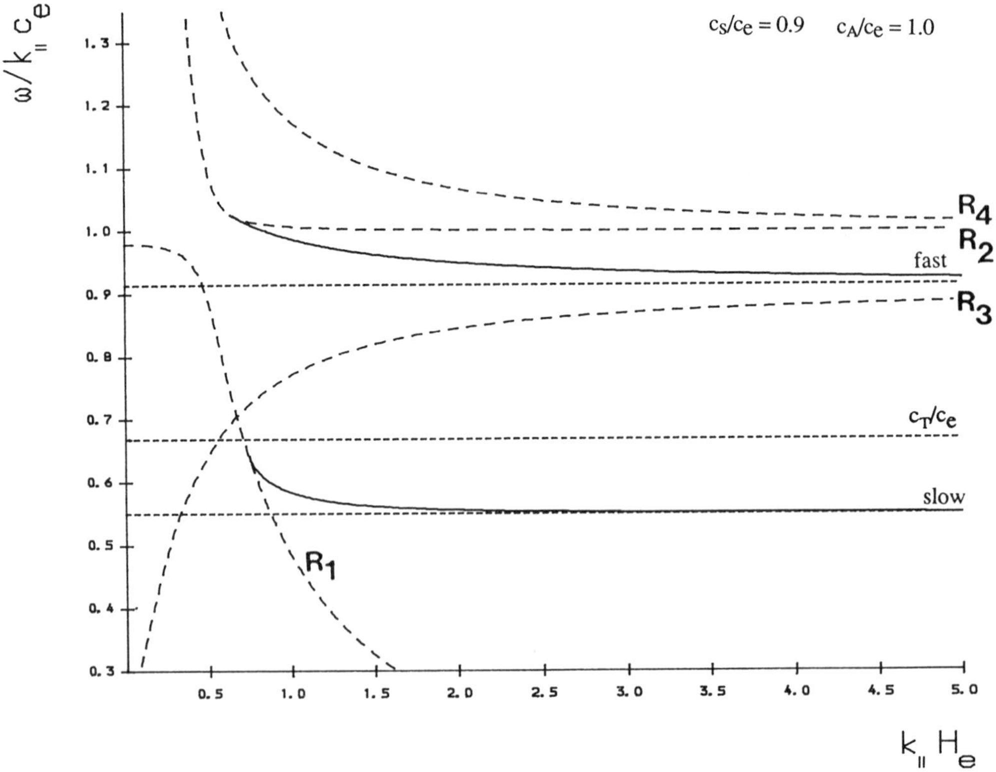

Figure 4. The phase-speed $\omega/k_\parallel$ (in units of sound speed c_e) as a function of dimensionless horizontal wavenumber, $k_\parallel H_e$, for surface waves propagating on a magnetic interface between two isothermal gases. The magnetic field is confined to the upper medium, which is cooler than the field-free gas below: $c_s = 0.9c_e$. If the Alfvén speed c_A is less than c_e only a slow surface wave exists. But for $c_A \geq c_s$, a fast surface wave exists in addition to the slow mode. The modes are illustrated for the case $c_A = c_s$. The dashed curves, labelled R_1–R_4, are cutoffs determined by M_0 and M_e. (After Miles and Roberts (1991b).)

and Roberts (1991b)):

$$(\omega^2 - k_\parallel^2 c_s^2)^2(\omega^2 - k_\parallel^2 c_e^2)^2(\Omega^4 - 1)F(\Omega^2) = 0, \tag{51}$$

where we have found it convenient to introduce the dimensionless squared frequency $\Omega^2 \equiv \omega^2/(gk_\parallel)$, and

$$F(\Omega^2) \equiv \Omega^8 - 2S\Omega^6 + (S^2 + 2\gamma - 1)\Omega^4 - 2(\gamma - 1)S\Omega^2 - D^2 \tag{52}$$

with

$$S = k_\parallel(c_s^2 + c_e^2)/g, \quad D = k_\parallel(c_s^2 - c_e^2)/g. \tag{53}$$

Now equation (51) is arrived at by a process of squaring, to remove the various roots occurring in M_e and M_0, and so not all of its roots are permitted solutions of the dispersion relation (49). For example, the roots $\omega^2 = k_\parallel^2 c_s^2$ and $\omega^2 = k_\parallel^2 c_e^2$ (Lamb modes) of equation (51) do not in fact satisfy the dispersion relation and so are not permitted modes of the system.

The root $\Omega^2 = 1$ (i.e., $\omega^2 = gk_\parallel$) is the so-called f-mode. The f-mode has been observed in some detail in the sun (Libbrecht, Woodard and Kaufman (1990)), though its relationship, if any, with sunspots is presently unclear. The f-mode is clearly a solution of equation (51), but is it a solution of equation (50)? An examination of equation (50) reveals that the f-mode is a solution provided (Miles and Roberts (1991a, b))

$$\frac{1}{2H_e} < k_\parallel < \frac{1}{2H_0}, \tag{54}$$

which requires that $\rho_0 > \rho_e$. This restriction on the horizontal wavenumber $k_\parallel$ is a consequence of the assumption of an exponential profile coupled with the requirement of a kinetic energy density that declines far from the interface. For it turns out that the vertical (incompressible) motion in the f-mode is simply

$$v_z = A\exp(-k_\parallel z), \tag{55}$$

and so gives a declining energy density $1/2\rho(z)v_z^2$ for wavenumbers $k_\parallel$ satisfying the inequality (54). An exponential profile in density is a reasonable representation of the believed structure in the sun's atmosphere but is less satisfactory as a model of the internal density structure. A slower growing density profile, such as that in a polytrope, is a better represention of the sun's internal density structure and satisfies finiteness of kinetic energy density as $z \to \infty$ for all $k_\parallel$. This suggests that only the upper limit on wavenumber is likely to be physically important. In terms of spherical harmonic degree l, arising in the helioseismologist's description of the sun's global modes of oscillations, the upper limit on $k_\parallel$ translates into an upper limit on degree l of about 3500, which is roughly of the order out to which the f-mode is observed. An added complication in all this is the magnetic field in the upper atmosphere. For theoretical calculations show that the f-mode is influenced by magnetism (Campbell and Roberts (1989), Evans and Roberts (1990b, 1991)). The effect is not large but within the limits of accuracy currently attained in helioseismology. Miles and Roberts (1991b) give a detailed discussion of the f-mode for the model under discussion here. Whether the f-mode's dispersion relation is different in the region of a sunspot is not known observationally. It is known, however, that global oscillations are of reduced amplitude in active regions as opposed to quiet regions.

Consider now the roots arising from the polynomial $F(\Omega^2) = 0$. In the limit $k_\parallel \to \infty$, for which $S \to \infty$ and $D \to \infty$, the 4th-degree polynomial in Ω^2 yields the reduced form

$$S^2\Omega^4 = D^2, \tag{56}$$

giving

$$\omega^2 = \left(\frac{\rho_e - \rho_0}{\rho_e + \rho_0}\right) gk_\parallel. \tag{57}$$

Thus we recover equation (38) with $c_A = 0$.

Figure 4 displays (after Miles and Roberts (1991b)) the results of a numerical solution of the dispersion relation (47) with parameters appropriate for penumbral conditions, namely $c_s = 0.9c_e$ (corresponding to a penumbral atmosphere with a temperature that is 81% of that in the field-free gas below) and $c_A = c_e$. The figure shows both fast and slow surface waves. Both of these modes have cutoffs at low $k_\parallel$. The fast surface wave, for example, is able to propagate only for $k_\parallel H_e$ greater than about 0.6. Taking H_e=150 km, for example, gives $k_\parallel H_e$ =900 km/λ, for a horizontal wavelength $\lambda(= 2\pi/k_\parallel)$. Accordingly, a fast surface wave of wavelength greater than about 10^3 km would leak into the non-magnetic region and so be unlikely to survive as a surface mode for very long. However, this feature of cutoff is model dependent and so may not have any serious physical consequences for penumbral surface waves. An investigation of more complicated models of the penumbral magnetic field and the field-free medium is necessary to examine this feature in greater detail.

5. Concluding Remarks

The structured and stratified nature of a sunspot, with an unknown distribution of magnetic field in its sub-photospheric layers, an uncertain degree of inhomogeneity within the visible umbra, and an evidently complex and fibrous distribution of magnetic field and/or gas density in the penumbra, makes for a complex description of the magnetohydrodynamic modes of oscillation of a spot. We have here concentrated on outlining the conclusions to be reached from a simple-minded view of a spot, emphasising its properties as, on the one hand, an elastic tube and, on the other hand, as a magnetic interface. Such a study, when complemented by more detailed investigations of tubes and tube-like structures in stratified atmospheres, hopefully provides a first step towards understanding the fascinating array of magnetohydrodynamic phenomena displayed by the sunspot.

Acknowledgement

It is a pleasure to thank Professors N. O. Weiss and J. H. Thomas for hosting such a pleasant and informative workshop and for providing me with the opportunity to attend it. I am grateful to Dr T. B. Bogdan for constructive comments on an earlier draft of this manuscript. Finally, I wish to acknowledge the financial support of NATO enabling me to attend the Workshop.

6. References

Abdelatif, T. E. (1988) 'Surface and body waves in magnetic flux tubes', Astrophys. J. **333**, 395-406.

Abels-van Maanen, A. E. P. M. and Weenik, M. P. H. (1979) Radio Sci. **14**, 301.

Adam, J. A. (1977) 'On the occurrence of critical levels in solar magnetohydrodynamics', Solar Phys. **52**, 293-307.

Adam, J. A. (1982) 'Asymptotic solutions and spectral theory of linear wave equations', Physics Reports **86**, 217-316.

Adam, J. A. (1986) 'Critical layer singularities and complex eigenvalues in some differential equations of mathematical physics', Physics Reports **142**, 263-356.

Antia, H. M. and Chitre, S. M. (1979) 'Waves in the sunspot umbra', Solar Phys. **63**, 67-78.

Appert, K., Gruber, R. and Vaclavik, J. (1974) 'Continuous spectra of a cylindrical magnetohydrodynamic equilibrium', Physics Fluids **17**,1471-1472.

Beckers, J. M. and Tallant, P. E. (1969) 'Chromospheric inhomogeneities in sunspot umbrae', Solar Phys. **7**, 351-365.

Bernstein, I. B. and Book, D. L. (1983) 'Effect of compressibility on the Rayleigh-Taylor instability', Phys. Fluids **26**, 453-458.

Bodo, G., Rosner, R, Ferrari, A. and Knobloch, E. (1989) 'On the stability of magnetized rotating jets: The axisymmetric case', Astrophys. J. **341**, 631-649.

Bogdan, T. B. (1992) 'Sunspot seismology: the interaction of a sunspot with solar p-modes', these proceedings.

Braun, D. C., Duvall, T. L. and Labonte, B. J. (1987) 'Acoustic absorption by sunspots', Astrophys. J. **319**, L27-31.

Braun, D. C., Duvall, T. L. and Labonte, B. J. (1988) 'The absorption of high-degree p-mode oscillations in and around sunspots', Astrophys. J. **335**, 1015-1025.

Braun, D. C., Labonte, B. J. and Duvall, T. L. (1990) 'The spatial distribution of p-mode absorption in active regions', Astrophys. J. **354**, 372-381.

Braun, D. C., Labonte, B. J., Duvall, T. L. and Ryutova, M. P. (1991) 'A possible mechanism for enhanced absorption of p-modes in sunspot and plage regions', Astrophys. J., preprint.

Browning, P. K. (1991) 'Mechanisms of solar coronal heating', Plasma Phys. and Controlled Fusion **33**, 539-571.

Cadez, V. M. and Okretic, V. K. (1989) 'Leakage of MHD surface waves in structured media', J. Plasma Phys. **41**, 23-30.

Cally, P. S. (1985) 'Magnetohydrodynamic tube waves: Waves in fibrils', Australian J. Phys. **38**, 825-837.

Cally, P. S. (1986) 'Leaky and non-leaky oscillations in magnetic flux tubes', Solar Phys. **103**, 277-298.

Cally, P. S. (1991) 'Phase-mixing and surface waves: a new interpretation', J. Plasma Phys. **45**, 453-479.

Cally, P. S. and Adam, J.A. (1983) 'On photospheric and chromospheric penumbral waves', Solar Phys. **85**, 97-111.

Campbell, W. R. and Roberts, B. (1989) 'The influence of a chromospheric magnetic field on the solar p- and f-modes', Astrophys. J. **338**, 538-556.

Chandrasekhar, S. (1961) Hydrodynamic and Hydromagnetic Stability, Oxford University Press, Oxford.

Chen, L. and Hasegawa, A. (1974) 'Plasma heating by spatial resonance of Alfvén wave', Phys. Fluids **17**, 1399-1403.

Chitre, S. M. (1992) 'Theory of umbral oscillations and penumbral waves', these proceedings.

Chitre, S. M. and Davila, J. M. (1991) 'The resonant absorption of p-modes by sunspots with twisted magnetic fields', Astrophys. J. **371**, 785-792.
Coulson, C. A. (1955) Waves, Oliver and Boyd, Edinburgh.
Cowling, T. G. (1976) Magnetohydrodynamics, Adam Hilger, Bristol.
Davila, J. M. (1987) 'Heating of the solar corona by the resonant absorption of Alfvén waves', Astrophys. J. **317**, 514-521.
Davila, J. M. (1990) 'Sunspot seismology theory', in E. R. Priest and V. Krishan (eds.), Basic Plasma Processes on the Sun, Kluwer Academic Publishers, Dordrecht, pp. 149-157.
Davila, J. M. (1991) 'Resonance absorption heating', in P. Ulmschneider, E. R. Priest and R. Rosner (eds.), Mechanisms of Chromospheric and Coronal Heating, Springer-Verlag, Heidelberg, pp. 464-479.
Edwin, P. M. and Roberts, B. (1983) 'Waves in a magnetic cylinder', Solar Phys. **88**, 179-191.
Einaudi, G. and Mok, Y. (1987) 'Alfvén wave dissipation in the solar atmosphere', Astrophys. J. **319**, 520-530.
Evans, D. J. (1990) The Effects of Magnetic Fields on Oscillations in the Solar Atmosphere, Ph. D. thesis, St Andrews University, pp. 194.
Evans, D. J. and Roberts, B. (1990a) 'The oscillations of a magnetic flux tube and its application to sunspots', Astrophys. J. **348**, 346-356.
Evans, D. J. and Roberts, B. (1990b) 'The influence of a chromospheric magnetic field on the frequencies of solar p- and f-modes: II. Uniform chromospheric field', Astrophys. J. **356**, 704-719.
Evans, D. J. and Roberts, B. (1991) 'The sensitivity of a chromospherically induced p- and f-mode frequency shift to the height of the magnetic canopy', Astrophys. J. **371**, 387-395.
Ferraro, V. C. A. and Plumpton, C. (1958) 'Hydromagnetic waves in a horizontally stratified atmosphere', Astrophys. J. **127**, 459-476.
Giovanelli, R. G. (1972) 'Oscillations and waves in sunspots', Solar Phys. **27**, 71-79.
Goedbloed, J. P. (1971) 'Stabilization of magnetohydrodynamic instabilities by force-free magnetic fields I. Plane plasma layer', Physica **53**, 412-444.
Goedbloed, J. P. (1983) Lecture Notes in Magnetohydrodynamics, Rijnhuizen Report 83-145, Assoc. Euratom - FOM, pp. 1-289.
Goedbloed, J. P. (1984) 'Plasma-vacuum problems in magnetohydrodynamics', Physica **12D**, 107-132.
Goedbloed, J. P. and Hagenbeuk, H. J. L. (1972) 'Growth rates of instabilities of a diffuse linear pinch', Phys. Fluids **15**, 1090-1101.
Gonzalez, A. G. and Gratton, J. (1991) 'Magnetoacoustic surface gravity waves', Solar Phys. **134**, 211-232.
Goossens, M (1991) 'Magnetohydrodynamic waves and wave heating in non-uniform plasmas', in E.R. Priest and A.W. Hood (eds.), Advances in Solar System Magnetohydrodynamics, Cambridge University Press, Cambridge, pp. 137-172.
Grossmann, W. and Tataronis, J. (1973) Z. Physik. **261**, 217.
Grossmann, W. and Smith, R. A. (1988) 'Heating of solar coronal loops by resonant absorption of Alfvén waves', Astrophys. J. **332**, 476-498.
Hain, K. and Lüst, R. (1958) Z. Naturforsch. **13a**, 936.

Heyvaerts, J. and Priest, E. R. (1983) 'Coronal heating by phased-mixed Alfvén waves', Astron. Astrophys. **117**, 220-234.

Hollweg, J. V. (1982) 'On the origin of solar spicules', Astrophys. J. **257**, 345-353.

Hollweg, J. V. (1986) 'Energy and momentum transport by waves', in B. Buti (ed.) Advances in Space Plasma Physics, World Scientific, Singapore, p.77.

Hollweg, J. V. (1987a) 'Resonant absorption of magnetohydrodynamic surface waves: Physical discussion', Astrophys. J. **312**, 880-885.

Hollweg, J. V. (1987b) 'Resonant absorption of magnetohydrodynamic surface waves: Viscous effects', Astrophys. J. **320**, 875-873.

Hollweg, J. V. (1988) 'Resonance absorption of solar *p*-modes by sunspots', Astrophys. J. **335**, 1005-1014.

Hollweg, J. V. (1990a) 'Heating of the solar corona', Computer Phys. Reports **12**, 205-232.

Hollweg, J. V. (1990b) 'MHD waves on solar magnetic flux tubes', in C. T. Russell, E. R. Priest and L. C. Lee (eds.), Physics of Magnetic Flux Ropes, AGU: Washington, Geophys. Mono. **58**, pp. 23-31.

Hollweg, J. V. (1991) 'Alfvén waves', in P. Ulmschneider, E. R. Priest and R. Rosner (eds.), Mechanisms of Chromospheric and Coronal Heating, Springer-Verlag, Heidelberg, pp. 423-434.

Ionson, J. A. (1978) 'Resonant absorption of Alfvénic surface waves and the heating of solar coronal loops', Astrophys. J. **226**, 650-673.

Jain, R. and Roberts, B. (1991a) 'Magnetoacoustic surface waves at a single interface', Solar Phys. **133**, 263-280.

Jain, R. and Roberts, B. (1991b) 'Magnetoacoustic surface waves', in P. Ulmschneider, E. R. Priest and R. Rosner (eds.), Mechanisms of Chromospheric and Coronal Heating, Springer-Verlag, Heidelberg, pp. 511-513.

Kalkofen, W. (1990) 'The heating of the quiet solar chromosphere', in E. R. Priest and V. Krishan (eds.), Basic Plasma Processes on the Sun, Kluwer Academic Publishers, Dordrecht, pp. 197-206.

Kruskal, M. and Schwarzschild, M. (1954) 'Some instabilities of a completely ionized plasma', Proc. Roy. Soc. (London) **A223**, 348-360.

Lee, M. A. and Roberts, B. (1986) 'On the behaviour of hydromagnetic surface waves', Astrophys. J. **301**, 430-439.

Libbrecht, K. G., Woodward, M.F. and Kaufman, J. M. (1990) 'Frequencies of solar oscillations', Astrophys. J. Suppl. **74**, 1127-1149.

Lighthill, M. J. (1960) 'Studies on magneto-hydrodynamic waves and other anisotropic motions', Phil. Trans. Roy. Soc. **A252**, 397-430.

Lites, B. W. (1986a) 'Photoelectric observations of chromospheric sunspot oscillations. III. Spatial distribution of power and frequency in umbrae', Astrophys. J. **301**, 992-1004.

Lites, B. W. (1986b) 'Photoelectric observations of chromospheric sunspot oscillations. IV. The Ca II H line and He I λ10830', Astrophys. J. **301**, 1005-1017.

Lites, B. W. (1988) 'Photoelectric observations of chromospheric sunspot oscillations. V. Penumbral oscillations', Astrophys. J. **334**, 1054-1065.

Lites, B. W. (1992) 'Observations of oscillations in sunspots', these proceedings.

Lites, B. W., White, O. R. and Packman, D. (1982) 'Photoelectric observations of propagating sunspot oscillations', Astrophys. J. **253**, 386-392.

Lou, Y.-Q. (1988) 'Viscous magnetohydrodynamic modes and *p*-mode absorption by sunspots', in Proc. Symp. Seismology of the Sun and Sun-like Stars, ESA SP-286, Noordwijk, pp. 305-310.

Lou, Y.-Q. (1990) 'Viscous magnetohydrodynamic modes and *p*-mode absorption', Astrophys. J. **350**, 452-462.

McKenzie, J. F. (1970) 'Hydromagnetic oscillations of the geomagnetic tail and plasma sheet', J. Geophys. Res., Space Phys. **75**, 5331-5339.

Miles, A. J. (1991) Magnetohydrodynamic Surface Waves in the Solar Atmosphere, Ph. D. thesis, St Andrews University, pp. 197.

Miles, A. J., Allen, H. R. and Roberts, B. (1991) 'Magnetoacoustic-gravity surface waves. II. Uniform magnetic field', Solar Phys., in press.

Miles, A. J. and Roberts, B. (1989) 'On the properties of magnetoacoustic surface waves', Solar Phys. **119**, 257-278.

Miles, A. J. and Roberts, B. (1991a) 'The *f*-mode and MHD surface waves', in P. Ulmschneider, E. R. Priest and R. Rosner (eds.) Mechanisms of Chromospheric and Coronal Heating, Springer-Verlag, Heidelberg, pp. 508-510.

Miles, A. J. and Roberts, B. (1991b) 'Magnetoacoustic-gravity surface waves. I. Constant Alfvén speed', Solar Phys., in press.

Moore, R. and Rabin, D. (1985) 'Sunspots', Ann. Rev. Astron. Astrophys. **23**, 239-266.

Moreno-Insertis, F. and Spruit, H. C. (1989) 'Stability of sunspots to convective motions. I. Adiabatic instability', Astrophys. J. **342**, 1158-1171.

Nocera, L., Leroy, B. and Priest, E. R. (1984) 'Phase mixing of propagating Alfvén waves', Astron. Astrophys. **133**, 387-394.

Nye, A. H. and Thomas, J. H. (1974) 'The nature of running penumbral waves', Solar Phys. **38**, 399-413.

Nye, A. H. and Thomas, J. H. (1976a) 'Solar magneto-atmospheric waves. I. An exact solution for a horizontal magnetic field', Astrophys. J. **204**, 573-581.

Nye, A. H. and Thomas, J. H. (1976b) 'Solar magneto-atmospheric waves. II. A model for running penumbral waves', Astrophys. J. **204**, 582-588.

Parker, E. N. (1974) 'The nature of the sunspot phenomenon I: Solutions of the heat transport equation', Solar Phys. **36**, 249-274.

Parker, E. N. (1979a) 'Sunspots and the physics of magnetic flux tubes. I. The general nature of the sunspot', Astrophys. J. **230**, 905-913.

Parker, E. N. (1979b) Cosmical Magnetic Fields, Oxford University Press, Oxford.

Parker, E. N. (1983) 'The propagation of torsion along flux tubes subject to dynamical nonequilibrium', Geophys. Astrophys. Fluid Dynamics **24**, 245-272.

Parker, E. N. (1987) 'Why do stars emit X-rays?', Physics Today **40**, 36-42.

Parker, E. N. (1991a) 'Heating solar coronal holes', Astrophys. J. **372**, 719-727.

Parker, E. N. (1991b) 'The phase mixing of Alfvén waves, coordinated modes, and coronal heating', Astrophys. J. **376**, 355-363.

Poedts, S., Goossens, M. and Kerner, W. (1990) 'On the efficiency of coronal loop heating by resonant absorption', Astrophys. J. **360**, 279-287.

Rae, I. C. and Roberts, B. (1981) 'Surface waves and the heating of the corona', Geophys. Astrophys. Fluid Dynamics **18**, 197-226.

Rae, I. C. and Roberts, B. (1982) 'Pulse propagation in a magnetic flux tube', Astrophys. J. **256**, 761-767.

Rayleigh, Lord (1877) The Theory of Sound, Dover, New York (reprinted by Dover, 1945), vol. 1, chap. X.

Roberts, B. (1980) 'Developpement recent dans la theorie des ondes MHD parfaites dans les milieux presentant une structure magnetique', Ann. De Physique **5**, 453-481.

Roberts, B. (1981a) 'Wave propagation in a magnetically structured atmosphere; I. Surface waves at a magnetic interface', Solar Phys. **69**, 27-38.

Roberts, B. (1981b) 'Wave propagation in a magnetically structured atmosphere; II. Waves in a magnetic slab', Solar Phys. **69**, 39-56.

Roberts, B. (1981c) 'Waves in magnetic structures', in L. E. Cram and J. H. Thomas (eds.) Physics of Sunspots, Sacramento Peak Observatory, Sunspot (New Mexico), pp. 360-383.

Roberts, B. (1985) 'Magnetohydrodynamic waves', in E. R. Priest (ed.), Solar System Magnetic Fields, Reidel, Dordrecht, pp. 37-79.

Roberts, B. (1986) 'Dynamical processes in magnetic flux tubes', in W. Deinzer, M. Knölker and H. Voight (eds.), Small-Scale Magnetic Flux Concentrations in the Solar Photosphere , Vandenhoeck & Ruprecht, Gottingen, pp. 169-190.

Roberts, B. (1987) 'On MHD solitons in jets', Astrophys. J. **318**, 590-594.

Roberts, B. (1988) 'Solar magnetohydrodynamics', in B. Buti (ed.), Cometary and Solar Plasmas Physics, World Scientific, Singapore, pp. 1-80.

Roberts, B. (1990a) 'Properties and models of photospheric flux tubes', in C. T. Russell, E. R. Priest and L. C. Lee (eds.), Physics of Magnetic Flux Ropes, AGU: Washington, Geophys. Mono. **58**, pp. 113-132.

Roberts, B.(1990b) 'Waves in magnetic flux tubes', in E. R. Priest and V. Krishan (eds.), Basic Plasma Processes on the Sun, Kluwer Academic Publishers, Dordrecht, pp. 159-174.

Roberts, B. (1990c) 'Oscillations in the sun', in B. Buti (ed.), Solar and Planetary Plasma Physics, World Scientific, Singapore, pp. 1-31.

Roberts, B. (1991a) 'Magnetohydrodynamic surface waves', in P. Ulmschneider, E. R. Priest and R. Rosner (eds.), Mechanisms of Chromospheric and Coronal Heating, Springer-Verlag, Heidelberg, pp. 494-507.

Roberts, B. (1991b) 'Magnetohydrodynamic waves in the sun', in E.R. Priest and A.W. Hood (eds.), Advances in Solar System Magnetohydrodynamics, Cambridge University Press, Cambridge, pp. 105-136.

Roberts, B., Edwin, P. M. and Benz, A. O. (1984) 'On coronal oscillations', Astrophys. J. **279**, 857-865.

Roberts, B. and Webb, A. R. (1978) 'Vertical motions in an intense magnetic flux tube', Solar Phys. **56**, 5-35.

Roberts, B. and Webb, A. R. (1979) 'Vertical motions in an intense magnetic flux tube: On the slender flux tube approximation', Solar Phys. **64**, 77-92.

Rosenthal, C. (1990) 'Absorption of acoustic waves in monolithic and fibril sunspot models', Solar Phys. **130**, 313-335.

Ryutov, D. D. and Ryutova, M. P. (1976) 'Sound oscillations in a plasma with magnetic filaments', Sov. Phys. J.E.T.P. **43**, 491-497.

Ryutova, M. P. (1990a) 'Waves and oscillations in magnetic fluxtubes', in J. O. Stenflo (ed.), Solar Photosphere: Structure, Convection and Magnetic Fields , IAU Symp. **138**, Reidel, pp. 229-249.

Ryutova, M. P. (1990b) 'Nonlinear waves in flux tubes', in E. R. Priest and V. Krishan (eds.), Basic Plasma Processes on the Sun, Kluwer Academic Publishers, Dordrecht, pp. 175-186.

Ryutova, M. P. and Persson, M. (1984) 'Dispersion properties and enhanced dissipation of MHD-oscillations in a plasma with random inhomogeneities', Physica Scripta **29**, 353-359.

Sakurai, T., Goossens, M. and Hollweg, J. V. (1991a) 'Resonant behaviour of MHD waves on magnetic flux tubes; I. Connection formulae at the resonant surfaces'Solar Phys. **133**, 227-245.

Sakurai, T., Goossens, M. and Hollweg, J. V. (1991b) 'Resonant behaviour of MHD waves on magnetic flux tubes; II. Absorption of sound waves by sunspots', Solar Phys. **133**, 247-262.

Scheuer, M.A. and Thomas, J.H. (1981) 'Umbral oscillations as resonant modes of magneto-atmospheric waves', Solar Phys. **71**, 21-38.

Sedlacek, Z. (1971) 'Electrostatic oscillations in a cold inhomogeneous plasma. I. Differential equation approach', J. Plasma Phys. **5**, 239-263.

Small, L. M. and Roberts, B. (1984) 'On running penumbral waves', in Hydromagnetics of the Sun, ESA SP-220, pp. 257-259.

Southwood, D. J. (1974) 'Some features of field line resonances in the magnetosphere', Planet. Space Sci. **22**, 483.

Spruit, H. C. (1981) 'Magnetic flux tubes', in S. Jordan (ed.), The Sun as a Star, NASA SP-450, Washington, pp. 385-412.

Spruit, H. C. (1982) 'Propagation speeds and acoustic damping of waves in magnetic flux tubes', Solar Phys. **75**, 3-17.

Spruit, H. C. and Roberts, B. (1983) 'Magnetic flux tubes on the sun', Nature **304**, 401-406.

Summers, D. (1976) 'Gravity modified sound waves in a conducting stratified atmosphere', Quart. J. Mech. Appl. Math. **29**, 117-126.

Tataronis, J. A. and Grossmann, W. (1973) 'Decay of MHD waves by phase mixing. I. The sheet pinch in plane geometry', Z. Physik **261**, 203-216.

Tayler, R. J. (1957) 'The influence of an axial magnetic field on the stability of a constricted gas discharge', Proc. Phys. Soc. (London) **B70**, 1049-1063.

Thomas, J. H. (1983) 'Magneto-atmospheric waves', Ann. Rev. Fluid Mech. **15**, 321-343.

Thomas, J. H. (1985) 'Hydrodynamic waves in the photosphere and chromosphere', in H .U. Schmidt (ed.), Theoretical Problems in High Resolution Solar Physics, Max Planck Institute: Munich, MPA 212, pp. 126-149.

Thomas, J. H., Cram, L. E. and Nye, A. H. (1982) 'Five-minute oscillations as a subsurface probe of sunspot structure', Nature **297**, 485-487.

Thomas, J. H., Cram, L. E. and Nye, A. H. (1984) 'Dynamical phenomena in sunspots. I. Observing procedures and oscillatory phenomena', Astrophys. J. **285**, 368-380.

Uberoi, C. (1982) 'A note on the existence of Alfvén surface waves', Solar Phys. **78**, 351-354.

Wentzel, D. G. (1979) 'Hydromagnetic surface waves on cylindrical fluxtubes', Astron. Astrophys. **76**, 20-23.

Wilson, P. R. (1980) 'The general dispersion relation for the vibration modes of magnetic flux tubes', Astron. Astrophys. **87**, 121-125.

Wittmann, A. (1969) 'Some properties of umbral flashes', Solar Phys. **7**, 366-369.

Yu, C. P. (1965) 'Magneto-atmospheric waves in a horizontally stratified conducting medium', Phys. Fluids **8**, 650-656.

Zirin, H. and Stein, A. (1972) 'Observations of running penumbral waves', Astrophys. J. Lett. **178**, L85-87.

Zhugzhda, Y. D. and Dzhalilov, N. S. (1984) 'Magneto-acoustic-gravity waves on the sun; III. The theory of running penumbral waves', Aston. Astrophys. **133**, 333-340.

Zhugzhda, Y. D., Locans, V. and Staude, J. (1983) 'Seismology of sunspot atmospheres', Solar Phys. **82**, 369-378.

Zhugzhda, Y. D., Locans, V. and Staude, J. (1987) 'The interpretation of oscillations in sunspot umbrae', Astron. Nachr. **308**, 257-269.

Zhugzhda, Y. D., Staude, J. and Locans, V. (1984) 'A model of the oscillations in the chromosphere and transition region above sunspot umbrae', Solar Phys. **91**, 219-234.

THEORY OF UMBRAL OSCILLATIONS AND PENUMBRAL WAVES

S. M. CHITRE
Tata Institute of Fundamental Research
Homi Bhabha Road, Bombay 400 005, India

ABSTRACT. Theoretical work on umbral oscillations and penumbral waves is reviewed. The 3-minute umbral oscillations are identified with the lowest fast magnetoatmospheric mode excited in the presence of the vertical magnetic field, in the shallow subphotospheric layers of the sunspot, while the running penumbral waves are gravity-modified magnetoacoustic fast modes generated in the horizontal penumbral magnetic field.

1. Introduction

An outstanding feature of any theory of sunspots is the role played by the strong spot magnetic field ($B \sim 2000G$) in influencing the mode of energy transport. It is now generally accepted that modified convection in some form must take place in a spot region, as there are convincing reasons to believe that sunspots cannot be in completely radiative equilibrium (cf. Chitre, 1963; Deinzer, 1965; Yun, 1970). In fact, the convectively unstable subphotospheric region of a sunspot, which is endowed with a strong magnetic field and which is thermally stratified in the vertical direction, is a highly suitable site for the generation and propagation of magnetohydrodynamic waves.

Observations have revealed a whole range of velocity fields in sunspots: these include large-scale wave motions like umbral oscillations and penumbral waves, filamentary structures and small-scale unresolved features. Nearly thirty years ago, Danielson (1961) reported observations of the penumbra composed of long filamentary structures of thickness ~ 300 km and length ~ 2000 km arranged radially around the umbra, with the characteristic lifetime of a penumbral filament of order 30 min. The remarkable observations of sunspot oscillations began in the late sixties with the discovery of umbral flashes in Ca II H and K lines by Beckers and Tallant (1969). This was followed by a succession of most interesting detection of wave motions: (i) 3-min umbral velocity and intensity oscillations reported in the photospheric and chromospheric spectral lines by Giovanelli (1972), Bhatnagar and Tanaka (1972), Beckers and Schultz (1972), (ii) Running penumbral waves discovered independently by Giovanelli (1972) and Zirin and Stein (1972), (iii) 5-min oscillations in the photosphere of the spot-umbra reported by Bhatnagar, Livingston and Harvey (1972), Solatu, Schröter and Wöhl (1976) and Thomas, Cram and Nye (1986).

Information about vertical propagation of umbral oscillations has been obtained by measuring the oscillations simultaneously in two or more spectral lines formed at different heights. Giovanelli, Harvey and Livingston (1978) found a phase lag between oscillations observed in Fe I $\lambda 5233$ (photospheric) and $H\alpha$, indicating upward phase propagation in the chromosphere. Lites and Thomas (1985) observed umbral oscillations simultaneously in Ti $\lambda 6304$ (low photospheric line) and Fe I $\lambda 5434$ (low chromospheric line) to show from

J. H. Thomas and N. O. Weiss (eds.), Sunspots: Theory and Observations, 333–343.

the peaks in power and coherence, and from small phase differences at 6 mHz, that the 3-min umbral oscillation has indeed the character of a vertically trapped mode in the photosphere. Thus, wave observations indicate that the dominant 3-min umbral oscillations have the character of a coherent, vertically standing wave in the photosphere and low chromosphere. Another recent development has been the measurement of 3-min umbral oscillations in the chromosphere-corona transition region above a sunspot umbra, using the UVSM aboard the SMM satellite. Gurman *et al.* (1982) have measured the oscillation in the UV line C IV $\lambda 1548$ (which is formed at $T_{eff} \sim 10^5$ K) to find significant velocity oscillations with periods in the range 129-173s. Thomas *et al.* (1987) have also reported simultaneous measurements of umbral oscillations in the photosphere, chromosphere and transition region in four spectral lines, including the C IV line from SMM.

A thorough review of the observations of sunspot oscillations is given in the paper by Lites in this volume. In what follows we shall summarise theoretical ideas concerning umbral oscillations and penumbral waves.

2. Umbral 3-min oscillations

The stability of a fluid layer in the presence of a vertical magnetic field has been discussed by Chandrasekhar (1961), Savage (1969) and Parker (1974) amongst others. But all these investigations were performed in the framework of the Boussinesing approximation which ignores the effects of compressibility, thus filtering out the magnetoacoustic modes. It is just these modes which turn out to be important in understanding the velocity oscillations in the sunspot umbra and penumbra, as demonstrated by the work of Nye and Thomas (1974; 1976), Uchida and Sakurai (1975), Antia, Chitre and Kale (1978), Scheuer and Thomas (1981) and Thomas and Scheuer (1982).

The 3-min oscillations are generally believed to be a resonant mode of the sunspot magnetic flux tube. In fact, there are two types of resonant modes in a sunspot that have periods around 3 minutes: these two types are the photospheric resonance and the chromospheric resonance. The photospheric resonance corresponds to fast magnetoatmospheric waves that are almost trapped in the subphotospheric layers; the resonance is caused by the trapping of the fast mode by downward reflection due to the rapidly decreasing density (hence increasing Alfvén speed) with height in the overlying layers and upward refraction due to the increasing sound speed in the convection zone. Following Moore (1973), Mullan and Yun (1973) we shall assume that the basic driving mechanism responsible for umbral oscillations is overstable oscillatory convection in a relatively shallow subphotospheric layer in the umbra. We demonstrate that the observed 3-min umbral oscillations are indeed a resonant response to overstable convection in a shallow subphotospheric umbral layer and the wave mode may be identified with a fast magnetoatmospheric wave.

The chromospheric resonance, on the other hand, corresponds to the slow magneto-acoustic wave that is nearly trapped in the chromosphere (Zhugzhda *et al.* 1983, 1984; Gurman and Leibacher, 1984). The slow modes are basically unmodified acoustic waves with motions along the vertical magnetic field and the trapping of these waves is caused by downward reflection at the chromosphere-corona transition region due to the rapid increase in the temperature (hence the sound speed) and upward reflection at the temperature minimum due essentially to an increase in the acoustic cut-off frequency resulting from the decreasing temperature towards the temperature minimum (Thomas, 1985).

We analyse the linear eigenmodes of a plane-parallel, thermally conducting, inviscid fluid layer of thickness d that is stratified under constant gravity in the presence of a uniform vertical magnetic field. We shall further assume the fluid to be an ideally conducting perfect gas which is governed by the following equations:

Momentum:

$$\rho \frac{d\mathbf{v}}{dt} = -\nabla p + \rho \mathbf{g} + \frac{\mathbf{j} \wedge \mathbf{B}}{c}$$

Mass:

$$\frac{d\rho}{dt} + \rho \nabla . \mathbf{v} = 0$$

Energy:

$$\rho c_v \frac{dT}{dt} - \Re T \frac{d\rho}{dt} = \nabla . (K_R \nabla T)$$

State:

$$P = \Re \varrho T$$

supplemented by Maxwell's equations:

$$\nabla \times \mathbf{H} = \frac{4\pi}{c} \mathbf{j}$$

$$\nabla \wedge \mathbf{E} = -\frac{1}{c} \frac{\partial \mathbf{B}}{\partial t}$$

$$\nabla . \mathbf{B} = 0$$

$$\mathbf{E} + \frac{\mathbf{v} \wedge \mathbf{B}}{c} = 0.$$

We assume the basic static state to be polytropic where the unperturbed pressure, density and temperature are related by $p_0 \propto T_0^{m+1}$ and $\rho_0 \propto T_0^m$, with the polytropic index $\Gamma = \frac{m+1}{m}$. The gas constant $\Re$ and the specific heat at constanct volume C_v are taken to be constant. We adopt the scale-height $H = \Re T_{base}/g$ as the unit of length and sound travel time $t = \sqrt{\Re T_{base}/g}$ as the unit of time and express the physical quantities in terms of the density, ρ_{base} and temperature, T_{base} at the base of the fluid layer. The governing equations are linearized by Fourier analyzing the physical variables as:

$$q = q_0(z) + q_1(z) \, exp \, i(\omega t + k_x x + k_y y)$$

We introduce two dimensionless parameters in the analysis:

$$G_B = \beta - 1 = \frac{B_0^2}{4\pi \Re \, T_{base} \, \rho_{base}} = \gamma \left(\frac{Alfv\acute{e}n \ speed}{sound \ speed} \right)^2$$

and

$$G_k = \frac{K_R \, g}{C_v \, \rho_{base} (T_{base})^{3/2}} = \left(\frac{sound \ travel \ time}{radiation \ diffusion \ time} \right).$$

The set of linearized equations is of sixth order in the z-derivatives and hence for a complete specification of the generalized eigenvalue problem we require in all six boundary conditions. We adopt the following boundary conditions at the upper and lower bounding surfaces:

At the upper boundary we assume free boundary conditions, namely, vanishing of the Lagrangian pressure (thermal + magnetic) perturbation, vanishing of the horizontal magnetic field component and the radiation flux condition; the lower boundary is taken to be maintained at constant temperature and there is no momentum flux across the boundary together with the vanishing of the horizontal component of the magnetic field (cf. Antia and Chitre, 1979). The linearized equations are solved numerically adopting a method described by Antia (1979) in order to evaluate the complex eigenvalues ω for specified horizontal wavenumbers.

For the sunspot umbra we adopt the following typical parameters: $B_0 = 3000$ G, $g = 2.47 \times 10^4$ cm s^{-2}, $d = 850$ km, $\rho_{base} = 2.67 \times 10^{-6}$ g cm^{-3}, $T_{base} = 15,000$ K, $\Gamma = 1.66, \gamma = 1.1, \kappa = 2 \times 10^{12}$ cm^2 s^{-1} and the dimensionless complex values so obtained are shown in Table 1 for three values of dimensionless wavenumbers k_x.

TABLE 1. Complex eigenvalues for different values of k_x corresponding to the choice of parameters $G_K = 0.035$ and $G_B = 0.215$. The fast and slow modes are designated by F and S respectively, while the f-mode has no nodes in pressure and velocity perturbations.

k_x	*Mode*	*Growth rate*	*Frequency*
	f	-1.45×10^{-2}	0.468
	S1	1.93×10^{-2}	0.388
0.5	F1	6.71×10^{-3}	1.666
	F2	8.44×10^{-3}	3.125
	f	-1.35×10^{-2}	0.929
	S1	5.03×10^{-2}	0.374
1.0	F1	8.49×10^{-3}	1.834
	F2	7.25×10^{-3}	3.216
	f	4.46×10^{-4}	1.827
2.0	S1	9.57×10^{-2}	0.359
	F1	4.61×10^{-3}	2.387
	F2	9.07×10^{-5}	3.550

It can be seen that the maximum growth rate (0.0085) is attained by the F1 mode around $k_x = 1$. The time-scale of 155s corresponding to the most unstable fast mode is in reasonable agreement with the observed periods of umbral oscillations. Note that the growth rates of slow modes decrease with G_B, while those of fast modes increase with G_B. We thus expect the fast modes to dominate over the slow modes for larger values of the magnetic field prevailing in the spot umbra.

Scheuer and Thomas (1981) have considered a more realistic three-layer model for the umbra with an overlying chromospheric layer and an underlying convection zone. The resonant lowest fast mode (F1), having a period of 153s is identified with a typical umbral oscillation; this is in essential agreement with the conclusion of Antia and Chitre (1979). The analysis of Scheuer and Thomas (1981) considers only adiabatic perturbations, but this should not essentially affect the computation of frequencies, although the growth rates are in the main determined by dissipative effects.

We believe that the driving mechanism for 3-min umbral oscillations is the overstable oscillatory convection in a shallow superadiabatic layer in the subphotospheric region of the umbra (Moore, 1973, Mullan and Yun, 1973). There will naturally be a leakage of the wave energy into the overlying chromosphere in the form of vertically propagating acoustic waves. In fact, Zhugzhda and his co-workers (1983, 1984) attribute the 3-min umbral oscillations to a resonant mode of slow magnetoacoustic wave excited by broadband acoustic noise from the underlying convection zone, which is very nearly trapped in the chromospheric region. It should perhaps be stressed that a certain amount of care must be exercised while classifying the fast and slow modes in a stratified atmosphere. The character of the wave mode is apt to change with wavenumber and with height: thus, the resonant mode may have a fast mode character in the subphotospheric layers, but when there is an escape of this wave energy into the overlying region, the same wave may display a slow mode character in the chromosphere. Fig. 1 shows the $(k-\omega)$ plot displaying the frequencies for the fast, slow and f-modes for the spot magnetic field $B_0 = 3000$ G.

The avoided crossings are clearly seen and one can, for example, switch from the f-mode to an $S1$ mode with increasing value of (k_x). Likewise, the $F1$ mode can go over to an f-mode. On the other hand, when $\beta \ll 1$, which is the case in the chromospheric layers, it can be readily seen from Fig. 2 that the $(k-\omega)$ plot is predominantly occupied by the slow modes, with the fast modes wedged in a very narrow region around $k_x \simeq 0$. In any case, because of the sharply declining matter density with height above the photosphere, the kinetic energy density of the 3-min oscillations is some five times larger in the photosphere than in the chromosphere (Abdelatif, Lites and Thomas, 1986). Moreover, in the strong magnetic field prevailing in the umbral photosphere the fast modes dominate, while in the chromosphere where the magnetic field is weaker, the slow modes dominate over the fast modes; here the growth rate of the most unstable mode is taken to be an indication of the strength. We should, of course, remember that the resonant frequencies of the photospheric and chromospheric resonators are nearly the same and this is going to lead to a coupling of the resonators. (cf. Zhugzhda, Staude and Locans, 1984)

3. Running penumbral waves

Observations of the spot penumbra reveal elongated structures (penumbral filaments) of approximate thickness 300 km and length 2000 km arranged radially around the umbra, with characteristic life-time of about 30 min (Danielson, 1961; Bray and Loughhead, 1964). Besides these, observations have also detected a remarkable pattern of running penumbral waves, which propagate radially outward from the umbra across the spot penumbra with largely vertical motions in Hα (Zirin and Stein, 1972; Giovanelli, 1972). The penumbral waves are found to have periods in the range 180-240s with associated horizontal wavelengths $\sim 2350-3800$ km and they travel outward in the penumbra at a typical speed of 15 km s^{-1}.

Nye and Thomas (1974, 1976) studied a piecewise linear model of the vertical structure of a typical sunspot penumbra. Initially, they gave an analytical solution for magneto-atmospheric waves in an isothermal atmosphere pervaded by a uniform horizontal magnetic field. The analysis was later extended to a two-layer penumbral model with the upper layer taken to be isothermal having a uniform horizontal magnetic field and the lower layer with a constant temperature gradient in the absence of a magnetic field. Small adiabatic perturbations in the lower layer were considered to identify the penumbral waves with fast magneto-atmospheric modes trapped at the photospheric level due to the increasing Alfvén

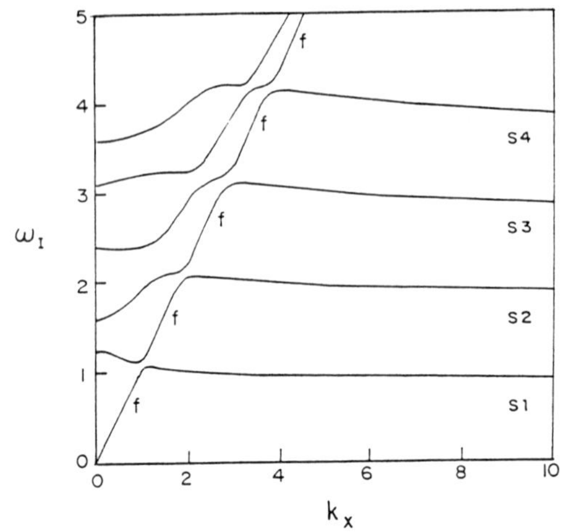

Figure 1. Frequency vs wavenumber plot corresponding to f, fast and slow modes in the presence of vertical magnetic field for which $\beta \simeq 1$.

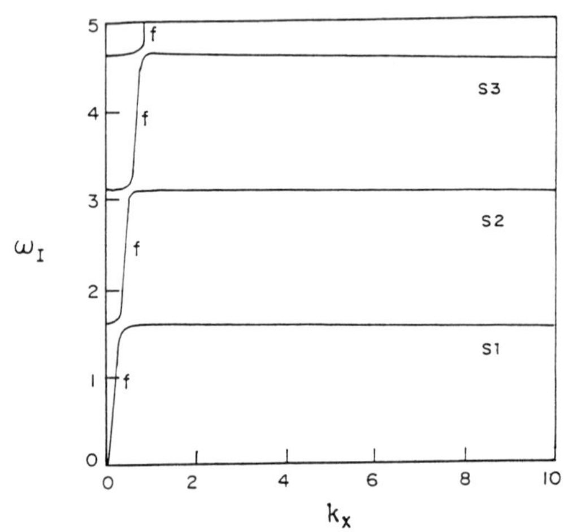

Figure 2. Frequency vs wavenumber plot corresponding to f, fast and slow modes in the presence of vertical magnetic field for which $\beta \ll 1$.

speed in the overlying atmosphere and increasing sound speed with depth in the convection zone.

Antia, Chitre and Gokhale (1978) considered a simplified problem with a view to studying the stability of a single plane-parallel polytropic inviscid, infinitely electrically conducting fluid layer, but with finite thermal conductivity, in the presence of a uniform horizontal magnetic field. For the range of parameters appropriate to the penumbra we choose a single layer of thickness $d = 850$ km, possessing a uniform horizontal magnetic field $B_0 = 1000$ G, $g = 2.47 \times 10^4$ cm s^{-2}, $\rho_{base} = 2.67 \times 10^{-6}$ gm cm^{-3}, $T_{base} = 15000$ K, $\Gamma = 1.66$ and $\gamma = 1.1$. The dimensionless complex numerical eigenvalues for the lowest fast (F1) mode are shown in Table 2.

TABLE 2. Complex eigenvalues for different values of k_x for the lowest fast mode (F1) corresponding to the choice of parameters: $G_K = 0.035, G_B = 0.024$.

k_x	*Mode*	*Growth rate*	*Frequency*
0	F1	$9.81 \times 10\ -3$	1.681
1	F1	$1.25 \times 10\ -2$	1.889
2	F1	$6.72 \times 10\ -3$	2.399

The maximum growth rate is obtained for a period of order 150 s and the corresponding wavelength is $\sim$ 3173 km. Thus, running penumbral waves may be identified with the fast magneto-atmospheric (F1) mode which turns out to be the most unstable. The corresponding most unstable slow mode, for the same choice of the basic parameters, yields a period $\sim$ 1700 s and it seems unlikely that the running penumbral waves have any relationship with the slow modes, although it will be interesting to identify this longer period observationally in the spot penumbra. Nye and Thomas (1974) have treated a more realistic three-layer model for the penumbra with adiabatic perturbations to conclude that the penumbral waves should be identified with gravity-modified magnetoacoustic fast modes that are vertically trapped at the photospheric levels. The source of excitation for penumbral waves is again oscillatory convection in the superadiabatic layer below the spot photosphere (Moore, 1973). Galloway (1978) has also suggested that the running penumbral waves are generated within the penumbral layer by overstable convection in the presence of the horizontal magnetic field. We conclude that the running penumbral waves are basically a photospheric phenomenon, since the phase velocity ($\sim$ 20 km s^{-1}) is typical of photospheric conditions rather than chromospheric. It will be interesting, however, to investigate the character and stability of slow modes in the chromospheric magnetic canopy above the penumbra.

4. Five-minute umbral oscillations

The observations of solar p-mode oscillations have been effectively used to infer the physical condition prevailing in the solar interior. Thus, acoustic modes of low-degree spherical harmonics ($\ell \lesssim 10$) sample a large portion of the solar body, while the high-degree ($\ell \lesssim 150$) acoustic modes probe only the outer regions of the solar convection zone. In fact, these high-degree acoustic modes are very well suited to yield information about the inhomogeneities that might be present in the subsurface layers – e.g. there has been some interest in studying the interaction of solar p-modes with the sunspot magnetic field in order to probe the structure of a sunspot with depth. Earlier attempts to use oscillations as a diagnostic tool of sunspot structure were concerned with umbral velocity oscillations. Thomas (1981) suggested that the 5-min oscillations observed in a sunspot umbra may be regarded as the response of the spot flux tube to the 5-min p-mode. Thomas, Cram and Nye (1982) argued that the 5-min oscillations detected in the umbra might be used as a probe of the sunspot structure, thus ushering in the era of sunspot seismology. Subsequently, Abdelatif, Lites and Thomas (1986) demonstrated that the spot umbra acts as a selective filter in admitting certain frequencies of p-mode oscillations more than others. The horizontally propagating p-modes incident on a sunspot flux tube are expected to be partially transmitted into the tube as magnetoacoustic waves. It is also found that there is a shift of power to longer horizontal wavelengths in the power spectrum of umbral oscillations compared to the power spectrum of oscillations in the surrounding convection zone; this could be attributed to the difference between the magnetoacoustic speed of a wave with a frequency ω and the sound speed of the wave in the ambient convection zone (cf. Abdelatif and Thomas, 1987).

The recent work of Braun, Duvall and LaBonte (1988) has quantified how the presence of a sunspot affects externally impinging p-mode waves. It is shown that for each of the sunspot observed there is a deficit of outgoing power relative to the incoming power and this large excess of power in the incoming acoustic waves over the outgoing waves may be interpreted as being due to some absorption process occurring in the sunspot. The observed absorption of the p-mode flux incident on the spot flux tube provides a valuable diagnostic probe of the interior of a sunspot in much the same manner as the study of an atomic structure by its bombardment with electromagnetic radiation and observation of the scattered radiation field.

It is quite possible that the incoming waves shake the magnetic field of the sunspot flux tube and energy which is initially in sound waves is transformed into magnetoacoustic wave moves propagating move or less parallel to the magnetic field of the spot. If these waves can then escape from the system, say, upward through the chromospheric layers or downward through the base of the sunspot, it would then be possible to explain the observed absorption by a completely non-dissipative process involving simply mode conversion in the presence of the spot magnetic field. It is also conceivable that part of the incoming wave flux that is transmitted into the flux tube could get converted into heat within the sunspot by some irreversible dissipative processes. The most frequently invoked mechanism is resonance absorption (Davila, 1987; Hollweg, 1988). An essential feature of the process of resonance absorption is that large-scale motions which are very nearly dissipationless can resonantly excite small-scale waves inside the flux tube whenever the global wave frequency equals the local Alfvén wave frequency. Thus, basic to the resonance theory is the existence of a global mode and its subsequent coupling to the small-scale dissipative waves.

The observations of Braun, LaBonte and Duvall (1990) show a variation of the absorption of incoming acoustic waves both with wavelength and azimuthal mode number. The absorption is found to be almost negligible for small transverse wavenumbers, but appears to exceed 50% for large values of the transverse wavenumber. Physically, we expect this variation of the absorption with wavenumbers to be due to some characteristic length of the system, such as the radius of the absorber or some characteristic gradient scale. It is, therefore, reasonable to expect that the amplitudes of scattered waves from the flux tube contain information about the sunspot structure in the subsurface layers. This clearly could become a powerful diagnostic tool for probing the structure of sunspots, once the mechanism responsible for the observed absorption of the incident acoustic flux is identified.

Chitre and Davila (1991) considered an inhomogeneous sunspot model with a vertical and azimuthal magnetic field to investigate the absorption of incoming acoustic modes in a narrow resonance layer inside the sunspot flux tube. It was demonstrated that in the presence of a twisted magnetic field any weak dissipative process (e.g. radiative or viscous dissipation) leads to the resonant absorption of both axisymmetric and nonaxisymmetric acoustic waves, as is observed (see also Rosenthal, 1990). The resultant heating rate turns out to be consistent with the observed absorption of the p-mode power incident on an isolated, inhomogeneously structured sunspot. If the calculated absorption rate is used in conjunction with the observed wavelength dependence of the absorption coefficient, it is found that the spot magnetic field must have a significant amount of twist in the portion of the flux tube below the photosphere. Furthermore, the presence of a twist in the magnetic field leads naturally to the observed dependence on the azimuthal wave mode number.

5. Conclusions

The unstable oscillatory modes in a thermally conducting inviscid polytropic fluid layer are computed separately for a uniform vertical and horizontal magnetic field. The main conclusions from the numerical computations are that for the range of parameters appropriate to the sunspot umbra, the overstable fast modes generated in the vertical magnetic field are in reasonable agreement with the observed features associated with 3-min umbral oscillations. Likewise, the computations of an idealized penumbral model also suggest that running penumbral waves could be identified with the fast magnetoacoustic modes propagating along the nearly horizontal magnetic field of the penumbra.

The study of instabilities for the sunspot umbra and penumbra reported here are of necessity applicable only to simplified spot models. It is naturally desirable to consider more realistic magnetic field configurations in the subphotospheric layers of the sunspot, rather that considering merely uniform horizontal or vertical magnetic field components. The stability analysis has also been done hitherto in the linear framework and there is need to undertake nonlinear calculations. There have been some beginnings of investigations into magnetoconvection in the nonlinear regime in the work of Knobloch et al. (1981) and Proctor and Weiss (1982). There are recent studies of nonlinear compressible magnetoconvection by Hurlburt and Toomre (1988), Hurlburt et al (1989) and Weiss et al (1990). (See also reviews by Hughes and Proctor, 1988, Weiss, 1991 and Proctor in this volume).

The stratification due to gravity is likely to be an important physical effect that must be included in any realistic sunspot model, as also is the incorporation of two-dimensional models for the sunspot flux tube.

Acknowledgement. It is a pleasure to express my gratitude to H. M. Antia and J. M. Davila for several valuable discussions and assistance with numerical computations. I should also like to thank Jack Thomas for stimulating my renewed interest in this field and for supplying me with useful background material. The support provided by Nigel Weiss and his comments on the manuscript are gratefully acknowledged.

References

Abdelatif, T. E., Lites, B. W. and J. H. Thomas (1986), *Ap. J.* **311**, 1015.

Abdelatif, T. E. and Thomas, J. H. (1987), *Ap. J.* **320**, 8 84.

Antia, H. M. (1979) *J. Comp. Phys.* **30**, 283.

Antia, H. M., Chitre, S. M. and D. M. Kale (1978), *Solar Phys.* **56**, 275.

Antia, H. M., Chitre, S. M. and M. H. Gokhale (1978), *Solar Phys.* **60**, 31.

Antia, H. M. and Chitre, S. M. (1979), *Solar Phys.* **63**, 67.

Bhatnagar, A. and Tanaka, K. (1972), *Solar Phys.* **24**, 87.

Bhatnagar, A., Livinston, W. C. amd Harvey, J. W. (1972), *Solar Phys.* **27**, 80.

Braun, D. C., Duvall, T. L. and B. J. LaBonte (1988), *Ap.J.* **335**, 1015.

Braun, D. C., LaBonte, B. J. and Duvall, T. L. (1990), *Ap.J.* **354**, 372.

Beckers, J. M. and Tallant, P. E. (1969) *Solar Phys.* **7**, 351.

Beckers, J. M. and Schultz, R. B. (1972), *Solar Phys.* **27**, 61.

Bray, R. J. and Loughead, R. E. (1964), *Sunspots*: Chapman and Hall, London.

Chandrasekhar, S. (1952), *Phil. Mag.* **43**, 501.

Chitre, S. M. (1963), *M.N.R.A.S.* **126**, 431.

Chitre, S. M. and Davila, J. M. (1991), *Ap.J.* **371**, 785.

Danielson, R. E. (1961), *Ap.J.* **134**, 275.

Davila, J. M. (1987), *Ap.J.* **317**, 514.

Deinzer, W. (1965), *Ap.J.* **141**, 548.

Galloway, D. J. (1978), *M.N.R.A.S.* **184**, 49P.

Giovanelli, R. G. (1972), *Solar Phys.* **27**, 71.

Giovanelli, R. G., Harvey, J. W. and Livingston, W. C. (1978), *Solar Phys.* **58**, 347.

Gurman, J. B., Leibacher, J. W., Steine, R. A., Woodgate, B. E. and Henze, W. (1982), *Ap.J.* **253**, 939.

Gurman, J. B. and Leibacher, J. W. (1984), *Ap.J.* **283**, 85 9.

Hollweg, J. V. (1988), *Ap.J.* **335**, 1005.

Hughes, D.W. and Proctor, M.R.E. (1988) *Ann. Rev. Fluid Mech.* **20**, 187

Hurlburt, N.E. and Toomre, J. (1988), *Ap. J.* **327**, 920.

Hurlburt, N.E., Proctor, M.R.E., Weiss, N.O. and Brownjohn, D.P. (1989) *J.Fluid.Mech.* **209**, 587.

Knobloch, E., Weiss, N. O., DaCosta, L. N. (1981), *J. Fluid. Mech.* **113**, 153.

Lites, B.W. and Thomas, J.H. (1985), *Ap. J.* **294**, 682

Moore, R. L. (1973), *Solar Phys.* **30**, 403.

Mullan, D. J. and Yun, H. S. (1973), *Solar Phys.* **30**, 83.

Nye, A. H. and Thomas, J. H. (1974), *Solar Phys.* **38**, 399.

Nye, A. H. and Thomas, J. H. (1976), *Ap.J.* **204**, 582.

Parker, E. N. (1974), *Solar Phys.* **37**, 127.

Proctor, M. R. E. and Weiss, N. O. (1982), *Rep. Prog. Phys.* **45**, 1317.

Rosenthal, C.S. (1990) *Solar Phys.* **130**, 313.

Savage, B. D. (1969), *Ap.J.* **156**, 707.

Scheuer, M. A. and Thomas, J. H. (1981), *Solar Phys.* **71**, 21.

Solatu, D., Schröter, E. H. and Wöhl, H. (1976), *Astron. Ap.* **50**, 367.

Thomas, J. H. (1981) in *The Physics of Sunspots* (ed. L. E. Cram and J. H. Thomas) 345.

Thomas, J.H. and Scheuer, M.A. (1982), *Solar Phys.* **79**, 19.

Thomas, J. H. (1985), *Aust. J. Phys.* **38**, 811.

Thomas, J. H., Cram, L. E. and Nye, A. H. (1984), *Ap.J.* **285**, 368.

Thomas, J.H., Lites, B.W., Gurman, J.B. and Ladd, E.F. (1987), *Ap. J.* **312**, 457.

Uchida, Y. and Sakurai, T. (1975), *P.A.S.J.* **27**, 259.

Weiss, N.O., Brownjohn, D.P., Hurlburt, N.E. and Proctor, M.R.E. (1990), *M.N.R.A.S.* **245**, 434.

Weiss, N.O. (1991), *Geophys. Ap. Fluid Dyn.* (in press)

Yun, H. S. (1970), *Ap.J.* **162**, 975.

Zhugzhda, Yu, D., Locans, V. and Staude, J. (1983), *Solar Phys.* **82**, 369.

Zhugzhda, Yu, D., Staude, J. and Locans, V. (1984), *Solar Phys.* **91**, 219.

Zirin, H. and Stein, A. (1972), *Ap.J. Lett.* **173**, L85.

SUNSPOT SEISMOLOGY: THE INTERACTION OF A SUNSPOT WITH SOLAR P-MODES

T. J. BOGDAN
High Altitude Observatory
*National Center for Atmospheric Research**
P.O. Box 3000
Boulder, Colorado 80307, U.S.A.

ABSTRACT. This article surveys developments in "sunspot seismology" from the decade following the original suggestion of Thomas *et al* (1982) that observations of solar acoustic oscillations in and about sunspots could be used to probe the internal structure of the sunspot with depth. Both the theoretical and observational components of sunspot seismology are considered at length, but considerably less is said about the presently ill-defined "inverse problem" for sunspot seismology. Much of our present understanding of the interaction of acoustic oscillations with magnetic flux tubes has been developed through careful consideration of idealized but instructive examples. A large part of this review is accordingly devoted to a discussion of these important sunspot seismology paradigms. Undoubtedly the highlight of the last ten years of research is the observational program carried out by Braun, Duvall and La Bonte (1987,1988,1990) demonstrating (i) that the interaction of p-modes with sunspots is sufficiently robust to be measured, and moreover, (ii) that sunspots appear to absorb up to 50% of the acoustic energy flux that is incident upon them.

1. Introduction

In the course of their travels through the solar interior, the trapped acoustic oscillations (p-modes) frequently encounter regions of magnetized plasma (magnetic flux tubes). Acoustic waves propagate because of the interplay of the fluid inertia (ρ, the fluid density) and the restoring force due to the fluid compressibility ($1/\gamma p$, p is the pressure and γ is the ratio of specific heats). Therefore, p-modes sense magnetic flux tubes because the magnetic field, (i) alters the compressibility $[1/(\gamma p + B^2/4\pi)]$, (ii) introduces an additional restoring force, the magnetic tension, and (iii) can (indirectly) cause the density of the magnetized plasma to differ from that of its surroundings. Most of the interactions between p-modes and flux tubes occur below the visible solar surface and consequently go undetected except insofar as their cumulative effects can slightly shift p-mode frequencies and/or split degenerate frequency multiplets. However, because the magnetic field does thread the solar surface in the form of long-lived, isolated, flux concentrations, and because the acoustic oscillations with frequencies between 2.5 and 4.5 mHz have upper reflection points within a density scale height or so of the $\tau_{5000} = 1$ layer, it becomes possible to observe *in detail* some aspects of the encounters between p-modes and magnetic flux

* *The National Center for Atmospheric Research is sponsored by the National Science Foundation.*

J. H. Thomas and N. O. Weiss (eds.), Sunspots: Theory and Observations, 345–367.

tubes in the solar envelope.

The study of the interaction of p-modes with the solar surface magnetic flux concentrations is termed "sunspot seismology" (Thomas *et al* 1982; Abdelatif & Thomas 1987), and it encompasses three related areas of investigation. The first area is purely observational and its concerns are the acquisition and analysis of high spatial and temporal resolution information about the p-modes in the quiet Sun surrounding the sunspot, as well as the magnetohydrodynamic (MHD) waves within the sunspot proper. Operationally this is achieved by recording time series of Doppler images taken in absorption lines formed fairly low in the photosphere. Data analysis then centers on assigning amplitudes and phases to the incoming and outgoing acoustic waves in the surrounding quiet Sun, while trying to isolate those waves within the sunspot that originate in response to the external acoustic forcing. The second line of inquiry is purely theoretical and seeks to identify, and understand, the nature of the physical processes that mediate the interaction of acoustic waves with magnetic flux tubes. The basic underlying question concerns the type of information about the structure of the magnetic flux tube which *in principle* can be encoded in the amplitudes and phases of the external p-modes and the forced internal magnetoacoustic-gravity waves. The third area of investigation examines a synthesis of theory and observation. The key issue here is what *in practice* can be learned about flux tube structure from less than perfect measurements compromised by unavoidable uncertainties and interpreted through idealized theoretical models.

This contribution surveys these three key area of active research in sunspot seismology in their broader context through a concise overview of the principal developments in theory and observation. No attempt is made to provide an inclusive resumé of all the relevant literature; the interested reader is urged to consult the previous reviews of the subject by Davila (1990) and Rosenthal (1991), who emphasize perspectives which are complementary to the one adopted here. The related review articles by Chitre (1992), Lites (1992), and Roberts (1992) that appear in this volume are also particularly valuable for their overlapping treatment of some common topics from significantly different points of view.

The remainder of this review is structured as follows. Theoretical investigations are treated in §§2,3. Section 2 addresses the interaction of sound waves with magnetic equilibria that are translationally invariant along a principal axis [see Fig. 1]. In §2.1 the basic apparatus for treating the interaction of acoustic waves with discrete structures is developed, and the nomenclature that provides the basis for the subsequent sections is introduced. Sections 2.2-2.4 proceed with a discussion of our current understanding of the interaction of p-modes with flux tubes in unstratified atmospheres. Novel facets of this interaction in gravitationally stratified atmospheres are highlighted in §3. Section 4 summarizes the present status of the observational component of sunspot seismology. The striking development here is the analysis of incoming and outgoing acoustic waves outside of a sunspot that was carried out by Braun *et al* (1987, 1988), which indicates a deficit of up to 50% in the outgoing acoustic power. This fascinating and largely unanticipated property of sunspots has provided an impetus for much of the ensuing theoretical developments. A synthesis of theory and observation which comprises the "inverse problem" in sunspot seismology is touched upon briefly in the concluding section (§5).

2. Interaction of Acoustic Waves with Magnetic Flux Tubes in Unstratified Atmospheres

2.1. THE T-MATRIX FORMULATION OF ACOUSTIC SCATTERING

In the surrounding homogeneous ideal fluid in which the magnetic flux tube is embedded, it is convenient to work with the velocity potential, $\phi(\mathbf{x}, t)$, from which the pressure and velocity

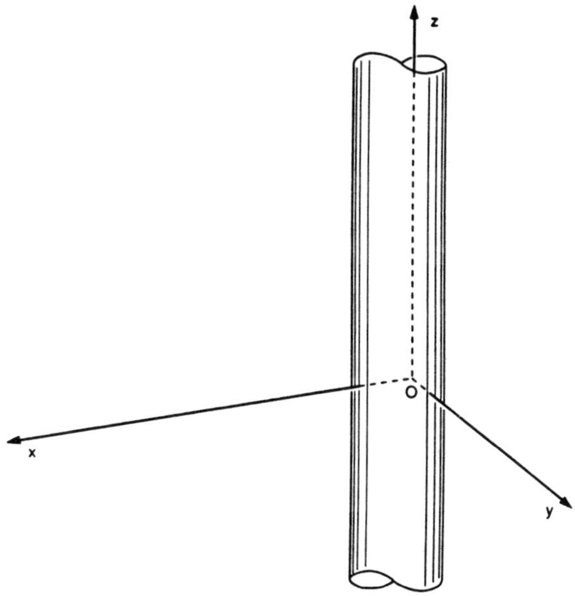

Figure 1. A cylindrical magnetic flux tube embedded in an homogeneous unmagnetized atmosphere.

fluctuations are given by the usual expressions:

$$\delta p(\mathbf{x},t) = \rho_0 \, \partial\phi/\partial t, \quad \delta \mathbf{v}(\mathbf{x},t) = -\nabla\phi \ . \tag{2.1}$$

The potential ϕ satisfies the scalar wave equation $\Box^2\phi = A(\mathbf{x},t)$, where $A(\mathbf{x},t)$ accounts for the sources of sound. In the unmagnetized fluid, the solution of the inhomogeneous wave equation is:

$$\phi(\mathbf{x},t) = \frac{1}{4\pi^2}\int d\omega \int dk_\parallel \sum_{m=-\infty}^{\infty} i^m e^{i[m\theta+k_\parallel z-\omega t]} \sum_{p=-\infty}^{\infty} A_p(k_\parallel,\omega)$$

$$\left[\delta_{p,m}\, J_m(k_\perp^{(a)} r) - T_{pm}(k_\parallel,\omega) H_m^{(1)}(k_\perp^{(a)} r)\right] \ , \tag{2.2}$$

where r, θ, z are cylindrical coordinates aligned with the axis of the magnetic flux tube (see Fig. 1), and $\delta_{p,m}$ is the Kronecker delta. The wavevector has been resolved into components parallel to, $k_\parallel$, and perpendicular to $k_\perp^{(a)} = k_\perp^{(a)}(k_\parallel,\omega) \equiv [\omega^2/C_0^2 - k_\parallel^2]^{1/2}$ ("a" = acoustic), the $[\hat{e}_z]$ flux tube axis, $J_m(x)$ and $H_m^{(1)}(x)$ are the Bessel and Hankel functions of the first kind, and $C_0 = (\gamma p_0/\rho_0)^{1/2}$ is the adiabatic sound speed. The branch defining $k_\perp^{(a)}$ is chosen such that $k_\perp^{(a)} = i[k_\parallel^2 - \omega^2/C_0^2]^{1/2}$ when $k_\parallel^2 C_0^2 > \omega^2$.

The coefficients $A_p(k_{\|},\omega)$ describe the acoustic field that is obtained in the absence of the magnetic flux tube, while the T-matrix, with components $T_{pm}(k_{\|},\omega)$, accounts for the additional components of the acoustic field resulting from the presence of the flux tube (Bates 1980). An alternate form for the term in square brackets in equation (2.2) is $\frac{1}{2}[\delta_{p,m} H_m^{(2)}(k_\perp^{(a)} r) + S_{pm}(k_{\|},\omega)\ H_m^{(1)}(k_\perp^{(a)} r)]$, where the components of the S-matrix are related to those of the T-matrix by $S_{pm} = \delta_{p,m} - 2\ T_{pm}$. When the magnetic flux tube is axisymmetric both the T- and S-matrices are diagonal and the shorthand rotation $T_{pm} \rightarrow \delta_{p,m} \cdot T_m$ is employed. Except in §2.4 where the cluster model is discussed, only axisymmetric flux tube equilibria will be considered here [for flux tubes with elliptic cross sections consult Morse & Feshbach (1953)], and hence only diagonal T- and S-matrices are necessary.

The T-matrix contains all the essential information necessary to describe both the transient and long-term interaction of acoustic disturbances with a particular magnetic flux tube. The poles in the complex $k_{\|} - \omega$ plane of the $T_m(k_{\|},\omega)$ describe the characteristic modes of oscillation of the flux tube, the so-called *tube modes*, and the residue provided by the remainder of the integrand at these poles gives the excitation amplitude of these modes for a given distribution of external acoustic sources. Additionally, the elements of the T-matrix may possess branch points and essential singularities which will typically account for the complicated transient response of the flux tube and its environs. The residue contributed by the elements of the T-matrix at the poles of $A_m(k_{\|},\omega)$ describes the steady-state scattered acoustic field that results when an acoustic disturbance of prescribed frequencies and wavenumbers insonifies the flux tube. In this limit one can readily construct the scattering and absorption cross sections (per unit flux tube length) by evaluating the T-matrix at the specified frequency and wavenumber (Morse & Ingard 1986):

$$\left.\begin{matrix} \sigma_s(k_{\|},\omega) \\ \sigma_a(k_{\|},\omega) \end{matrix}\right\} = \frac{C_0}{|\omega|} \sum_{m=-\infty}^{\infty} \left\{\begin{matrix} 4\ |T_m(k_{\|},\omega)|^2 \\ \alpha_m(k_{\|},\omega) \end{matrix}\right. \qquad [2.3]$$

where

$$\alpha_m(k_{\|},\omega) \equiv 1 - |1 - 2\ T_m(k_{\|},\omega)|^2 = 1 - |S_m(k_{\|},\omega)|^2 \qquad [2.4]$$

is the absorption coefficient employed by Braun *et al* (1988). The absorption coefficient vanishes when the outgoing wave is simply phase-shifted relative to the incoming wave, and then one writes $S_m(k_{\|},\omega)$ =exp$[2i\delta_m(k_{\|},\omega)]$, with $\delta_m(k_{\|},\omega)$ denoting the m-th partial wave phase shift. If $\sigma_a(k_{\|},\omega) = 0$ then the scattering is elastic and no acoustic energy is absorbed by the flux tube in the steady-state.

The T-matrix is obtained by suitably matching the exterior solution [equation (2.2)] to a solution of the linearized MHD equations pertinent to the interior of the magnetic flux tube. This matching generally requires enforcing the continuity of the normal displacement and the total pressure perturbation across the interface bounding the flux tube. In practice such a procedure is complicated by the fact that the wave modes of a spatially-varying magnetized plasma can be quite complex, and that the interface may be irregular in shape. Presently, only a few very special equilibria are known for which this matching can be carried out and for which T-matrices are readily obtained. These equilibria are of necessity quite idealized, but they nevertheless prove to be invaluable for the insights they offer for understanding more realistic and complex configurations. These special equilibria are discussed in §§2.2-2.4 below.

2.2. THE UNIFORMLY MAGNETIZED FLUX TUBE

The simplest equilibrium model for sunspot seismology consists of an isothermal uniformly magnetized cylinder of radius a, embedded in a homogeneous ideal fluid. The internal structure of such a flux tube is described by two additional parameters: the sound speed C_S and the Alfvén speed C_A. Lateral force balance across the current sheet bounding the flux tube sets the ratio of internal to external fluid density $\eta = 2C_0^2/(2C_S^2 + \gamma C_A^2)$, where for the sake of simplicity it has been assumed that γ is the same in the magnetized and unmagnetized media.

The linearized equations valid within the flux tube admit plane-wave solutions, but unlike the surrounding fluid, these motions have nonvanishing vorticity and an anisotropic dispersion relation with $k_\perp^{(ma)}(k_\parallel,\omega) = [(\omega^2 - k_\parallel^2 C_S^2)(\omega^2 - k_\parallel^2 C_A^2)/(C_S^2 + C_A^2)(\omega^2 - k_\parallel^2 C_T^2)]^{1/2}$ ("ma" = magnetoacoustic), with $C_T \equiv C_S C_A/(C_S^2 + C_A^2)^{1/2}$ denoting the tube, or cusp, speed. The matching between the internal and external solutions is readily carried out and one obtains (Abdelatif & Thomas 1987; Ryutov & Ryutova 1976; Wilson 1980; Bogdan 1989):

$$T_m(k_\parallel,\omega) = \frac{\omega^2 k_\perp^{(ma)} J_m'(k_\perp^{(ma)}a) J_m(k_\perp^{(a)}a) - \eta[\omega^2 - k_\parallel^2 C_A^2] k_\perp^{(a)} J_m'(k_\perp^{(a)}a) J_m(k_\perp^{(ma)}a)}{\omega^2 k_\perp^{(ma)} J_m'(k_\perp^{(ma)}a) H_m^{(1)}(k_\perp^{(a)}a) - \eta[\omega^2 - k_\parallel^2 C_A^2] k_\perp^{(a)} H_m^{(1)\prime}(k_\perp^{(a)}a) J_m(k_\perp^{(ma)}a)} . \qquad [2.5]$$

When ω and $k_\parallel$ are both real and satisfy $|\omega/k_\parallel| < C_0$, the denominator of equation (2.5) is real and may have isolated zeroes. These zeroes are the poles of the T-matrix and describe tube modes that are confined to propagate along the flux tube and decline exponentially away from the flux tube into the unmagnetized fluid (McKenzie 1970; Defouw 1976; Abdelatif 1988; Evans & Roberts 1990). On the other hand, when $|\omega/k_\parallel| > C_0$ the numerator of equation (2.5) is the real part of the denominator, implying (*i*) each $\alpha_m(k_\parallel,\omega)$ vanishes identically, and (*ii*) that the maximal value $|T_m(k_\parallel,\omega)|$ can attain is unity, which occurs when the imaginary component of the denominator vanishes. The first property indicates that in the steady-state interaction of an acoustic plane-wave with a uniformly magnetized flux tube no energy is absorbed: the scattering is purely elastic. The second property indicates that when a flux tube is insonified at one of its scattering resonances, where a particular $T_m(k_\parallel,\omega) = 1$, the scattering cross section exceeds the wavelength of the incident acoustic wave *independent of the radius of the flux tube* (Bogdan 1989). These scattering resonances—or alternatively leaky tube modes (Spruit 1982; Cally 1985, 1986)—result from poles of the T-matrix that lie slightly off the real $k_\parallel$ and ω axes. The offset distance gives the quality factor of the resonance, which measures the ease with which the tube mode can couple to outgoing acoustic waves.

Acoustically compact flux tubes (diameter $\ll$ wavelength) couple very weakly to outgoing acoustic waves and consequently their scattering resonances have extremely high quality factors. As a result, the level of perturbations in the near-field ($|\mathbf{r}| \gtrsim a$) are enhanced by a factor $\simeq |H_m^{(1)}(k_\perp^{(a)}a)/J_m(k_\perp^{(a)}a)|$ relative to their off-resonance values. The scattering resonances occur for incident acoustic waves with parallel phase velocities in the neighborhood of $\omega/k_\parallel \simeq \pm C_k \equiv \pm C_A/(1+\eta^{-1})^{1/2}$, for $m \neq 0$, and $\omega/k_\parallel \simeq \pm C_T$. The former resonances are associated with the excitation of the surface kink, or bending, oscillations (Ryutov & Ryutova 1976), while the latter are associated with the excitation of slow body modes (Defouw 1976). Since resonance scattering requires that these characteristic parallel phase velocities be in excess of the external sound speed, only highly evacuated ($\pm C_k$), or very warm ($\pm C_T$), slender flux tubes will possess scattering resonances.

During the transient phase of the insonification of a slender flux tube at one of its scattering resonances, energy is extracted from the incident waves to build this enhanced near-field response, a process which Ryutova (1986; Ryutov & Ryutova 1976) calls acoustic absorption via the Landau resonance. At late times, however, when the near-field motions have saturated at their asymptotic level, there is a detailed balance between absorption and re-emission, leading to the zero absorption cross section predicted by equations (2.3) and (2.5). For sufficiently compact flux tubes, the near-field motions will be so large that nonlinear effects enter. The potentially interesting consequences of this phenomenon, especially for the cluster sunspot model, have not yet been fully explored.

When the flux tubes are large (diameter $\geq$ wavelength) the scattering resonances become plentiful but are offset at such large distances from the real axes that their individual quality factors are small. Abdelatif and Thomas (1987) find that in this limit, the superposition of these many resonances leads to a gentle sinusoidal variation of the scattering cross section with the internal horizontal wavenumber, $k_\perp^{(ma)}$, on the characteristic scale scale $\sim \pi/a$.

A uniform background field-aligned flow, $\mathbf{U} = U\hat{\mathbf{e}}_z$, within the flux tube still results in an analytic solution for the scattering problem. One simply makes the replacement $\omega \to \omega - k_\parallel U$ throughout equation (2.5) *with the exception* of the ω^2's which multiply the first of the two terms in the numerator and the denominator, and of course in the definition of the external perpendicular wavenumber (McKenzie 1970; Bogdan 1989). The additional physical effects entailed by such a procedure are not as straightforward. In particular, the flux tube is now bounded by a vortex sheet, and with the appropriately modified T-matrix, equation (2.2) must now describe the vagaries of the magnetized Kelvin-Helmholtz instability in the relevant parameter range (Birkinshaw 1984; Wu & Wang 1991). The flow shifts the locations of the scattering resonances of a slender flux tube according to (Bogdan 1989): $\omega/k_\parallel \simeq \eta U/(1+\eta) \pm [C_k^2 - \eta U^2/(1+\eta)^2]^{1/2}$ and $\omega/k_\parallel \simeq U \pm C_T$. A more intriguing and subtle manifestation of the vortex sheet is the so-called negative energy wave (NEW) phenomenon (McKenzie 1972; Craik 1985), noted in the sunspot seismology context by Rytuova (1988). For flow speeds well below the threshold for the magnetized Kelvin-Helmholtz instability, the presence of a slight amount of viscosity allows the negative energy waves to grow in amplitude by extracting energy and momentum from the background flow field. Including nonlinear effects, via mode-mode coupling, for example, appears to achieve basically the same end, although the time scales can be quite different. The consequences of this process for the interaction of acoustic waves with a magnetic flux tube situated in an organized downflow have yet to be spelled out in any detail, but by analogy with wave interactions with shear flows (Miles 1957; McKenzie 1972; Jones 1977), one might conceivably find both under-reflection $[\alpha_m(k_\parallel,\omega) > 0]$ and over-reflection $[\alpha_m(k_\parallel,\omega) < 0]$.

Another interesting extension of the uniformly magnetized cylinder model, which is still computationally feasible, takes into account the influence of the radiation field in p-mode/sunspot scattering (Bogdan & Knölker 1991). The radiation is treated as gray and is assumed to satisfy local thermodynamic equilibrium. In addition, the Eddington approximation is used to relate the perturbed radiation pressure tensor to the perturbed mean intensity. The equilibrium configuration has the flux tube in both mechanical and radiative equilibrium with its surroundings, and one must supply a constitutive relation, like Kramer's law, to relate the jump in opacity at the flux tube boundary to the density contrast η.

The resulting linearized equations of radiation hydrodynamics vitiate only the first of equations (2.1), and still permit plane-wave solutions with, however, two independent choices for the lateral wavenumber: $k_\perp^{(a)}(k_\parallel,\omega)$ corresponding to a radiatively damped acoustic mode, and $k_\perp^{(r)}(k_\parallel,\omega)$ representing a material-modified radiation-diffusion mode (Mihalas & Mihalas 1984; Bogdan & Knölker 1991). To account for the possibility of two types of scattered wave modes, one need only replace the second term in square brackets appearing in equation (2.2) by the

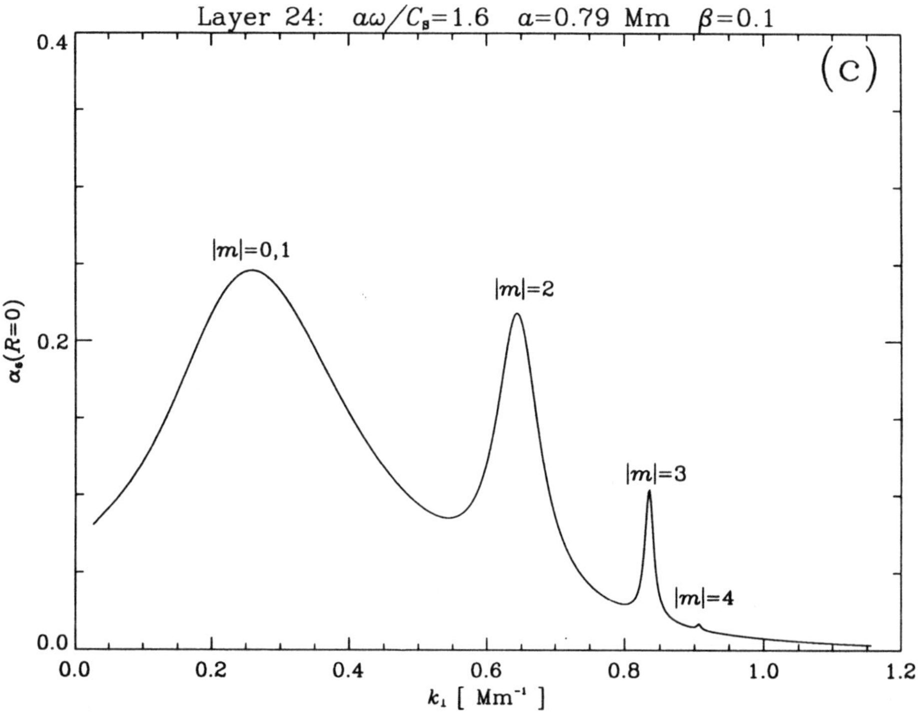

Figure 2. Absorption coefficient averaged over azimuthal orders between -5 and +5 as a function of the perpendicular p-mode wavenumber at fixed frequency. The external atmosphere represents conditions at a depth of 425 km in the quiet Sun. The azimuthal orders responsible for each absorption resonance are indicated [from Bogdan & Knölker 1991].

sum $T_{pm}^{(a)}(k_{\|},\omega)\ H_m^{(1)}(k_{\perp}^{(a)}r) + T_{pm}^{(r)}(k_{\|},\omega)\ H_m^{(1)}(k_{\perp}^{(r)}r)$. The conversion of acoustic modes to radiation-diffusion modes gives the appearance of acoustic absorption [*e.g.*, $|1 - 2\ T_m^{(a)}(k_{\|},\omega)| < 1$] because the radiation-diffusion modes carry off some of the energy that would otherwise have resided in the outgoing acoustic waves. On physical grounds this mode conversion results from the different levels of temperature (and hence intensity) perturbations associated with the external acoustic and internal magnetoacoustic wave modes, coupled with the jump in opacity at the flux tube surface. For optically thick disturbances, the radiation-diffusion wave is essentially a thermal relaxation wave and the damping of the acoustic disturbance results from the dissipation inherent in the thermal conduction across the flux tube boundary. Ryutova & Persson (1984) point out that this process can be greatly enhanced for magnetic flux tubes with a large circumference to cross sectional area ratio. In particular, a cluster sunspot composed of N identical constituent flux tubes has $N^{1/2}$ times more surface area to support such an acoustic dissipation mechanism than does a single monolithic sunspot of the same cross sectional area.

With the radiative effects included, the matching between the internal and external solutions is carried out as in the nonradiative case, only with the additional constraints that the continuity of the normal radiation flux and the perturbed radiation pressure must be maintained at the flux tube boundary. When the external acoustic waves are optically thick (optical depth of a wavelength $> 10^2$) this mode conversion is exceedingly weak except near the scattering resonances where absorption coefficients approaching unity are possible. Figure 2, taken from Bogdan and Knölker (1991), illustrates a best-case-scenario for acoustic absorption due conversion to radiation-diffusion modes. For this example the ratio of gas to magnetic pressure inside the flux tube is 1:10, the flux tube radius is 0.79 Mm, and the p-mode frequency and optical depth of one wavelength are 3.34 mHz, and 10^5, respectively, and the sound speed in the external atmosphere is 10.4 km s^{-1}.

On physical grounds, more interesting effects are expected when the optical depth of the incident wave is of order unity. However, the homogeneous external atmosphere fails to provide any insight in this case because the *in situ* radiative damping of the acoustic wave is so strong that the basic assumptions underlying the concepts of scattering and absorption cross sections become untenable. Here, one is really forced to look beyond this simple example towards stratified external atmospheres (Spiegel 1977) where a p-mode can spend most of its time deep in the atmosphere where radiative damping is inconsequential, while interacting with magnetic structures in surface layers where efficient mode conversion is a distinct possibility. In such studies, the Eddington approximation must necessarily be discarded in favor of the full linearized equation of transfer (Delache & Froeschlé 1972; Dzhalilov *et al* 1992).

2.3. FLUX TUBES WITH RADIAL VARIATIONS: RESONANCE ABSORPTION

The next level of sophistication in modeling the interaction of acoustic waves with magnetic flux tubes replaces the uniformly magnetized flux tube by one in which radial variations of the equilibrium are permitted. In addition to its radius a, the flux tube is now parameterized by three functions: $\Pi(r)$ the total pressure, which generates the azimuthal component of the magnetic field via $B_\theta^2(r)/4\pi = -rd\Pi/dr$; the axial magnetic field profile $B_z(r)$, which then gives the gas pressure $p(r) = B_z^2(r)/8\pi + d(r^2\Pi)/dr^2$; and the temperature, $T(r)$. The total pressure satisfies the constraints $\Pi(r) \geq 0$, $d\Pi/dr \leq 0$, $d(r^2\Pi)/dr^2 \geq 0$, while $\Pi(r=a)$ must equal the external gas pressure (Parker 1979a).

Under these more general circumstances, the linearized equations valid within the flux tube remain separable but no longer permit plane-wave solutions. In ideal MHD, the radial variations of the internal perturbations follow from the solution of a second-order ODE first derived by Hain and Lüst (1958). A striking feature of this equation is that singular points may arise at certain finite values of r. These cylindrical critical surfaces occur at radii $r_A(k_\|,\omega,m)$ and $r_c(k_\|,\omega,m)$ satisfying:

$$\omega^2 = \omega_A^2(r_A) \equiv \frac{1}{4\pi\rho(r_A)}\left\{\frac{m}{r_A}B_\theta(r_A) + k_\| B_z(r_A)\right\}^2 \qquad [2.6]$$

$$\omega^2 = \omega_c^2(r_c) \equiv \frac{C_S^2(r_c)\omega_A^2(r_c)}{C_S^2(r_c) + C_A^2(r_c)}, \qquad [2.7]$$

and in ideal MHD some [but not all (Hollweg 1988; Sakurai *et al* 1991a)] components of the perturbations will diverge at the critical surface. A consequence of this singular behavior of the internal eigenfunctions is that the resulting components of the T-matrix satisfy the inequality $|1-2T_m(k_\|,\omega)| < 1$, indicating a net flow of acoustic energy into the flux tube. This phenomenon is called "resonance absorption" (although in ideal MHD true absorption of the acoustic wave is impossible).

The physical underpinnings of resonance absorption become clear if one considers the approach to such a steady-state via an initial value problem formulation. Such calculations (Booker & Bretherton 1967; Kapraff & Tataronis 1977; Poedts *et al* 1990; Poedts & Kerner 1992) indicate that outside of the immediate neighborhood of the critical surface, the perturbed fields rapidly approach their steady-state behavior predicted by the singular solution, consistent with the steady inflow of energy into the neighborhood of the critical surface—the critical layer—dictated by the T-matrix. In the ever shrinking neighborhood of the critical surface an oscillatory steady-state is not achieved, but instead, amplitudes and radial gradients continue their monotonic growth. The analogy with the event horizon of a black hole is not uncompelling—information and energy approach, but are never quite able to reach, the critical surface after being trapped in the critical layer (Acheson 1973; McKenzie 1973). The "absorption" in ideal MHD is simply an artifact of the nonexistence of a global steady-state solution, and the deficit of outgoing acoustic energy is continually stored in the critical layer. In a practical sense this growth of amplitudes and radial gradients in the critical layer must eventually cease because (*i*) nonlinear terms are no longer negligible in the ideal MHD equations, or (*ii*) small but finite levels of dissipation can convert the energy flowing into this layer into heat. In the former case, it is not presently clear what (if any) steady-state is available to the problem (Margolis & Su 1978; Clark & Peltier 1984). In the latter case the ensuing steady-state can be calculated *ab initio* by solving the nonsingular ODE that results from linearizing the viscid and/or resistive equations. Such calculations (Lou 1990; Goossens & Poedts 1992) show that for sufficiently weak dissipation, the acoustic absorption is given by the T-matrix calculated in the singular ideal MHD limit *independent* of the level or nature (viscosity versus resistivity) of the dissipation mechanism.

Although resonance absorption has been a popular coronal heating mechanism since the late 1970's, it has only recently been considered as a possible explanation of p-mode absorption by sunspots. Hollweg's (1988) initial estimates assumed a cartesian slab geometry and invoked a sufficiently thin critical layer $[|d\omega_A^2(r_A)/dr| \gg \omega^2 k_\perp^{(a)}]$ that an approximate analytic solution of the singular ODE could be employed. In a subsequent investigation (Sakurai *et al* 1991a) a more realistic cylindrical geometry was used, but the restriction to thin critical layers was retained. This restriction permits instructive analytic formulae for the $\alpha_m(k_\|,\omega)$ (Sakurai *et al* 1991b), but it has the attendant disadvantage that the large radial gradient of $\omega_A^2(r)$ at the critical layer precludes interesting levels of acoustic absorption.

Lou (1988, 1990) numerically integrated the nonsingular viscid equations in an untwisted magnetic flux tube and was therefore not constrained to treating sharp critical layers. Lou's equilibrium is characterized by the generating functions:

$$\left\{\begin{aligned} \Pi(r) &= p_0 \\ \frac{B_z^2(r)}{8\pi} &= p_0\left[1-\frac{1}{3}\left[1+\frac{1}{2}\exp\left(-\lambda(r/a-1)^2\right)\right]\left[1+\exp\left(-\lambda(r/a-1)^2\right)\right]\right] \\ T(r) &= \frac{1}{3}\,T_0\left[2+\exp\left(-\lambda(r/a-1)^2\right)\right] \end{aligned}\right.$$

which contain an adjustable "sharpness" parameter λ, in addition to the pressure p_0 and temperature T_0 of the external atmosphere. The limit $\lambda \gg 1$ gives equilibria similar to those considered by Sakurai, Hollweg and Goossens (1991a,b). For typical values of $a, \omega, k_\|$ and $m = 1$ Lou showed that absorption coefficients as high as 0.5 could be obtained for the optimal value $\lambda = 5$, while significantly lower absorption levels were found at both larger and smaller values of λ. Since $B_\theta = 0$, no r_A exists when $m = 0$.

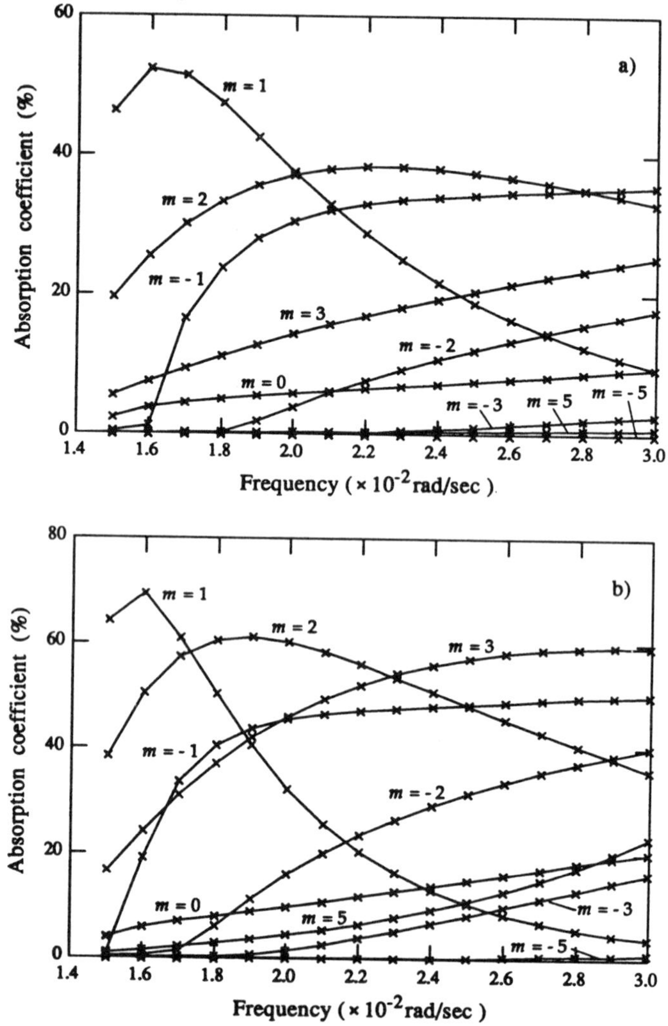

Figure 3. Absorption coefficients at fixed horizontal wavenumber as a function of p-mode frequency for a twisted magnetic flux tube for various azimuthal orders as indicated. In the top panel the flux tube radius is 4.2 Mm, while in the bottom panel it is 6.3 Mm [from Goossens & Poedts 1992].

Using the same equilibrium as Lou, Goossens and Poedts (1992) integrated the nonsingular resistive equations. They verified Lou's results and carried out an extensive survey of the relevant parameter space. They went on to include twist in Lou's equilibrium by setting

$$\Pi(r) = p_0 + \frac{4B^2_{\theta\max}}{\pi} \int_{r/a}^{1} ds\ s\ (1-s)^2$$

while leaving $p(r)$ and $T(r)$ unchanged. They pointed out that the presence of a significant amount of twist induces more efficient p-mode absorption. This is illustrated by Fig. 3, which is reproduced from their paper, and corresponds to an equilibrium configuration with $\lambda = 5$ and $B_{\theta\max}/B_z(r=0) = 0.1$, which is insonified by an acoustic plane-wave with a fixed lateral wavenumber $k_\perp^{(a)} = 1.0\ \mathrm{Mm}^{-1}$.

Rosenthal (1991, 1992) has also computed absorption coefficients by integrating the nonsingular viscid equations for diffuse equilibria, both with and without twist. He also considers randomly nested concentric magnetic flux rings given by the generating functions $\Pi(r) = p_0$, $T(r) = T_0[1 - \sigma B_z^2(r)/8\pi p_0]$, and

$$B_z(r) = \sum_{j=1}^{N} \sum_{\pm} \frac{B_j}{1 + |(r \pm r_j)/a|^{n_j} + |2(r \pm r_j)/3a|^{10}} \ ,$$

containing several adjustable parameters, some of which describe the location (r_j), strength (B_j), and sharpness in definition (n_j) of the flux rings. Although such an equilibrium model is admittedly contrived, it arguably captures some of the physics of the cluster sunspot model without the severe (albeit possible important) complications that arise from treating multiple scattering in two—as opposed to one—spatial dimensions (see §2.4 below). When N is large an incident p-mode can find many ($\leq 2N$) nested critical layers upon which it can be absorbed, and Rosenthal's calculations indeed indicate that much-enhanced absorption coefficients are obtained under these circumstances.

One suspects that large absorption coefficients should arise when the incoming acoustic wave can excite a global mode of oscillation of the flux tube while simultaneously coupling to a critical layer. The kink mode is a promising global resonance for untwisted flux tubes with $\omega_A^2(r=0) \gg \omega^2$ and a sharp transition from $\omega_A^2 = \omega_A^2(0)$ to zero over a thin boundary layer at the flux tube surface. The discrete Alfvén modes (Goedbloed 1984) are promising candidates for highly twisted flux tubes (Goossens & Poedts 1991). Resonance absorption in the presence of steady equilibrium velocity fields presents yet another interesting possibility for enhanced absorption through the coupling of the negative energy waves to the critical layer (Hollweg *et al* 1990; Yang & Hollweg 1991; Goossens *et al* 1992).

2.4. THE CLUSTER MODEL: MULTIPLE SCATTERING

The cluster (or spaghetti) sunspot model (Parker 1979b) can be treated by extending the models of §§2.2,2.3 to consider an equilibrium of several parallel magnetic flux tubes placed in close proximity to one another in a homogeneous unstratified atmosphere. The internal structure of the j-th flux tube can be specified independent of the others and is subject only to the usual constraint that it must be in mechanical equilibrium with its immediate unmagnetized environment.

Without loss of generality, one may assume that the scattering properties of the j-the flux tube in isolation are known, so that its T-matrix, $T_{pm}^{(j)}(k_\parallel, \omega)$, can be calculated along the lines set out in §§2.1-2.3 above. Equation (2.2) continues to be a valid representation of the acoustic field *outside* the flux tube cluster, with the unsuperscripted $T_{pm}(k_\parallel, \omega)$ now interpreted as the T-matrix for the composite scatterer. This T-matrix will, in general, be nondiagonal since the flux tube cluster is not expected to be axisymmetric about any particular origin. The dependence of $T_{pm}(k_\|,\omega)$ on the properties of the individual scatterers is illustrated by recasting equation (2.2) in the form:

$$\phi(\mathbf{x},t) = \frac{1}{4\pi^2}\int d\omega \int dk_\| \; e^{i[k_\| z-\omega t]} \sum_{m=-\infty}^{\infty} i^m \Big\{ A_m(k_\|,\omega)J_m(k_\perp^{(a)}r)e^{im\theta} - \\ - \sum_{j=1}^{N} B_m^{(j)}(k_\|,\omega)T_m^{(j)}(k_\|,\omega)H_m^{(1)}(k_\perp^{(a)}|\mathbf{r}-\mathbf{r}_j|)e^{im\theta_j} \Big\} \qquad [2.10]$$

where $T_{pm}^{(j)}(k_\|,\omega) = \delta_{p,m} \cdot T_m^{(j)}(k_\|,\omega)$ is the (assumed) diagonal T-matrix of the j-th flux tube, the axis of which threads the $x-y$ plane at $\mathbf{r}_j$, and θ_j is the local cylindrical coordinate angle, $\cos\theta_j = (\mathbf{r}-\mathbf{r}_j)\cdot\hat{\mathbf{e}}_x/|\mathbf{r}-\mathbf{r}_j|$. As before, the coefficients $A_m(k_\|,\omega)$ give the acoustic field that would be present (relative to the origin from which the individual flux tube position vectors, $\mathbf{r}_j$, are reckoned) in the absence of all flux tubes, and, unlike the direct generalization of equation (2.2), equation (2.10) is valid even in the unmagnetized lanes which separate the flux tubes.

The complications of multiple scatter are hidden in the vector of local excitation coefficients, $B_m^{(j)}(k_\|,\omega)$. These are related to the acoustic field that would be present in the absence of all flux tubes, represented relative to the *local* coordinate systems centered on the individual flux tube axes $\mathbf{r}_j$ [*viz.*, $A_m^{(j)}(k_\|,\omega)$], via an infinite coupled set of linear equations of the form (Bogdan & Fox 1991):

$$(\mathrm{I}-\mathrm{M})\cdot\mathbf{B} = \mathbf{A} \qquad [2.11]$$

where I is the identity matrix and M is the multiple scatter matrix. The multiple scatter matrix accounts for the component of the local excitation at a given flux tube that is due to waves scattered by the other flux tubes in the cluster. If multiple scatter is weak, then $(\mathrm{I}-\mathrm{M})^{-1} = \mathrm{I}+\mathrm{M}+\mathrm{M}\cdot\mathrm{M}+\cdots$, and the familiar first Born approximation amounts to retaining only the first term in this expansion.

One can translate (Bogdan & Cattaneo 1989; Bogdan & Fox 1991) the outgoing scattered waves from the individual constituents of the cluster into outgoing waves centered on the origin if the second term in braces in equation (2.10) is replaced by $A_m(k_\|,\omega)\; T_m(k_\|,\omega)\cdot H_m^{(1)}(k_\perp^{(a)}|\mathbf{r}|)\exp(im\theta)$, where the diagonalized cluster T-matrix is:

$$T_m(k_\|,\omega) \equiv \frac{1}{A_m(k_\|,\omega)} \sum_{j=1}^{N} \sum_{p=-\infty}^{\infty} (-i)^{p-m}\; B_p^{(j)}T_p^{(j)}J_{p-m}(k_\perp^{(a)}|\mathbf{r}_j|)e^{i(p-m)\varphi_j} \;, \qquad [2.12]$$

and $\cos\varphi_j = \hat{\mathbf{r}}_j\cdot\hat{\mathbf{e}}_x$. The resulting expression for $\phi(\mathbf{x},t)$ is again only valid outside of the flux tube cluster, but it permits the absorption and scattering cross sections to be calculated from the standard expression, equation (2.3), using the diagonalized T-matrix given above. If the individual flux tubes within the flux tube cluster are elastic scatterers, then the absorption cross section for the cluster as a whole must also vanish, although the individual absorption coefficients constructed according to equation (2.4) need not vanish themselves.

The crux of the problem lies in successfully solving for the local excitation, by inverting I$-$M, and projecting it onto $\mathbf{A}$. The singularities of the inverse matrix in the complex $k_\| - \omega$

plane describe the novel collective oscillations of the flux tube cluster. Such a computation has been carried out (Bogdan & Fox 1991) for the simplest cluster consisting of a pair of identical uniformly magnetized flux tubes of separation d and radius $a(< d/2)$. In this example there is a highly localized mode of oscillation which is indicated by the scattering resonance shown in figure 4. This mode consists of transverse undulations of the thin fluid channel separating the two flux tubes and of the neighboring magnetized plasma inside the flux tubes, while the far sides of the flux tubes exhibit little significant motion. For acoustically compact flux tubes, the phase velocity of this collective mode is $\omega/k_{\|} \simeq \pm C_A[1 + (1 - 4a^2/d^2)^{1/2}\eta^{-1}]^{1/2}$, which approaches the Alfvén speed as the flux tubes are brought closer together. In figure 4, this phase speed is finally larger than the external sound speed when $d/2a = 1.05$, allowing this collective mode to be resonantly excited by an obliquely incident acoustic plane-wave. This equilibrium is found to have other novel collective modes: one example consists of a pair of surface waves that propagate in unison on the surfaces of the individual flux tubes along helical trajectories.

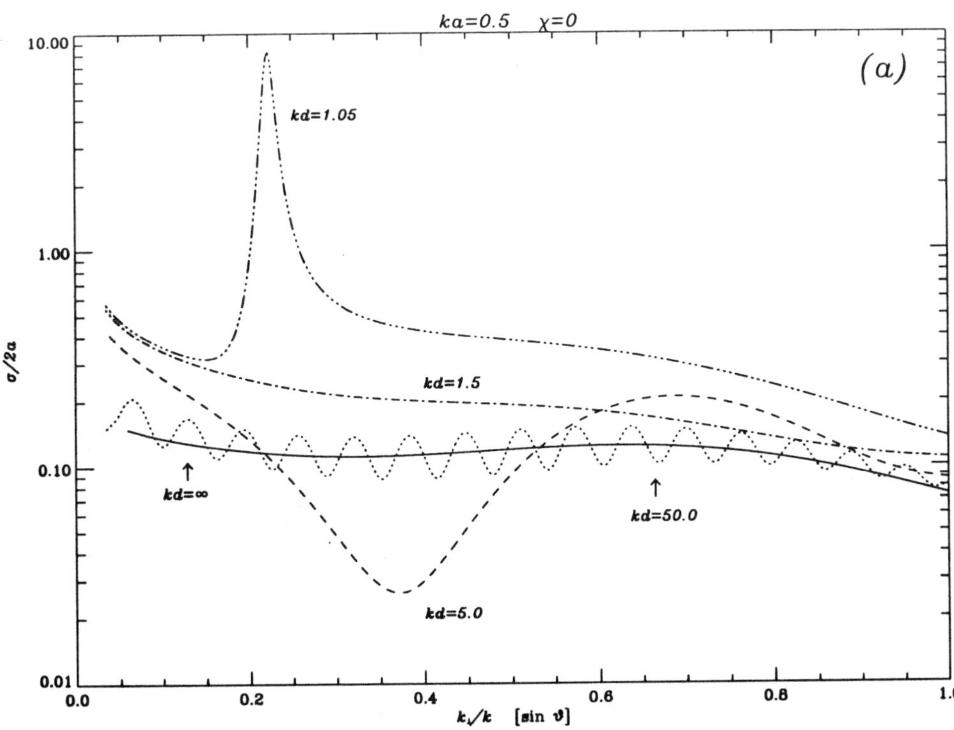

Figure 4. Scattering cross section in flux tube diameters as a function of perpendicular p-mode wavenumber at fixed frequency. The different curves correspond to different flux tubes separations. The lateral wavevector of the p-mode is aligned with the flux tube separation vector, and the wavelength is twice the circumference of the flux tubes. The ratios of the internal sound speed and Alfvén speeds to the external sound speed are 5/6 and 4/3 respectively [from Bogdan & Fox 1991].

One would like to extend this investigation to pairs of flux tubes possessing internal critical layers, as well as clusters containing numerous constituent flux tubes. The former extension poses no additional difficulty, but the latter entails increased computational efforts in solving equation (2.11). Alternative approaches which attempt to circumvent a direct solution of equation (2.11) seem, for the present, fraught with their own difficulties. The Born approximation (Rosenthal 1991) fails when the flux tubes lie well within one lateral acoustic wavelength of one another ($k_\perp^{(a)}|\mathbf{r}_i - \mathbf{r}_j| \leq 1$), which is likely to be the case in most sunspot seismology applications. Ensemble averaging, or effective medium approximations, are also of dubious value because (i) there may not be a sufficient number of individual flux tubes in the cluster for a statistical treatment to be accurate, and (ii) one must overcome the spurious damping that results from phase cancellation when averaging the linear equations (Bogdan & Cattaneo 1989). One means to avoid this latter problem is to employ an "acoustic transfer equation" for the intensity $|\phi(\mathbf{x},t)|^2$ (Ishimaru 1978), instead of working with the usual wave equation for the velocity potential $\phi(\mathbf{x},t)$. The advantage here is that even after averaging over some ensemble of equilibrium configurations, the equation for the intensity conserves energy, while the averaged wave equation for the velocity potential yields decaying solutions. The physical explanation is obvious: at distances of order one acoustical depth or greater into the fibril magnetic field the intensity of the acoustic radiation varies little from one configuration to the next, but the phase of the individual waves varies dramatically in recognition of the fact that most of the acoustic signal has been previously scattered by other magnetic flux tubes.

3. The Interaction of Acoustic Waves with Magnetic Flux Tubes in Gravitationally Stratified Atmospheres

When stratification is taken into account, the waves scattered by the magnetic flux tube need no longer have the same variation with depth as the incident wave. This occurs primarily because of a contrast in the stratification between the interior and the exterior of the magnetic flux tube, and, as we shall presently see, because of the existence of the additional slow mode (s-mode) which is present only in the magnetic flux tube.

To illustrate these issues, suppose, for example, that the magnetic flux tube is embedded in an adiabatically stratified polytropic atmosphere which extends infinitely far ($z \to -\infty$) below a rigid surface set at some finite distance z_0 below the surface where the complete polytrope would have terminated. In this case equations (2.1) continue to be valid and equation (2.2) is modified to:

$$\begin{aligned}\phi(\mathbf{x},t) = \frac{1}{2\pi}\int d\omega \sum_{l=1}^{\infty}\sum_{m=-\infty}^{\infty}\sum_{p=-\infty}^{\infty} i^m\, e^{i(m\theta-\omega t)+k_\perp z}\\ U(-l-\Delta_l; m_*; -2k_\perp z)\sum_{n=1}^{\infty} A_p(n,\omega)[\delta_{p,m}\,\delta_{n,l}\,J_m\,(k_\perp r) -\\ -T_{pm}(n \to l;\omega)H_m^{(1)}(k_\perp r)]\end{aligned} \qquad [3.1]$$

where $k_\perp(l,\omega) \equiv m_*\omega^2/g(m_* + 2l + 2\Delta_l)$, $m_* = 1/(\gamma - 1)$ is the polytrope index, $\mathbf{g} = -g\hat{\mathbf{e}}_z$ is the gravitational acceleration, $\Delta_l = \Delta_l(z_0)$ is a correction to the dispersion relation accounting for the truncation of the polytrope, and $U(a;b;x)$ is one of the two linearly independent solutions of the confluent hypergeometric equation (Abramowitz & Stegun 1964). The T-matrix now permits the outgoing scattered wave to be composed of different azimuthal multipole components ($m \neq p$), and different radial orders ($l \neq n$), than those of the exciting p-mode (Rosenthal 1991).

As in the unstratified models $T_{pm}(n \rightarrow l, \omega)$ is once again calculated by matching the exterior solution to an interior solution by requiring continuity of the perturbed pressure and normal displacement everywhere on the flux tube surface. Only the uniformly magnetized flux tube is known to permit a separation of the linearized equations, with the two attendant physical liabilities that (*i*) the field lines must be anchored from above on a rigid surface at depth z_0 in the external polytrope where $B^2/8\pi$ is less than the external pressure, and (*ii*) the run of density with the depth is the same both inside and outside the flux tube.

The equation (Ferraro & Plumpton 1958; Scheuer & Thomas 1981) which describes the variation with depth of the perturbations within the uniformly magnetized flux tube is a fourth-order differential equation (neglecting purely torsional Alfvén waves to which the external p-modes will typically not couple). In the high-beta ($\beta = 8\pi p/B^2$) plasma found deep in the flux tube, this fourth-order equation factors into a pair of weakly coupled second-order equations. One of this pair has the external p-mode (and its exponentially growing counterpart) as its homogeneous solution. The other operator has the quasi-Alfvénic s-modes which propagate freely in either direction along the magnetic flux tube, having the form (Roberts & Soward 1983; Zhugzhda & Dzhalilov 1982):

$$\delta v_\perp(z) \sim \begin{cases} (-z)^{\frac{1}{2}} \cdot H_\nu^{(1)}(-2\nu\omega z/C_A(z)) \\ (-z)^{\frac{1}{2}} \cdot H_\nu^{(2)}(-2\nu\omega z/C_A(z)) \end{cases} \quad [3.2]$$

where $\nu \equiv 1/(m_* + 2)$, and $C_A(z) \propto (-z)^{-m_*/2}$ is the Alfvén speed, as its homogeneous solutions. Because the s-modes have such a short wavelength, they are confined to, and effectively guided by, the flux tube at large depth.

The s-modes and p-modes couple strongly where the plasma beta is of order unity. Indeed, here neither the p-modes of the external atmosphere nor the s-modes given by equation (3.2) are individually solutions of the fourth order ODE. One consequence, therefore, of the interaction of the impinging external p-mode with the $\beta \lesssim 10$ region of the magnetic flux tube will be the generation of a downward flux of s-modes (in this model an upward flux is ruled out by the rigid plate at $z = -z_0$) and presumably some attendant disbursement of the outgoing scattered p-modes over various radial orders.

In order to quantify these notions one needs to solve the fourth-order ODE down to sufficiently large β, and match this solution onto the external p-modes, in order to determine the $T_{pm}(n \rightarrow l, \omega)$, and ultimately the efficiency of the conversion of p-modes to s-modes as a function of frequency and lateral wavenumber. A somewhat simpler proposition, which might still provide reasonably accurate results, hinges on using the Born approximation to estimate *both* the distribution of scattered p-modes over all available radial orders (Rosenthal 1991), *and* the flux of downward propagating s-modes guided along the magnetic flux tube (Spruit 1991; Spruit & Bogdan 1992). Future developments along these lines might include bold initiatives which attempt to deal with flux tubes that flare open with increasing height (Cally 1983; Poedts & Goossens 1988).

4. Observations of the Interaction of P-Modes with Sunspots

4.1 RESPONSE OF THE SUNSPOT INTERIOR TO EXTERNAL P-MODE FORCING

Thomas *et al* (1982) laid the conceptual groundwork for observational sunspot seismology with their spatially-unresolved measurements of velocity fields in sunspot umbrae. They measured the

Doppler shift in the Stokes-V profile of the Fe I λ6302.5 line, ensuring that the signal originated in the highly magnetized umbra, and used a nearby telluric line of O_2 as a reference. Power spectra of the time series for the sunspot taken on two different days yielded three reproducible power peaks at $\nu \simeq 2.8$, 3.3, and 3.8 mHz. The authors speculated that the strong selective response of the umbra at these particular frequencies in response to the broadband external acoustic forcing should in principle provide information about the variation of the sunspot radius (perhaps amongst other things) with depth.

Abdelatif, Lites, and Thomas (1986) carried out spatially resolved observations of umbral velocity fields, using the umbral line Ti I λ6303.8. Two-dimensional images were constructed every 60 s by raster-scanning the spectrograph slit across the field of view containing the sunspot. A comparison of $k_\perp - \omega$ power spectra for the entire field of view and that of the umbra alone indicates suppressed power in the umbra for $\omega/k_\perp \lesssim 25$ km s^{-1} while no such suppression is evident in the surrounding quiet Sun. When integrated over wavenumbers, the power in the p-mode band (2.5 mHz $< \nu <$ 4.5 mHz) is found to be suppressed by a factor of 2-3 in the umbra relative to the quiet Sun, suggesting that the umbra acts as a selective p-mode filter. However it needs to be kept in mind that such comparisons can be complicated owing to differences in the height of line formation and in the structure of the oscillation eigenfunctions between magnetized and unmagnetized plasmas.

In a subsequent paper Abdelatif and Thomas (1987) suggested a physical interpretation for the suppressed umbral power for $\omega/k_\perp \lesssim 25$ km s^{-1}. They noted that in an *unstratified* uniform vertical magnetic field, the usual dispersion relation gives $\omega/k_\perp = \{(C_S^2 + C_A^2)\omega^2(\omega^2 - k_\parallel^2 C_T^2)/[(\omega^2 - k_\parallel^2 C_A^2)(\omega^2 - k_\parallel^2 C_S^2)]\}^{1/2}$, where $k_\parallel$ is the vertical component of the wavenumber, C_S is the adiabatic sound speed C_A is the Alfvén speed, and $C_T = C_S C_A/\sqrt{C_S^2 + C_A^2}$. Since $C_T \ll C_A$, C_S the minimal value $\omega/k_\perp$ can attain occurs when $k_\parallel$ is close to zero. They suggest that this minimal value should be associated with the figure of 25 km s^{-1} obtained from the comparison of their power spectra.

4.2 THE INTERACTION OF P-MODES AND SUNSPOTS VIEWED AS A CLASSICAL SCATTERING PROBLEM

Braun, Duvall, and LaBonte (1987, 1988) suggested that with spatially-resolved Doppler images one could use *only* data from the surrounding quiet Sun, and treat the interaction of p-modes with a sunspot as a classical scattering problem. To illustrate the idea consider a spherical polar coordinate system with the sunspot situated at the north pole ($\theta = 0$). The temporal sequence of Doppler images then yields a three-dimensional array $\delta v_\parallel(\theta_j, \phi_j, t_k)$, where j indexes the useful pixels, and $k = 0, 1, \ldots, N$ the time series ($t_k = k \cdot \Delta t$). The useful pixels are those that satisfy $\theta_{\rm inner} < \theta_j < \theta_{\rm outer}$, where $\theta_{\rm inner}$ is the smallest polar angle that excludes the sunspot, and $\theta_{\rm outer}$ is the largest angle still within the field of view. Now in the unmagnetized quiet Sun surrounding the sunspot, the standard equations governing non-radial stellar oscillations are separable (Unno *et al* 1989), and the angular variations are $\propto [AP_l^m(\cos\theta) + BQ_l^m(\cos\theta)]e^{im\phi}$. Since the field of view defined by the useful pixels contains neither $\theta = 0$, nor $\theta = \pi$, it is not necessary to discard the associated Legendre function of the second kind $Q_l^m(\cos\theta)$ (Abramowitz & Stegun 1964), and so one seeks a spectral decomposition:

$$\delta v_{\parallel}(\theta_j,\phi_j,t_k) = \sum_{n=-\frac{N}{2}}^{\frac{N}{2}} \sum_{m=-M}^{M} \sum_{p=0}^{L-1} V(p,m,n)\{P_{l_p}^m(\cos\theta_j) - \frac{2i}{\pi} Q_{l_p}^m(\cos\theta_j)\} \exp\{i[m\phi_j + 2n\pi t_k/N\Delta t]\} + c.c. \,, \tag{4.1}$$

in terms of a discrete set of L orders $\{l_0, l_1, l_2, \ldots, l_{L-1}\}$ which may be chosen to yield quasi-orthogonality of the polar-angle basis functions over the useful pixels. The largest order $l_{L-1} \simeq \pi/\Delta\theta$ is set by the spatial Nyquist sampling frequency, where $\Delta\theta$ is the pixel spacing.

For $l >> 1$ (Abramowitz & Stegun 1964),

$$P_l^m(\cos\theta) - \frac{2i}{\pi} Q_l^m(\cos\theta) \sim \frac{(-1)^m \Gamma(l+m+1)}{\Gamma(l+\frac{1}{2})} \sqrt{\frac{2}{\pi\theta}} \exp\left\{i[(l+\frac{1}{2})\theta - \frac{\pi}{4} - \frac{m\pi}{2}]\right\} \,, \tag{4.2}$$

so, for negative values of n, $V(p,m,n)$ gives the amplitudes of the outgoing waves, while positive values of n describe the incoming waves. One can then read off from equation [4.1] the absorption coefficient,

$$\alpha_m(p,|n|) = 1 - \left|\frac{V(p,m,-|n|)}{V(p,m,|n|)}\right|^2 \,, \quad n = 1,2,\ldots,N/2 \tag{4.3}$$

and the partial-wave phase shift,

$$\delta_m(p,|n|) = \frac{1}{2} \arg\left[V(p,m,-|n|)/V(p,m,|n|)\right], \tag{4.4}$$

which, together, provide a complete description of the interaction of p-modes with sunspots, subject to the usual limitations of finite spatial and temporal sampling.

In their analysis Braun, Duvall, and LaBonte (1988) used the Fe I λ8688 line, and employed Hankel functions of argument $k_\perp R_\odot \sin\,\theta$ [$\simeq (l+1/2)\theta$ for $\theta << 1$], in lieu of the associated Legendre functions. For large $k_\perp R_\odot \sin\theta$,

$$J_m(k_\perp R_\odot \sin\theta) + iY_m(k_\perp R_\odot \sin\theta) \sim \sqrt{\frac{2}{\pi k_\perp R_\odot \sin\theta}} \exp\left\{i[k_\perp R_\odot \sin\theta - \frac{\pi}{4} - \frac{m\pi}{2}]\right\} \,, \tag{4.5}$$

and a comparison of equation [4.5] with [4.2] shows that the two approaches give (up to a multiplicative factor) the same variation with θ in the limit of large l where the curvature of the solar surface is presumably unimportant. Figure 5 is reproduced from their paper and documents vividly the significant absorption of p-modes by sunspots. They also studied the dependence of the absorption parameter on the azimuthal order and concluded that the absorbing region appeared to be somewhat larger than the penumbra. In later studies (Braun *et al* 1990; Braun & Duvall 1990), they applied the scattering analysis point-by-point throughout an active region to map out the variation of absorption with position and/or magnetic field strength. Their maps of p-mode absorption show maxima coincident with the visible sunspots, but extended regions of absorption

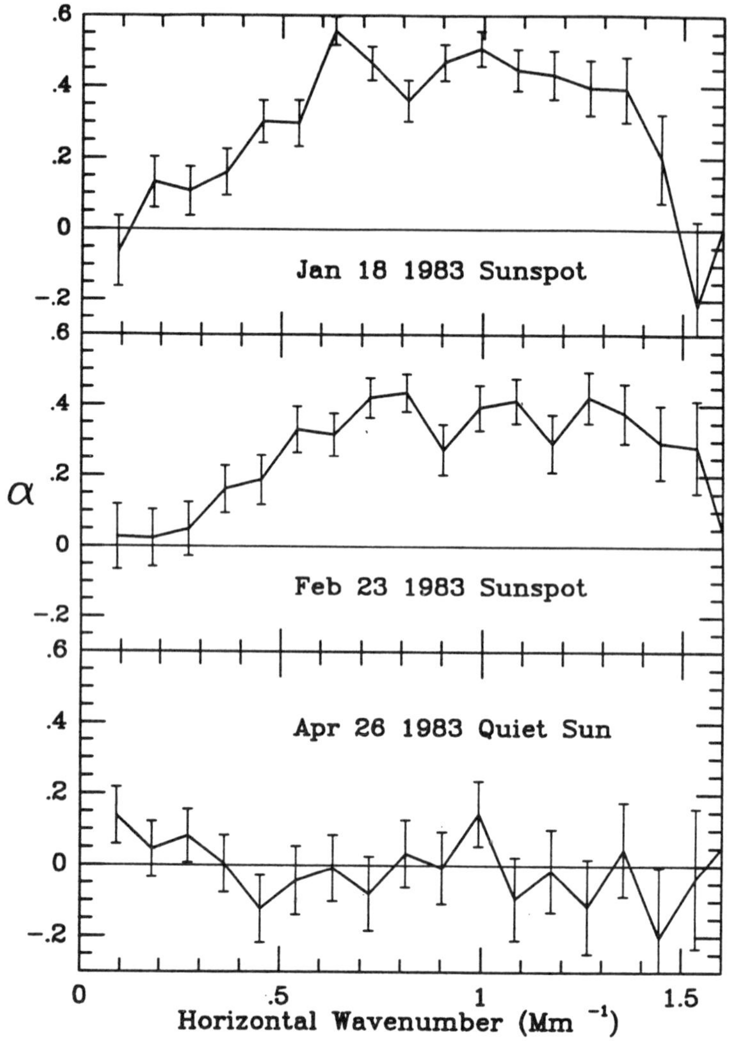

Figure 5. Absorption coefficient averaged over azimuthal orders from -5 to +5 and frequencies as a function of lateral wavenumber. The top two panels are for two different sunspots, while in the bottom panel the analysis has been carried out for a patch of quiet Sun [from Braun *et al* 1988].

are also associated with the neighboring plage. Absorption coefficients as large as 0.7 were obtained for a particular sunspot group.

5. Towards an Inverse Problem for Sunspot Seismology

Despite the promising developments which have occurred over the last decade in both the theoretical and observational components of sunspot seismology, it is evident that a sense of maturity has yet to emerge from either line of inquiry. One ultimately requires absorption coefficients *and* partial-wave phase shifts measured as functions of the order and degree of the external p-modes for a range of isolated magnetic concentrations ranging from sunspots downward in size to knots and pores. Likewise, the variation of p-mode absorption with mean magnetic field strength in plage deserves more attention in order to answer such questions as whether the level of absorption is more sensitive to the peak field strength in the magnetic flux tubes or the relative concentration of the flux tubes. Theoretical models need to meet the challenge of quantifying the prominent role that strong vertical stratification (density scale height$\sim$ 150 km) is likely to play in mediating the interaction of p-modes with solar surface magnetic flux concentrations. How such interactions translate into perturbations to the emergent intensity, and how these perturbations in turn influence the interaction of p-modes with sunspots are also issues that probably merit deeper thought and more sophisticated modeling efforts. A thorough understanding of the propagation of p-modes in discretely structured magnetic fields, although presently not well-understood, appears to be essential for further progress.

A straightforward synthesis of observations and deductive theory in the context of an "inverse problem" approach to sunspot seismology is, for the present, premature, and consequently there is a dearth of such endeavors in the literature. A notable exception is the contribution by Brown (1990) which advocates a different—essentially phenomenological—point of view. The idea is basically to treat the interaction between the p-mode and the sunspot in the Born approximation, where the scattered acoustic field due to an element of magnetized fluid is related by a linear operator acting on the local acoustic field generated by the external sources (so multiple scatter is completely ignored). One then adopts a simple parameterization of this operator that involves perhaps one or at most a few variable parameters. For example, one could assume a simple acoustic source with strength proportional to the local pressure excess, and a dipole source with strength and directionality proportional to the local fluid displacement in each fluid element. The inverse problem would then consist of solving for the variation of these parameters from fluid element to fluid element, given the observed incoming and outgoing wave fields. Fluid elements which, through such a process, acquire sufficiently small parameter values are then deemed operationally to reside outside of the magnetic flux tube. This approach offers hope for significant and rapid progress in sunspot seismology, particularly if one can ascribe some physical interpretation to the chosen parameterization of the linear operator.

This last caveat is crucial, and it deserves a final comment. In the example given above—where the local fluid element can emit monopole and dipole radiation in response to the exciting acoustic field—one suspects that the implied two constants of proportionality can be related to the degree of contrast in compressibility and density between the magnetized and surrounding unmagnetized fluid elements. One possibility is to visualize the fluid element as an acoustically compact sphere of radius a ($a \ll \lambda$ the acoustic wavelength) embedded in a homogeneous atmosphere. The scattering of an acoustic plane-wave,

$$\delta p_{inc}(r,\theta,t) \;=\; \exp[i(kr\cos\theta - \omega t)], \qquad [5.1]$$

(r, θ, ϕ are now the usual spherical polar coordinates) from such a sphere results in the scattered field:

$$\delta p_{scat}(r,\theta,t) \;\sim\; \frac{e^{ikr}}{r}\frac{k^2a^3}{3}\left[\kappa - 1 + \frac{3\eta - 3}{2\eta + 1}\cos\theta\right], \quad r \to \infty\ , \qquad [5.2]$$

where κ is the ratio of internal to the external fluid compressibility [$\sim \gamma p_0/(\gamma p + B^2/4\pi)$], and η the ratio of internal to external fluid densities (Morse & Ingard 1986). The relation between the two proportionality parameters and the physical conditions within the sunspot is now explicit. A more realistic approach might be to embed the acoustically compact sphere in a stratified, presumably polytropic, atmosphere. The pair of relations analogous to equations (5.1) and (5.2) for such a situation would be useful to work out; the work of Kato (1966), Brown (1990), and Rosenthal (1991), provides a useful point of departure for such an endeavor.

Acknowledgements

I am grateful to T. Brown, M. Goossens, S. Poedts and H. Spruit for allowing me to state results of their research prior to publication, and to D. Braun and S. Poedts for supplying figures from their papers in a timely fashion. To D. Degenhardt and T. Duvall Jr., I express my thanks for their instructive comments on this manuscript. Many of the viewpoints expressed here have been shaped by conversations with colleagues, in this regard, particular thanks are due to T. Brown, F. Cattaneo, V. Ferrari, M. Goossens, J. Hollweg, M. Knölker, M. Lee, I. Lerche, J. McKenzie, D. Mihalas, S. Poedts, A. Richmond, B. Roberts, C. Rosenthal, R. Rosner, A. Skumanich, H. Spruit, J. Thomas, and E. Zweibel.

Literature Cited

Abdelatif, T.E. 1988, Surface and body waves in magnetic flux tubes, *ApJ*, **333**, 395.

Abdelatif, T.E., Lites, B.W., & Thomas, J.H. 1986, The interaction of p-modes with a sunspot. I. Observations, *ApJ*, **311**, 1015.

Abdelatif, T.E., & Thomas, J.H. 1987, The interaction of p-modes with a sunspot. II. Simple theoretical models, *ApJ*, **320**, 884.

Abramowitz, M.A., & Stegun, I.A. 1964, *Handbook of Mathematical Functions*, (Washington: National Bureau of Standards).

Acheson, D.J. 1973, Valve-effect of inhomogeneities on anisotropic wave propagation, *J. Fluid Mech.*, **58**, 27.

Bates, R.H.T. 1980, General introduction to the extended boundary condition, in *Acoustic, Electromagnetic and Elastic Wave Scattering—Focus on the T-Matrix Approach*, ed.: V.K. Varadan & V.V. Varadan, (New York: Pergamon), p.21.

Birkinshaw, M. 1984, The Kelvin-Helmholtz instability for relativistic particle beams—I. Stability analyses in the time and space domains for vortex-sheet flows, *MNRAS*, **208**, 887.

Bogdan, T.J. 1989, On the resonance scattering of sound by slender magnetic flux tubes, *ApJ*, **345**, 1042.

Bogdan, T.J., & Cattaneo, F. 1989, The normal modes of a resonant cavity containing discrete inhomogeneities: The influence of fibril magnetic fields on the solar acoustic oscillations, *ApJ*, **342**, 545.

Bogdan, T.J., & Knölker, M. 1991, Scattering of acoustic waves from a magnetic flux tube embedded in a radiating fluid, *ApJ*, **369**, 219.

Bogdan, T.J., & Fox, D.C. 1991, Multiple scattering of acoustic waves by a pair of uniformly magnetized flux tubes, *ApJ*, **379**, 758.

Booker, J.R., & Bretherton, F.P. 1967, The critical layer for internal gravity waves in a shear flow, *J. Fluid Mech.*, **27**, 513.

Braun, D.C., Duvall Jr., T.L., & LaBonte, B.J. 1987, Acoustic absorption by sunspots, *ApJ*, **319**, L27.

Braun, D.C., Duvall Jr., T.L., & LaBonte, B.J. 1988, The absorption of high-degree p-mode oscillations in and around sunspots, *ApJ*, **335**, 1015.

Braun, D.C., Duvall Jr., T.L., & LaBonte, B.J. 1990, The spatial distribution of p-mode absorption in active regions, *ApJ*, **354**, 372.

Braun, D.C., & Duvall Jr., T.L. 1990, P-mode absorption in the giant active region of 10 March 1989, *Solar Phys.*, **129**, 83.

Brown, T.M. 1990, An inverse method for p-mode scattering measurements, *Solar Phys.*, **128**, 133.

Cally, P.S. 1983, Umbral oscillations in the presence of a spreading magnetic field, *Solar Phys.*, **88**, 77.

Cally, P.S. 1985, Magnetohydrodynamic tube waves: Waves in fibrils, *Austr. J. Phys.*, **38**, 825.

Cally, P.S. 1986, Leaky and non-leaky oscillations in magnetic flux tubes, *Solar Phys.*, **103**, 277.

Chitre, S.M. 1992, (these proceedings).

Clark, T.L., & Peltier, W.R. 1984, Critical level reflection and resonant growth of nonlinear mountain waves, *J. Atmos. Sci.*, **41**(21), 3122.

Craik, A.D.D. 1985, *Wave Interactions and Fluid Flows*, (Cambridge: Cambridge Univ. Press).

Davila, J.M. 1990, Sunspot seismology theory, in *Basic Plasma Processes on the Sun*, ed.: E.R. Priest & V. Krishan, (Dordrecht: Reidel), p. 149.

Defouw, R.J. 1976, Wave propagation along a magnetic flux tube, *ApJ*, **209**, 266.

Delache, P., & Froeschlé, C. 1972, Time dependent radiative transfer. Damping of a temperature fluctuation I. Near LTE grey approximation, *Astr. Ap.*, **16**, 348.

Dzhalilov, N.S., Zhugzhda, Y.D., & Staude, J. 1992, Radiation—hydrodynamic waves in an optically thick gray atmosphere. I. Homogeneous model, *Astr. Ap*, (in press).

Evans, D.J., & Roberts, B. 1990, The oscillations of a magnetic flux tube and its applications to sunspots, *ApJ*, **348**, 346.

Ferraro, V.C.A., & Plumpton, C. 1958, Hydromagnetic waves in a horizontally stratified atmosphere, *ApJ*, **127**, 459.

Goedbloed, J.P. 1984, Plasma-vacuum interface problems in magnetohydrodynamics, *Physica D*, **12**, 107.

Goossens, M., Hollweg, J.V., & Sakurai, T. 1992, Resonant behaviour of MHD waves on magnetized flux tubes III. Effect of equilibrium flow, *Solar Phys.*, (in press).

Goossens, M., & Poedts, S. 1992, Linear resistive MHD computations of resonant absorption of acoustic oscillations in sunspots, *ApJ*, (in press).

Hain, K., & Lüst, R. 1958, Zur Stabilität zylindersymmetrischer Plasmakonfigurationen mit Volumenströmen, *Zeit. f Naturforschg.*, **13**a, 936.

Hollweg, J.V. 1988, Resonance absorption of solar p-modes by sunspots, *ApJ*, **335**, 1005.

Hollweg, J.V., Yang, G., Cadez, V.M., & Gakovic, B. 1990, Surface waves in an incompressible fluid: Resonant instability due to velocity shear, *ApJ*, **349**, 335.

Ishimaru, A. 1978, *Wave propagation and Scattering in Random Media. II. Multiple Scattering, Turbulence, Rough Surfaces, and Remote Sensing*, (New York: Academic), §14.

Jones, D.S. 1977, The scattering of sound by a simple shear layer, *Phil. Trans. R. Soc. (London) A*, **284**, 287.

Kapraff, J.M., & Tataronis, J.A. 1977, Resistive effects in Alfvén heating, *J. Plasma Phys.*, **18**, 209.

Kato, S. 1966, The response of an unbounded atmosphere to point disturbances, *ApJ*, **143**, 893.

Lites, B.W. 1992, (these proceedings).

Lou, Y.-Q. 1988, Viscous magnetohydrodynamic modes and p-mode absorption by sunspots, in *Proc. Symp. Seismology of the Sun and Sun-like Stars*, (ESA SP-286, Dec. 1988), p. 305.

Lou, Y.-Q. 1990, Viscous magnetohydrodynamic modes and p-mode absorption by sunspots, *ApJ*, **350**, 452.

Margolis, S.B., & Su, C.H. 1978, Boundary-value problems in stratified shear flows with a nonlinear critical layer, *Phys. Fluids*, **21**(8), 1247.

McKenzie, J.F. 1970, Hydromagnetic oscillations of the geomagnetic tail and plasma sheet, *J. Geophys. Res.*, **75**, 5331.

McKenzie, J.F. 1972, Reflection and amplification of acoustic-gravity waves at a density and velocity discontinuity, *J. Geophys. Res.*, **77**(16), 2915.

McKenzie, J.F. 1973, On the existence of critical levels, with application to hydromagnetic waves, *J. Fluid Mech.*, **58**, 709.

Mihalas, D., & Mihalas, B.W. 1984, *Foundations of Radiation Hydrodynamics*, (New York: Oxford Univ. Press).

Miles, J.W. 1957, On the reflection of sound at an interface in relative motion, *J. Acoust. Soc. Am.*, **29**(2), 226.

Morse, P.M., & Feshbach, H. 1953, *Methods of Theoretical Physics, Part II*, (New York: McGraw-Hill), §11.2.

Morse, P.M., & Ingard, K.U. 1986, *Theoretical Acoustics*, (Princeton: Princeton Univ. Press).

Parker, E.N. 1979a, *Cosmical Magnetic Fields*, (New York: Oxford Univ. Press).

Parker, E.N. 1979b, Sunspots and the nature of magnetic flux tubes. I. The general nature of the sunspot, *ApJ*, **230**, 905.

Poedts, S., & Goossens, M. 1988, The continuous spectrum of MHD waves in 2-D solar loops and arcades. Parametric study of poloidal mode coupling for poloidal magnetic fields, *Astr. Ap.*, **198**, 331.

Poedts, S., Goossens, M., & Kerner, W. 1990, Temporal evolution of resonant absorption in solar coronal loops, *Computer Phys. Commun.*, **59**, 95.

Poedts, S., & Kerner, W. 1992, Time-scales and efficiency of resonant absorption in periodically driven resistive plasmas, *J. Plasma Phys.*, (in press).

Roberts, B. 1992, (these proceedings).

Roberts, P.H., & Soward, A.M. 1983, The effect of a weak magnetic field on stellar oscillations, *MNRAS*, **205**, 1171.

Rosenthal, C.S. 1991, Sunspots and sunquakes: Acoustic oscillations as a probe of magnetic structure in the Sun, *Ph.D. Dissertation*, University of Cambridge.

Rosenthal, C.S. 1992, Absorption of acoustic waves by sunspots II: Resonance absorption in axisymmetric fibril models, *Solar Phys.*, (in press).

Ryutov, D.A., & Ryutova, M.P. 1976, Sound oscillations in a plasma with "magnetic filaments", *Sov. Phys. JETP*, **43**(3), 491.

Ryutova, M.P. 1986, Generation of plasma flows by oscillations of magnetic flux tubes, in *Proc. of the Joint Varenna-Ambastumani Intl. School & Workshop on Plasma Astrophys.*, (ESA SP-251, Aug. 1986), p.71.

Ryutova, M.P. 1988, MHD-waves and their nonlinear interaction in a plasma with magnetic flux tubes, in *Proc. of the Joint Varenna-Ambastumani Intl. School & Workshop on Plasma Astrophys.*, (ESA SP-285, Dec. 1988), p.77.

Ryutova, M.P., & Persson, M. 1984, Dispersion properties and enhanced dissipation of MHD-oscillations in a plasma with random inhomogeneities, *Physica Scripta*, **29**, 353.

Sakurai, T., Goossens, M., & Hollweg, J.V. 1991a, Resonant behaviour of MHD waves on magnetic flux tubes. I. Connection formulas at the resonant surface, *Solar Phys.*, **133**, 227.

Sakurai, T., Goossens, M., & Hollweg, J.V. 1991b, Resonant behaviour of MHD waves on magnetic flux tubes. II. Absorption of sound waves by sunspots, *Solar Phys.*, **133**, 247.

Scheuer, M.A., & Thomas, J.H. 1981, Umbral oscillations as resonant modes of magneto-atmospheric waves, *Solar Phys.*, **71**, 21.

Spiegel, E.A. 1977, Photoconvection, in *Lecture Notes in Physics, 77, Problems of Stellar Convection*, ed.: E.A. Spiegel & J.P. Zahn, (Berlin: Springer-Verlag), p. 267.

Spruit, H.C. 1982, Propagation speeds and acoustic damping of waves in magnetic flux tubes, *Solar Phys.*, **75**, 3.

Spruit, H.C. 1991, Absorption of p-mode waves by magnetic fields, in *Lecture Notes in Physics, 388, Challenges to Theories of the Structure of Moderate Mass Stars*, ed.: J. Toomre & D.O. Gough, (Berlin: Springer-Verlag), p. 121.

Spruit, H.C., & Bogdan, T.J. 1992, The conversion of p-modes to slow-modes and the absorption of acoustic waves by sunspots, *ApJ*, (submitted).

Thomas, J.H., Cram, L.E., & Nye, A.H. 1982, Five-minute oscillations as a subsurface probe of sunspot structure, *Nature*, **297**, 485.

Unno, W., Osaki, Y., Ando, H., Saio, H., & Shibahashi, H. 1989, *Nonradial Oscillations of Stars*, (Tokyo: Univ. of Tokyo Press).

Wilson, P.R. 1980, The interaction of acoustic waves with flux tubes, *ApJ*, **237**, 1008.

Wu, D., & Wang, D. 1991, The Kelvin-Helmholtz instability of a cylindrical flow with a shear layer, *MNRAS*, **250**, 760.

Yang, G., & Hollweg, J.V. 1991, The effects of velocity shear on the resonance absorption of MHD surface waves: Cold plasma, *J. Geophys. Res.*, **96**(A8), 13,807.

Zhugzhda, Y.D., & Dzhalilov, N.S. 1982, Transformation of magnetogravitational waves in the solar atmosphere, *Astr. Ap.*, **112**, 16.

VI. THE RELATION OF SUNSPOTS TO THE GLOBAL SOLAR MAGNETIC FIELD

Cyclic magnetic activity on the Sun is traditionally monitored by counting the incidence of sunspots, which are its most conspicuous products. These spots are surface features but they show systematic patterns of behaviour which imply that their magnetic flux remains coherent through most of the convection zone. So it is important to relate them to the deep-seated global field involved in the 22-year magnetic cycle. It is widely believed that this cycle is generated by a dynamo located at, or just below, the base of the convection zone, where helioseismology has revealed a strong radial gradient in angular velocity. Although the intricacies of dynamo theory, along with detailed properties of the solar cycle, lie beyond the scope of this volume, we are concerned with the formation and evolution of the flux tubes that emerge in active regions to form sunspots.

It seems likely that the dynamo produces a predominantly toroidal magnetic field, with a strength between 10^4 and 10^5 G, in a region of convective overshoot that may only be about 10^4 km thick. Such a layer, with an abrupt decrease in the field strength at its upper boundary, can become unstable to modes driven by magnetic buoyancy. These modes are expected to have a much longer wavelength along the field than across it. In suitable circumstances their nonlinear development may result in the formation of isolated loops of flux that are free to rise up through the convection zone.

An isolated flux tube cannot remain in static equilibrium with its surroundings: instead it will float upwards owing to magnetic buoyancy. So loops that are anchored at the base of the convection zone develop an Ω-shaped configuration and eventually break through the surface. The dynamics of buoyant flux tubes has been examined in considerable detail. Their rise is affected by Coriolis forces and by ambient turbulence, and a rising loop will be diverted or distorted unless the field is sufficiently strong.

In the first chapter of this section D. W. Hughes reviews the strengths and weaknesses of solar dynamo theory; he then goes on to discuss Rayleigh–Taylor instabilities of a stratified magnetic layer and presents recent nonlinear computations which show how such a layer can break up. Next, F. Moreno-Insertis describes the rise of individual flux tubes through the convection zone. Such tubes are unstable to kinks, which develop as gas drains downwards from their crests. A number of arguments suggest that the initial field strength must approach 10^5 G.

In a contributed paper at the Workshop A. R. Choudhuri considered mechanisms for preventing Coriolis forces from deflecting a rising flux tube, either by turbulent exchange or by Kelvin-Helmholtz instabilities. There were also three contributed papers concerned with aspects of the solar dynamo. P. R. Wilson argued that polar fields emerge from below rather than as a result of flux transport from lower latitudes. Then R. L. Jennings described an ambitious numerical simulation of convection in a rotating, compressible layer which acts as a turbulent dynamo. Finally, D. J. Galloway explained the astrophysical relevance of the fast dynamo problem and presented the first convincing numerical example of a fast dynamo.

THE FORMATION OF FLUX TUBES AT THE BASE OF THE CONVECTION ZONE

D.W. HUGHES
Department of Applied Mathematical Studies
The University
Leeds LS2 9JT
U.K.

ABSTRACT. Sunspots are the surface manifestation of a cyclical magnetic field within the Sun. Here we discuss some of the current theories of this field and how it may give rise to sunspots.

1. Introduction

The title of this review is of course rather a leading one, presupposing as it does that magnetic flux tubes do indeed need to be formed and that the site of their formation is the base of the convection zone. In the first half of this review I shall attempt to show why the idea is plausible; in the second half I shall describe attempts at modelling the process. The actual rise to the solar surface of buoyant flux tubes is described elsewhere in these proceedings (Moreno-Insertis 1992); I shall thus try and limit discussion to the earlier stages of the field's evolution.

Although, naturally, most of the talks at this workshop have focussed on the detailed structure of sunspots it should always be borne in mind that sunspots are the (complicated) surface manifestation of a global solar magnetic field. After all, sunspots do not exist in isolation; somehow they must be coupled to a large-scale sub-surface magnetic field. As shown by the many beautiful observations presented at the meeting, sunspots are objects of considerable, even bewildering complexity. However, although the picture of a sunspot as a piece of a toroidal flux rope pushing its way through the surface is seen to be drastically over-simplified, at the crudest level the idea remains valid. Hale's laws concerning the bipolarity of sunspots and their approximate alignment with the equator are typically obeyed, strongly suggesting the emergence of an underlying predominantly toroidal flux tube. It is obviously then of interest to ask whether the field exists in such a tubular form throughout the solar interior or, alternatively, whether part of the flux is in the form of a large-scale diffuse field. The distinction is an important one with the physics of the two types of field configuration being rather different.

An isolated tube of magnetic flux will adjust rapidly to having the same total pressure (gas + magnetic) as its non-magnetic surroundings. Accordingly, the gas pressure inside the tube is always less than that of the external atmosphere. Whether this reduced pressure leads to the tube being buoyant depends crucially on the temperature difference between the tube and its surroundings. For example, it follows from the perfect gas law that a tube in thermal equilibrium with its surroundings will necessarily be less dense and so

J. H. Thomas and N. O. Weiss (eds.), Sunspots: Theory and Observations, 371–384.

will be buoyant. Indeed, only for one particular temperature difference between the tube and the external atmosphere will the tube be in mechanical equilibrium, with the same density as its surroundings. In general, therefore, isolated flux tubes are to be regarded as non-equilibrium phenomena.

For a diffuse field, on the other hand, things can be rather different. By a diffuse field I mean, ideally, one that is homogeneous in the horizontal direction; for example in Cartesian geometry, fields of the form $(B_x(z), B_y(z), 0)$. In contrast to isolated flux tubes, such fields can exist in hydrostatic equilibrium simply by assuming some of the loading of the overlying atmosphere; of course these equilibria may be unstable but that is another matter. In spherical geometry, where the field topology typically is more complicated, I shall use the term 'diffuse fields' rather loosely, simply to refer to fields with horizontal variation on a scale large compared to flux tube dimensions.

2. The Global Solar Magnetic Field

There are two key questions regarding the magnetic field that participates in the solar cycle. One is 'What is the mechanism responsible for producing the cyclical field?'; the other is 'What are the strength and location of the field?'. These are not strictly separate issues since a full knowledge of the driving mechanism would of course include a knowledge of the field's location. However, the distinction is useful since it is possible to obtain an approximate answer to the second question without having to answer the first. Calculating the distribution of the sub-surface field is a tricky business, but an estimate of the magnetic flux at depth can be obtained by consideration of the observed surface magnetic features (Galloway & Weiss 1981; Parker 1987). The total flux appearing at the surface at any one time can be measured and, when coupled to estimates of how efficient the Sun is in recycling used flux, gives an idea of the total (unsigned) flux beneath the surface. Both Galloway & Weiss (1981) and Parker (1987) arrive at a value of about $10^{22} - 10^{23}$ Mx. The corresponding field strength of course then depends on how tightly packed is the flux. For reasons that we shall discuss shortly, Galloway & Weiss propose that the bulk of the flux is located in a thin layer, of thickness 10^4 km, just beneath the convection zone, with a field strength of the order of 10^4 G; Parker, on the other hand, suggests that the field has a rather larger filling factor, occupying the lower half of the convection zone with a strength of about 10^3 G. Simply from the surface observations it is not possible to say with any certainty what the field looks like at depth — further insight must be gained from a detailed consideration of the physical processes taking place in the solar interior.

The first of the two questions above addresses the fundamental problem of solar magnetohydrodynamics. There are two possible candidates for the driving mechanism of the field; either via dynamo action or by a hydromagnetic oscillator. The debate on this issue is of long standing and is still not fully resolved, although at the moment dynamo action is certainly the more fancied of the two. Nevertheless, in the open spirit of the workshop it seems appropriate to discuss both possibilities. The pros and cons of both mechanisms have been elegantly expounded in a review article written ten years ago by Cowling (1981) and so I shall therefore focus mainly on more recent developments.

2.1 THE CURRENT STATE OF SOLAR DYNAMO THEORY

In a dynamo the magnetic field is maintained against Ohmic dissipation by the inductive motions of an electrically conducting gas or fluid; in a cyclic dynamo both the poloidal and toroidal fields change sign every half-cycle. Since the Earth has maintained its field for something like 10^5 Ohmic decay times there is absolutely no doubt that a rotating fluid body is capable of regenerating a magnetic field — the issue is whether such a process is taking place within the Sun. A full treatment of the dynamo problem, in which the velocity and magnetic fields are combined in a self-consistent way, is extremely difficult to attain. Consequently, most attention has been focussed on the kinematic problem, in which the magnetic field evolves under the influence of a *prescribed* velocity field, with no account taken of the back-reaction of the Lorentz force on the fluid motions. A most elegant formulation of the kinematic problem is possible when there is a separation of scales between a large-scale magnetic field and small-scale turbulence. The mean-field theory describing the field's evolution in this case was developed by Steenbeck, Krause & Rädler (1966 and subsequent papers, translated into English by Roberts & Stix 1971), following on from an earlier idea of Parker (1955) (for further descriptions of mean-field theory see Moffatt 1978, 1983; Krause & Rädler 1980). In the simplest example of an oscillatory dynamo the velocity field is characterised by four (possibly space-dependent) parameters; α, some measure of the helicity ($\mathbf{u} \cdot \nabla \times \mathbf{u}$) of the turbulence; $\nabla\omega$, the shear or differential rotation; U, the meridional circulation; and η, the turbulent magnetic diffusivity. These so-called $\alpha\omega$-dynamos admit travelling wave solutions, with propagation towards the equator when $\alpha\partial\omega/\partial r < 0$ (in the northern hemisphere) (Parker 1955). Simple, though not rigorous, considerations of how the convecting gas is influenced by compressibility and rotation suggest that the net helicity in the northern hemisphere should be negative (corresponding to *positive* α) and that therefore $\partial\omega/\partial r$ should be negative (see Moffatt 1978). Stix (1976) advanced independent arguments for this choice of the sign of α.

By suitable choices of the free parameters, solutions to the kinematic dynamo problem can be obtained that agree rather well with the observed surface solar magnetic field (Roberts & Stix 1972; Yoshimura 1975). Furthermore, the inclusion of plausible nonlinear terms leads to complicated temporal behaviour reminiscent of the solar cycle (Weiss, Cattaneo & Jones 1984). However, although kinematic mean-field dynamo theory has proved instructive, it is not the final word. Indeed, with regard to the solar magnetic field, there are two types of difficulty with this approach. The first concerns the very validity of the mean-field formulation of the solar field (Moffatt 1978; Layzer, Rosner & Doyle 1979; Cowling 1981; Hoyng 1990). The second, on which I shall concentrate here, concerns problems that arise when dynamical behaviour is taken into account. In particular, we must address the fact that (a) the field is, at least sometimes, of an intermittent nature, (b) the field will interact with the convective motions in a complicated fashion, and (c) in a non-kinematic description of a dynamo, quantities such as the helicity and the differential rotation cannot be chosen arbitrarily, but must emerge in a self-consistent manner from the hydrodynamics. Such dynamical considerations also provide information as to the location of the solar dynamo. From kinematical considerations, which highlight the importance of differential rotation and of helical motions, the most promising site for dynamo action is the convection zone, the scene of the most vigorous gaseous motions; however, the dynamical aspects of the magnetic field lead to certain difficulties with this idea, which I shall discuss below.

Parker (1975) pointed out that isolated flux tubes of strength $\gtrsim 10^2$G at the base of the convection zone will rise to the surface, by magnetic buoyancy, on a time-scale of a few months. He argued that this rapid escape of flux tubes could cause problems for the solar dynamo, which presumably has a regeneration timescale of the order of the solar cycle period. Parker's (1975) study of a rising horizontal flux tube in which buoyancy is balanced by aerodynamic drag has been extended by a number of authors to include other forms of retarding force and also the important effects arising from undulatory perturbations of the tube (see Hughes 1991, Moreno-Insertis 1992 for complete references); the outcome of most of these calculations is again that flux tubes can traverse the convection zone in a matter of months.

A related issue dealing with the confinement of magnetic flux is that of the interaction between convective motions and magnetic fields. For two-dimensional convection, magnetic flux can be expelled from convective cells (Weiss 1981a,b; Proctor & Weiss 1982). In three dimensions things naturally become more complicated. Kinematic calculations show that at low values of the magnetic Reynolds number R_m the field can be pumped preferentially downwards, say, by an array of cells in which the downward motions are connected and the upward cells disconnected (Drobyshevski & Yuferev 1974). However, at higher R_m, even in the kinematic regime, the picture is not so straightforward, with some downward pumping but also with the formation of a layer of reversed flux near the top of the domain (Arter, Proctor & Galloway 1982). In the compressible regime, even for two-dimensional flows, yet more complications arise (see the reviews of Hughes & Proctor 1988 and Proctor 1992). Thus, although great strides have been made in the study of magnetoconvection in recent years we are still not yet in a position to say with any certainty just what is going on in the solar convection zone. On theoretical grounds we might expect that, for the bulk of the convection zone, rotation would enforce 'cartridge belt convection', cells with axes aligned parallel to the rotation axis; numerical simulations of rotating spherical convection seem to confirm this (e.g. Gilman & Miller 1986) as do some of the Spacelab experiments of Hart, Glatzmaier & Toomre (1986) (although others though show considerably less ordered motions). However, for the Sun there is no observational evidence of such flows; indeed, the originators of giant cells (Simon & Weiss 1968) have recently laid them to rest, arguing that there is no organised convective structure on scales larger than that of supergranules (Simon & Weiss 1991). Further support for the case against cartridge belt convection comes from helioseismology observations (Brown *et al.* 1989), which are incompatible with such motions. The case for cartridge belt cells relies on the Coriolis force being dominant; if it is not, such as when the magnetic field is sufficiently strong as to be dynamically significant, then the convection pattern can be very different (Zeldovich, Ruzmaikin & Sokoloff 1983; Jones, Roberts & Galloway 1990). Indeed, Ribes, Mein & Mangeney (1985) have argued that the movement of surface magnetic features suggests the presence of azimuthal ('doughnut') rolls encircling the rotation axis. The detailed study of nonlinear, compressible magnetoconvection represents a considerable computational challenge, and so is still in its early stages. Nordlund *et al.* (1992) have investigated the interactions between rotating turbulent convection and a dynamo-generated magnetic field; their main finding, of considerable interest in the solar context, is that the mean flux is pumped downwards while strong ropes of field are formed around swirling downflows.

The most ambitious dynamo calculations are those of Gilman (1983) and Glatzmaier (1985a), who modelled the interaction between convection and magnetic fields in a rotating spherical shell. Their simulations were dynamical, in that there was no prescribed helicity distribution or differential rotation profile; these emerged self-consistently as the flow evolved, governed by the induction, momentum and energy equations. The calculations are of considerable importance in that they demonstrate clearly the possibility of a self-consistent cyclic dynamo; they are also of interest in that there are significant differences between the calculated velocity and magnetic fields and those observed on the Sun. The simulations show an equatorial surface acceleration (as on the Sun) with the convection adopting a pattern of drifting cartridge belt cells; the angular velocity then must decrease inwards. The net helicity in the northern hemisphere is negative and the magnetic field migrates towards the poles. This is actually in agreement with the prediction of kinematic mean-field theory, although this may just be coincidental since the assumption of two distinct scales is not really valid. In addition, the cycle period is rather short compared to that of the Sun.

The preceding three paragraphs have seen a brief discussion of some of the problems that have arisen from considerations of a dynamo operating within the convection zone; in the light of these problems we are now in a position to consider the current ideas regarding the solar dynamo. As mentioned in the introduction, one of the important features of the observed surface field is its regularity, as described by Hale's polarity laws. Consequently, the flux tubes that emerge as sunspots must be of sufficient strength as to resist distortion and shredding by the turbulent convection. This would seem to imply that the tubes should always be of at least equipartition strength (with their magnetic energy equal to the external kinetic energy); if they start their journey near the base of the convection zone this gives a field strength there of the order of 10^5G (Moreno-Insertis 1992). This represents a strong argument against the magnetic field always being in the convection zone only in the form of discrete flux tubes since it seems unlikely that a tube could be amplified to equipartition strength before succumbing to a combination of magnetic buoyancy and turbulent shredding. In addition it is difficult to explain how the processes of diffusion and reconnection operate in a flux tube dynamo (see Schüssler 1983 for a discussion of fibril and flux tube dynamos). Van Ballegooijen (1982) has considered the possibility that flux tubes may be anchored in the overshoot zone, the convectively stable transition region between the convective and radiative zones. Although the magnetic buoyancy of isolated flux tubes is suppressed in a subadiabatic atmosphere, van Ballegooijen nevertheless concludes that the longest containment times are only of the order of one year. He goes on to suggest that flux tubes may however be confined to the overshoot region for longer times by an inward Coriolis force arising from advection by an equatorward mean flow.

The alternative scenario to the field always being in a tubular state is that the solar dynamo regenerates a *diffuse* field and that the flux tubes we see at the surface arise from some secondary process. From considerations of two-dimensional magnetoconvection, Galloway & Weiss (1981) argued that the bulk of the magnetic flux should be contained in the convective overshoot region (see also Spiegel & Weiss 1980); they suggested in addition that the overshoot region would also be the site of the solar dynamo. A related possibility is that the dynamo operates in the convection zone itself, with a strong element of downward pumping. A layer of strong, reasonably ordered magnetic field would then be built up, either at the base of the convection zone or in the overshoot region, which could subsequently give rise, through some instability mechanism, to the strong flux tubes that emerge at the surface. In the light of the most recent three-dimensional simulations (Nordlund *et al.* 1992) the idea of a strong field in the overshoot region would seem to stand on even firmer ground.

For rather different reasons, the idea of a dynamo operating in the overshoot zone has arisen from considerations of the dynamo calculations of Gilman (1983) and Glatzmaier (1985a). They find that the helicity changes sign towards the base of the convection zone and that, at least on kinematic grounds, it is of the right sign as to produce equatorward migration of magnetic features. Glatzmaier (1985b) attempted to model an overshoot zone dynamo but, for numerical reasons, was unable to simulate a complete magnetic cycle. DeLuca & Gilman (1986, 1988) have also considered overshoot dynamos, though the effects of the overshooting were just encapsulated in the α term.

The most recent support for a deep-seated dynamo comes from the helioseismology measurements. They indicate that there is little radial variation in angular velocity for most of the convection zone and that the interior rotates nearly as a solid body, at a rate equal to that of the surface at some mid-latitude. If this is so then there must be a transition region with considerable radial gradients in angular velocity; $\partial\omega/\partial r$ would be positive at low latitudes, negative at high latitudes. As pointed out by Gilman, Morrow & DeLuca (1989), such a region of rapid variation in ω is potentially favourable for dynamo action; from considerations of the helicity in global convection models, the product of the differential rotation and the helicity at the base of the convection zone is such as to produce equatorward migration of magnetic features in low latitudes, as observed on the Sun, and poleward migration at high latitudes.

Since the seismology results indicate little radial variation in angular velocity throughout the convection zone, this means that the surface latitudinal variation is maintained with depth. Babcock (1961), in one of the earliest models of the solar dynamo, postulated that poloidal field would be transformed to toroidal field via latitudinal shear. The differing signs of the toroidal field in the two hemispheres then follows very naturally; however, the prediction of kinematic theory that the dynamo waves propagate vertically has meant that the idea has received relatively little attention. Lerche & Parker (1972), however, considered such a model and concluded that vertically propagating waves could still give rise to the observed surface equatorward migration. The idea of latitudinal shear playing some role in the solar dynamo would still seem to be an interesting possibility.

2.2 AN ALTERNATIVE TO THE DYNAMO PROCESS?

As mentioned earlier, the alternative to the idea of dynamo action is that the observed solar field is produced by some sort of hydromagnetic oscillator. This is quite a long-standing suggestion, dating back to Walén (1946). It is envisaged that a cyclical toroidal field is generated by repeated winding and unwinding of a *non-reversing* poloidal field. To remain coherent, the poloidal field would need to be located in the radiative interior, away from the turbulence of the convection zone. The notion of such a poloidal field in the radiative interior is not inconceivable since the Ohmic decay time is comparable to the Sun's lifetime on the main sequence. However, this supposes that the Sun, on arriving on the main sequence, possessed a sizeable poloidal field; whether such a field would have survived the fully convective Hayashi phase is hard to say without a better knowledge of the turbulent diffusion of magnetic fields (see Cowling 1981; Mestel & Weiss 1987). The generation of toroidal field by shearing a poloidal field is essentially the same as occurs in a dynamo; however, besides this similarity there are two big differences between the two mechanisms. The first is that in an oscillator there is no regeneration of the poloidal field. The second concerns the role played by magnetic diffusion; in a dynamo, diffusion is essential both for dissipation and reconnection; in an oscillator, which is, very crudely, a torsional Alfvén wave, diffusion is not of fundamental importance. Although the idea of an oscillator within the Sun has been around for a long time there is still no realistic model of such a mechanism (although see Piddington 1976 and earlier papers). At present it is certainly not possible to provide answers to the questions of how the oscillations are driven, why the field propagates towards the equator, and why the period is 22 years.

Deciding, observationally, between an oscillator and a dynamo – especially when there is no satisfactory model of either – is a delicate matter. The observational evidence to date strongly favours dynamo action. There is certainly evidence of poloidal field reversals (Howard & LaBonte 1981) and, in addition, the torsional waves with a period of 11 years detected by Howard & LaBonte (1980) are most readily interpreted as being due to the (quadratic) Lorentz force arising from a dynamo-generated field of period 22 years (Schüssler 1981; Yoshimura 1981). However, although such evidence is compelling it is not conclusive. For example, it *could* be argued that, since the poloidal field at the heart of any oscillator must be situated within the radiative zone, it would give no clue at the surface as to its structure; poloidal field reversals would then be accounted for simply by the dispersal of surface flux within the convection zone.

Another observation that may be able to distinguish between the two mechanisms is that of the phase maintenance of the cycle. Dicke (1970) argued that that because the early arrivals of the sunspot maxima of 1778 and 1788 were followed by a compensating long cycle, the mechanism responsible for the cycle must keep time; considerable differences in cycle lengths must then be accounted for by changes in the convection zone dynamics. Dicke (1978) and Gough (1990) have used the sunspot data for the past three hundred years to see if the best fit is found by a process with no long-term memory (a model of a turbulent dynamo) or by a process that keeps perfect time subject to some stochastic element (an oscillator filtered through the turbulent convection zone). Unfortunately the results are tremendously inconclusive. This is partly due to the sunspot record being too short; however there is the additional problem that any oscillations could be highly nonlinear, with no clear maintenance of phase. A further observation of interest is that the p-mode frequencies vary with the solar cycle; this though is due to changes in the structure of the very outer layers of the Sun (Libbrecht & Woodard 1990) and so sheds no light on the problems concerning us here. Despite tantalising hints – from both helioseismology and neutrino measurements – there is no firm evidence of any cyclic variation of the deeper interior (see the review by Gough & Toomre 1991).

To return to the topic in the title of the paper, it is interesting to note that the idea for the formation of sunspots is essentially the same whether the magnetic field is produced by an oscillator or by a dynamo that creates a diffuse field in the overshoot zone. In both cases it is postulated that the toroidal field is built up to sufficient strength that it becomes unstable. A stitch of flux, or possibly an entire ring, is thus released into the convection zone, through which it rises to appear at the surface as sunspots. We shall deal with this issue in the following section.

3. The Nonlinear Breakup of a Magnetic Layer

In Section 2 we discussed several factors pointing to the creation of a strong, predominantly toroidal magnetic field in the overshoot zone. In this section we consider how the field can escape, as it must, eventually to appear at the surface as sunspots and other magnetic features. Magnetic fields are of course susceptible to a whole host of instabilities. Since the radius at the base of the convection zone considerably exceeds the critical radius at which the buoyancy and tension of a flux tube are in balance (Jensen 1955; Weiss 1964) we expect that the most important instabilities will be buoyancy-driven. The precise structure of the field in the overshoot zone is not known and so it makes sense to examine, at least initially, the stability properties of reasonably simple field configurations. The obvious starting point is to consider, in Cartesian geometry, a purely horizontal field that depends only on the vertical coordinate. Buoyancy instabilities of such fields fall into two categories: *(i)* instabilities due to a continuously stratified field; *(ii)* instabilities arising from a discontinuity in the field. For smoothly varying fields there is a wide range of possible instability mechanisms; the linear theory of such modes has been explored in some detail (e.g. Acheson 1978, 1979; Hughes 1991), but little progress has been made into the nonlinear regime. Instabilities driven by a discontinuity in the field are less complicated; the nonlinear evolution has been studied and, since it is this that is of most interest in the context of flux tube formation, I shall concentrate here on these modes. By virtue of the magnetic pressure, a magnetic field can support denser field-free gas, in equilibrium; instabilities that ensue are of the Rayleigh-Taylor type, resulting from the release of gravitational potential energy.

This problem was first tackled by Kruskal & Schwarzschild (1954), who examined the stability of an isothermal plasma supported above a vacuum by a uniform magnetic field $\mathbf{B} = B_0\hat{\mathbf{y}}$. They found, by an essentially local analysis, that in the absence of all diffusion, instability occurs when

$$2k_y^2 < |k_x|/H_p\,, \tag{1}$$

where k_x and k_y are the wavenumbers in the x and y directions and H_p is the pressure scale height of the plasma. The most readily destabilised modes are interchanges ($k_y \equiv 0$), which do no work against magnetic tension. Parker (1979) dealt with the related problem of a magnetic gas of uniform density ρ and uniform field $\mathbf{B} = B_0\hat{\mathbf{y}}$ supporting a non-magnetic gas of uniform density $\rho + \Delta\rho$, where $\Delta\rho/\rho = V_A^2/2u^2$ (V_A = Alfvén speed, u = characteristic thermal speed). For incompressible perturbations, in the absence of all diffusion, the growth rate s is given by

$$s^2 = \frac{g\Delta\rho(k_x^2+k_y^2)^{1/2} - \rho V_A^2 k_y^2}{2\rho + \Delta\rho}\,, \tag{2}$$

with instability thus occurring when

$$k_y^2/(k_x^2+k_y^2)^{1/2} < (g/V_A^2)\Delta\rho/\rho\,. \tag{3}$$

As in the Kruskal & Schwarzschild problem, the most readily destabilised modes are the interchanges ($k_y \equiv 0$). The most rapidly growing mode has infinitesimally small transverse dimensions ($k_x \to \infty$), although this of course is no longer true when diffusion is included. It can be seen from (2) that for interchange modes the only role of the magnetic field is to provide the destabilising density jump $\Delta\rho$; the dispersion relation is identical to that obtained by considering two uniform superposed fluids whose densities differ by $\Delta\rho$ (Chandrasekhar 1961).

Cattaneo & Hughes (1988) considered the linear and nonlinear evolution of these interchange modes, incorporating diffusion and the full effects of compressibility. The equilibrium state in their model consisted of a finite slab of uniform, unidirectional horizontal

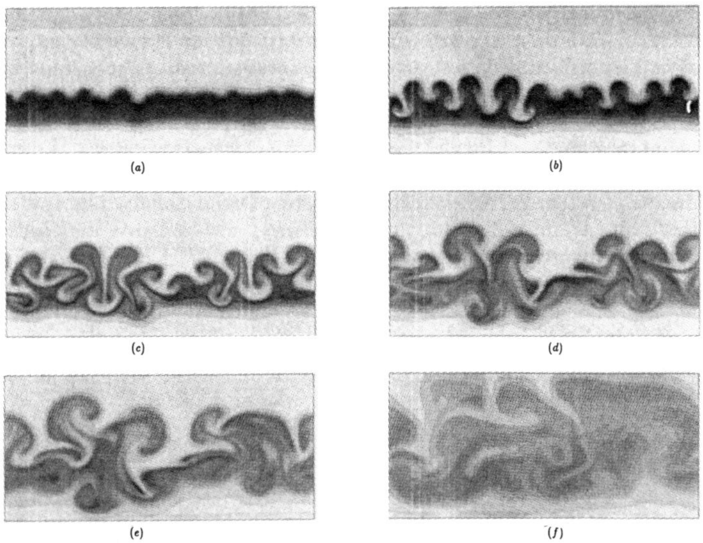

Figure 1. Time evolution of an unstable magnetic layer. The field lines remain straight, into the plane of the paper. Darker regions correspond to stronger field (from Cattaneo & Hughes 1987).

magnetic field embedded in a convectively stable, piecewise polytropic atmosphere. The total pressure and temperature were continuous, leading to a destabilising density jump at the upper magnetic interface. As shown in Figure 1a,b the initial stage of the instability is marked by the rise of the buoyant magnetic gas and the fall of the denser, field-free gas above. Consequently, a strong shear develops, causing the onset of Kelvin-Helmholtz instabilities that wrap the gas into regions of strong vorticity, located in the wings of the mushrooms of Figure 1b. Once most of the available gravitational potential energy has been released, the motions are dominated by the vorticity distribution. The preferred horizontal scale is such that the strongest interactions are between adjacent vortices on neighbouring mushrooms, acting so as to move the vortex pair downwards; indeed, the interactions can be of sufficient strength as to pull down pockets of strong field, even though these are less dense than the neighbouring gas. The field is rapidly dispersed by the interacting vortices, on a timescale short compared to the magnetic diffusion time.

Cattaneo, Chiueh & Hughes (1990a) extended these calculations to include the effects of a weak poloidal magnetic field, as must be present in the Sun. The initial field was still horizontal, but no longer unidirectional; in Cartesian geometry $\mathbf{B} = (B_x(z), (B_0^2 - B_x^2)^{1/2}, 0)$ (also the lowest region of field-free gas was excluded). As in the earlier calculations, only y-independent motions were considered, but now with non-zero components of the velocity and magnetic field in all three spatial directions. As we shall see below, a key feature of this field configuration is the possible existence of a *resonant surface* on which the poloidal field vanishes. (Although the existence of a resonant surface in the Cartesian model might seem rather arbitrary, since a rotation about the z-axis mixes the 'poloidal' and 'toroidal' ingredients, the uniqueness of the resonant surface is ensured by noting that the Cartesian model is constructed as a local representation of the system in spherical geometry, where the poloidal and toroidal components *are* uniquely defined.) From the energy principle of ideal MHD it can be shown that the equilibrium is stable if

$$4\beta\frac{(\gamma m - m - 1)}{\gamma} > \frac{(m+1)}{(\delta B_x)^2} + \frac{2}{\theta D}\,. \tag{4}$$

Here $\beta = \mu_0 p_0 / B_0^2$, with p_0 the gas pressure at the top of the domain, γ is the usual ratio of specific heats, m is the polytropic index, δB_x is the shear at the magnetic interface, θ is the dimensionless temperature gradient and D is the distance between the interface and the resonant layer. From (4) it can be seen that, as expected, stability is enhanced by large magnetic shears and large subadiabatic gradients. Furthermore we see that instability is most likely to occur when the resonant layer is close to the magnetic interface (D small); in such cases the field is untwisted close to the interface and the instability proceeds rather as for a purely toroidal field.

Figure 2 shows the nonlinear development of the instability from two different equilibrium states, both of which have B_x varying quadratically with height, but which differ in the location of the resonant layer. For the equilibrium of Figure 2a the resonant layer is situated just beneath the magnetic interface. As predicted by linear theory, the instability remains confined to the upper reaches of the field, where the poloidal component is weak, and is similar in character to the evolution of the purely toroidal field shown in Figure 1. The poloidal field increases with depth below the resonant layer, resisting the instability and allowing part of the original layer to avoid disruption. The equilibrium of Figure 2b, on the other hand, has a resonant surface close to the bottom boundary; the evolution of the instability is seen to be markedly different. The poloidal field is strongest at the magnetic interface and so, to avoid excessive work against magnetic tension, the instability adopts a long horizontal lengthscale. The poloidal field decreases with depth down to the resonant surface, where it vanishes; the preferred mode assumes a long vertical scale, utilising the untwisted field at depth. In consequence, most of the magnetic layer is consumed by the instability, leaving little behind. Besides the considerable difference in scale (both horizontal and vertical), the ejected flux of the two cases differs in one other very important respect. When the resonant layer is close to the magnetic interface the escaping field is predominantly toroidal; Kelvin-Helmholtz instabilities set in (as for the purely toroidal field of Figure 1), strong vortices are formed and the field is rapidly distorted. In contrast, when the resonant surface is deep-seated, the ejected field has a strong poloidal component at its outer edge; Kelvin-Helmholtz instabilities are suppressed, the fragments thus resist deformation and rise unscathed.

Although, from the idealised calculations described above, it is not possible to make any definitive statements about the solar magnetic field, some interesting features of possible significance do arise. Cattaneo, Chiueh & Hughes (1990b) have speculated that the variation with the solar cycle of the scale of emerging flux (Golub *et al.* 1981) could be accounted for by a temporal variation in the location of the resonant layer. At solar minimum the resonant layer would be situated close to the magnetic interface. The emerging flux would then be small-scale and virtually untwisted; it would thus be easily distorted, influenced both by the vorticity distribution and by the turbulence in the overlying convection zone. Such flux would appear at the surface as small-scale ephemeral regions with no systematic pattern of emergence. At solar maximum, the resonant layer would be located deeper

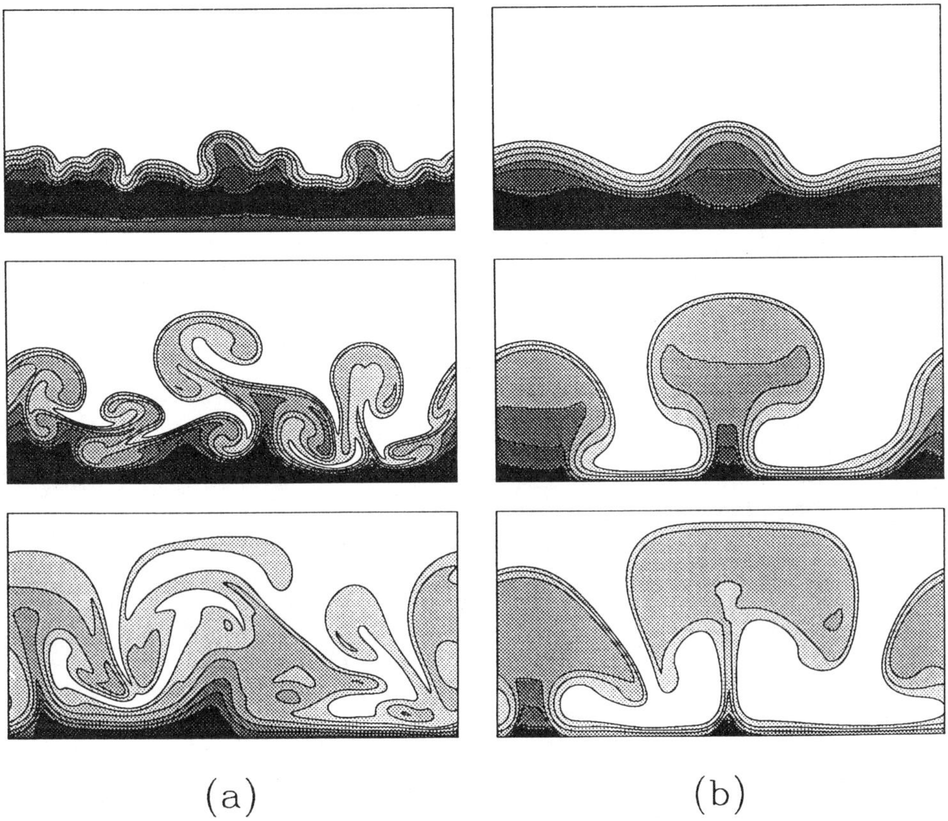

Figure 2. Time evolution of a predominantly toroidal field. The solid lines are poloidal field lines; the grey-scale represents the toroidal field (into the page). The equilibrium in (a) has the poloidal field vanishing near the magnetic interface and increasing with depth; in (b) it vanishes near the lower boundary and increases with height. The motions are independent of the direction into the page (from Cattaneo *et al.* 1990b).

down, giving rise to coherent tubes of flux with a helical field structure. These would resist deformation by the convection and appear at the surface as sunspots, with well-defined polarity laws.

4. Summary and Discussion

The pattern of emergence of sunspots certainly suggests the existence of an ordered large-scale magnetic field in the solar interior; furthermore, the flux tubes that appear as sunspots must be of at least equipartition strength, having resisted distortion during their journey through the convection zone. As discussed in Section 2, several independent factors (studies of magnetoconvection, self-consistent dynamo models, helioseismology measurements) point to the idea of a strong, predominantly toroidal diffuse field being stored in the region of convective overshoot, situated beneath the convection zone proper. The evidence is certainly suggestive, although not conclusive. The possibility also exists that the field always takes a tubular form (and is never diffuse). Van Ballegooijen (1982) has suggested the overshoot zone as the most satisfactory storage location for such tubes. Parker (1987) has advanced the alternative idea that the tubes are located towards the base of the convection zone, held down against buoyancy by what he terms 'thermal shadows' — cold regions on top of the tubes, formed by the magnetic field suppressing the heat flow. However, it seems that there are more severe difficulties to be overcome with the notion of the field only taking the form of discrete tubes. The fundamental problem is that flux tubes are really rather slippery objects, and consequently somewhat difficult to pin down. Although it is possible, by a certain conspiracy of forces, to construct flux tube equilibria, it is by no means clear that these will be stable. Of course, all is not plain sailing with the diffuse field approach either. Just as for the case of isolated tubes, the dynamo process is not fully understood. In addition, a solution is needed to the problem of building up a strong field that then releases flux tubes of equipartition strength into the convection zone. There is certainly no shortage of instabilities available; the question is whether they can act so as to release large fragments of strong field. If indeed this is how the Sun operates then then the rapid rise of isolated flux tubes is not a matter for concern.

Both theory and observations should allow progress to be made in determining the structure of the interior field. Even with the computing power available today, a high resolution self-consistent dynamo calculation is still a massive computational undertaking. Such computations are indeed valuable; however it is also true that much can be learnt by focussing on specific aspects of the whole problem. The results described in Section 3 showed that a layer of field could become unstable and release magnetic fragments of a range of sizes. In these calculations the equilibrium state possessed a fairly strong field; it would be of considerable interest to see whether, when the field strength is gradually increased from an initially low value, the instability can be suppressed, by overshooting convection, until the field strength reaches some critical value. In addition it is necessary to try and understand the three-dimensional evolution of such instabilities and to investigate further the effects of non-uniform field distributions. An increased understanding of the dynamics of the convection zone itself is being obtained from numerical studies of three-dimensional compressible convection (Stein & Nordlund 1989; Cattaneo *et al.* 1991). Over the past few years helioseismology has shed new light on ideas of the internal rotation of the Sun. Better observations in the future (via the GONG network) are bound to add to our knowledge of both the rotation profile and, possibly indirectly, the magnetic field.

Acknowledgments

I am grateful to D.J. Galloway and N.O. Weiss for helpful comments on the paper.

References

Acheson, D.J. (1978) *Philos. Trans. R. Soc. London A* **289**, 459.

Acheson, D.J. (1979) *Solar Phys.* **62**, 23.

Arter, W., Proctor, M.R.E. & Galloway, D.J. (1982) *Mon. Not. R. Astron. Soc.* **201**, 57p.

Babcock, H.W. (1961) *Astrophys. J.* **133**, 572.

Brown, T.M., Christensen-Dalsgaard, J., Dziembowski, W.A., Goode, P., Gough, D.O. & Morrow, C.A. (1989) *Astrophys. J.* **343**, 526.

Cattaneo, F. & Hughes, D.W. (1987) in *Theoretical Problems in High Resolution Solar Physics II* (eds. G. Athay and D.S. Spicer) NASA-CP-2483, pp. 101-104.

Cattaneo, F. & Hughes, D.W. (1988) *J. Fluid Mech.* **196**, 323.

Cattaneo, F., Chiueh, T. & Hughes, D.W. (1990a) *J. Fluid Mech.* **219**, 1.

Cattaneo, F., Chiueh, T. & Hughes, D.W. (1990b) *Mon. Not. R. Astron. Soc.* **247**, 6p.

Cattaneo, F., Brummell, N.H., Toomre, J., Malagoli, A. & Hurlburt, N.E. (1991) *Astrophys. J.* **370**, 282.

Chandrasekhar, S. (1961) *Hydrodynamic and Hydromagnetic Stability*, Clarendon.

Cowling, T.G. (1981) *Ann. Rev. Astron. Astrophys.* **19**, 115.

DeLuca, E.E. & Gilman, P.A. (1986) *Geophys. Astrophys. Fluid Dyn.* **37**, 85.

DeLuca, E.E. & Gilman, P.A. (1988) *Geophys. Astrophys. Fluid Dyn.* **43**, 119.

Dicke, R.H. (1970) in *Stellar Rotation* (ed. A. Stetteback) pp. 289-317 Reidel, Dordrecht.

Dicke, R.H. (1978) *Nature* **279**, 676.

Drobyshevski, E.M. & Yuferev, V.S. (1974) *J. Fluid Mech.* **65**, 33.

Galloway, D.J. & Weiss, N.O. (1981) *Astrophys. J.* **243**, 945.

Gilman, P.A. (1983) *Astrophys. J. Suppl.* **53**, 243.

Gilman, P.A. & Miller, J. (1986) *Astrophys. J. Suppl.* **61**, 585.

Gilman, P.A., Morrow, C.A. & DeLuca, E.E. (1989) *Astrophys. J.* **338**, 528.

Glatzmaier, G.A. (1985a) *Astrophys. J.* **291**, 300.

Glatzmaier, G.A. (1985b) *Geophys. Astrophys. Fluid Dyn.* **31**, 137.

Golub, L., Rosner, R., Vaiana, G.S. & Weiss, N.O. (1981) *Astrophys. J.* **243**, 309.

Gough, D.O. (1990) *Phil. Trans. R. Soc. London A* **330**, 627.

Gough, D.O. & Toomre, J. (1991) *Ann. Rev. Astron. Astrophys.* **29**, 627.

Hart, J.E., Glatzmaier, G.A. & Toomre, J. (1986) *J. Fluid Mech.* **173**, 519.

Howard, R. & LaBonte, B.J. (1980) *Astrophys. J.* **239**, L33.

Howard, R. & LaBonte, B.J. (1981) *Solar Phys.* **74**, 131.

Hoyng, P. (1990) in *Solar Photosphere: Structure, Convection and Magnetic Fields. Proc. IAU Symp. 138* (ed. J.O. Stenflo) Kluwer, pp. 359-374.

Hughes, D.W. (1991) in *Advances in Solar System Magnetohydrodynamics* (eds. E.R. Priest and A.W. Hood) Cambridge University Press, pp. 77-104.

Hughes, D.W. & Proctor, M.R.E. (1988) *Ann. Rev. Fluid Mech.* **20**, 187.

Jensen, E. (1955) *Ann. Astrophys.* **18**, 127.

Jones, C.A., Roberts, P.H. & Galloway, D.J. (1990) *Geophys. Astrophys. Fluid Dyn.* **53**, 183.

Krause, F. & Rädler, K.-H. (1980) *Mean-Field Magnetohydrodynamics and Dynamo Theory*, Pergamon.

Kruskal, M. & Schwarzschild, M. (1954) *Proc. R. Soc. London A* **223**, 348.

Layzer, D., Rosner, R. & Doyle, H.T. (1979) *Astrophys. J.* **229**, 1126.

Lerche, I. & Parker, E.N. (1972) *Astrophys. J.* **176**, 213.

Libbrecht, K.G. & Woodard, M.F. (1990) *Nature* **345**, 779.

Mestel, L. & Weiss, N.O. (1987) *Mon. Not. R. Astron. Soc.* **226**, 123.

Moffatt, H.K. (1978) *Magnetic Field Generation in Electrically Conducting Fluids*, Cambridge University Press.

Moffatt, H.K. (1983) *Rep. Prog. Phys.* **46**, 621.

Moreno-Insertis, F. (1992) These proceedings.

Nordlund, Å., Brandenburg, A., Jennings, R.L., Rieutord, M., Ruokolainen, J., Stein, R.F. & Tuominen, I. (1992) submitted to *Astrophys. J.*

Parker, E.N. (1955) *Astrophys. J.* **122**, 293.

Parker, E.N. (1975) *Astrophys. J.* **198**, 205.

Parker, E.N. (1979) *Cosmical Magnetic Fields*, Clarendon.

Parker, E.N. (1987) *Astrophys. J.* **312**, 868.

Piddington, J.H. (1976) in *Basic Mechanisms of Solar Activity. Proc. IAU Symposium 71* (eds. V. Bumba and J. Kleczek) pp. 389-407.

Proctor, M.R.E. (1992) These proceedings.

Proctor, M.R.E. & Weiss, N.O. (1982) *Rep. Progr. Phys.* **45**, 1317.

Ribes, E., Mein, P. & Mangeney, A. (1985) *Nature* **318**, 170.

Roberts, P.H. & Stix, M. (1971) *The turbulent dynamo: a translation of a series of papers by F. Krause, K.-H. Rädler and M. Steenbeck*, Tech. Note 60, NCAR, Boulder, Colorado.

Roberts, P.H. & Stix, M. (1972) *Astron. Astrophys.* **18**, 453.

Schüssler, M. (1981) *Astron. Astrophys.* **94**, L17.

Schüssler, M. (1983) in *Solar and Stellar Magnetic Fields. Proc. IAU Symp. 102* (ed. J.O. Stenflo) Reidel, pp. 213-236.

Simon, G.W. & Weiss, N.O. (1968) *Z. Astrophys.* **69**, 435.

Simon, G.W. & Weiss, N.O. (1991) *Mon. Not. R. Astron. Soc.* **252**, 1p.

Spiegel, E.A. & Weiss, N.O. (1980) *Nature* **287**, 616.

Steenbeck, M., Krause, F. & Rädler, K.-H. (1966) *Z. Naturforsch.* **21a**, 369.

Stein, R.F. & Nordlund, Å. (1989) *Astrophys. J.* **342**, L95.

Stix, M. (1976) *Astron. Astrophys.* **47**, 243.

van Ballegooijen, A.A. (1982) *Astron. Astrophys.* **113**, 99.

Walén, C. (1946) *Ark. Mat. Astr. Fys* **33a**, No. 18.

Weiss, N.O. (1964) *Mon. Not. R. Astron. Soc.* **128**, 225.

Weiss, N.O. (1981a) *J. Fluid Mech.* **108**, 247.

Weiss, N.O. (1981b) *J. Fluid Mech.* **108**, 273.

Weiss, N.O., Cattaneo, F. & Jones, C.A. (1984) *Geophys. Astrophys. Fluid Dyn.* **30**, 305.

Yoshimura, H. (1975) *Astrophys. J. Suppl.* **29**, 467.

Yoshimura, H. (1981) *Astrophys. J.* **247**, 1102.

Zeldovich, Ya.B., Ruzmaikin, A.A. & Sokoloff, D.D. (1983) *Magnetic Fields in Astrophysics*, Gordon and Breach, New York.

THE MOTION OF MAGNETIC FLUX TUBES IN THE CONVECTION ZONE AND THE SUBSURFACE ORIGIN OF ACTIVE REGIONS

F. MORENO-INSERTIS
Instituto de Astrofísica de Canarias
38200 La Laguna (Tenerife)
Spain

ABSTRACT. The rise of kink-unstable magnetic flux tubes through the solar convection zone is discussed with a view to explaining the subsurface origin of active regions. In spite of the advances and insights obtained by analytical and numerical means in the past decade(s), there are still important uncertainties as to the viability of a mechanism of this kind to explain the origin of sunspots and active regions. In this paper, the main features of the rise of kink unstable tubes and some open questions regarding this process are presented.

1. Introduction

A widespread opinion about the origin of active regions is that a strand of magnetic flux detaches itself from the general solar toroidal field at the bottom of the convection zone and rises to the surface to produce the panoply of phenomena generally called magnetic activity. Although in various forms and degrees of development, this idea has repeatedly appeared in the literature of the past thirty years (Parker 1955, Babcock 1961, Schmidt 1968, Ponomarenko 1970, Zwaan 1978, Spiegel & Weiss 1980, Golub *et al* 1981, van Ballegooijen 1982a, Parker 1984, Moreno-Insertis 1984, 86). Substantial progress in understanding and modelling a process of this kind has been achieved in the past decade. However, I think that a complete proof that this process is indeed a viable mechanism for producing active regions has not been given yet.

From an observational point of view there are clear hints that the surface magnetic activity reflects the presence of a toroidal system of magnetic flux buried deep in the convection zone: suffice it to mention the approximate east-west orientation of AR's, Hale's polarity law or the appearance of active longitudes. As to the field distribution in the toroidal system, one contemplates two alternative forms: (a) the magnetic field may be in the form of a continuum or *sea* of field. One envisages this situation specially underneath the convection zone proper, in the overshoot region. (b) the magnetic field may be in the form of individual flux tubes, possibly also in the overshoot region. Both views need not be incompatible. In fact a continuum of field may be subject to instabilities of various kinds (magnetic Rayleigh- Taylor, etc. See Hughes 1991, 92 and references therein), so that flux tubes may form out of it. The intermittency of the field distribution is, at any rate, a common feature of MHD turbulence problems (e.g., Kraichnan 1976). Additionally, flux tubes are seen to form in 3D magnetoconvection simulations (Brandenburg *et al* 1991).

J. H. Thomas and N. O. Weiss (eds.), Sunspots: Theory and Observations, 385–410.

In this paper we are interested in the behavior of individual magnetic flux tubes originally in the deep convection zone, *in as far as they may give rise to active regions.* First of all we will consider the possible states of equilibrium of toroidal flux tubes which allow a *sea-serpent* evolution, i.e., such that, after a sinusoidal perturbation is superimposed on them, the crest of the tube rises to the photosphere while the valleys remain buried underneath the convection zone. Secondly, the results of numerical simulations of such a process will be briefly shown (Sect. 5). The second part of the paper will be devoted to individual aspects of the rise of kink-unstable tubes which seem to be at odds with the idea of having the AR's at the surface originate from flux tubes which had equipartition field strength in the deep convection zone. Sect. 6, in particular, will treat the influence of rotation on the rise of the tubes, Sect. 7 the problem of the extreme weakening of the field strength along the rise and Sect. 8 will deal with the possible fragmentation of the tube in its interaction with the surrounding medium. Finally, the problem of the field strength of the magnetic flux tubes in the deep convection zone is treated in Sect. 9.

2. The acceleration terms for a flux tube in the slender tube approximation

The description of the motion of the magnetic flux tubes (mft) in the present context is generally done within the framework of the slender flux tube approximation (Defouw 1976, Roberts and Webb 1978, Spruit 1981,Ferriz-Mas and Schüssler 1989). In this approximation, only the values of the physical variables on the axis of the tube appear in the final equations. Further, permanent pressure equilibrium across the tube is assumed ($p_e = p_i + p_m$; p_e and p_i stand for the gas pressure outside and inside the tube, respectively, and p_m is the magnetic pressure in the tube). By subtracting the equation of motion of the surrounding medium (component parallel to the field lines) from the equation of motion of the flux tube one obtains the following acceleration terms for the motion of the mass elements of the tube (for details, see the literature quoted above):

(a) **Buoyancy Term:**

$$\boldsymbol{g}\,\frac{\Delta\rho}{\rho_i}\,, \tag{1}$$

with

$$\Delta\rho \stackrel{\text{def}}{=} \rho_i - \rho_e\,. \tag{2}$$

(b) **Magnetic pressure gradient** along the tube:

$$\frac{1}{\rho_i}\,\frac{\partial p_m}{\partial s}\,\boldsymbol{l}\,. \tag{3}$$

This term comes from the subtraction of the internal and external gas pressure gradients along the tube. In (3), s is the arc-length parameter and $\boldsymbol{l}$ is the unit tangent vector, both along the central field line of the tube.

There are also some terms in the direction perpendicular to the tube's axis:

(c) **Magnetic curvature force** (R_c = radius of curvature of the field lines)

$$\frac{1}{R_c}\frac{B^2}{4\pi\rho_i}\,\boldsymbol{n}\,, \tag{4}$$

where $\boldsymbol{n}$ is the unit vector in the direction normal to the field lines and toward the center of curvature. Finally,
(d) **drag force** (aerodynamic drag),

$$-C_D\,\frac{\rho_e}{\rho_i}\,\frac{|V|\boldsymbol{V}}{\pi R}\,, \tag{5}$$

where $\boldsymbol{V}$ is the relative transversal velocity of the tube with respect to the surroudings, R is the radius of the tube and C_D is an empirical coefficient of order unity which is normally taken from wind-tunnel experiments (see discussion in Moreno-Insertis 1986).

In more complicated situations, for instance when the whole system is rotating, one has to include additional acceleration terms (Coriolis, etc).

3. Equilibrium of the flux tubes in the deep convection zone

As a starting point for the calculation of the time evolution of a magnetic tube, an equilibrium condition of some sort is normally assumed (thermal equilibrium, force equilibrium). Mechanical equilibrium, i.e. the vanishing of the sum of all forces acting on the tube, requires a degree of fine tuning (a precise value of $\Delta\rho$ to compensate for the magnetic curvature force, for instance) which at first sight seems difficult to be reached by the tube spontaneously. One thus prefers to assume a situation in which the internal temperature equals the external temperature ('thermal equilibrium'), mainly because one knows of a mechanism, radiative exchange, which does the job of equalizing temperatures. However, it is well known that a flux tube with $T_e = T_i$ will be buoyant because of the reduced internal pressure (magnetic buoyancy, Parker 1955). In fact it can be shown that a toroidal flux tube in thermal equilibrium with its surroundings *in the convection zone* cannot get rid of its buoyancy and will rise *as a whole* to the surface on a fairly short time-scale as compared to the activity cycle period (Parker 1975, Moreno-Insertis 1983). This time-scale is also substantially shorter than the time-scale for radiative exchange (see Sect. 8 below), so that, after all, a complete thermal equilibrium may be difficult to reach if the tube was formed with a temperature imbalance with respect to the surroundings.

Assume, now, that the originally buoyant tube was not in the convection zone but, rather, at its lower boundary, namely in the overshoot region. If so, there is a chance that the tube will lose its magnetic buoyancy when rising before leaving that region. How great a chance? A blob of plasma rising adiabatically in a subadiabatic environment tends to become cooler than its surroundings. This effect may be small, since the overshoot region is only slightly subadiabatic, but may suffice to make the inner density close enough to the density outside so that the tube comes into the vicinity of a mechanical force equilibrium with $|g\,\Delta\rho| = |B^2/(4\pi R_c)|$.

A simple method to obtain a quantitative estimate for the buoyancy reduction can be devised by realizing that once the tube goes through a level in which buoyancy and magnetic curvature force compensate each other, it may thereafter execute oscillations about that position. The oscillations are damped because of the effect of the drag force; the frequency of oscillation, which may be called the *magnetic Brunt-Väisälä frequency*, is given by

$$\omega_{MBV}^2 = \frac{{v_A}^2}{2{H_p}^2}\left(\frac{1-2\nabla_a}{\gamma} - \beta\delta\right), \tag{6}$$

with

$$\delta = \nabla - \nabla_{ad}. \tag{7}$$

In (6) and (7), ∇ is the logarithmic temperature gradient, $\nabla = d\log T/d\log P$, ∇_{ad} its adiabatic value, γ is the specific heat ratio, v_A and H_p are the Alfvén speed and pressure scale-height, respectively, and β stands for the ratio of gas to magnetic pressure. For simplicity we have used in (6) the expression valid for a horizontal tube in a cartesian geometry. The more complete treatment including, e.g., rotation and spherical geometry can be found in Moreno-Insertis (1992). (6) is the magnetic counterpart to the customary Brunt-Väisälä frequency known from the theory of gravity waves (Lighthill 1978). The necessary condition for the oscillations to take place is that $\omega_{MBV}^2 > 0$; otherwise, the tube will simply run away up into the convection zone.

The flux tube can only be held within the overshoot region through this mechanism if the amplitude of these oscillations is smaller than the depth of the overshoot region. An estimate for the amplitude can be obtained by equating the force driving the oscillation at maximum elongation, Δz, to the buoyancy force experienced by the tube in the initial state of thermal equilibrium, $|g\Delta\rho| = g\rho/\beta$:

$$|\omega_{MBV}^2\,\Delta z| = \left|g\frac{\Delta\rho}{\rho}\right| = \frac{g}{\beta}. \tag{8}$$

With (6) and (8) one can easily obtain Δz for a variety of values of the magnetic field strength and of the subadiabaticity of the ambient stratification, as shown in Table 1 .

The dashed line in the table separates those amplitude values that are smaller than (or at most roughly comparable to) the depth of the overshoot region as calculated in the existing models (van Ballegooijen 1982b, Schmitt, Rosner and Bohn 1984, Pidatella and Stix 1986,

Table 1. Amplitude of oscillation around the position of mechanical equilibrium of a flux tube of field strength B_0 initially in thermal equilibrium with its surroundings for various values of the superadiabatic coefficient δ of the external stratification.

	Δz (km)	δ			
		-10^{-4}	-10^{-5}	-10^{-6}	-10^{-7}
B_0	$5\,10^3 G$	8.5	$8.5\ 10^1$	$8.4\ 10^2$	$8.3\ 10^3$
	$10^4 G$	$3.4\ 10^1$	$3.4\ 10^2$	$3.4\ 10^3$	$3.2\ 10^4$
	$2\,10^4 G$	$1.4\ 10^2$	$1.3\ 10^3$	$1.3\ 10^4$	$1.1\ 10^5$
	$10^5 G$	$3.4\ 10^3$	$3.2\ 10^4$	$2.0\ 10^5$	$4.2\ 10^5$

Skaley and Stix 1991) from those that are clearly larger than that. As can be seen, even very strong fields ($B_0 \sim 10^5$ G) can lose a substantial part of their initial buoyancy by rising in the overshoot region if the superadiabatic coefficient δ is as negative as -10^{-5}. The models mentioned above, however, predict something like $\delta = O(-10^{-6})$ in a large part of the overshoot region. If so, only flux tubes with field strength up to a few times 10^4 G could get rid of their buoyancy before leaving the overshoot region. The actual value of δ present in the overshoot region is difficult to ascertain; the existing models of the overshoot region can only serve as a rough guide to the actual stratification present underneath the convection zone: note, for instance, that if there is a (possibly thick) magnetic field layer in those depths, it will have a noticeable effect on the stratification which has not been taken into account in those models.

The necessary condition $\omega^2_{MBV} > 0$ commented above substitutes the condition $\delta < 0$ of the non-magnetic case. Thus, depending on the magnetic field strength, the loss of buoyancy through rise may continue to operate in the bottom layers of the convection zone proper. By calculating the corresponding oscillation frequency, the method followed above can be readily generalized to more complicated situations, including rotation, etc.

Flux tubes that have reached a state of mechanical equilibrium will not maintain it forever, even if they are stable to all kinds of mechanical perturbations. In fact, through radiative heat inflow they will become slightly hotter and will rise slowly on a long thermal time-scale, similarly to the phenomenon described by Parker (1975). All in all, we can say that, assuming that their field strength is not too high, flux tubes initially in thermal equilibrium in the overshoot region will reach a state of mechanical equilibrium higher up in that region or in the lowermost layers of the convection zone proper.

An alternative way of holding buoyant flux regions has been devised by Parker (1987a) based on the reduction of heat transport through a strong magnetic field and the consequent formation of a cool *thermal shadow* above the magnetized region which could counteract the buoyancy force. This mechanism applies to horizontally extended magnetic regions or sheets rather than to flux tubes. For details the reader is referred to the original series of papers (Parker 1987a,b,c, 1988a,b,c). A further possibility for equilibrium in the deep convection zone is given if one assumes the existence of a meridional flow that holds the flux tubes in place against their buoyancy. This has been studied by van Ballegooijen and Choudhuri (1988) but has the drawback of requiring an additional ingredient, the meridional flow, whose existence and characteristics are as yet unknown.

4. Stability

4.1 Basic stability analysis

Consider a flux tube that is in a position of mechanical equilibrium, or, as explained in the foregoing section, that is carrying out oscillations about that position. In the latter case, the equilibrium is obviously stable against perturbations which move the tube as a whole keeping its toroidal shape. Yet, it may be unstable to perturbations which bend the field lines. Considering, for simplicity, perturbations which keep the flux tube within a cone of constant latitude (θ = const in spherical coordinates r, θ, ϕ) then matter may flow from the crests of the perturbed tube to the valleys, making the crests lighter and the valleys heavier, thereby destabilizing the original equilibrium condition (as in the Parker instability for a continuum of field, Parker 1966. See also Schüssler 1980). If this mechanism is effective enough, it may not be overcome by the restoring effect of the magnetic curvature force (assuming the wavelength is not too short) nor by the stabilizing effect of the outside stratification (if $\delta < 0$). If the instability develops into the nonlinear regime, the rising loop of the unstable tube may reach the photosphere while the valleys remain buried underneath the convection zone, the whole adopting a sea-serpent shape and, perhaps, producing active regions as a result. For lack of a better name this kind of instability is customarily called the 'kink instability' of magnetic flux tubes, but it should not be confused with the well known kinking or 'buckling' instability of magnetic field distributions containing an azimuthal component around the main axis of the tube, which corresponds to a different physical mechanism.

The first study of the linear stability of toroidal flux tubes in the solar convection zone was carried out by Spruit and van Ballegooijen (1982). They obtained a dispersion relation for perturbations of circular wavenumber m in the form of a quadratic polynomial in ω^2, with coefficients which basically depended on $\beta\delta$ and m only. The resulting stability diagram is shown in Fig. 1, with dots marking the values of marginal stability. As can be seen, all modes with $m < 4$ are unstable in a convection zone ($\delta > 0$). In fact, the mode with $m = 0$ (perturbation of the tube as a whole keeping its toroidal shape) is *less* unstable than the modes with positive m (for m not too high). This is due to the destabilizing effect of the downflow along the field lines explained above. The consequence of this is that whenever the infinite wavelength mode ($m = 0$) is stable (or unstable with a growth rate well below that of the unstable $m > 0$-modes), the tube can develop a sea-serpent shape while remaining anchored underneath the convection zone.

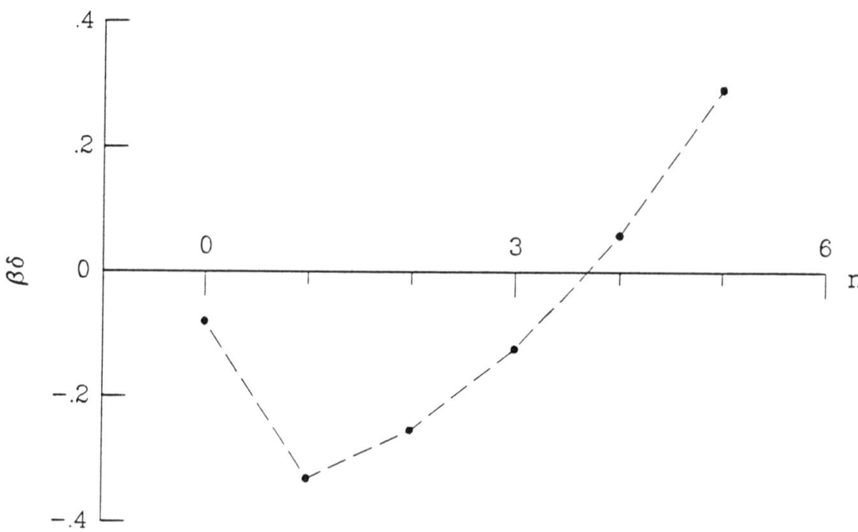

Fig. 1 – Stability of a toroidal flux tube against perturbations that bend the field lines as derived from the analysis of Spruit and van Ballegooijen (1982). The dots mark the position of marginal stability: values of $\beta\delta$ above them correspond to instability.

4.2 Stabilization through rotation

Similarly to the case of a continuum of field (Gilman 1970, Acheson 1979), the rotation of the physical system can strongly affect the stability character of a magnetic flux tube. The corresponding stability analysis was carried out by van Ballegooijen (1983; see also Ferriz-Mas and Schüssler, 1992). Rigid rotation ($d\Omega/dr = 0$), has a generally stabilizing effect on all unstable modes, either in the sense that it makes them stable, or, at least, because it lowers the growth rate of those that remain unstable; the boundary between stable and unstable modes, in any case, is shifted toward higher $\beta\delta$ values. Using $\Omega = 3\,10^{-6}$ sec^{-1} (which should be a rough approximation for the rotation rate in the deep layers of the convection zone, see Libbrecht 1988, Goode *et al* 1991), the mode with infinite wavelength, for example, turns out to be stable in the whole convection zone except its uppermost 50,000 km. The stabilizing effect (in the sense explained above) also applies for the kink modes ($m > 0$), but to a lesser extent. For example, the $m = 1$ and $m = 2$ modes are still unstable in the whole convection zone. As a result of all this, the range of values of the field strength for which a sea-serpent evolution is possible is larger in the rotating case than in the non-rotating case. The growth-rates of the unstable modes for the given value of Ω and $\delta = 10^{-8}$ (the value adequate to the bottom layers of the convection zone, cf. Spruit 1977) can be seen in Fig. 2. Note that the first two harmonics are unstable for all field strengths shown in the figure whereas the mode with infinite wavelength is stable except for very large values of B, $B \gtrsim 3\ 10^5$ G.

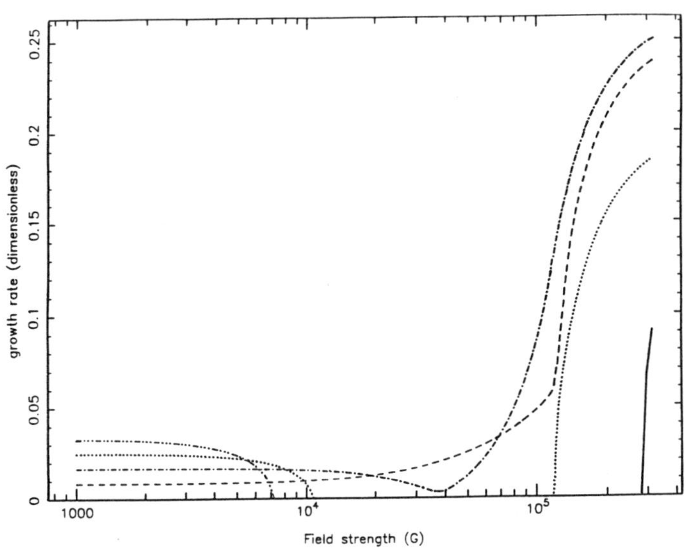

Fig. 2 – Growth rate of the unstable modes taking into account the rotation of the Sun as derived from the stability analysis of van Ballegooijen (1983) for $\delta = 10^{-8}$ (bottom of the convection zone). Shown are the modes $m = 0$ (solid line), $m = 1$ (dashed line), $m = 2$ (dot-dash), $m = 3$ (dotted line) and $m = 4$ ($- \cdot \cdot \cdot$).

4.3 Poleward slip

Magnetic flux tubes may be unstable also to perturbations which move the tube outside of the cone of constant latitude. Displacing a toroidal tube along a surface of constant solar radius so that the toroidal shape of the tube is preserved, for instance, will give rise to an unstable motion in which the tube keeps moving toward the solar axis (poleward slip) (Spruit and van Ballegooijen, 1982, Pneumann and Raadu, 1972). The driving force for this instability is the magnetic curvature force, which tends to diminish the radius of the flux ring. The buoyancy force may modify the unstable evolution but, since it is radially directed, it cannot balance the latitudinal component of the magnetic curvature force; hence, it cannot prevent the displacement of the tube toward the pole.

In fact, the rotation of the Sun does modify this unstable evolution. The Coriolis force may prevent the migration of the tube toward the pole if the field strength is not too high ($B \lesssim 10^5$G), similarly to the case of kink-unstable tubes explained below (Sect. 6): it effectively opposes any flux ring motion transversal to the rotation axis, so that the poleward-slip instability can be suppressed. If, additionally, the tube is trapped in the overshoot region, e.g. through the mechanism explained in Sect. 3, then it will not be able to move in any direction and will stay in its original position. (see Moreno-Insertis, 1992).

5. Numerical modelling of the nonlinear evolution of a kink-unstable flux tube

The linear analyses mentioned in previous paragraphs can only serve as a guideline as to which ranges of field strength and depths are interesting to select possible candidates of flux tubes that end up being rooted underneath the convection zone while their upper stretches reach the photosphere. A limitation of the linear approach is given, for example, by the fact that the stratification changes fairly abruptly from mildly superadiabatic to strongly subadiabatic along the overshoot region, i.e., within a fraction of a scale-height. It may then easily happen that parts of a toroidal tube originally contained within the overshoot region are brought up to the convection zone by a perturbation while the rest remains below it. The applicability of the linear analyses in such cases is limited. At any rate, once the flux tube has begun its unstable evolution it will enter the nonlinear regime long before it reaches the upper convection zone.

To study the evolution of the flux tubes before they give rise to active regions, one has to have recourse to numerical means to solve the system of partial differential equations that govern this process. Numerical simulations of this kind have been carried out by several authors in the past decade (Moreno-Insertis 1984, 1986, Chou and Fisher 1989, Choudhuri 1989). In all these simulations it was shown that, for a variety of initial and boundary conditions, kink-unstable flux tubes lying at the bottom of the convection zone or slightly below it do indeed develop a sea-serpent shape with the top of the tube reaching the photosphere and the bottom remaining anchored underneath the convection zone. Two instances of this evolution can be seen in Fig. 3, one for a cartesian geometry and the other for a spherical case. In spite of the different geometry, the spherical calculation reproduces the basic features of the cartesian one. Among the latter, one can mention:

a) the rise of the tubes occurs within a few months for values of the initial field strength between 10^4 G and 10^5 G and magnetic flux above 10^{20} Mx. Since the drag force increases

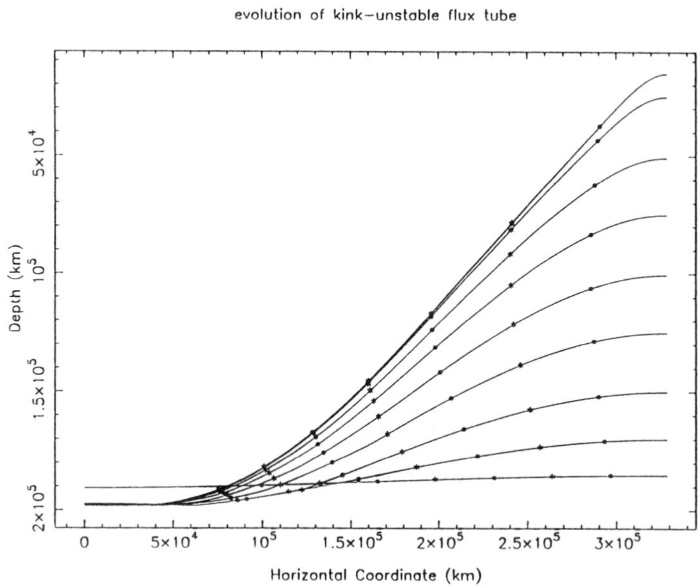

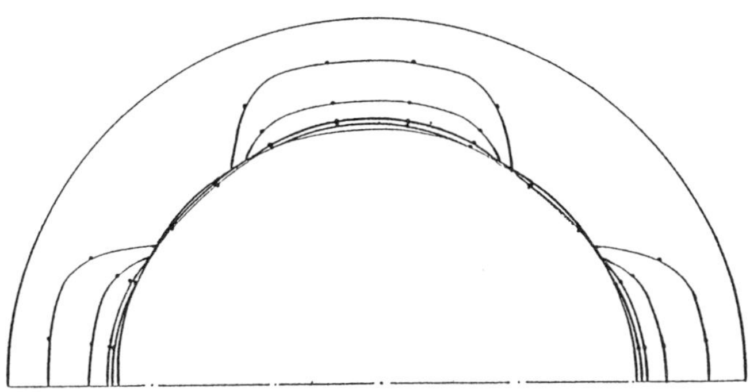

Fig. 3 – Time evolution of an unstable magnetic flux tube originally lying at the bottom of the convection zone in the absence of rotation. (a) cartesian geometry with an initial field strength $B_0 = 10^4$ G as calculated with the numerical code of Moreno-Insertis (1986) (b) spherical geometry with $B_0 = 1.7\,10^4$ G, taken from Choudhuri (1989). Dots and crosses in the figures mark the time evolution of fixed mass elements of the tube.

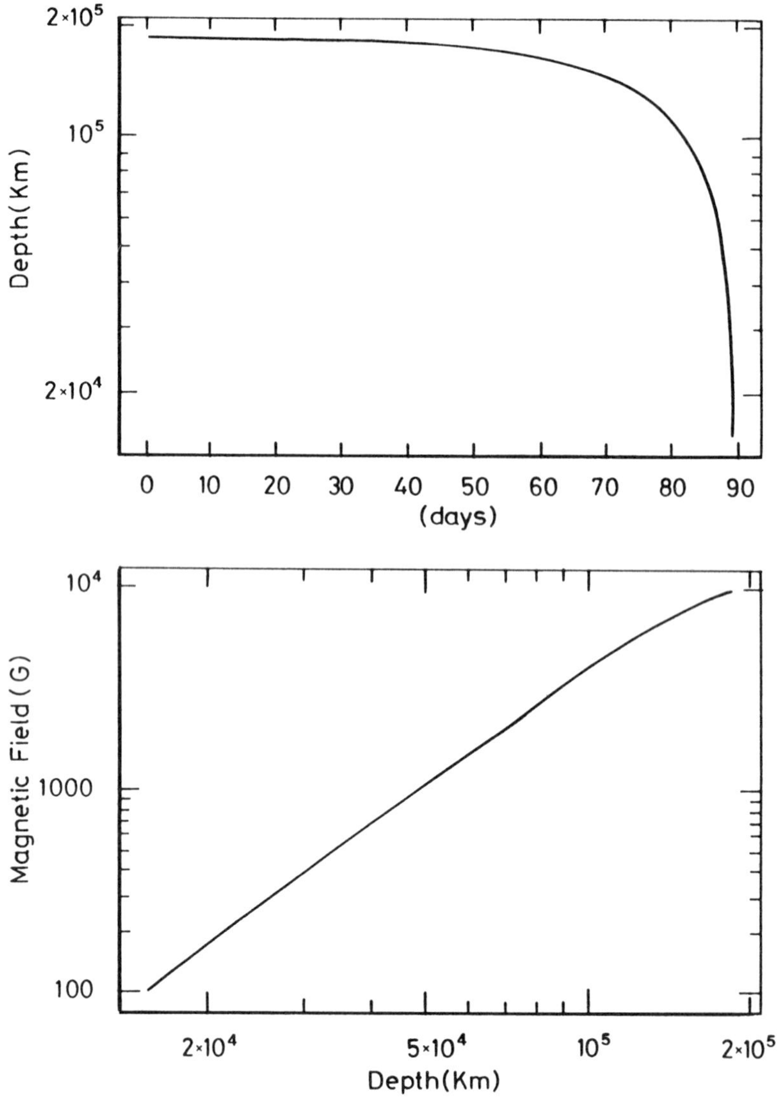

Fig. 4 – Details of the calculation shown in Fig. 3(a):
(a) Position of the uppermost mass element of the tube as a function of time.
(b) Magnetic field strength of the crest of the tube as a function of instantaneous position.

with decreasing radius (see eqn. 5), lower values of the magnetic flux imply longer rise times. In Fig. 4a, the duration of the rise is shown in the form of a diagram depicting the depth of the crest of the tube as a function of time. The horizontal section of the curve corresponds to the initial phase of development of the instability. This phase can be made arbitrarily long by considering initial perturbations of very small amplitude. The rise time quoted above can be taken to mean the rise once the amplitude of the perturbation in position had grown above 1/10 of a scale-height.

b) the downflow along the field lines that is essential for the development of the instability at the linear stage causes the stretching of the crest of the rising tube. However, as can be seen through the Lagrange markers in Fig. 3, this effect is not very pronounced along the rise; the original separation of two Lagrange elements of the tube does not increase by more than about a factor 2 if $B_0 \simeq 10^4$ G. The reason for this is the stratification of the convection zone, which is increasingly superadiabatic in higher layers. The evolution of the mass elements in the flux tube being adiabatic, they become increasingly buoyant when reaching higher levels along their rise. The component of the buoyancy force along the tube axis can then effectively counteract the downflow and even turn it into an upflow along the field lines near to the crest of the rising loop (as can be seen in the two final tube configurations of Fig. 3a). This effect is less pronounced the higher the original field strength, but, even for $B_0 = 10^5$ G the stretching does not go beyond a factor 7.

c) the downflow is also responsible for the enhanced stability of the section of the tube below the convection zone. Given that a symmetric boundary condition was used, the density per unit length of the lowermost mass elements monotonically increases with time. This causes the field to become weaker and makes the tube sink somewhat to find a new equilibrium in deeper, denser levels. Periodic boundary conditions also allow the tube to become anchored in the overshoot region (Caligari 1991).

d) there is a continuous weakening of the field strength due to lateral expansion of the tube when rising across successive scale-heights toward the photosphere (see Fig. 4b). This weakening can be countered by the stretching of the tube associated with the downflow along the field lines, but, as implied by what has been said in b), only to a very limited extent. The consequences of this fact are discussed in Sec. 7 below.

The physical behavior of the magnetic flux tubes in these simulations is well understood (see the original literature quoted above). For the application of such calculations to explain the origin of active regions, however, it is necessary to consider in some detail their limitations and the open questions that are posed by them. In the following sections we discuss what we think are the most important ones, namely 1) the effect of rotation, 2) the weakening of the magnetic field along the rise and 3) the fragmentation of the tubes into thinner strands.

6. Rotation

If the time-scale of the flux tube evolution is long enough, the Coriolis force may effectively counteract the forces in the direction transversal to the rotation axis (radial direction in cylindrical coordinates) with the result that the rising tube reaches the photosphere at high solar latitudes. This has been shown by Choudhuri and Gilman (1987) and Choudhuri (1989). The basic physical reason underlying this effect is easy to understand and is directly related to the phenomenon of inertial oscillations in a rotating system (e.g. Tritton 1988,

Greenspan 1968). Consider a particle at rest at time $t = 0$ in a rotating system with cylindrical coordinates (ϖ, ϕ, z) and angular velocity vector $\mathbf{\Omega} = (0, 0, \Omega)$. If the particle is acted on by a constant non-fictitious acceleration a in the ϖ–direction, then, due to the action of the Coriolis force, it will execute an oscillatory motion of the form:

$$\begin{cases} \dot{\varpi} &= \frac{a}{2\Omega} \sin 2\Omega t \,, \\ \dot{\phi} &= \frac{a}{2\Omega\varpi_0}(\cos 2\Omega t - 1) \,. \end{cases} \tag{9}$$

In (9) the symplifying 'quasi-cartesian' condition $|\dot{\phi}| \ll |\Omega|$, i.e., $|a/4\Omega^2\varpi_0| \ll 1$, has been assumed. (9) tells us that on a time-scale $1/(2\Omega)$ the rotational effects impose an inertial oscillation on the particle that limits the motion toward the axis or away from it. Applied to the rise of magnetic flux tubes, when the rise time is of order π/Ω or longer, the Coriolis force is going to counteract the buoyancy force in the (cylindrical) radial direction. The only remaining component of the buoyancy force is in the z-direction so that, if the inertial oscillation has a small amplitude, the tube will rise in a direction parallel to the rotation axis.

To obtain an estimate of the relevance of this phenomenon for the rise of flux tubes, we substitute in the foregoing expressions the acceleration a with g/β, the magnetic buoyancy acceleration corresponding to thermal equilibrium ($T_i = T_e$), as in equation (8). For Ω, the value corresponding to the bottom of the convection zone in low latitudes (as deduced from helioseismological studies, Libbrecht 1988) is used, so that the time-scale π/Ω is roughly 2 weeks. The amplitude of the oscillatory motion turns out to be

$$\text{max. elongation:} \qquad \frac{a}{4\Omega^2} = \frac{{v_A}^2}{8\Omega^2 H_p} \approx B_4^2\; 9\,10^2\text{km}\,, \tag{10}$$

$$\text{max. velocity:} \qquad \frac{a}{2\Omega} = \frac{{v_A}^2}{4\Omega H_p} \approx 5.3\; 10^2 B_4^2 \text{cm sec}^{-1}\,. \tag{11}$$

B_4 above means the magnetic field in units of 10^4 Gauss. The numerical values in (10) and (11) were calculated for the lower convection zone. Also there, the quasi-cartesian condition $|\dot{\phi}| \ll |\Omega|$ is fulfilled whenever $3\,10^{-3} B_4^{-2} \ll 1$.

Consider now the consequences of the foregoing for the rise of *toroidal* flux tubes, a problem considered by Choudhuri and Gilman (1987).

a) if $B_0 \lesssim 10^4$ G, then the rise time of the flux tube in the absence of rotation is longer than 30 days (Moreno-Insertis 1983). The Coriolis force will thus be able to impose inertial oscillations on the tube motion. On the other hand, the amplitude (10) is a small fraction of a scale-height up to the highest levels of the convection zone. The tube, therefore, can only rise along a direction parallel to $\mathbf{\Omega}$ with an oscillation with velocity amplitude (11) superimposed on the motion. The main deviation from this behavior can be expected to occur in the upper convection zone, where thermal equilibrium is no longer applicable, so that the buoyancy force will be stronger than as used in (10).

b) if $B_0 \gtrsim 10^5$ G, the rise time is of the order of several days or less in the absence of drag. The amplitude of the inertial oscillation is of the order of the depth of the convection zone.

In this case one can then expect the effect of the Coriolis force to be limited so that the tube may not be deflected to higher latitudes.

The results of the numerical calculation of Choudhuri and Gilman for the rise of tubes in the absence of drag can be seen in Fig. 5, taken from their original paper. In it the rise of flux rings with initial field strength 1.7 10^4 G (left), 5.4 10^4 G (middle) and 1.7 10^5 G (right) is shown for tubes starting at various latitudes. The behavior predicted by (9)-(11) is fulfilled in most of the convection zone in the two cases with weaker field strength (left and middle figures). In the upper convection zone, as expected, the increasingly superadiabatic stratification makes the buoyancy force dominant: the trajectories then turn more radial at those heights. In any case, even if the tube starts close to the equator, it reaches the photosphere in high latitudes. The case with stronger initial field is much less affected by the Coriolis force, as expected. The trajectories are not deflected toward the poles and the tube goes out of the convection zone at latitudes comparable to those of the non-rotating problem. The inclusion of drag causes the oscillation of the weaker flux tubes to be damped. The latitude of emergence of the tubes is anyway as in the drag-free case (high if $B_0 \lesssim 10^4$ G, roughly equal to the original latitude if $B_0 \gtrsim 10^5$ G; see Choudhuri and Gilman, Sect. IV)

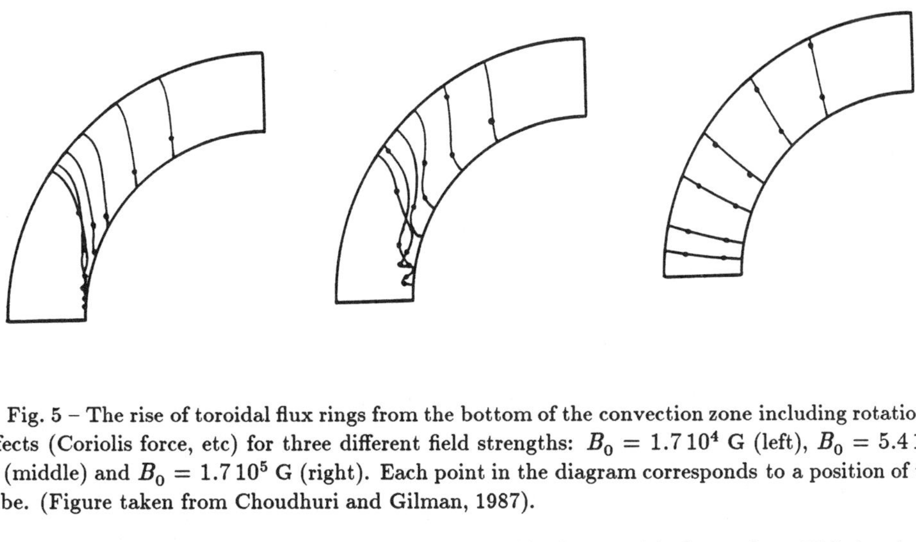

Fig. 5 – The rise of toroidal flux rings from the bottom of the convection zone including rotational effects (Coriolis force, etc) for three different field strengths: $B_0 = 1.7\,10^4$ G (left), $B_0 = 5.4\,10^4$ G (middle) and $B_0 = 1.7\,10^5$ G (right). Each point in the diagram corresponds to a position of the tube. (Figure taken from Choudhuri and Gilman, 1987).

A similar study can be carried out for the rise of kink-unstable flux tubes. This has been done by Choudhuri (1989). The most noticeable difference as compared to the non-rotating case of Fig. 3(b) is that, again for $B_0 = 1.7\,10^4$ G, the crest of the tube now suffers the same kind of polar drift as the toroidal rings of Fig. 5 (leftmost diagram), so that the tube appears at high latitudes in the photosphere, outside the activity belt.

Summarizing, if one takes these results at face value one would conclude that (a) either some fundamental physical ingredient of the actual solar problem is missing in these idealized models or (b) the flux tubes that one sees erupting at the photosphere had a high field strength [at least $O(10^5\ G)$] at the bottom of the convection zone. The problem is further complicated by the possible development of the Kelvin-Helmholtz instability arising from the shear between the longitudinal flow in the tube caused by the Coriolis force and the ambient flow close to the tube. Order-of-magnitude estimates for this problem can be found in a recent paper by D'Silva and Choudhuri (1991).

7. Tube expansion and magnetic field weakening along the rise

Although modified by the magnetic forces, the mass elements of the rising flux tube are subject to a process of expansion similar to that suffered by unmagnetized plasma parcels in the updraft of a convective cell. Any blob of plasma moving across successive pressure and density scale-heights must continuously adapt its pressure and density to those of the surroundings; as a result, any infinitesimal volume of an unmagnetized plasma parcel increases its size as an inverse power of the ambient density, namely $R \propto \rho(z)^{-1/3}$.

When rising across the convection zone, a flux tube also adapts its density and pressure to the surroundings. Since $\beta \gg 1$, the condition $\rho_e \approx \rho_i$ is fulfilled during the rise. Magnetic flux conservation, on the other hand, requires that

$$B\,\frac{\lambda}{\rho} = \text{const} \tag{12}$$

with λ = mass per unit length. λ does not vary much along the rise, as seen above, at most by a factor of a few units. Thus, the following proportionality relations should be approximately fulfilled:

$$\begin{aligned} B &\propto \rho\,, \\ R &\propto \rho^{-1/2}\,. \end{aligned} \tag{13}$$

The rise of a flux tube across the convection zone should then be accompanied by a decrease (by orders of magnitude) of the field strength with a concurrent growth of the tube's cross section.

The consequences of these facts for the flux tube evolution are explored in Figs. 6 and 7. In Fig. 6, the field strength of a rising tube is shown as a function of depth for tubes with initial field strength 10^5 G (upper pair of solid and dotted lines) and 10^4 G (lower pair). The equipartition field is also shown (dashed line) as taken from Spruit's model (1977): here equipartition is used in the customary sense that the magnetic energy equals the kinetic energy of the convection. The solid lines correspond to the field strength calculated using the proportionality law (13). As can be seen, if the initial field strength is close to equipartition (lower solid line) *it will be below equipartition during the whole rise.* In fact the field strength will be orders of magnitude below equipartition in the upper convection zone. The result of an actual numerical calculation of the evolution of kink-unstable tubes

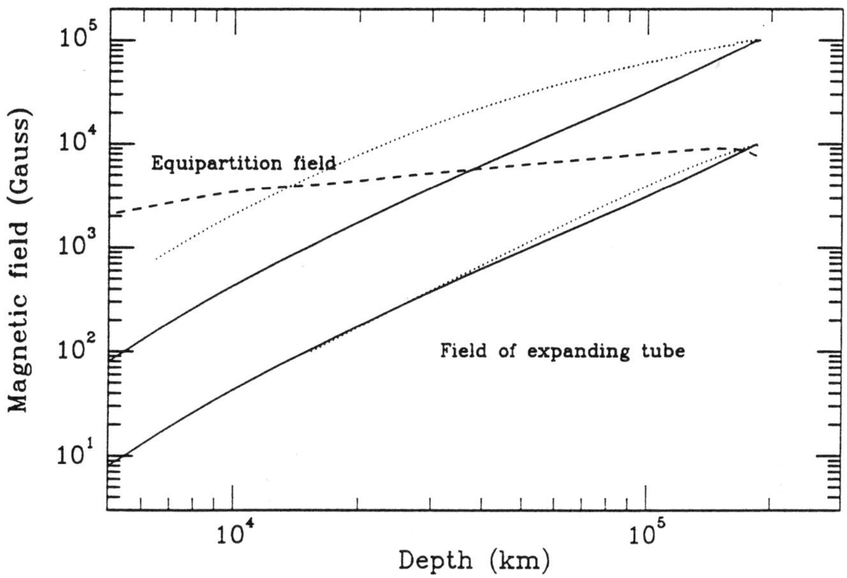

Fig. 6 – Magnetic field strength of a rising flux tube that follows the approximation $B(z) \propto \rho(z)$ (solid lines) compared to the equipartition field strength as given in Spruit (1977) (dashed line) for tubes with initial field strength $B_0 = 10^4$ G (lower curve) and $B_0 = 10^5$ G (upper curve). For comparison, the calculated field strength of the crest of kink-unstable tubes is also shown (dotted lines).

is shown in the form of dotted lines (as calculated with the code of Moreno-Insertis 1984). Solid and dotted lines almost coincide for the flux tube originally with equipartition field: the reason for this is the limited amount of stretching of the crest of the tube (cf. Sec.5 and fig. 3a). A flux tube with an initial field roughly 10 times the equipartition value (upper solid and dotted lines), on the other hand, will remain above equipartition for most of the rise. In the case of a kink-unstable tube (upper dotted line), in particular, the field is above equipartition up to about 15,000 km, i.e., roughly speaking, until the depth of the bottom of the supergranules.

The lateral expansion of the rising flux tube can be seen in Fig. 7. In Fig.7a the ratio between the radius of the tube cross section and the local pressure scale-height (upper curves) and between the radius and the depth of the tube axis below the surface (lower curves) are shown for the case $B_0 = 10^4$ G, $\Phi = 3\ 10^{22}$ Mx. The originally thin tube becomes an extremely thick object in the upper convection zone: for the magnetic flux used in the figure, the upper boundary of the tube reaches the surface at the latest when its lower boundary is still about midway between top and bottom of the convection zone!

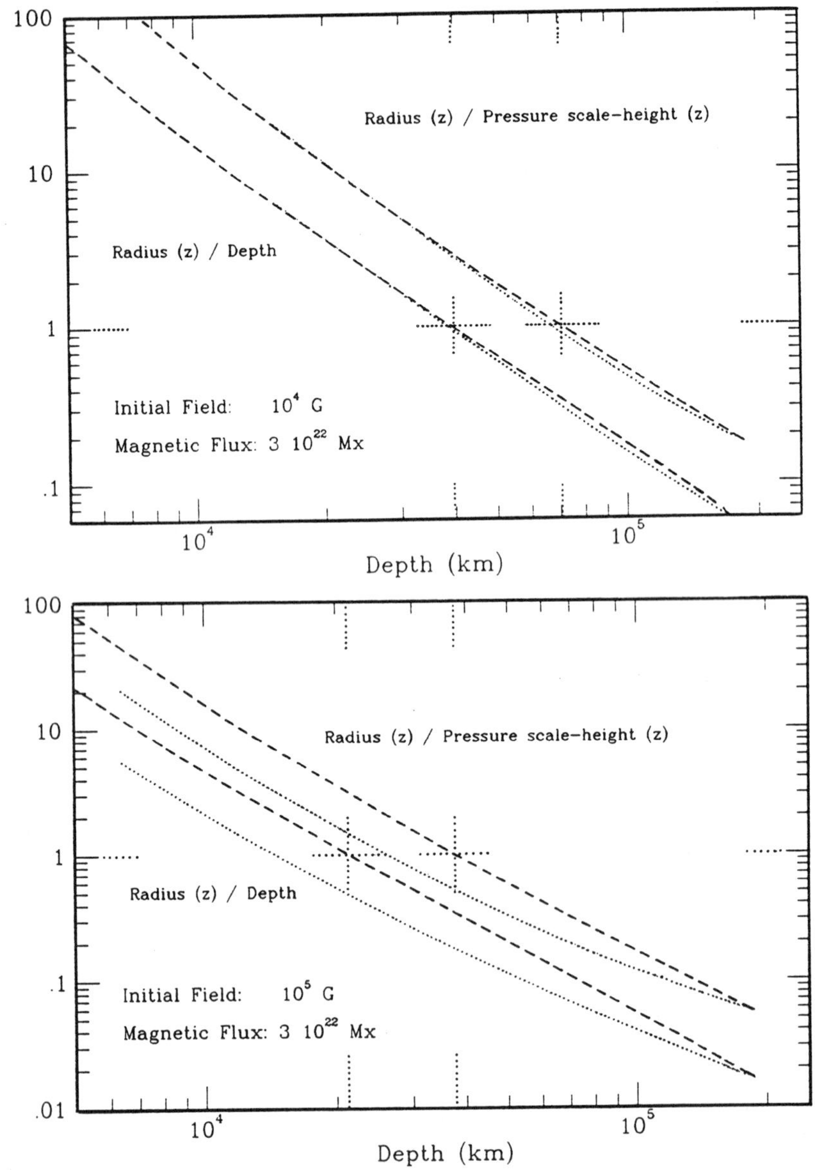

Fig. 7 – Radius of a rising flux tube following the approximation $R \propto \rho(z)^{-1/2}$ (dashed lines) and as calculated for the uppermost mass element of a kink-unstable tube (dotted lines). Shown are the ratios of the radius to the local pressure scale-height and to the instantaneous depth of the crest of the tube. (a) $B_0 = 10^4$ G; (b) $B_0 = 10^5$G.

The flux tube becomes an extended region of weakly magnetized plasma (as measured by the equipartition field) occupying a large range in depths below the surface. This conclusion also holds if the tube becomes fragmented along the rise (Sect. 8): the resulting tubules would expand and their field strength decrease in the manner described above. In the final phases before reaching the surface, the *bunch* of tubes extends across the upper half of the convection zone.

The corresponding curves for $B_0 = 10^5$ G can be seen in Fig.7b. Now (cf. dotted lines) the radius of the flux tube (or tube ensemble) is still moderately small ($R < H_p(z)$) until about the depth of the bottom of the supergranules. $R(z) \sim z$ is reached in the thin flux tube calculation when $z = 20,000$ km. This is probably a lower bound, since the tube is no longer thin in those heights.

As pointed out by several authors (Galloway and Weiss 1981, Moreno-Insertis 1986), a magnetic flux tube with $B^2 \ll {B_{eq}}^2$ is not able to withstand the convective flows around it, but, rather, may be severely distorted and shredded by them. A simple estimate can be made as follows: if the tube does find itself inmersed in a convective eddy with typical speed v_c, then the magnitude of the turbulent pressure fluctuations will be

$$p_{turb} \sim \mathrm{O}\left(\frac{1}{2}\rho v_c^2\right) . \tag{14}$$

When a pressure difference of that order is applied at the boundaries of the tube, the resulting gradient will be

$$\frac{p_{turb}}{2R} \sim \frac{1}{4}\frac{\rho v_c^2}{R} = \frac{{B_{eq}}^2}{8\pi}\frac{1}{2R} . \tag{15}$$

We now compare (15) to the buoyancy force in the tube along the rise. Except in the uppermost layers of the convection zone, one can use as an upper bound for the latter $\rho g/\beta$. The ratio between the two forces is therefore

$$\frac{{B_{eq}}^2/16\pi R}{\rho g/\beta} \gtrsim \frac{1}{2}\frac{H_p}{R}\left(\frac{B_{eq}}{B}\right)^2 . \tag{16}$$

The function on the rhs coincides (up to factors of order unity) with the ratio of drag force over buoyancy force for the case when the relative velocity between the tube and its surroundings is v_c; it has been represented in Fig. 8 for the two cases considered in Figs. 6 and 7.

The upper curve ($B_0 = 10^4$ G), shows the ratio (16) to be well above 1 during most of the rise, while the opposite is true for $B_0 = 10^5$ G. In the former case, therefore, the flux tube will be passively carried around by the convective flows. The outcome of the interaction between the rising loop of an unstable tube and the surrounding flows, in particular, will actually depend on the ratio between the typical size of the locally dominating convective flows, L_{conv}, and the typical length-scale of the tube along the field lines, Λ; Λ can be, e.g., the separation of the footpoints of a rising tube or the initial wavelength of the perturbation.

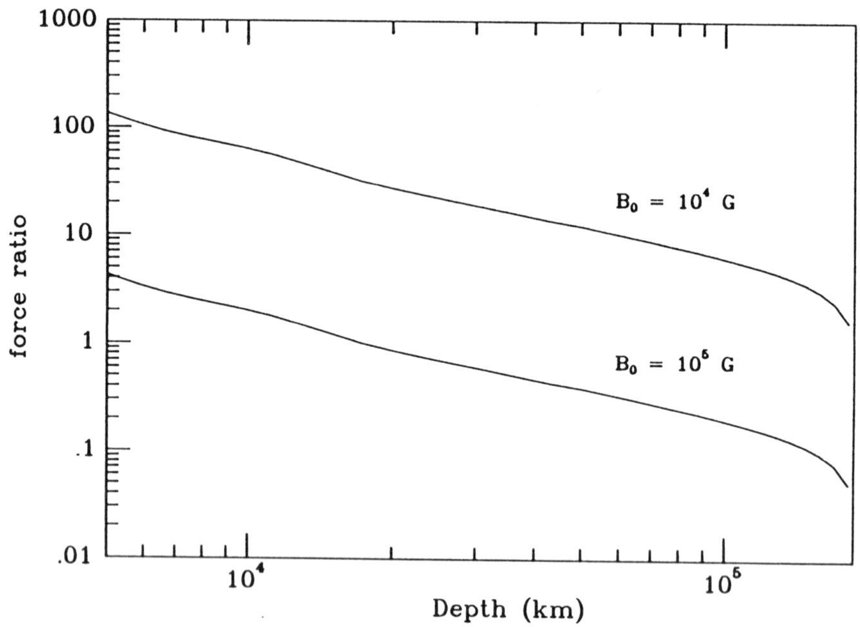

Fig. 8 – The expression (16) as a function of depth, which approximates the ratio between the force exerted on the tube by the surrounding convective flows and the buoyancy force of the tube if it were in thermal equilibrium with the surroundings for $B_0 = 10^5$ G (lower curve) and $B_0 = 10^4$ G (upper curve). The approximate proportionality laws (13) have been used.

If, for example, $\Lambda \lesssim L_{conv}$, then the tube, although passively moved about, may still keep its coherence; it may reach high layers of the convection zone in the updraft of the convective eddy. In the opposite case, $L_{conv} \ll \Lambda$, the disrupting effect of the turbulent flows will be fully felt by the tube. It will be distorted, it may be shredded and its coherent East-West orientation may be lost.

The phenomenon described in this paragraph (tube expansion and field weakening) must be specially noticeable in the final 10,000 km prior to arrival at the surface. The number of pressure scale-heights in that range of depths is very high, so that if the tube were to behave the way it does in deeper layers, it would end up having an extremely weak field, say a fraction of 1 Gauss; its radius should be correspondingly large. In fact the evolution of the flux tube in those heights can no longer be described using the thin flux tube approximation; to properly figure out the physical evolution of the tube, one has to use the full set of MHD equations. The most obvious possibility is to solve them using a numerical code, perhaps along the lines of the work of Nozawa *et al* (1992), possibly with an initial condition of non-zero vertical velocity. Until a solution of this kind is obtained, it will be difficult to predict any values for field strength or other physical variables of the rising tube close before outbreak at the surface.

8. Fragmentation of the rising tube

A variety of physical processes may cause the splitting of the tube into thinner strands or tubules as it rises. Although still largely unknown, these processes may play a non-negligible role in the evolution of the magnetic flux tubes in the convection zone. A recent and interesting discussion on the fragmentation of flux tubes can be found in Schüssler (1991); the reader is referred to that publication for details and further examples.

As a first instance of such an unstable process, the flows surrounding the flux tube may give rise to a dynamical instability in which the original circular cross section adopts a mushroom-like profile and finally splits into two separate regions. This has been studied for the rise of bubbles in a liquid (Tsinganos 1980 and refs. therein). Examples of rising tubes showing that behavior have been produced by Schüssler (1979). Further, the mushroom-shaped objects resulting from instabilities at the upper boundary of a magnetized layer (Cattaneo and Hughes 1990) are also suggestive of a process of this kind.

The *fluting instability* may also cause the disruption of the tube at the 'inner' (or concave) side of a curved section of the tube. This process could affect especially the crest of the rising loop. The reason for the appearance of this instability is that energy can be gained by exchanging magnetized and unmagnetized plasma because the magnetic curvature force tends to straighten the field lines of the tube. On the other hand, the stabilizing effect of gravity that prevents the sunspots from breaking apart at the surface (Meyer, Schmidt, Weiss 1977) does not work in our case, since the density inside and outside the tube are very close to each other. The time-scale for the fluting instability can be estimated to be (Schüssler 1991, Cap 1976),

$$\tau \lesssim \frac{(2RR_c)^{1/2}}{v_A}. \tag{17}$$

Putting $R_c \sim H_p$, one obtains for a tube in the bulk of the convection zone:

$$\tau \lesssim \Phi_{22}^{1/4}\, B_4^{-5/4}\, 1.5\mathrm{d}, \tag{18}$$

with Φ_{22} the magnetic flux in units of 10^{22} Mx and B_4 the field strength in units of 10^4 G. It is not clear at present what effect this instability may have on the structure of the crest of the tube once it has developed into the nonlinear phase and (possibly) saturated; the time-scale (18) at any rate is short in comparison to the expected rise times of the tube.

Through these two mechanisms or other possible unstable processes the rising magnetized region may become a bunch of thin tubules. This may affect the rise in (at least) two different ways:

a) the thermal exchange between the tube and its surroundings will be enhanced and

b) the drag force exerted by the surrounding flows, which grows as R^{-1}, may become dominant for the motion in the direction normal to the tube axis.

A rough estimate for the time-scale in which the temperature difference $T_i - T_e$ tends to be levelled off is given by

$$\tau_{rad} = \frac{R^2}{\kappa_T} = \frac{\Phi_{18}}{B_4\,\kappa_{T6}}\ 1\ \mathrm{yr}, \tag{19}$$

in which κ_T is the thermal conductivity adequate for an optically thick medium (κ_{T6} in units of 10^6 cm^2 sec^{-1}). If, now, $B_0 = 10^4$ G, then, in the bulk of the convection zone we can use $B_4\, \kappa_{T6} < 10$. We then have

$$\tau_{rad} \gtrsim \Phi_{19}\ 1\ \mathrm{yr}\,. \tag{20}$$

We deduce that only for $\Phi \lesssim 10^{18}$ Mx can the thermal exchange with the surroundings make the evolution of the tube deviate from adiabatic. Note that a flux tube departing from the bottom of the convection zone with the total flux of an active region, say $\Phi \approx 10^{22}$ Mx, should be split into 10,000 tubules for the resulting fibrils to have as small a value of the magnetic flux as 10^{18} Mx. In other words, unless the individual strands resulting from the fragmentation processes are very small, the rise of the tube ensemble will be basically adiabatic.

It is more difficult to assess the importance of the fragmentation processes in relation to the enhancement of the drag force. Small radius means strong coupling to the surrounding flows. An argument along the lines of the discussion in Sec. 7 can be applied. In fact, going back to eq. 16, if $R \ll H_p$ then the drag force can become dynamically dominant. Assuming that the individual elements of the tube ensemble are fully exposed to the ambient flows they encounter in their journey, then the drag force will make the relative velocity transversal to the tube small and the tubes will be passively carried around by the surrounding flows. As explained in Sec. 7, the ratio between the length-scale of the turbulent elements and the wavelength of the rising tube will determine whether the coherence of the tube ensemble is maintained or not.

The fragmentation processes treated in this section could be prevented to some extent if the magnetic field of the tube had an azimuthal component, i.e., if the field were *twisted* around the tube axis (see, e.g., Tsinganos 1980, Parker 1979). Yet, the evolution of a rising tube with twisted field, even the simple case of a rising toroidal flux ring, poses a number of interesting problems and questions which are still unanswered. The weakening associated with the lateral expansion of the tube, for example, affects differently the axial and azimuthal components of the field. In fact, while in the absence of longitudinal stretching $B_{axial} \propto R^{-2}$, if the lateral expansion of the tube proceeds uniformly along its radius, $B_{azim} \propto R^{-1}$. The ratio B_{azim}/B_{axial}, therefore, steadily increases with time and with it also the pitch angle of the twisted field. As the tube radius grows, that ratio becomes very large, and the pitch angle may reach values close to 90 degrees. The possibility and development of such a process in a tube rising across the convection zone and the ensuing instabilities (e.g., buckling) are topics which merit further study.

9. Field strength at the bottom of the convection zone

One of the interesting aspects of the subject matter of this paper is the possibility of establishing a relationship between the existing observations of active regions and the properties of the magnetic flux system buried deep in the convection zone. Both the time dependent calculations treated in the foregoing paragraphs as well as the stationary models of flux tubes extending across the convection zone (as in van Ballegooijen 1982a) aim at providing information in this respect. The failure of the slender tube approximation for the treatment of the magnetic flux tubes in the uppermost layers of the convection zone prevents a direct comparison of the observations with the results of the time-dependent calculations. The latter, however, can provide valuable guidance and insights, e.g. for the problem of deducing the field strength of the magnetized region at the bottom of the convection zone, B_0.

A reasonable first guess for B_0 is to assume a value not far from the equipartition field in those depths, i.e., $B_0 \sim 10^4$ G. A magnetic flux tube of that field strength, however, is not likely to reach the photosphere in concentrated form. Assuming a higher field strength, say $B_0 = 10^5$ G, solves that difficulty but brings in problems of its own. We can summarize and list advantages and problems of those field strength values as follows (see also Petrovay 1991):

a) $B_0 = O(B_{eq}) \sim 10^4$G at the bottom of the convection zone.

> The main reason in favor of choosing this value is of course its closeness to the only fiducial value available at that depth, namely equipartition with turbulent convection. Apart from that, as seen in Sect. 3, a flux tube with that field strength can easily get rid of its initial buoyancy by rising a short distance within the overshoot region. The tube does not rise as a whole to the surface and the choice of an initial thermal condition is no longer critical.

In spite of the foregoing, flux tubes with that field strength are not likely to produce active regions as a result of a sea-serpent evolution. Here are some reasons seen in previous sections of this paper:

> The Coriolis force can easily make the tube appear at high latitudes at the surface instead of within the activity belt.

> The diameter of the tube becomes extremely large as it reaches the upper convection zone (may even become 1/2 the depth of the convection zone).

> The field strength is below the equipartition value by 1 or 2 orders of magnitude during most of the rise.

> The tube may be easily carried away and/or deformed and shredded by the surrounding turbulent flows.

b) $B_0 \sim 10^5$G. In this case there are two large uncertainties:

> It is difficult to imagine how such a large field strength may arise at the bottom of the convection zone if it is to be generated by a turbulent dynamo mechanism in which the turbulent motions have an energy comparable to the magnetic energy of the equipartition field (by definition).

> Except if $\delta \lesssim -10^{-5}$, it is difficult to hold the tubes within the overshoot region if they are initially in thermal equilibrium with the surroundings. This problem may be less critical than it appears, for various reasons: first, the δ–distribution in the overshoot region is not well known; at its lower boundary, δ becomes very negative, $\delta = O(-10^{-2})$. Second, the time-scale for thermal exchange with the surroundings is very long (see Sect. 8), so that it is not clear which thermal condition is the most reasonable choice for the initial state of the flux tubes.

On the other hand, for the understanding of the rise of the tube there are several advantages in assuming a high initial field strength:

> The Coriolis force may not be able to make the tube deviate from the original latitude it was at originally

> Large radii (greater than the local pressure scale-height) are only reached in the uppermost few ten thousand km in the convection zone.

> The field strength stays above the local equipartition value also roughly up to that height.

As a corollary to the foregoing, if flux tubes with $B_0 \sim 10^4$ G are to give rise to the observed activity at the surface, then some mechanism must reprocess and concentrate the magnetic flux in the upper convection zone.

The last remark provides a clue of a possible way out of the dilemma presented in this subsection which also fits with the suggestions presented by Golub *et al* (1981). The field strength of the tubes formed at the bottom of the convection zone possibly shows a distribution of values around the equipartition field. A certain fraction (possibly most) of the magnetic flux in that depth could be of equipartition strength or close to it. If those flux tubes undergo a process of kink-instability development, then the rising loop will become weak, be shredded and reprocessed by the local turbulent motions in the upper convective zone and, perhaps, give rise to small flux concentrations (say, inner network fields, possibly also ephemeral active regions). The appearance of *active longitudes* in the pattern of small-scale flux emergence at the surface rotating with an angular velocity adequate to the deep convection zone (Stenflo 1991) could be directly related to this possibility. On the other hand, a small fraction of the flux in the submerged toroidal flux system could have a field strength noticeably above average. Kink-unstable flux tubes initially with that field strength could rise to the surface in a concentrated form and within the activity belt and thus give rise to normal active regions.

10. Conclusion

Although commonly accepted as a mechanism for the formation of active regions, the process of rise of a kink-unstable magnetic flux tube from the bottom of the convection zone to the surface is still not fully understood. Existing simulations show the possibility of anchoring of the flux tubes as a result of the development of the instability while the crest of the unstable tube rises toward the surface so that a 'sea-serpent' shape ensues. Many physical details of such a process have also been explained. However, a difficulty when trying to draw the whole picture of the 'prehistory' of active regions is the effect of the strong stratification and turbulent flows of the convection zone on the rising tube. They may prevent the arrival of the tube at the surface in concentrated form if its original field strength is close to the equipartition value at the bottom of the convection zone or below it: such a tube becomes weakly magnetized (in comparison to the local equipartition value) and very thick (as measured by the local scale-height) in the upper half of the convection zone. This effect becomes extreme in the final few ten thousand km below the photosphere. In the face of the existing evidence, one would be tempted to conclude that the rise of flux tubes originally with equipartition field strength just serves as a mechanism to feed the upper convection zone with unconcentrated magnetic flux, which then would suffer the effect of the local convective motions and perhaps give rise to small-scale magnetic features at the surface (ephemeral AR's, inner network tubes, etc). Normal active regions, on the other hand, could directly result from the rise of flux tubes with a field strength of at least about one order of magnitude above equipartition at the bottom of the convection zone.

The structure of the convection zone and of its lower boundary is becoming much better known thanks to helioseismological studies (e.g., Christensen-Dalsgaard, Gough and Thompson, 1991, Goode *et al* 1991). Many existing uncertainties about the properties of convection and magnetic fields in those levels (extent and structure of magnetized layer, existence and size of large-scale convective cells, size of overshoot region, etc) may perhaps be resolved in the coming years. This would allow, in particular, to overcome the uncertainties concerning the original state of the flux tubes, one of the stumbling blocks when trying to understand the prehistory of active regions. Time-dependent calculations of the evolution of unstable tubes must also be complemented with a 2-D or 3-D study of the outbreak of a rising magnetized region at and close below the photosphere. Finally, the evolution of the submerged part of the flux tubes after formation of the active region is also an interesting topic for further study.

Acknowledgements The author is grateful to the editors of this book and organizers of the workshop on 'The Theory of Sunspots', Profs. J.H.Thomas and N.O.Weiss, for their invitation to participate in this excellent conference and to Drs. H.U.Schmidt and M. Schüssler for their useful comments on this manuscript. Interesting discussions with Prof. E. N. Parker and Drs. M. Vázquez, A.R. Choudhuri and H.C. Spruit are also acknowledged.

11. References

Acheson, D.J. (1979), 'Instability by magnetic buoyancy', *Solar Phys.* **62**, 23

Babcock, H.W. (1961), 'The topology of the Sun's magnetic field and the 22 year solar cycle.', *Astrophys. J.* **133**, 572

Brandenburg, A., Jennings, R.L., Nordlund, Å, Stein, R.F., Tuominen, I. (1991), 'The role of overshoot in solar activiry: a direct simulation of the dynamo', in I. Tuominen, D. Moss, G. Rüdiger (eds.), The Sun and cool stars: activity, magnetism, dynamos, IAU Coll. 130, Springer Verlag, Berlin.

Caligari, P. (1991), 'Diplomarbeit', Univ. Freiburg (Germany).

Cap, F. (1976), 'Handbook on Plasma Instabilities', Academic Press, New York

Cattaneo, F., Hughes, D.W. (1988), 'The nonlinear breakup of a magnetic layer: instability to interchange modes', *J. Fluid Mech.* **196**, 323-344

Chou, D.-Y., Fisher, G.H. (1989), 'Dynamics of anchored flux tubes in the convection zone. I. Details of the model', *Astrophys. J.* **341**, 533-548.

Choudhuri, A.R. (1989), 'The evolution of loop structures in flux rings within the solar convection zone', *Solar Phys.* **123**, 217-239

Choudhuri, A.R., Gilman, P.A. (1987), 'The influence of the Coriolis force on flux tubes rising through the solar convection zone', *Astrophys. J.* **316**, 788-800

Christensen-Dalsgaard, J., Gough, D.O., Thompson, M.J. (1991), 'The depth of the solar convection zone', *Astrophys. J.* **378**, 413-437.

Defouw, R.J. (1976), 'Wave propagation along a magnetic tube', *Astrophys. J.* **209**, 266

D'Silva, S.Z., Choudhuri, A.R. (1991), 'The effect of Kelvin-Helmholtz instability on rising flux tubes in the convection zone', *Solar Phys.* to appear

Ferriz-Mas, A., Schüssler, M. (1989), 'Radial expansion of the magnetohydrodynamic equations for axially symmetric configurations', *Geophys.Astroph.Fluid Dyn.***48**, 217-234.

Ferriz-Mas, A., Schüssler, M. (1992), 'Instabilities of magnetic flux tubes in a stellar convection zone. I: Toroidal flux tubes in differentially rotating stars', *Geophys.Astroph.Fluid Dyn.*, submitted.

Galloway, D.J., Weiss, N.O. (1981), 'Convection and magnetic fields in stars', *Astrophys. J.* **243**, 945-953

Gilman, P.A. (1970), 'Instability of magnetohydrostatic stellar interiors from magnetic buoyancy. I', *Astrophys. J.* **162**, 1019-29

Golub, l., Rosner, R., Vaiana, G.S., Weiss, N.O. (1981), 'Solar Magnetic fields: the generation of emerging flux', *Astrophys. J.* **243**, 309-316

Goode, Ph.R., Dziembowski, W.A., Korzennik, S.G., Rhodes, E.J. (1991), 'What we know about the Sun's internal rotation from solar oscillations', *Astrophys. J.* **367**, 649-657

Greenspan, H.P. (1968), 'The theory of rotating fluids', Cambridge Univ. Press, Cambridge.

Hughes, D.W. (1991), 'Magnetic Buoyancy', in E.R.Priest (ed.), *Advances in Solar System MHD*, Cambridge Univ. Press, Cambridge.

Hughes, D.W. (1992), this volume.

Kraichnan, R.H. (1976), 'Diffusion of passive-scalar and magnetic fields by helical turbulence', *J. Fluid Mech.* **77**, 753-768

Libbrecht, K.G. (1988), 'solar p-mode frequency splittings', in V. Domingo, E.J.Rolfe (eds.), Seismology of the Sun and Sun-like Stars, ESA SP-286.

Lighthill, J. (1978), 'Waves in Fluids', Cambridge Univ. Press, Cambridge

Meyer, F., Schmidt H.U., Weiss, N.O (1977), *Monthly Notices Roy. Astron. Soc.* , **179**, 741.

Moreno-Insertis, F. (1983), 'Rise-times of magnetic flux tubes in the convection zone of the Sun', *Astron. & Astrophys.* **122**, 241

Moreno-Insertis, F. (1984), 'Ph.D.Thesis', Univ. of Munich.

Moreno-Insertis, F. (1986), 'Nonlinear time-evolution of kink-unstable magnetic flux tubes in the conveciton zone of the Sun', *Astron. & Astrophys.* **166**, 291.

Moreno-Insertis, F. (1992), in preparation

Nozawa, S., Shibata, K., Matsumoto, R., Sterling, A.C., Tajima, T., Uchida, Y. Ferrari, A., Rosner, R. (1992), 'Emergence of magnetic flux from the convection zone into the solar atmosphere. I. Linear and nonlinear adiabatic evolution of the convective-Parker instability', *Astrophys. J.* Suppl. **78**.

Parker, E.N. (1955), 'The formation of sunspots from the solar toroidal field', *Astrophys. J.* **121**, 491

Parker, E.N (1966), 'The dynamical state of the interstellar gas and field', *Astrophys. J.* **145**, 811-833.

Parker E.N. (1975), 'The generation of magnetic fields in astrophysical bodies, X. Magnetic buoyancy and the solar dynamo', *Astrophys. J.* **198**, 205

Parker E.N. (1979), 'Cosmical Magnetic Fields', Clarendon Press, Oxford

Parker E.N. (1984), 'Depth of origin of solar active regions', *Astrophys. J.* **280**, 423.

Parker, E.N. (1987a), 'the dynamical oscillation and propulsion of magnetic fields in the convective zone of a star. I. General considerations', *Astrophys. J.* **312**, 868-879

Parker, E.N. (1987b), 'The dynamical oscillation etc. , II. Thermal Shadows', *Astrophys. J.* **321**, 984

Parker, E.N. (1987c), 'The dynamical oscillation etc., III. Accumulation of heat and the onset of the Rayleigh-Taylor instability', *Astrophys. J.* **321**, 1009 Ref: *Thermal shadows*

Parker, E.N. (1988a), 'The dynamical oscillation etc., IV. Eruption to the surface', *Astrophys. J.* **325**, 880

Parker, E.N. (1988b), 'The dynamical oscillation etc., V. Instability and propulsion of flux bundles', *Astrophys. J.* **326**, 395

Parker, E.N. (1988c), 'The dynamical oscillation etc., VI. Small flux bundles, network fields and ephemeral regions', *Astrophys. J.* **326**, 407

Petrovay, K. (1991), 'On the properties of toroidal flux tubes in the solar dynamo', *Solar Phys.* **134**, 407.

Pidatella, R.M., Stix, M. (1986), 'Convective overshoot at the base of the Sun's convection zone', *Astron. & Astrophys.* **157**, 338-340

Pneumann, G.W., Raadu, M.A. (1972), 'Forced diffusion of solar magnetic fields', *Astrophys. J.* **172**, 739

Ponomarenko, Y.B. (1970), 'The formation mechanism of active regions', *Astr. Zh.* **47**, 98-102 (*Soviet Astr.* **14**, 78)

Roberts, B., Webb, A.R. (1978), 'Vertical motions in an intense magnetic flux tube', *Solar Phys.* **56**, 5-35

Schmidt, H.U. (1968), 'Magnetohydrodynamics of an active region', in K.O. Kiepenheuer (ed.), structure and development of solar active regions, IAU Symp.35, p.95-107, Reidel, Dordrecht

Schmitt, J.H.M.M., Rosner, R., Bohn, H.U. (1984), 'the overshoot region at the bottom of the solar convection zone', *Astrophys. J.* **282**, 316-329

Schüssler, M. (1979), 'Magnetic buoyancy revisited. Analytical and numerical results for rising flux tubes', *Astron. & Astrophys.* **71**, 79-91.

Schüssler, M. (1980), 'Flows along magnetic flux tubes. I.Equilibrium and buoyancy of a slender magnetic loop in the interior of a star', *Astron. & Astrophys.* **89**, 26-32

Schüssler, M. (1991), 'Comments on the structure and dynamics of magnetic fields in stellar convection zones', Habilitationsschrift, Univ. Göttingen.

Skaley, D., Stix, M. (1991),, 'The overshoot layer at the base of the solar convection zone', *Astron. & Astrophys.* **241**, 227-232.

Spiegel and Weiss (1980), 'magnetic activity and variations in solar luminosity', *Nature* **287**, 616.

Spruit, H.C. (1977), Ph.D.Thesis, Univ. Utrecht

Spruit, H.C. (1981), 'Motion of magnetic flux tubes in the solar convection zone and chromosphere', *Astron. & Astrophys.* **98**, 155-160

Spruit, H.C., van Ballegooijen, A.A. (1982), 'Stability of toroidal flux tubes in stars', *Astron. & Astrophys.* **106**, 58-66

Stenflo, J.O. (1991), 'Relation between small-scale flux emergence and global evolution of the Sun's magnetic field', in A.V.Alvensleben (ed.), JOSO Annual Report 1990.

Tritton, D.J. (1988), 'Physical Fluid Dynamics', 2nd ed., Clarendon Press.

Tsinganos, K.C. (1980), 'Sunspots and the physics of magnetic flux tubes. X. On the Hydrodynamic instability of buoyant fields', *Astrophys. J.* **239**, 745-760

van Ballegooijen, A.A. (1982a), 'The structure of the solar magnetic field below the photosphere', *Astron. & Astrophys.* , **106**, 43-52 van Ballegooijen, A.A. (1982b), 'The overshoot layer at the base of the solar convective zone and the problem of magnetic flux storage', *Astron. & Astrophys.* **113**, 99-112.

van Ballegooijen, A.A. (1983), 'On the stability of toroidal flux tubes in differentially rotating stars', *Astron. & Astrophys.* **118** 275-284

van Ballegooijen, A.A., Choudhuri, A.R. (1988), 'The possible role of meridional flows in suppressing magnetic buoyancy', *Astrophys. J.* **333**, 965

Zwaan, C. (1978), 'On the appearance of magnetic flux in the solar photosphere', *Solar Phys.* **60**, 213-240

VII. CONCLUDING SUMMARY

Many fascinating results – both observational and theoretical – were presented at the Workshop. It had been decided that no attempt would be made to summarize them all. Instead, E. N. Parker, who has written many papers on the theory of sunspots over more than 30 years, was invited to present his own impressions of the current state of the subject. His stimulating survey of the challenges posed by the sunspot phenomenon provides a fitting conclusion to this volume.

THE SUNSPOT PHENOMENON: A COMMENTARY

E.N. PARKER
Enrico Fermi Institute and
Depts. of Physics and Astronomy
University of Chicago
Chicago, Illinois 60637 USA

ABSTRACT. The present state of detailed knowledge of the sunspot phenomenon presents a host of unanswered questions. This summary is an attempt to pose some of the more outstanding problems, with the basic question remaining as to why a star is obliged to form sunspots at all.

1. Background Considerations

This writing is intended as a summary of impressions and questions at the end of five days of presentations on the diverse aspects of the sunspot phenomenon. The primary scientific question is why the Sun is compelled to force magnetic flux bundles into the local concentration that makes a sunspot? The next questions are what aspects of the observed sunspot phenomenon are a direct consequence of the forcing process, and what aspects are a consequence of the concentrated field, irrespective of the forcing? We may also ask what causes other stars, including some giants and a variety of subdwarfs, to develop enormous cool patches on their surfaces, suggestive of the sunspot phenomenon on the Sun? We may conjecture that a cool patch on a star is a single enormous sunspot, or a close cluster of many smaller sunspots, or quite a different phenomenon altogether.

It is an observed fact that magnetic fields continually emerge through the surface of the sun, forming east–west bipolar magnetic regions, whose sense of orientation reverses at approximately eleven year intervals. From these facts we infer that the magnetic field is generated by the fluid motion in the convective zone. The simplest viable dynamo process invokes a nonuniform rotation of the Sun to shear the weak poloidal field ($\sim$ 10 gauss) to produce a vast azimuthal, or toroidal, field of several kilogauss in the lower regions of the convective zone. Somewhere in the convective zone the convection produces local Ω and U bulges in the toroidal field which are rotated into meridional planes to provide the magnetic circulation that makes up the weak poloidal field. More complicated processes are possible, and the uncertainty in present ideas on the precise form of the dynamo arises from the inference from helioseismology that there is no significant radial variation in the

J. H. Thomas and N. O. Weiss (eds.), Sunspots: Theory and Observations, 413–423.

rotation rate across the convective zone. There is, of course, a variation with latitude, which can also produce the migratory bands of field that are implied by the emergence of bipolar magnetic regions at the surface. But for the present the subject waits on improved precision and decreased uncertainty in the helioseismological results.

It should be recalled that the Sun does not always exist in its present "normal" state. The ^{14}C record shows that the 15th and 17th centuries were periods of greatly reduced solar activity – the Spörer and Maunder Minima, respectively – while the 12th century was a time of extraordinary activity – the Medieval Maximum. We may ask whether the solar dynamo functioned only weakly in the 15th and 17th centuries, while going into a hyperactive state in the 12th century. Or was the dynamo functioning normally while the emergence of field at the surface of the Sun was suppressed for some reason? Direct observations of sunspots during the second half of the 17th century indicate the normal 11–year sunspot period, so at least some aspect of the dynamo was functioning normally. But this is a digression from the topic of the physics of the sunspot itself.

It is a curious fact that a region of magnetic field appears at the surface of the sun in fibril form, arriving at the surface as a collection of magnetic flux bundles in an updraft from the "dynamo region," believed to lie somewhere in the lower half of the convective zone or in the overshoot region below the convective zone. Field boils up to the surface in the form of Ω loop's with as much as 3×10^{22} Maxwells emerging at an active site in a couple of days, repeated at intervals of a week or so at the same location to provide an "activity nest." A total flux in excess of 10^{24} Maxwells may emerge in a single activity nest over a period of 6–12 months. So there must be an enormous quantity of magnetic flux in the lower convective zone. Some conjectures as to the strength of the azimuthal field run well in excess of 10^4 gauss, although the maximum kinetic energy density obtained in mixing–length models of the convective zone is substantially less, corresponding to the energy density $B^2/8\pi$ of $3-10\times10^3$ gauss. But this unsolved dynamo problem need not concern us. The observed fact is that flux bundles, presumed to come from the azimuthal field deep in the convective zone, emerge in batches to form an Ω loop or bipolar magnetic region at the surface of the Sun. The emergence of fresh flux at the surface goes on for a period of a day or two and then ceases, often repeated at the same position in a week or so.

It has been pointed out by several speakers that the intrinsic buoyancy of strong magnetic fields in the Sun is expected to bring a horizontal flux bundle of several kilogauss up out of the lower half of the convective zone in a month or so, far too fast to permit the accumulation of so strong a field in the first place. Thermal shadows tend to suppress this buoyant rise and various Rayleigh–Taylor and thermal convective instabilities are expected to produce the Ω loops that extend upward to the surface. Bipolar magnetic regions appear at the visible surface of the Sun, representing the apex of an Ω loop rising through the surface.

The strong tendency of the major magnetic eruptions to recur at special fixed locations – the activity nests – is not understood, nor is it clear why an eruption stops after a couple of days. What anchors the feet of the Ω loop so that greater lengths of flux tube do not rise over a period of time? Finally, there is the question of the disappearance of magnetic flux from the surface of the Sun once it has arrived there as recurring bursts of emergence of magnetic field. The flux at the surface continually fades away, while continually replenished by fresh eruptions. But what is the mechanism of the fading? Retraction below the surface, the opposite of emergence, is the only simple possibility, but observations show evidence of only an occasional retraction, which appears at the surface of the Sun as the merging and disappearance of patches of opposite field.

The same dilemma exists for the field of a sunspot, which, when the time comes for it to die, fades away faster than can be accounted for by the escape of magnetic flux bundles from its periphery. The fading occurs over dimensions of 10^4 km in a characteristic time of the order of 3×10^5 sec and is without adequate explanation at the present time.

2. Formation of Sunspots

The curious feature of the emerging flux, leading to the formation of sunspots, is the tendency for neighboring flux bundles to cluster together, exhibiting a mutual attraction that overcomes their basic mutual repulsion. For it must be remembered that the individual flux bundles observed in the photosphere expand to fill all the available volume in the tenuous chromosphere above. Hence, when they crowd together, the expanded field in the chromosphere is compressed, in opposition to its pressure $B^2/8\pi$. At the photospheric level the newly emerged flux bundles contain fields of the order of 0.5 kilogauss, which soon compresses to 1.5 kilogauss, evidently as a result of the downflow of gas in the interior of the flux bundle and/or radiative cooling at the surface. It is these 1.5 kilogauss flux bundles that are observed to cluster tightly together. The accumulated flux bundles become dark (cool), forming what is called a pore, when the overall diameter reaches about 1500 km. Continued clustering accrues more flux bundles into the pores and brings the pores together to produce larger magnetic regions. When bundles and pores accumulate to a diameter of 4000 km, the peculiar and complex penumbral phenomenon appears at one or more locations around the boundary of the pore, quickly filling in around the periphery wherever fresh flux is not streaming in. At this point the phenomenon is considered to be a fully developed sunspot, of which the original pore becomes the umbra. Growth by continuing accumulation of flux bundles and pores may continue to diameters of 10^4 km and more.

The astonishing property of the phenomenon of sunspot formation is that the clustering tendency continues only as long as there is fresh magnetic flux emerging at the center of the region. When emergence ceases, clustering ceases, and the disintegration of the sunspot, or sunspots, begins in earnest. In fact flux bundles may begin to break away from the outerside (relative to the other end of the magnetic bipole) even before the emergence of flux and the accumulation of bundles on the inner side has ceased. In the dispersal phase the magnetic flux fades away, with some of the loss to be attributed to the breaking away of small bundles of flux, reversing the earlier formation process. Some spots disintegrate as quickly as they were formed, while large circular spots develop a general outflow of gas at the visible surface, sweeping away other magnetic fields to form the encircling field–free moat, which may last for days, and occasionally for weeks or months.

What unseen forces cause this peculiar clustering of magnetic flux bundles? The clustering is in opposition to the magnetic stresses, so the explanation must lie in the hydrodynamics. There is no observational clue at the surface of the Sun as to what is happening below the surface. We can only assume that somewhere below the dense carpet of granules there are converging hydrodynamic flows that sweep the flux bundles together during the period of flux emergence. Thus the converging flow is the basic cause of the sunspot. But what initiates the converging flow? If the emerging flux is borne to the surface on a convective updraft, how is it that there is a converging flow at either end of the emerging Ω loop? Since neither a converging nor a diverging flow is observed at the surface, we can only guess that the convergence at either end is required to feed the downdraft associated with the convective updraft. But that is only a conjecture. We note that the later development of the moat around a long–lived sunspot indicates an updraft to provide the surface outflow. Meyer, Schmidt, Weiss, and Wilson in 1974 emphasized that a powerful converging flow ($\sim$ 1 km/sec at a depth of several thousand km) is required as a collar around the sunspot field to provide confinement and stability of the spot.

So what is the form of the converging flow that creates the observed clustering of the flux bundles and then maintains the stability of the sunspot created by the clustering? What initiates the converging flow? And why does the converging flow survive only for a short period after cessation of the emergence of fresh flux, although in some cases that period evidently stretches out to weeks or days? One can sketch the upwelling that brings the flux bundles to the surface, placing the return flow (the downdrafts) along the flux bundles that make up each leg of the Ω loop, as already noted. In that case the inertia of the downdrafts may establish converging flows at their upper ends, when the downdrafts become horizontally separated from the upwelling as the bipolar magnetic field spreads out. The circulation is presumably maintained by the convective forces. But in that case why does the flow terminate in most cases soon after the original upflow ceases? It is evident that we really have no solid ideas here.

There seems to be no direct way to get at these primary questions by observations of the surface of the Sun or by theoretical and numerical prowess at the present time. We may hope that inversion of helioseismological data may one day provide information on systematic flows beneath the surface.

The alternative approach is to study the formation of sunspots in great detail and study the sunspots themselves in great detail. For the sunspot is a complex dynamical phenomenon that is of interest in its own right. The sunspot lies at the visible surface, as the conspicuous cap to the mysterious doings below. The penumbra represents an intriguing magnetohydrodynamic frill whose peculiar structure reflects in some way the unknown processes below. The placid umbra is less flamboyant but, with its bright points and internal oscillations, is worthy of serious scientific consideration. There is also the hope that helioseismology can be developed one day as a probe so sensitive as to provide indications of the structure beneath the visible surface.

In the meantime the task is to explore what can be seen of the sunspot, which is already a challenge to our incomplete theoretical knowledge of collective hydrodynamics, radiative transfer, and magnetohydrodynamics.

3. The Penumbra Phenomenon

Consider the penumbra. This phenomenon appears only after substantial clustering of flux bundles has enlarged the pore to 4000 km or more. Therefore, it would seem that the penumbra is not an essential part of the primary forces that cause the clustering. However, it is apparent that the penumbra is, in all probability, a direct consequence of the special properties of the clustering forces, so that we might one day infer something about the nature of the clustering forces from a careful consideration of the penumbra. For the penumbra is a peculiar structure indeed. If it is superfluous for the initial clustering, it contains usually more than half of the total flux of the region once it is developed, i.e. the penumbra usually exhibits a greater total flux than the umbra, spread over a region 3–5 times larger than the umbra. To put this in simple terms, when the pore reaches 4000 km an effect arises which causes more of the arriving flux to be added in less concentrated form around the outside of the concentrated pore, now called the umbra. The formation of the penumbra is a rather abrupt occurrence and it is not clear whether the penumbra is constructed from magnetic flux pirated from the umbra or from newly clustering flux from outside, or both. It would be interesting to study the growth of the umbra as a function of time from a time well before the appearance of the penumbra to a time well after to understand the source of the magnetic flux.

The astonishing aspect of the penumbra is that it appears to be composed of radial magnetic flux tubes and sheets that are alternately nearly horizontal and inclined to the horizontal by $\sim 40°$, providing the observed dark and bright radial penumbral filaments, respectively. The outward Evershed flow ~ 6 km/sec observed at the outer ends of the dark filaments may be a siphon flow along the nearly horizontal flux bundles. In fact it has been pointed out that the horizontal flux bundles form a canopy extending outward from the umbra, rising to a height of a couple of hundred km above the level of the normal photosphere. One observes normal granules streaming out from under the outer edge of the canopy as the origin of the outward flow in the moat. On the other hand, the canopy is optically thick and less bright than the normal photosphere, so things cannot be entirely normal under the canopy. We presume that the steeply inclined field in the penumbra extends down through the photosphere, inhibiting the normal convection and hence accounting for the reduced surface brightness of the canopy.

The inner parts of a dark filament, which seem to hang over into the umbra, show rippling motions that travel inward to the (often) bright bead marking the inner end of the filament in the umbra. What is it that supplies energy to maintain the bright bead?

The bright filaments in the umbra show little or no systematic streaming. Their steep inclination ($\sim 40°$) suggests a relatively distant connection back into the Sun well beyond the penumbra. X–ray pictures of the Sun show that some of this highly inclined flux from the inner penumbra connects off to more distant active regions.

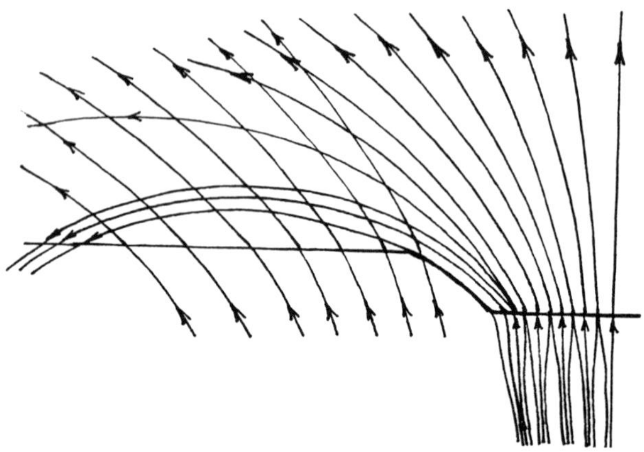

Figure 1. A sketch of the field lines form the umbra and penumbra, showing the interlacing of field in the penumbra.

We may ask ourselves how this peculiar array of interfaced horizontal and inclined fields came into being. It is sketched in Fig. 1. It is certainly no accident because it seems to be a universal property of all penumbrae. The close packing of alternate layers of horizontal and inclined fields is immune to rapid reconnection insofar as the layers are broad sheets. However, if the dark filaments represent ribbons (lying on edge) of horizontal field of finite width, then there may be some reconnection. If so, the reconnection is compensated by suitable motions continually rotating the magnetic flux back to the horizontal and inclined orientations in order to maintain the penumbra for a few days. It is perhaps just such activity that gives the penumbra its dynamical vigor.

But why is the penumbra created in the first place, with the principal add–on flux bundles going into the strongly inclined orientation outside the umbra, while flux bundles from the umbra are pulled down between layers of inclined field to form the canopy? Presumably most of the magnetic flux from the umbra extends upward (and away to distant points) surrounded by the inclined portion of the penumbral field, as sketched in Fig. 1. The effect arises only when the umbral diameter is sufficiently large ($\gtrsim$ 4000 km) and hence is flaring out rapidly from the vertical.

A related question is at what point does the penumbra disappear during the disintegration? Does it disappear when the diameter falls below 4000 km. And is there an abrupt increase in the rate of loss of magnetic flux from the region at the time of disappearance, or does the rate of flux loss continue smoothly. Does the umbra cease to shrink while the penumbra disappears? These are all the reverse of the obvious questions concerning the appearance of the penumbra.

4. Sunspot Structure

The existence of a sunspot by whatever cause raises some interesting questions concerning the flow of heat upward through the umbra and the diversion of heat around the umbra. The ACRIM data show that a sunspot blocks the heat flow, and, for a few days at least, causes the brightness of the Sun to diminish by an amount generally in keeping with the darkness of the spot. No bright ring appears around the periphery of the spot no matter how long the spot survives. It is a curious fact that the darkness of the umbra of a sunspot is almost entirely independent of the radius R of a sunspot (over a factor of ten in the radius) and the magnetic field intensity in the umbra is also pretty much independent of R. These facts evidently bear on the subsurface structure of the spot. A single monolithic flux tube, with a radius $R \sim 10^4$ km extending downward into the depths of the convective zone of the Sun, does not provide sufficient heat transport to supply the umbral brightness (equal to about one fifth the normal photospheric heat output of 6×10^{10} ergs/cm^2 sec) without some form of internal convective transport, for radiative transport is weak. For this reason consideration has been given to vertical heat transport by the vertical oscillation of cells of fluid along the magnetic lines of force. The cells of gas oscillate up and down exchanging heat at the extreme ends of their oscillations by overlap and horizontal radiative transfer with their neighbors similarly oscillating from above and below. Such motions, if sufficiently vigorous, can supply the meager umbral heat flux. However, the model of vertical oscillations overlooks the problem of transferring the heat horizontally across the field into the center of the single grand flux bundle, which has to take place somewhere below the surface if the center of the umbra is to be supplied. We know of no mechanism for effective convective transport across a strong field. So the idea that a sunspot represents the upper end of a single monolithic flux tube leads to a serious difficulty, to account for the approximately uniform temperature T that is largely independent of R.

There is no reason to assume a monolithic column of field, and an obvious alternative is the assumption that the subsurface field of a sunspot is composed of many separate flux bundles, with field–free fluid streaming freely between, sketched in Fig. 1. Free convection then supplies heat from below to the underside of the umbra more or less independently of the radius. One would extend this idea to the field below the penumbra. The field strength within the individual bundle would be determined by the dynamics of the individual flux bundle and so is more or less independent of the number of bundles packed together to provide the radius R of the total cluster. It is difficult to escape the idea that the observed umbral dots represent the upper ends of projections of relatively field–free gas poking upward through the umbra from the region of separated flux bundles below. Therefore, a diminished magnetic field in the umbral dots would be expected, and the observation of that field would provide an important test of the concept. However, it must be remembered that the umbral dots lie deep, whereas most spectral lines are formed one or more scale heights above $\tau = 1$, so that strong Zeeman splitting is expected whether in the line of sight to an umbral dot or not. Bruce Lites mentioned some FeII lines that appear only in the hotter material of the umbral dot, thereby permitting an unambiguous measurement of the field in a dot.

In the meantime it might be interesting to perform some convective legerdermain in a monolithic model to see if it is a viable possibility. Theoretical studies so far indicate the necessity of vigorous, nonlinear convection, which is not easily reconciled with the generally placid umbra.

There is an interesting possibility that combines both the monolithic concept with the cluster concept, and this would be indicated if observations showed a strong field in the umbral dots. The idea is that the field is a single coherent monolithic flux bundle from the umbra downward a distance h to where the field breaks up into a cluster of separate flux bundles (of greater field strength) with field–free gas between. Thus free convection transports heat between the separate bundles to supply the center of the monolith, and vertical convection transports the heat upward from there. The umbral dots are then just the upper ends of rising oscillating elements of gas constrained to motion along the field, although again the necessary vigor of the convection seems to be excluded by the 20–30 minute life of the individual dots. We imagine that h is a few hundred km. This model is not particularly appealing because of its complexity. We shall see what observations force upon us.

5. General Theoretical Problems

We may ask what can be done on the purely theoretical side to explore the properties of dynamical convection in the presence of magnetic fields? For most of our knowledge of the elementary properties of flux bundles in a static or active fluid of low electrical resistivity has been gained from the construction of simple idealized theoretical examples. Is it not possible to assemble our present qualitative knowledge of many different effects into one gigantic numerical code that can simulate the convection in a stratified atmosphere and a substantial magnetic field? Might we not unlock the secrets of the subphotospheric origin of the sunspot by brute force simulation? The reviews of numerical simulation presented at this Workshop have shown how far the art of the numerical simulation has progressed, with a variety of interesting and important illustrations of the exclusion of field, the concentration of cold downdrafts, etc. However, the numerical analyst would be the first to point out that the Reynolds number and the magnetic Reynolds number, or Lundquist number, are too large to permit a complete simulation in 3–D or even in 2–D with any present computer or with any presently conceivable computer. Hence, it is necessary to "parametrize" the simulation, which is usually accomplished by declaring that the effect of turbulence on velocity, magnetic, and temperature fields of scale in excess of some value ℓ can be represented adequately by eddy diffusion coefficients based on all eddies smaller than ℓ.

This is just as much a necessity for analytical treatment which, like the numerical simulation, can deal only with the relatively simple topology of the mean fields.

Now the eddy diffusion of scalar fields, e.g. a minor contaminant or temperature, is based on the simple concept of mixing. A drop of ink in a bucket of water is rapidly and effectively mixed throughout the entire volume of the bucket by the random walk of the individual elements of fluid in a turbulent flow. The ink is soon deformed into a ribbon that ergodically fills the bucket. For isotropic turbulence the diffusion coefficient is a scalar, of the order of $0.1\, v\ell$ where v is the rms velocity of the dominant eddies of scale ℓ. However, a vector field provides complications because the fluid elements initially lying on a single field line continue to be connected by a field line as the length of that line grows exponentially, permitting the elements of fluid to be scattered to the far corners of the volume. Conservation of magnetic field in an incompressible fluid decrees that the strength of the field grows in proportion to the length of the line. The stress density and energy density of the field increase exponentially as the square of the field strength, so that the extended magnetic field quickly overpowers that turbulent velocity field in which it is carried. To put it in different words, the mean field is transported and mixed while the small–scale (large Fourier wave number) components of the field grow rapidly and exponentially, soon dominating the turbulence and the mean field. The usual assignment of an eddy diffusivity overlooks this catastrophic situation.

It is obvious, of course, that the small–scale fields normally cannot grow much beyond equipartition with the turbulence. But the fact is that there is no satisfactory theoretical representation of the turbulent diffusion of a vector field. We have pointed out that neighboring parallel bands of intense magnetic field may be freely intermixed in planes perpendicular to the bands. This is just two–dimensional mixing in the perpendicular planes without deformation of the individual elemental flux bundles, so it is a 2–D scalar diffusion problem, the theory of which was provided by G.I. Taylor and others almost seventy years ago. Then if the two bands are oppositely directed, the close small–scale association of their opposite elemental flux bundles leads to reconnection, driven now by the magnetic energy. Reconnection provides free loops, among other things, which can be freely rotated into planes perpendicular to the mean field direction. Vainshtein, Rosner, and I have suggested that this may be what happens in the solar dynamo.

Unfortunately most magnetic configurations are not so simply treated and the nature of eddy diffusion is not clear. I see little or no prospect of proceeding with definitive numerical or analytical simulations until the "eddy diffusivity" can be treated properly, because the eddy diffusivity plays an essential role in the outcome of most simulations.

It is interesting to note that since Leighton's work over twenty years ago on the random walk of flux bundles across the surface of the Sun, there have been increasingly detailed and quantitative studies of this 2–D, and essentially scalar, diffusion. The interesting outcome of these studies is an estimated eddy diffusion coefficient of about 1×10^{12} $cm^2/sec = 10^2$ km^2/sec in the central portions of active regions, 2×10^{12} cm^2/sec in the outer portions of active regions, and 6×10^{12} cm^2/sec between high and low latitudes. A basis for understanding these different estimates from different granule motions, etc. is not available.

The problem and confusion may be related to the fading problem mentioned earlier, where it is observed that the total flux in each half of a bipolar magnetic region, and the total flux in a sunspot, fade away some ten times faster than can be accounted for by the observed departure of individual patches of field, i.e. flux bundles, or by the merging and cancellation with patches of field of opposite sign, i.e. retraction of the flux bundle into the Sun, reversing the initial process of emergence. There must be more to it than meets the eye, because field lines have no ends ($\nabla \cdot \mathbf{B} = 0$) and Faraday's law of induction associates nonconservation of magnetic flux with enormous electromotive forces that would produce electric currents in the highly conducting fluid so as to cancel the decline of the total flux. That is to say, the low electrical resistivity and large dimensions of the fields in the Sun guarantee a close adherence to the well known principles of magnetohydrodynamics.

This brings us to the interesting theoretical possibility of the fast dynamo. Consider a flow of characteristic amplitude v and characteristic scale ℓ, so that the magnetic Reynolds number is $R_m = v\ell/\eta$ where η is the molecular resistive diffusion coefficient. A fast dynamo arises for a specified $\mathbf{v}(\mathbf{r}, t)$ if there is a growth of flux in the limit $R_m \to \infty$. The simple fast dynamos so far conceived do not involve velocity fields that one would expect to find in the Sun. However, one may ask whether a complete theory of turbulent eddy diffusivity in the context of a solar dynamo is not simply an example of a fast dynamo, because it generates magnetic flux in the presence of large R_m.

There is an obvious question that arises in connection with the fast dynamo and the possible correspondence of turbulent diffusion. Note the fact that current models of what may be fast dynamos involve drawing the magnetic field into progressively thinner sheets, of thickness ϵ, say, as R_m increases. Define the effective magnetic Reynolds number $R_\epsilon = \epsilon v/\eta$. What is the limiting value of R_ϵ as $R_m \rightarrow \infty$? I would guess that the limit of R_ϵ is of the order of unity, because somewhere there must be effective diffusion if the necessary topological alterations of the field are to take place. If this is the case, then the fast dynamo is equivalent to the kinematical view of turbulent diffusion, that the field is pulled and sheared into filaments whose thickness decreases more or less exponentially in time until molecular resistivity provides rapid diffusion. On the other hand, there is the obvious alternative that the fast dynamo is dynamical rather than kinematical. That is to say, it may be that large field gradients are formed in association with rapid neutral point reconnection. In the Petschek reconnection mode, for instance, the rate of cutting of field is of the order of the Alfven speed divided by $\ln R_m$. Technically, then, this would not provide a fast dynamo because the dynamo rate declines to zero as $R_m \rightarrow \infty$. But in nature R_m is large but finite and $\ln R_m$ does not exceed 50. This appears to be fast enough, and we have shown that a workable dynamo can be constructed from this fact. But there is little more that can be said without a better knowledge of what is really happening beneath the surface of the Sun.

Indeed, what is happening beneath the surface of the Sun? Zwaan's review describes in detail the marching in and out of the flux bundles that seem first to build and then to demolish the sunspot, respectively. Other presentations have described the remarkable structure of the sunspot, which is somehow a creation of the "magnetometeorology" of the Sun. The observational descriptions are vivid and baffling because (a) we do not have much idea what is happening below the surface and (b) we evidently do not see all the magnetic flux bundles at the surface. We may look to advances in helioseismology to provide some clues on the magnetic structure, and perhaps the fluid motions, beneath the sunspot. Substantial theoretical work has established the general nature of the effect of the motion and magnetic structure on the mode frequencies. The difficult tasks of acquiring the necessary observations and inverting the data still lie ahead.

High resolution and high sensitivity are the hope and expectation for picking up the "invisible" magnetic flux component at the surface. Recent work by Stenflo suggests a whole new world of weak small-scale fields. The necessary resolution of 100 km, or better, is probably achievable only from an instrument of 1 m (or larger) aperture in space. But with NASA's determination not to explore the vital physics of the Sun, we will have to make do with ground based alternatives for the foreseeable future.

Index